Process Control and Optimization Handbook

for the Hydrocarbon Processing Industries

Gulf Publishing Company • Book Division • Houston, London, Paris, Tokyo

Process Control and Optimization Handbook

for the Hydrocarbon Processing Industries

Compiled by
Les A. Kane
Mechanical/Electrical Editor
Hydrocarbon Processing

Process Control and Optimization Handbook

Library of Congress Cataloging in Publication Data

Main entry under title:

Process control and optimization handbook for the hydrocarbon processing industries.

Includes bibliographical references.

1. Chemical process control—Handbooks, manuals, etc. I. Kane, Les A. 1949–

TP155.75.P74 661′.81 80-15243

ISBN 0-87201-144-5

Contents

Part I.

Computers in Control and Optimization

Glossary of Computer and Control Terms

Actuator—Device which accepts the output from a control system and moves a final control element (usually a valve) to change a process condition.

A/D converter (analog to digital)—Device which converts analog signals from process sensors into digital form for use by computers and other digital equipment.

Adaptive control—Method of control whereby tuning (response) of the control system is varied with process conditions, unlike other control systems where response times are set manually and remain constant.

Algorithm—Mathematical representation of the action performance by a controlling device to achieve its task such as proportional integral and derivative.

Architecture—With reference to computers, it is how the computer is constructed or wired and the path of data through it.

Baud rate—Number of bits or discreet pieces of information transmitted each second.

BCD (Binary Coded Decimal)—Mathematical system using binary (zeros and ones) notation to represent decimal (base 10) numbers.

Bit—A single binary digit, 0 or 1.

Bubble memory—Memory system that relies on the creation and destruction of small magnetic areas (bubbles) within a special semiconductor material to store information. Its major advantages are: its contents can be changed online with the computer, the stored information will not be destroyed with loss off power and it has no moving parts.

BUS—The term used to categorize a number of wires that all perform a similar function. For example, a data BUS is the group of wires that carries the data.

Byte—A number of adjacent binary digits operated upon as a unit.

Cascade control—Method of control whereby the output of one controller becomes the input to a secondary controller. Its purpose is to eliminate the long response time of certain processes, usually temperature. For temperature control, the primary controller responds to the actual temperature whereas the secondary controller responds to another variable such as steam pressure or flow.

Configure—To program the control system interconnections and controller algorithms.

Controller—Device which receives a measurement of the process variable, compares that measurement with another representing the desired control point, and adjusts its output based on the selected control algorithm to minimize error between the two measurements.

Controller tuning—Procedure followed during process startup that adjusts the control system so that it provides the desired control response for process changes.

CPU (Central Processing Unit)—The part of a computer which performs the actual calculations. For a microcomputer, the CPU is the microprocessor.

CRT (Cathode Ray Tube)—Device used in televisions and some control systems to display data.

D/A converter (digital to analog)—Device which converts digital signals from the computer and other digital devices into analog signals for use by control or monitoring devices in the plant.

Data highway—The set of wires used for intercommunication between computers and other digital devices.

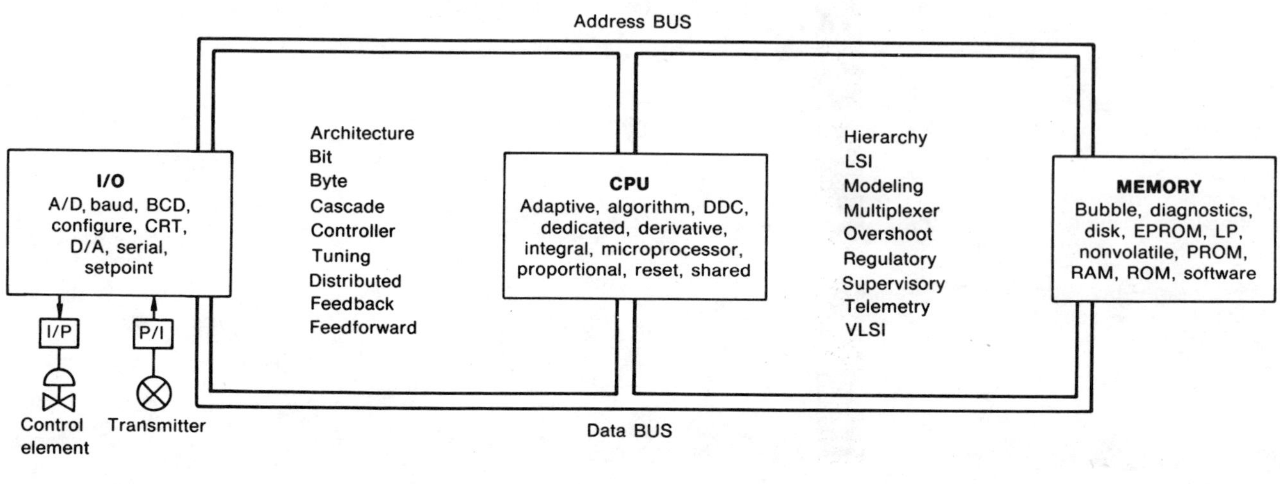

DDC (Direct Digital Control)—A method of computer control whereby the computer acts as the controller by performing the control algorithms and sending signals directly to the final control elements (usually valves).

Dedicated control—Using one controller to control one process variable.

Derivative control—A control algorithm which anticipates when a process variable will reach its desired control point by sensing its rate of change. This allows a control change to take place before the process variable overshoots the desired control point.

Diagnostics—A program contained within a control system to troubleshoot itself.

Disk—Magnetic memory storage device which contains no information during manufacture. Its contents are input and changed on line with the computer and will remain with loss of power. It does, however, require moving parts.

Distributed control—A control system design which allows control calculations to be performed in different parts of the control system. As opposed to centralized control whereby all control calculations are performed in one section of the control system, distributed control offers additional reliability.

EPROM (Erasable Programable Read Only Memory)—Memory device with information placed into it during manufacture that cannot be altered with the computer. It can be erased and reprogramed with special equipment.

Feedback control—A type of control whereby the controller receives a feedback signal representing the condition of the controlled process variable.

Feedforward control—Type of control which takes corrective action based on input disturbances before the process is upset.

Hierarchy—The term designates which computer/component in a multicomputer/component control system performs which level of control complexity.

Integral control—Control algorithm which resets the controller output to compensate for any offset between the process variable and the desired control point do to proportional control. This type of control is also called reset.

I/P transducer (current to pneumatic)—Device which takes an electrical current signal and converts it to a proportional pneumatic signal to interface electronic and pneumatic parts of a control system. It is primarily used as an interface between electronic control systems and pneumatically actuated valves.

I/O (Input/Output)—Devices used to put data into and receive data from a control system.

LSI (Large Scale Integration)—Technology of manufacturing small electronic circuits which contain an extraordinary number of components for their size. The microcompressor is a product of LSI technology.

LP (Linear Program)—A computational method used by process optimizing computers for solving arrays of linear inequalities.

Microcomputer—Small computing system containing, as a minimum, a microprocessor (CPU), a memory and some type of I/O device.

Microprocessor—Single-chip central processing unit.

Modeling—Writing a program to simulate process, economic or other conditions.

Multiplexer—Device which takes a number of input signals, encodes them and transmits them over fewer output signal wires. It is used to reduce the number of wires in long transmission systems. The original signals are regained by a demultiplexer at the receiving end.

Nonvolatile—A term which refers to memory that does not lose its contents if power is lost.

Overshoot—A condition where a control system allows a process variable to cross over its setpoint.

PID controller (Proportional, Integral and Derivative)—The letters in front of controller indicate which algorithms the controller uses to do its job.

P/I transducer (pneumatic to current)—Device used to interface pneumatic and electronic parts of a control system.

PLC (Programable Logic Controller)—Device used to control on-off process operations. The sequence of on-off executions and sometimes other control functions can be changed by programing instead of wiring changes.

Process variable—The process condition that is being controlled, such as temperature, flow, pressure, etc.

PROM (Programable Read Only Memory)—Device with information placed into it during manufacture that cannot be altered by the computer. It can, however, be reprogramed using special equipment.

Proportional control—A control algorithm whereby the output change of the controlling device is proportional to changes in the process variable signal.

RAM (Random Access Memory)—Memory which contains no information during manufacture. Its contents are input and changed at any memory location by the computer system. Its contents will be lost with loss of power.

Regulatory control—Control where the primary objective is to keep the controlled process variable constant in spite of upsets.

Reset—Another term for integral.

ROM (Read Only Memory)—Memory with information placed into it during manufacture that cannot be altered.

Serial communication—Sending bits of information in succession along a single wire.

Setpoint—Intended level of control for a process variable.

Shared processing—Controlling more than one process variable with a single controller or processing unit.

Software—The computer system program as opposed to hardware which is the physical components of the system.

Supervisory control—A method of computer control whereby the computer provides only setpoints to individual controllers which independently perform the actual control algorithms.

Telemetry—Measuring a quantity, transmitting the result to a distant station and there indicating or recording the measured quantity.

Transmitter—A device which measures the process variable and converts that measurement to a signal usable by the control system.

VLSI (Very Large Scale Integration).

Online Computer Optimization 1: What It Is and Where To Do It

A burst of activity is underway to implement closed-loop steady-state optimization for crude units, cat crackers, refinery distillations and many other process. Presented are the concepts of optimization, the mathematical techniques used and places where it really pays off for a variety of HPI processes

Dr. P. R. Latour, Setpoint, Inc., Houston

Computer optimization offers several advantages over conventional control. Primarily, it bridges the gap between a plant's true business objectives and its actual operation. In the early 1970s, with the advent of minicomputers, offline optimization of single process units was considered after the difficulties of online optimization were realized. The argument was that optimization is not required very often; hence, it is better done offline. However, more and more plants are turning to online optimization due to a lack of enforcement of action, a lack of accurate input information, the time lag involved in making changes and the advent of less expensive and increasingly reliable equipment. Presented are the concepts of optimization, the mathematical techniques used and places where it really pays off for a variety of HPI processes.

Operating objectives. Process control instruments and computers are installed in industrial plants to provide automatic control and make them run better. This equipment is used by operators and engineers to measure the state of the plant and regulate the operating conditions at specified targets.

The operating objectives of industrial plants can be summarized in a global way in terms of three factors listed in decreasing priority in Table 1: safety, product quality, and efficiency.

There are numerical specifications associated with each of these. For safety there are numerical limit values for metal temperatures, vessel pressures, tank inventories, and equipment speeds which specify the boundaries of safe process operating limits. These numerical values come from the equipment manufacturer or past experience. The numerical values relating to product quality come from specifications in contracts or from market customers. The numerical values associated with optimum operating efficiency include utility costs, product values, and maintenance costs.

Most process control systems indirectly contribute to the Table 1 operating objectives by logging, alarming, and regulatory control of operating conditions. The operators and unit engineers provide the interface for translating the general operating objectives of Table 1 into the specific command inputs to the process control system of the plant. They bridge the gap between business and the plant operation.

In the literature of modern control engineering there has been a great deal of discussion about the gap between theory and practice in the past 15 years. The implication is that sufficient advanced theory and technology are available. If it were merely implemented, all process control problems would be solved. Although there is no

TABLE 1—Operating objectives for industrial plants

1. **Safety.** Do not operate the process in any way which might harm people or damage equipment.
2. **Product.** Operate the plant such that product quality specifications and production rate targets (or inventories) are achieved.
3. **Efficiency.** Operate the process economically, to minimize long term operating costs.

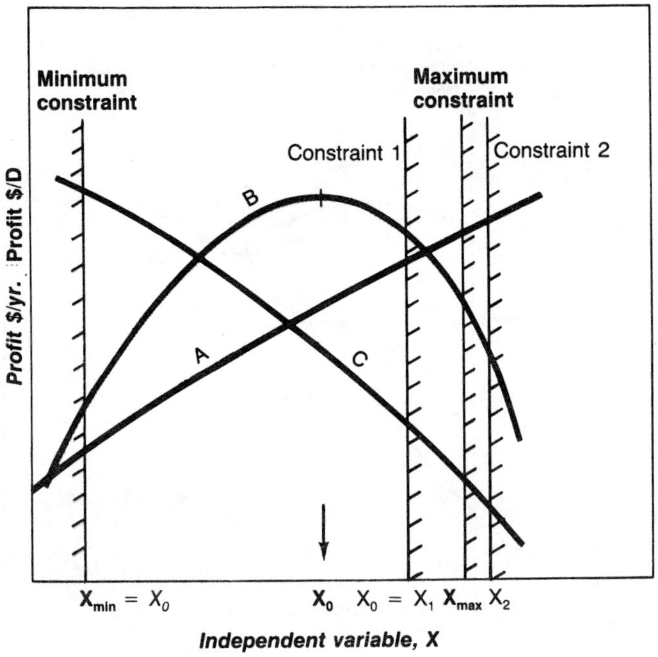

Fig. 1—Three different profit functions.

doubt the theory/practice gap exists, there is a different type of gap between the specific setpoint inputs into current process control systems and the true operating objectives of the industrial plant, Table 1. It seems apparent that genuine progress can be measured by the success of bridging this business/control gap.[1]

This tutorial article discusses the concept of steady-state optimization for online computer control and illustrates how it represents a direct, straightforward technique for bridging the business/control gap, allowing design of advanced computer control systems to automatically control industrial plants to meet realistic evolving business objectives.

WHAT IS OPTIMIZATION?

The concept of optimization is very broad and complex and connotes different meanings for people in different settings. We need to distinguish between two general control methods: dynamic regulation and steady-state optimization. The functions and objectives of each type of control are quite different. Also, the techniques used to implement these control methods are also different.

Dynamic regulation throttles manipulated variables to hold dependent variables near specified target values, which may be fixed or time varying. This control function is achieved by feedback control, sometimes embellished with feedforward control, adaptive control, nonlinear control, multivariable regulatory control, or modern optimal control theory. The word optimization is often used in the context of dynamic optimization. This occurs when the process is described by differential equations and the objective function is an integral over time. Optimization here should be viewed as the best algorithms and coefficients for the best type of dynamic regulation. The famous Linear Quadratic Gaussian problem (LQG) of modern control theory[2] provides the dynamic control algorithm which optimizes a particular quadratic performance index for linear differential equations' systems. This dynamic performance index does not directly relate to the steady-state economic business objectives of process industry plants.[1,3] This is the major reason that dynamic optimal control theory is not the final answer for complete automation of continuous processes.

Steady-state optimization is applied to processes and objectives which are described by algebraic equations rather than differential equations. Since optimum means best, either maximum or minimum, we must have a physical system to be optimized and an expression of an objective function measure of best performance.

There are three different types of optimization problems that are commonly encountered in industrial process operations: operating conditions, allocation and scheduling. Optimizing steady operating conditions include distillation reflux ratios, intermediate stream compositions and reactor conditions. Allocation problems have to do with the best distribution of a fixed resource among several parallel pieces of equipment. Scheduling problems refer to the proper timing or frequency of repeating a periodic operation for cleaning, maintenance, or refurbishing. These usually constitute the setpoints for the dynamic regulation control system. Allocations and operating conditions are most readily amenable to online optimization.

The job of any control system is to determine the proper values for the independent variables of the process. The independent variables must be identified and defined. Their ranges must be specified, and their influence on the process dependent variables and on the business objectives must be determined.

It is instructive to consider three general possibilities for the effect of the independent variables on plant profits. A maximum constraint profit function is illustrated in Fig. 1A. Here the profit increases monotonically as the independent variable is increased. However, as the independent variable becomes larger, a physical constraint is encountered, either on the independent variable itself or on one or more dependent variables. The most limiting constraint in Fig. 1A is constraint no. 1. If it were removed, the independent variable could be increased to constraint no. 2. It is clear that the optimum operating point for the independent variable is at the constraint's X_1. Note that to accurately determine the value of X_1, it is necessary to accurately determine the relationship between X and the dependent constraint available. Absolute values are important. However, once we assume the profit is maximum at X_1, there is no need to know the magnitude of the profit or the shape of the profit function. These concepts have a strong bearing on the model requirements and optimization procedure.

An unconstrained hill profit function is illustrated in Fig. 1B. Here the profit function reaches a smooth maximum within the allowable range of the independent variable. This represents a tradeoff between two or more competing or conflicting phenomena that contribute to the profit function. The optimum value of the independent variable occurs where the profit function has zero slope, or the slope of one phenomenon equals the negative slope of the other phenomenon. In this example, the location of physical constraints is immaterial. The crucial requirement is proper economic values and accurate slopes. Note also that the absolute value of the profit level is also immaterial. When the profit function is differentiated with respect to X, absolute terms or constants drop out. These considerations influence the modeling requirements as well as the optimization algorithm. Again, some parameters are crucial, others immaterial.

A minimum constraint profit function is illustrated in Fig. 1C. Now the slope is negative, but the concepts are similar to the maximum constraint profit function in Fig. 1A. The optimum value of the independent variable is at X_{Min}. Since the optimum feasible solution, X_0, must line between X_{Min} and X_{Max} these three figures cover all the general possibilities. Multiple hills are not encountered. Since the unconstrained hill situation is the most complex to handle, significant simplifications and cost savings can be realized for a constrained optimum.

All three possibilities are illustrated in Fig. 2 for optimization of FCC conversion.

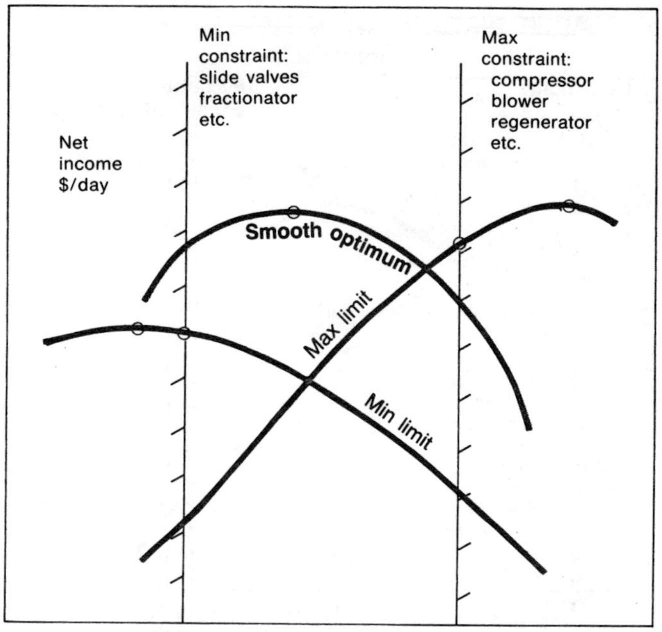

Fig. 2—Optimum FCC conversion.

Optimization techniques. There are a wide variety of mathematical optimization techniques employed for steady-state optimization computer control.

Linear programming is useful for simultaneously optimizing a number of independent variables when constraints are important and relationships are linear as illustrated in Figs. 1A and C. Sectional linear programming can handle problems where relationships are not linear.[4,5,6] An example flowchart for the sectional linear programming procedure is given in Fig. 3.[4]

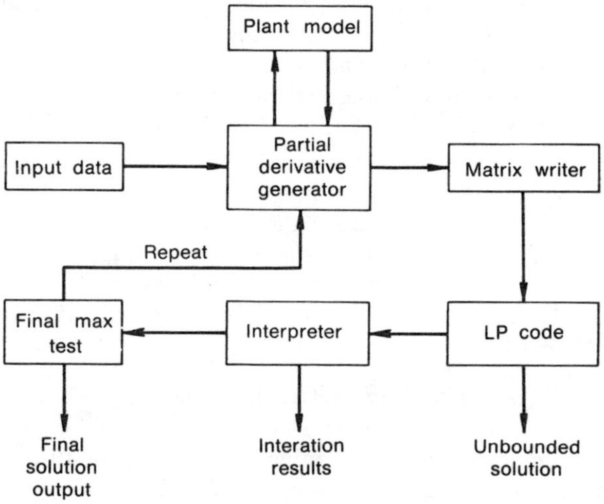

Fig. 3—Sectional linear programming procedure.

Constraint control methods are very useful when problems can be decomposed into single, independent variables and constraints predominate[7] as in Figs. 1A and C.

Gradient steepest-ascent methods are useful for finding multiple independent variables at the top of a hill. The property of quadratic convergence is very desirable to overcome numerical problems associated with sectional LP.

The sequential simplex method[8] is sometimes called the triangular method or the three-toed sloth approach. Its advantage is that it operates directly on the objective function rather than derivatives.

Evolutionary operation (EVOP) is suitable when there is no mathematical model of the process but the objective function can be directly measured.

For simpler problems, using **analytic derivatives** to give a closed form solution for the top of the hill provides an elegant solution for online implementation and circumvents convergence and numerical problems. It also clearly shows the model parameters that are critical to the solution.

The Lagrange Multiplier Method has proven suitable for multivariable allocation problems with a smooth hill.

Newton-Raphson methods have been employed for finding the top of an optimum hill, particularly for single variable problems.

The details of these algorithms are well documented in the literature and will not be covered in this article. Each has its advantages and disadvantages. Enough experience is available with these methods to allow selection of the proper algorithm or combination of algorithms to suit the nature of a particular problem.

OPPORTUNITIES FOR OPTIMIZING OPERATING CONDITIONS

The most important class of optimization problems for steady-state online control is the proper selection of operating conditions. This class of problems can be divided into three fundamental areas.

Material balance. The optimum material balance determines the proper routing of the main streams through the plant or refinery. We find it is important to analyze each main process flow or independent variable on the

TABLE 2—Examples of optimum operating conditions

1. **Distillation columns**
   - Reflux ratio
   - Hard target impurity
   - Soft target impurity
   - Pressure
   - Recycles
   - Multiple impurities
   - Feed rate, temperature
2. **Fluid catalytic cracking**
   - Reactor temperature, conversion
   - Recycles, selectivity
   - Preheat, heat balance
   - Fresh feed rate
   - Pressure profile
   - Catalyst replacement
3. **Olefin furnaces**
   - Temperature, severity
   - Steam hydrocarbon ratio
   - Hydrocarbon feed rate
4. **Crude distillation unit**
   - Pumparound refluxes
   - Stripping steam
   - Overflash
   - Charge rate
   - Vacuum tower steam, pressure, temperature
5. **Absorbers**
   - Liquid flow, temperature
6. **Furnaces**
   - Excess air
7. **Coal gasification**
   - Gas bypass

material balance to determine whether it is profitable to increase or decrease that flow. These usually relate to the primary specifications for products or the component losses and are not independent optimization variables. However, splits of feed to different parallel units may be optimized. Optimum inline blending of components determines the economic optimum yields and the feed capacity of the plant.

Energy balance. The second area for optimizing operating conditions is to view the energy balance of the process. The energy balance is usually affected by secondary flows which represent a type of recycle in the plant. Distillation tower reflux and unconverted product recycles affect the energy balance. Pressure settings within the plant also affect the energy balance, particularly in distillation towers and in utility supply headers. The optimum allocations that were indicated earlier also contribute to the energy balance. The control of air or oxygen for combustion operations also presents opportunities for energy balance optimization.

Reaction conditions. The third area of optimum operating conditions relates to chemical reactors. The optimum reaction conditions usually represent constraints or tradeoffs between competing economic forces. The proper operating temperature and hence reaction rate or conversion to desired products is a primary optimization variable on many reactors. Sometimes this is a tradeoff between reaction rate and equilibrium. Recycle variables that affect selectivity, catalyst activity, the amount of excess reactants, or reaction pressure, also represent independent variables for optimization.

The economic benefits from reactor optimization are believed to have far exceeded the benefits from other types of optimization and much work emphasis has been placed on reactor optimization. Unfortunately, the cost for successful implementation of reactor optimization is also the highest. It is not clear, in general, which of these three areas represents the opportunity for the fastest payout.

Optimization problems for these three areas can be decoupled and attacked in a stepwise fashion. It is possible to conceive and implement an optimizer for material balances that can be a profitable endeavor whether or not the energy balance and the reactor conditions are also optimized. Likewise, it is possible to work on specific optimizations of the energy balance which can be successfully implemented independently of whether or not the material balance and the reactors are optimized. Modular optimizers are easiest to implement and use.

A number of specific examples of optimum operating conditions that have been implemented with online control computers are listed in Table 2.

Distillation. There are quite a number of optimization opportunities in the operation of distillation column trains.[9] Before the oil embargo, fuel gas on the Gulf Coast was less than 20¢ per MMBtu (or 20¢/kilo Joule). The high differential values for components from petrochemical plants indicated that the proper strategy was to maximize the consumption of utilities to maximize separation within a distillation tower constraint. The limiting constraint may be the condenser, tower tray flooding, or reboiler duty. Distillation columns were commonly operated at maximum reflux ratio.

With the recent emphasis on energy conservation the new attitude is to minimize the reflux ratio. Now the question is, "How low can the reflux be set?" It is quite clear there is an economic minimum or optimum reflux ratio, below which the energy savings are less than product losses. The optimum reflux point is illustrated in Fig. 4.

A unique report[10] of online optimization of distillation reflux ratio describes computer control of a propylene upgrading unit. This application includes two propylene/propane splitter towers and two propylene/methylacetylene and propadiene distillation towers. Operators tend to over reflux towers because they are easier to operate within product specifications. Many columns are run on a fixed flow control for reflux, so reflux ratio is higher than needed when feed rates are low.

Optimization of distillation column operating pressure is also important. The general relationships are illustrated in Fig. 5. For most columns the optimum point is minimum pressure limited by the condenser cooling medium. This maximizes relative volatility and minimizes reflux for a given separation. As the condenser temperature limit changes, the tower pressure should also be changed. When heat of vaporization decreases with pressure, a smooth hill optimum can occur. This has been solved analytically.

There are often important recycles affiliated with distillation operations, and the composition levels of these recycles are prime candidates for optimization.

Sometimes a product from a series of distillation columns contains multiple impurities, both lighter components and heavier components. When the specification is only on total purity for that stream, the relative amounts of the light and heavy components are subject to optimization.[9]

Distillation columns really don't have a fixed upper limit for feed rate. Many times the feed capacity is rather soft and can be increased by decreasing reflux and hence, separation. This means that as peak capacity is increased there is a gradual decline in the tower's operating performance. If increasing feed rate is economically advantageous, there may be an economic tradeoff hill between capacity benefits and yield debits.

The normal rule-of-thumb for the feed temperature to a distillation column is that it is best to enter the feed as a liquid at or near its bubble point. There is a hill optimum feed temperature or enthalpy to tradeoff heat and recovery. This will provide both energy savings and potential capacity increases. This problem has not been adequately covered in the literature.

Large aromatics plants, separation sections for olefins plants, and refinery saturates or unsaturates gas plants are among the fundamental operations in the petrochemical industry. These plants provide many opportunities for optimum distillation control applications.

Fluid cat cracking. There are many optimization opportunities for fluid catalytic cracking. A primary variable is reactor temperature which determines the conversion of feedstock to lighter products. At low conversion the products are heavy and there is little gas and coke make. As conversion is increased each product sequentially increases to a maximum yield and then decreases: cycle oil, gasoline and light gases. Most cat crackers in North America are operated for maximum gasoline production, but some northern refineries and most European refineries tend to be run for higher middle distillate yield, varying with the season.

In the 1960s cracking for maximum gasoline usually meant running to a maximum conversion, limited by a physical constraint, typically the gas compressor, the air blower, or the regenerator burning temperature. In the last ten years there have been major process enhancements to this basic refinery process that have led to higher selectivity to gasoline with a corresponding reduction in gas and coke yields. These process enhancements include more selective zeolite catalysts, a reduction in fluid bed cracking with an increase in riser cracking, and higher temperature regeneration to lower the carbon-on-regenerated catalyst. These process improvements have tended to cause a smooth economic hill for optimum conversion rather than maximum conversion to a constraint. This situation is illustrated in Fig. 6. This is particularly true for European cracking operations at lower economic optimum conversions.

On the other hand, optimization of recycle has changed in the opposite way. In the 1960s, recycle was used to enhance the selectivity to gasoline and there was a smooth optimum amount of recycle at the top of a hill. This would trade off the value of selectivity versus the capacity

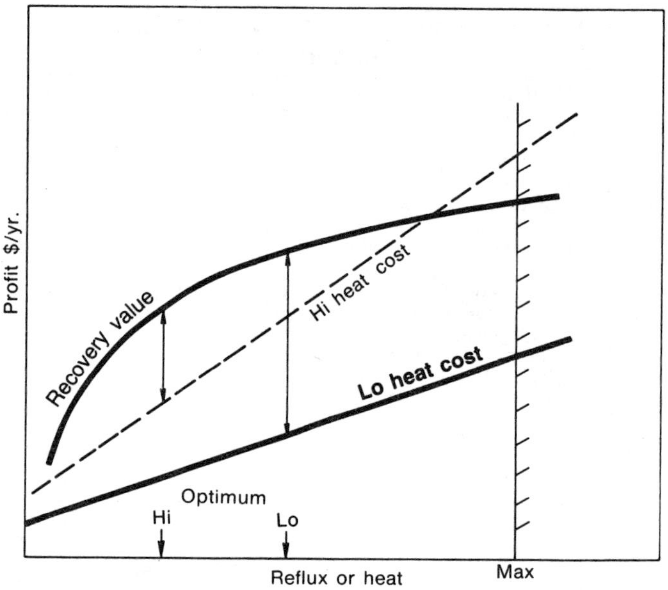

Fig. 4—Optimum distillation reflux.

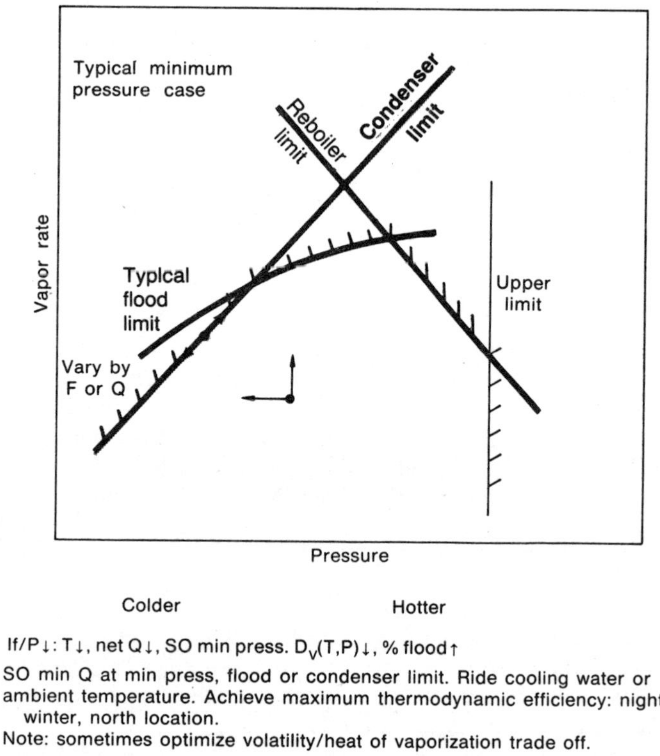

Fig. 5—Distillation tower pressure loading.

loss and heat cost for recycling. With the process enhancements just mentioned, the need for recycle has greatly diminished, and often the optimum gas oil recycle is zero, the minimum constraint. Whereas reactor temperature optimization has become more complex, recycle optimization has become simpler.

The optimum feed preheat or feed temperature to a cat cracker is an interesting optimization variable. Sometimes it is located at the top of a hill; other times the proper policy is maximum feed heat, and still other times the answer is minimum feed heat. Feed temperature has a complex influence on the heat balance of the reactor-

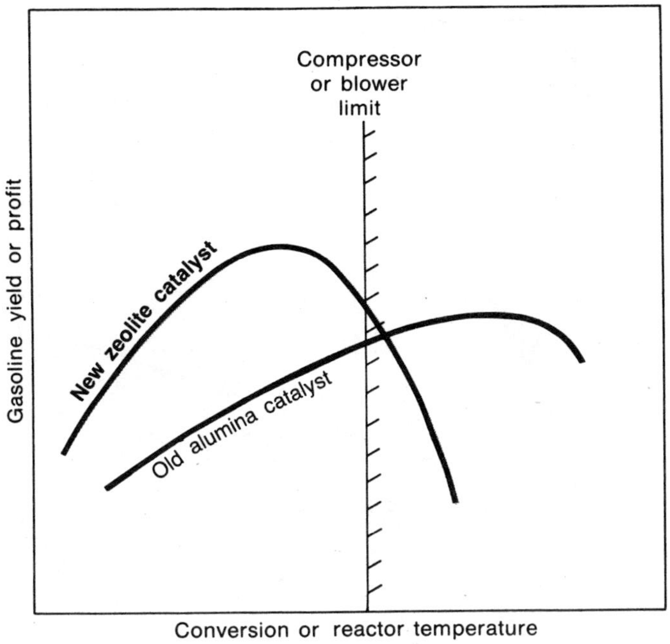

Fig. 6—FCC optimum conversion.

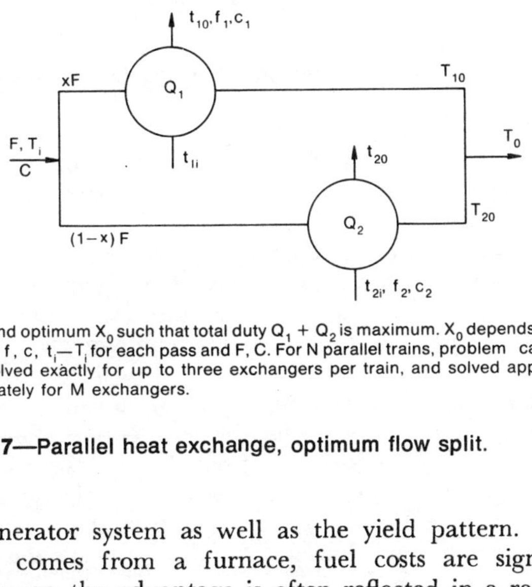

Find optimum X_0 such that total duty $Q_1 + Q_2$ is maximum. X_0 depends upon U, f, c, t_i—T_i for each pass and F, C. For N parallel trains, problem can be solved exactly for up to three exchangers per train, and solved approximately for M exchangers.

Fig. 7—Parallel heat exchange, optimum flow split.

regenerator system as well as the yield pattern. If feed heat comes from a furnace, fuel costs are significant. However, the advantage is often reflected in a reduction in catalyst circulation rate. This increases selectivity to gasoline and reduces the coke yield. The higher valued products and the increase in potential regenerator capacity may or may not offset the increased fuel cost.

Setting a fresh feed rate to a cat cracker is the primary material balance variable. There is often controversy as to whether this should be set by an online optimization for the unit or by an offline refinery-wide optimization that accounts for other units and inventory position. This represents an excellent example of the proper selection of independent variables and the determination of who or how they will be optimized, on what frequency, and on what basis. As with any process unit, processing more feed is usually accompanied by a decline of product yield or process performance. When increased capacity is desirable there is a smooth optimization for this tradeoff and it should be included in the online optimization.

The pressure profile through a fluidized bed reactor-regenerator system for a catalytic cracker is determined by the settings on the regenerator pressure controller and the wet gas compressor suction pressure controller. There is an optimizing strategy for these independent variables also.

There is another independent variable for FCC operations concerning catalyst replacement, both flow and frequency. The cost of fresh catalyst addition must be compared with the value of higher catalyst activity.[8]

Ethylene. Olefin furnaces for steam cracking have been the subject of online optimization control for several years. The optimum cracking temperature or severity depends on product values and downstream processing costs and constraints. The hydrocarbon feed rate and the steam flow are also optimization variables. The type of feed, from ethane through gas oil, the amount of coking in the furnace tubes, and the relative values of olefin products have profound effects on the optimum operating conditions for these furnaces.

Crude oil distillation. The crude distillation unit is the first operating unit in all petroleum refineries. A number of optimization problems have been solved on these important units.

Pumparound reflux streams improve the heat recovery to crude and hence fuel requirements. However, they do this at the expense of separation in the upper sections of the tower and in limiting cases can cause a decline in the yield of the more valuable products. This tradeoff presents an opportunity for optimization just like any distillation column reflux.

Stripping steam is used to improve product separation, but there is a point of diminishing returns. In fact, for some units it has been determined that the proper amount of steam is zero. Trading off the cost for steam versus its process credit is another optimization problem.

The proper amount of overflash in a crude unit can be an important variable for optimization. For some units where side draws and bottoms components are combined for cat cracker feed, the optimum point is at a minimum sufficient to prevent entrainment from escaping with side draw distillates. For other units more overflash provides refluxing to enhance products at the expense of energy. This tradeoff can be quite sharp and economically significant.

The crude charge to some crude units is split to dissimilar parallel heat exchangers as shown in Fig. 7. The flow and inlet temperature conditions for the other side of these exchangers can vary. Also tube fouling changes. An optimization question exists for the proper split of a given crude flow between these two parallel exchangers or exchanger networks. If all the crude goes to one or the other of the exchangers, the heat duty is low. The sum of the duties of both exchangers has a maximum at some intermediate split. The combined duty as a function of the fraction of the crude to one exchanger is illustrated in Fig. 8, showing the smooth optimum.

When refinery capacity is desirable, the charge rate to a crude unit becomes a very important optimization variable.

An important product cut in most refineries today is made on the vacuum tower, when clear distillates have intrinsically higher value than black vacuum pitch. The three main operating conditions that determine the yield and quality of vacuum gas oils are stripping steam, operating pressure, and temperature. The optimum amount of stripping steam usually occurs at the top of a hill. Small amounts are advantageous for reducing the partial pressure of oil and hence increasing the gas oil yield. However, excess steam increases the tower pressure drop and can lead to increases in base pressure and hence reductions in gas oil yield.

Absorbers. Absorber columns are widely used in refineries and gas processing plants to recover C_3 and heavier hydrocarbons from gases. The primary operating variable for a gas absorber is the liquid absorbent flow and temperature. A heavy lean oil is used as an absorption liquid. The liquid is recycled through the plant and regenerated for recovery of the heavy components. Although increasing liquid absorbent flow rate improves recovery of the valuable components, there is an operating debit for recycling costs. For many units there is an economic optimum for the proper amount of absorption liquid, based on the tradeoff between operating cost for recycle and the value for recovering the heavier components.

Combustion. Furnaces and boilers display a smooth optimum efficiency for the proper amount of excess air. Many methods have been proposed for optimizing this important variable.

Coal gasification. Close study of proposed coal gasification plants has revealed a number of variables for potential optimization. One example is the bypass of low Btu gas around the primary shift reaction section. This affects the plant energy balance and the overall distribution between primary shift, secondary shift and methanation. It influences the final product yield.

In industry today many of these optimizations are solved on a one time basis by process engineers, and the solution is used for extended periods of time. If the premises are deemed to be obsolete a month or two later, an engineering study is done to redetermine the optimum policy. Online optimization is the mathematical automation of this engineering effort. The online computer and its associated model contain the process technology and correlations that have been previously developed. It provides a tool to ensure that all this technology is continually employed to best optimize the process. It tends to illustrate and guide where model improvements should be made to enhance the various simplifying assumptions. It can play an integral role in process development.[11]

OPPORTUNITIES FOR OPTIMIZING ALLOCATION

Another steady-state optimization problem in industrial applications is the proper allocation of a limited resource among parallel processes. Some examples of successful implementation of this method follow.

Steam generators. In large chemical manufacturing complexes, high pressure steam is generated by a number of boilers of varying size and efficiency. The economic optimum load distribution among the different boilers to meet varying steam demand is reached when the incremental steam cost is identical for each boiler. This is commonly known as economic dispatch.

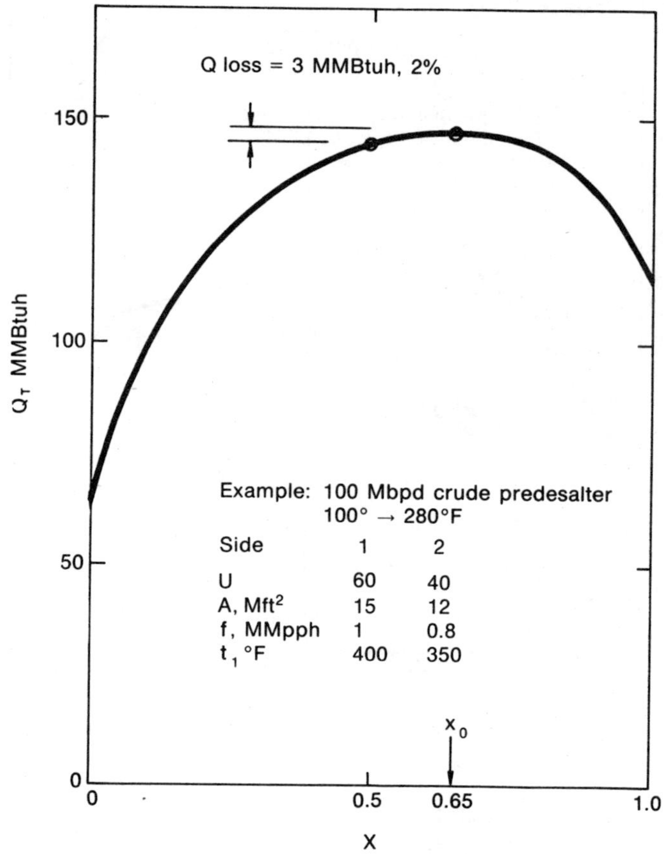

Fig. 8—Parallel heat exchange, optimum flow split example.

Refrigeration. Olefins plants require low temperature distillations for recovery of the light component products. The low temperature condensers are usually supported by ethylene and propylene refrigeration cycles. These cycles are usually run with a fixed supply of refrigeration duty, set by compressor capabilities. The distillation users are somewhat flexible in their demand requirements. There is an optimum allocation to distribute this fixed refrigeration capability among several competing users.

If each distillation user has a local optimization for reflux, it is easy to assign a price for refrigeration condensing. The optimum allocation of refrigeration is achieved when each independent user pays the same cost for refrigeration, and that cost equals or exceeds the true cost for producing the refrigerant. Optimum allocation of refrigerant duties not only conserves energy or maximizes yields but can sometimes contribute to increased plant capacity.

Parallel distillation columns. In some older plants where distillation columns or other process units are operated in parallel, an optimization problem for the proper allocation of feed to parallel equipment exists. This optimization has been successfully implemented for propylene distillations by online computer control.[10] This improves plant efficiency by reducing the total utilities requirements and component losses. It also provides an increased capacity capability.

Limited fuel. During winter emergencies, the government sometimes limits the fuel available to industrial plants in favor of residential customers. When a plant

TABLE 3—Examples of optimum allocation problems

1. Loading electric generating plants
2. Loading steam boilers
3. Refrigeration for olefins plant separations condensers
4. Limited fuel supplies to plant furnaces and boilers
5. Feed rate to dissimilar distillaton towers or process units operated in parallel
6. Flow split to parallel heat exchangers
7. Blending
8. Limited hydrogen to several refinery unit consumers
9. Dissimilar feeds (C_2, C_3, C_4, naphtha, gas oil) to olefin plant cracking furnaces
10. Balancing multiple pass furnaces
11. Load shedding electric power users

TABLE 4—Examples of optimum scheduling problems

1. **Catalytic reformer**
   Reactor regeneration frequency
   Catalyst replacement frequency
2. **Heat exchangers**
   Cleaning frequency
3. **Olefin furnaces**
   Tube decoking frequency

TABLE 5—Different objective functions for a propane absorber

1. Maximize C_3 recovery from gas
2. Maintain a specified C_3 content in residue gas
3. Optimize tradeoff of the incremental C_3 recovery value versus the cost of lean absorption oil.
4. Optimize tradeoff of the incremental C_3 recovery value versus the cost of lean absorption oil, but keep C_3 content in residue gas above the contract minimum specification.

Note 1. All four assume you can not exceed the maximum absorption oil rate, limited by absorber flooding, pump, cooling capacity, stripping capacity, etc.
Note 2. For the last two, you must model key component differential values, absorption oil regeneration costs, and effect of liquid flow on recoveries.

uses its fuel in a number of process heaters and boilers, there is a question of the optimum allocation of the limited fuel. Which heaters can make best use of the available energy? This is another example of an allocation optimization problem.

A number of other areas for optimum allocation are listed in Table 3.

OPPORTUNITIES FOR OPTIMIZING SCHEDULING

There are many periodic operations in hydrocarbon processing scheduled on a regular basis. These optimization problems invariably represent a tradeoff between increased maintenance costs and improving average process performance. These problems can be solved offline using online computer calculated input data, or solved by an online computer with advisory output to the operator, or solved online and closed loop. Some important examples are listed in Table 4.

Catalyst regeneration. Catalytic reformers and many other fixed bed catalytic reaction systems require periodic catalysts regeneration for coke removal or refurbishment of catalyst activity. Sometimes the catalyst must be replaced entirely. Since catalyst activity, and hence process performance, slowly declines with time, a long period between regenerations gives low maintenance cost but lower average process performance. On the other hand, a frequent regeneration gives high maintenance cost and lost production, but higher process performance when operating. The problem is to determine the optimum frequency schedule for regenerations to trade off these opposing objectives.

Heat exchangers. The same scheduling problem applies to cleaning heat exchanger bundles. The preheat train for petroleum refinery crude distillation units is subject to gradual fouling. As the exchangers become more and more fouled, heat recovery declines and the furnace fuel requirement increases. The problem is to determine an optimum cleaning frequency to trade off maintenance costs and fuel costs.

Furnace decoking. Olefin furnaces for steam cracking offer another important application for optimum scheduling. As the cracking operation continues for several months, carbon is deposited on the inside of the furnace tubes. This inhibits heat transfer, causing a decline in process performance with time. However, increased maintenance cost and lost production result if the furnaces are shut down for steam decoking. There is a continuing debate among olefin plant operators about the optimum frequency for furnace tube decoking, and how to determine that optimum.

DEFINING WHAT YOU WANT TO DO

The objective function for process optimization represents a type of model for the plant, expressing, "What you want to do." It is one of the two models required for optimization: the process model and the economic model. The process model describes the process behavior as viewed by the primary sensors for temperature, composition and flow. It mathematically expresses, "What you have." The economic model describes the effect of the process operation on the business. It describes the objective functions for the process. There may be a number of different potential objective functions that may apply at different times. They can generally be stated in words rather than specific numbers, although there are supporting economic data to augment verbal objectives. Some example process units illustrate a variety of objective functions.

Absorber. Examples of some different objective functions for a propane absorber column are listed in Table 5. The first is the simplest and most common. This objective implies that the C_3 value is very high with respect to the possible cost for increased absorption oil. The words maximize or minimize an independent variable generally imply going to a physical constraint. The word optimize generally implies a tradeoff at the peak of a hill. The second objective implies that the contract specification for product quality is of overriding importance. The third objective is probably more complete and realistic and embodies the first objective. The fourth objective is the most general and realistic of all and embodies all the three preceding objectives. If the control system is provided the capability to achieve the fourth objective, then it can achieve each of the others as well. Since the engi-

TABLE 6—Different operating objectives for a fluid cat cracker

1. Maximize gasoline yield, fixed feed rate
2. Fixed gasoline production, minimum feed rate
3. Maximum conversion to light products, load compressor/regenerator constraints
4. Optimize economic yields, fixed feed
5. Maximize gasoline production with fixed cycle oil production
6. Economic optimum yields and feed rate, product values vary stepwise with production
7. Maximum feed with fixed yield pattern
8. Maximum fuel oil destruction for refinery
9. Maximum FCC gasoline plus olefins for alkylate

TABLE 7—Different objective functions for distillation columns

1. Maximize yield of more valuable components from a given feed, within purity specifications
2. Maximize product purity at a given production rate from a given feed
3. Minimize energy consumption: reboilers, condensers, within purity specifications
4. Optimize energy consumption versus product recovery value, separation
5. Maximize distillate production, within specification
6. Optimize feed rate, tradeoff capacity versus recovery

neering work and software required to implement the fourth objective is considerably greater than the first, it is important that the objectives be clearly determined before the control system is designed.

Fluid cat cracking. A number of different operating objectives for fluid cat crackers are listed in Table 6. Each of these is a realistic business objective for operating a cracking unit for a particular period. The proper optimization technique and the model requirements for each of these objectives may differ. The cost for design implementation and maintenance of each of these may also differ.

A control system could be provided to achieve any or all these objectives, as selected by the operator. He could push a button calling for business objective number 3 to maximize conversion to light products, or business objective number 7 to maximize feed with a fixed yield pattern. This would be a control system that matches the operating objectives for industrial plants listed in Table 1. The operator no longer enters setpoint targets for temperature or composition. Instead he enters directly the desired business objective, and the optimizer achieves it.

Distillation. A number of different objective functions for operating distillation columns are listed in Table 7. Each of these calls for a different model and different optimizing strategy. To optimize energy consumption, it is obvious that the differential economic value of the light key as well as the heavy key components at the tower must be assigned and considered. Often there are economic models required to assign realistic values for these key components at each tower. This includes modeling of recycle costs, the cost for compression, vaporization, and condensation, and the cost for processing a component in a number of downstream distillation operations. These models for intermediate process streams ultimately relate to known values for plant products, fuel costs, electric power cost, or feedstock cost.

Aromatics plants have an example of this concept. The

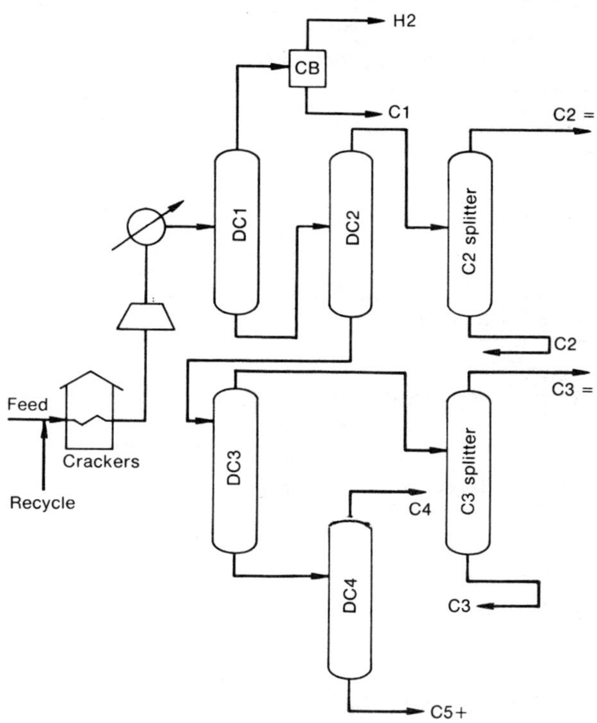

Fig. 9—Olefin plant separation train.

feed is a benzene-toluene rich stream to the benzene tower. High purity benzene product is taken overhead and the toluene rich bottoms is recycled through a hydrodealkylation reaction system for partial conversion to benzene. The reactor effluent mixture is combined with plant feed to the benzene tower. The question arises, "What is the proper setting of benzene tower reflux or reboiler energy consumption?" While material balance control is used to maintain benzene purity, refluxing determines the residue benzene content in the bottoms toluene which is recycled through the HDA reactor system. The differential value of benzene in the tower bottoms and tops must be determined in order to optimize tower reflux. This differential value depends on the cost for recycling benzene through the system. This cost must include pumping. heating and condensing, and any significant reactor effects.

An olefins plant is sketched in Fig. 9. The columns recover ethylene, propylene and butadiene products from unconverted ethane, propane, butane and other feed components. Material balance control can be used to maintain product purity but distillation reflux and energy balance is used to determine the level of olefins residue in the bottoms of each of the recycling streams. To optimize this reflux and refrigeration load, it is essential to determine the differential product values for olefins and paraffins in the tops and bottoms of the corresponding columns. This differential value depends upon the cost of recycling the olefin back through the cracking furnaces and through the process plant. This economic performance model includes processing costs which depend on compression, refrigeration, vaporization, condensation and reactor effects. These cost debits may also include capacity debits if recycle impinges upon a unit constraint that limits fresh feed rate. These situations can cause significant changes in the differential values at the column and

TABLE 8—Influence of coil outlet temperature on run time, yield, and profitability of naphtha cracking

Coil temp, °C	860	850	840
Yields, W%			
H_2	0.95	0.92	0.89
CH_4	15.09	14.64	14.19
C_2H_2	0.57	0.47	0.39
C_2H_4	29.0	28.15	27.26
C_2H_6	4.15	4.20	4.23
C_3H_6	13.33	14.09	14.73
C_4H_6	4.47	4.57	4.61
BTX	13.56	13.27	12.97
C_5+	28.39	28.29	28.38
NC_6 conversion, %	95.7	93.6	91.2
C_3H_6/C_2H_4 ratio	0.46	0.50	0.54
Gross margin, \$/MT naphtha	82.22	80.9	79.13
Onstream time, days	38	58	80
Furnace run time, %	94.5	96.5	97.5
Net Δ, MM\$/yr	—0.59	0	—1.47

Feed and economic basis in reference 12

have large effects on the distillation energy consumption requirement.

Nonlinearities. One of the characteristics of these economic performance models for process units is strong nonlinearities.

It is no longer adequate to think in terms of fixed product prices with highly flexible markets. Sometimes product prices vary stepwise with production volume or quality. On the day that new product prices or energy costs become effective, these numbers are entered into the optimizing computer and immediately affect the way the process is run.

A number of important applications display strong variations in utility cost depending upon amount and source of the utility. This is true both for low pressure steam, hydrogen supplies, switching among different fuels and electric power level.

Algebraic models that have been developed for intermediate stream values show some nonlinear variations with downstream processing.

Also, many complex tradeoffs compare tangible physical benefits against intangibles or social debits and display a highly nonlinear character. When these characteristics are present they may be difficult to handle, but they also suggest that online optimization could be beneficial when compared with simpler human methods.

Many complex tradeoffs are becoming important. They represent instances where more modeling work must be done to advance the application of optimization. Usually one side in this tradeoff is clear and can be modeled to cover tangible or certain physical effects. It is easy to predict the economic benefit from a closer approach tc quality specifications. For example, if blended gasoline octane average can be reduced, this gives a savings in lead. If the average value of the pour point of a fuel can be decreased within specifications, a yield credit is realized. The modeling difficulty arises on the other side of the tradeoff; in these examples, the possible loss of market volume because of poorer average quality.

Another example of a complex tradeoff is energy vs. the environment. The cost to remove sulfur from fuel oils or to operate cleanup processes can be clearly stated in terms of energy. The difficult tradeoff is with the cost or disadvantage for pollution, including SO_2 content in the atmosphere.

A third example of complex tradeoffs is short term versus long term. Usually the short term benefits from improved production rates or higher operating temperatures are clear, but the long term debits for poorer equipment life, coking, or metal fatigue are more nebulous and difficult to quantify.

An interesting example was recently published[12] on the effect of ethylene plant furnace outlet temperature on plant profitability. The results are summarized in Table 8. This example clearly shows the importance of furnace run time debits on tradeoffs for an optimum yield. If the outlet temperature differs from the optimum by 10° C, the economic loss can be from \$500,000 to \$1 million per year.

Another complex tradeoff is the classic economic question of the relationship between price levels and market volume. Increasing price levels give a clear economic benefit at constant market volume. But if market volume will decline, it is difficult to predict how much, as a function of price.

Complex tradeoffs and nonlinearities create modeling problems. A mathematical expression is required to quantify an intangible social debit. Once this expression is agreed upon, optimization can be done.

Part 2 of this article will discuss the benefits of online optimization along with how to estimate their magnitude, the steps required for a successful project and mistakes to avoid.

LITERATURE CITED

1 Latour, P. R., "Comments on Assessment and Needs—Process Control," *AIChE Symposium Series*, Vol. 72, No. 159, p. 161.

2 Athans, M. Ed, "Linear-Quadratic-Gaussian Problem," *IEEE Trans On Automatic Control*, Vol. AC-16, No. 6, December 1971.

3 Latour, P. R., "Optimal Control of Linear Multivariable Plants with Constant Disturbances," 1971 JACC, Washington University, St. Louis, Aug. 13, 1971.

4 Larmon, F. P., Van Reusel, J., Viville, L., "Online Optimization of an Ethylene Oxide Unit," 5th IFAC/IFIP International Conference on Digital Computer Application to Process Control, The Hague, Netherlands, June, 1977.

5 Baxley, R. A., "Treatment of Distillation Systems in Online Optimization," 5th IFAC/IFIP International Conference on Digital Computer Application to Process Control, The Hague, Netherlands, June, 1977.

6 Griffith, R. E., and Stewart, R. A., "A Nonlinear Programming Technique for Optimization of Continuous Processing Systems," *Management Sci.*, Vol. 7:379, 1961.

7 Baxley, R. A., "Local Optimizing Control for Distillation," *Instrumentation Technology*, October, 1969, p. 75.

8 Fiero, W. J., and Kelly, P. E., "To Optimize the FCC Unit," *Hydrocarbon Processing*, September, 1977, p. 117.

9 Latour, P. R., "Objectives and Strategies for Composition Control of Distillation Columns," *Instrumentation Technology*, Vol. 25, No. 7, July, 1978, p. 67.

10 Lamb, M. Y., "Computer Control of a Propylene Upgrading Unit," 85th National AIChE meeting, Philadelphia, Pa., June, 1978.

11 Latour, P. R., "Process Design Decisions Related to Process Control," Paper 60b, AIChE 74th National Meeting, New Orleans, La., March, 1973.

12 Goossens, A. G., Dente, M., Ranzi, E., "Improve Steam Cracker Operations," *Hydrocarbon Processing*, September, 1978, p. 227.

13 Harrison, T. J., and Ash, R. H., Minicomputers in Industrial Control, ISA, Pittsburgh, Pa., 1978, p. 43.

14 Latour, P. R., "Digital Control Techniques—Tutorial," 17th National ISA Aerospace Instrument Symposium, Las Vegas, Nev., May 12, 1971.

Online Computer Optimization 2: Benefits and Implementation

Impressive energy savings is one of the major potential benefits of online computer optimization. This and other benefits are explained along with how to estimate their magnitude, the steps required for a successful project and mistakes to avoid

Dr. P. R. Latour, Setpoint, Inc., Houston

PART 1 of this article presented the concepts of online computer optimization, the mathematical techniques used and pointed out specific places where online optimization has proven beneficial to HPI processing and operations. The next steps involve an investigation of all the benefits, estimating their magnitudes for justification purposes, and making sure you can satisfy the requirements for a successful project. These steps as well as mistakes to avoid are presented here.

BENEFITS OF ONLINE OPTIMIZATION

Benefits of steady-state optimization must come from improved plant and business performance. The specific sources can vary widely for a variety of reasons. In general terms, the revenues come from improved yields of valuable products, reduced energy consumption, and higher processing rates either through increased capacity of existing equipment or longer stream factors. Optimization may also influence reduction in a number of other operating costs including maintenance, equipment wear and staff utilization. There are longer term intangible benefits from interactions with particular plant groups including the operators, the engineering department and management.

The engineering benefits come from improved process trouble shooting and assistance in making quicker and more accurate decisions. This ultimately relates to improved process operations. There is also an interaction with process design of new units. The knowledge that comes from computer optimization can have a bearing on equipment sizing and upon assumed capacity factors.[1] If computer optimization can increase capacity, then a new plant design which includes a computer optimization can have smaller, less expensive equipment for a given thru-put.

The sources of steady-state optimization benefits are listed in Table 1.

Prediction methods. The concepts for economic justification of optimizing computer control are the same as for justifying any other engineering project investment.[2] They hinge on the economic performance model and a prediction of the additional benefits that will accrue as a result of the computer optimization investment over the life of the project. Hence, the basic approach is a base case and delta case method of predicting benefits. The starting point is determination of the base case for current operations, which requires an assessment of what's happening now, or the performance of the current process control system.

It is important to make a distinction between benefits that will accrue from dynamic regulation of processes and separate benefits that will accrue from steady-state optimization. Since each of these is a separate and distinct function, with a separate and distinct cost and physical implementation, each must have a separate and distinct economic benefit.

The common approach is to assume that dynamic regulation reduces variations in dependent variables so that the operator or the computer optimizer can move the setpoint closer to a specification target and hence reap steady-state benefits. This credits no benefit to the advanced regulatory controls for steadier operation alone, but specifies them as prerequisites for achieving all the steady-state benefit by the operators or optimization.

Frequently, optimization is not implemented online, and it is up to the operator to achieve the optimum targets and economic benefits. It is possible to assign true dollar incentives for improved regulatory control about the same steady-state average value setpoints.[3] These

TABLE 1—Benefits from steady-state optimization

- Improved product yields and/or quality
- Reduced energy consumption, operating costs
- Increased capacity of equipment, stream factors
- Less maintenance cost
- More efficient engineers, operators
- Tighter, lower cost process designs

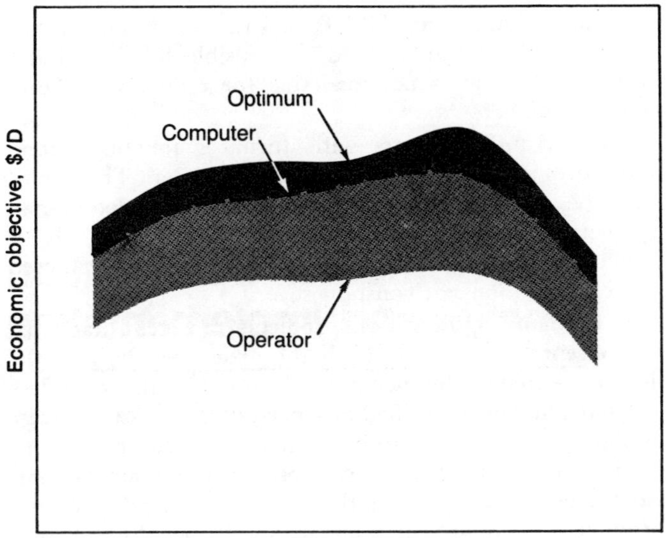

Fig. 1—Optimizer benefit illustration.

substantial dynamic benefits are invariably omitted in process control system justifications. Next, the steady-state optimization benefits can be independently assessed. Open-loop operator control might achieve some fraction of these, and automatic closed-loop optimization may achieve a higher percentage of these benefits.

Economic performance of any plant will vary with time. There are steady periods and upset periods. Actual performance achieved by the operator is illustrated in Fig. 1. His performance must necessarily be somewhat below the theoretical optimum or maximum objective function performance as shown. The integrated average difference between these curves is the potential that remains for computer optimization. A prediction must be made of the actual achievement with computer optimization. Sometimes, if the economic objective function is incomplete or poorly formulated, the operator's curve can exceed the optimum curve. Since this represents a contradiction, it means the operator exceeded some limit without incurring the same penalty as the optimization expression indicated. These situations must be resolved by more realistic economic models.

Although the general concepts for determining benefits are not difficult, the work is difficult and time consuming in practice. A number of methods for predicting these benefits are listed in Table 2. They invariably include the concept of documenting the base of current operations and predicting the improvement situation with the control system.

Sometimes for regulatory control and a linear profit function to a constraint, the method is to reduce the variations and then change the setpoint to achieve the same probability of exceeding the limits that was used in the base case. Since the variation distribution is narrower, this allows a closer approach to constraints and increased benefits. This is the simplest and most common method.[3]

Difficulties occur when the optimum is at the top of a hill. We now must model not only the curvature of the hill but the expected position of the peak with regard to current operating conditions.

There are generally two components of the benefits. One is a once-and-for-all directional improvement based on the fact that the discovered optimum point is normally above or below the current operating position. The operators now discover where at least one optimum point is, and that provides a one time credit for the discovery. If offline or engineering optimization has been frequently practiced, this component may be negligible. The second component of the benefits is derived from the capability for recalculating the optimum point and ensuring that the process follows it in a timely fashion. This component is typically handled by a simulation model with a sequence of test runs. A few typical spot points are determined and an assessment made of the difference in objective functions for these typical cases. This is sufficient to allow an extrapolation to the future.

A more rigorous approach is to simulate both the process and the computer optimizer together and subject this simulation to a series of random variations. This Monte-Carlo simulation approach constructs Fig. 1 with and without the optimizer, allowing an average economic benefit to be estimated. This rigorous and complete approach has rarely been used in practice.

One common method is to compare the best current operator with an average operator in the plant. This difference in performance represents a conservative estimate of what can be truly achieved in the plant and only credits the computer with the capability of matching the best operator. This method is highly conservative, but is believable and simple to document.

Another method is to build on the experience of past successful projects that incorporated test runs and post audits. Test runs at the beginning of a project are made to document the base case performance of the plant, including calculations of the true economic objective. After the project is complete, a series of post audit test runs are made both with and without the computer to document its performance. It may be possible to extrapolate this experience to similar processes in other

TABLE 2—Methods for predicting benefits of optimization

- Base case, delta cases
- Reduce variations, change setpoint to achieve same probability of exceeding limits
- Simulation model runs, typical spot points
- Simulate process and optimizer, inflict Monte Carlo variations
- Compare best operator vs. average operator
- Test runs and post audits

TABLE 3—Flexible nature of optimization Investments

For an existing plant with many possible varying constraints, there is a unique flexibility characteristic of an investment in computer optimization instead of in bigger or more efficient process equipment.

Each process equipment investment only contributes to one bottleneck, for the fraction of time that the bottleneck is limiting, and gives an improvement only up to the next limiting bottleneck.

The computer investment contributes a modest improvement against many bottlenecks, in a additive way, most of the time. Economic justification premises can be less rigid.

plants. In making these comparisons it is important that the two plants exhibit similar types of business environment variations. This would include changes in feedstock, product quality, economic pricing and operating conditions. Similarities in size of physical equipment are less important than similarities in the variations. Simple plants with a steady feedstock, uniform product quality, and a fixed business environment can be optimized once and for all, whereas complex processes subject to volatile business conditions present an opportunity for repeated reoptimization.

Avoid the pitfalls. Predicting benefits must be done with a great deal of care. In the past, benefits claims have been poorly documented and higher than achieved. Some justifications are based on the following incorrect philosophy: "This process has high thruput of valuable products, and a 1% improvement in yield provides $2,000 per day in benefit which will easily justify computer control, and surely a computer can improve the yield by 1%." A similar dangerous premise is, "Our fuel bill is $20,000 per day, so if we save five percent of that we could justify our project, and surely we can save five percent."

For distillation it is incorrect to sum the reboiler heat duties, and claim a certain percentage reduction in this total heat. The heat duty of a distillation column reboiler or condenser is used to generate both reflux and overhead distillate product. There is an absolute minimum heat required to run the plant to boil and condense the distillate products with zero reflux. If many of the columns are operating at rather low reflux ratios (like $R/D=1$), then only half of the total reboiler duty is available for reflux variations. If reboiler and condenser duties are reduced by ten percent this must mean a 20 percent reduction in reflux, which may not be feasible. Also, for some distillation columns there are large, sensible heat effects. A debutanizer with a large reboiler furnace has a feed at 40° C and a bottoms feed to a catalytic reformer at 110° C. About one half of the reboiler furnace heat is used for sensible heating of this liquid. That heat is not available for reduction by process control.

Another pitfall occurs with furnaces for the second and third reactors in the catalytic reformer. These provide the intrinsic heat of reaction which is determined by octane requirement and reformer severity. It is difficult to claim fuel savings on these heaters by altered process conditions, at constant severity.

Since many optimization problems represent tradeoffs, it is essential to include both the gross benefits and the debits to predict net benefits. To do this, the complete tradeoff must be quantified and defined. Impressive energy savings have been reported from process control of distillation columns. However, most of these are accompanied by a reduction in the average separation of the products. There is often an economic penalty for this. A fair justification should account for this penalty.

In summary, detailed and specific work is required to determine economic benefits from computer control and optimization. It is dangerous to predict a general percentage of improvement without highlighting the basis for that assumption and pinpointing the source of the benefit. This must be done by careful analysis of the material and energy balances for process units. It is possible to assign benefits on an individual control loop basis for advanced regulatory control functions and to assign benefits for optimization of each independent variable on a stand alone basis. This practice determines where the strong process sensitivities are, which variables should be included in the optimization, and which may be neglected as unimportant. For example, in catalytic cracking it is possible to conceive, design, implement and justify an optimization for recycle ratio alone. It is also possible to design, implement and justify optimization of temperature or conversion alone or combine optimization of both these variables together.

TABLE 4—Requirements for online optimization

- Select proper independent variables
- Formulate model requirements
- Formulate business objective functions
- Select optimization algorithms appropriate to nature of the problem (process and business). Always guarantee a feasible solution. Handle partial equipment failures.
- Specify sensor inputs, manual inputs
- Define human interface for objectives, defaults, economic parameters, physical limits, laboratory analyses. The system must inform people what it is doing and why.
- Specify the interface with regulatory controls, consider timing, move limits, output sequencing, process dynamics, stability and interactions.
- Specify computer hardware and software
- Consider procedures for maintaining model fidelity, process and economic

Nature of investment. Computer optimization is competing for specific investment funds with other methods for process enhancement. Many times the benefits of computer optimization might be achieved by alternate methods. For example, additional heat exchangers, larger or more efficient furnaces, more distillation trays or sophisticated stream analyzers. Computer optimization must compete with other engineering project investments on an earning power and risk basis.

There is an important difference in the character of investments in computer optimization because of its flexible nature. This is defined in Table 3. Although it requires premises about the variability of the process and business environment, it does not require premises about specific bottlenecks and limitations that must be embedded in investments for process equipment. It can often represent the best investment for making small incremental improvements against many constraints for complicated processes subject to highly varying conditions.

REQUIREMENTS FOR SUCCESS

There are a number of considerations in the design and implementation of successful online industrial optimization. The key requirements are listed in Table 4.

It is most important to select the proper independent variables. The most convenient independent variables can be selected by the structure of the regulatory control strategy. Also, the structure of the optimization must be determined. One extreme is a series of single variable local optimizers that are independent; the other extreme is a massive multivariable optimization of all independent variables simultaneously. The design of regulatory controls and online optimizers must be compatible.

Next, a formulation of the model requirements must be made to determine which dependent variables must be predicted from which independent variables.

Then the business objective functions must be written.

The next step is to select the appropriate algorithm for optimization, depending upon the nature of the problem. The model and algorithm should be robust and always report a feasible solution. The total system should handle partial equipment or instrument failures to keep the computer onstream time high.

All input variables from sensors and manual entries must be clearly specified. Important consideration should be given to the interface and displays between the optimizer and people. Since the computer optimization is a partner with process operators and management in achieving the plant operating objectives, a great deal of work is required to allow people to understand what the optimizer is doing and why, and also to allow the people to influence, interrupt and guide the functions and operation of the optimizer.

TABLE 5—What online optimizer does not do

- Determine business objectives of plant
- Specify model
- Select criteria for finding and defining the optimum point
- Select the best set of independent variables
- Specify mechanical or safety limits
- Determine market requirements for product quality
- Verify product values, fuel and feed costs, production rate limits

TABLE 6—Perspective: online optimization

Technology—Broad, complex, fundamental, evolving philosophy. Need more applications, software and expertise.

People—Involves operators, engineers, managers, planners, instrument people, programmers, maintenance groups.

Digital era—Adequate hardware and general software for automation of industrial plants. Costs are still dropping.

Measurements—Sensors generally adequate. Not a fundamental barrier, but optimization will justify further improvements.

Incentives—Control room consolidation and moderization of older plants; high capital integrated new plants; high costs of energy, raw materials, equipment, manpower; complex specifications and business demands; all contribute to escalating benefits.

Limitations—The number of doubters of digital control feasibility is in rapid decline, they will soon be extinct. However, skilled implementation people are in short supply.

The interface of the optimizer to the regulatory control system requires design attention.[4]

The particular computer hardware and software configuration for implementing the system must be specified. Large models and optimizers are commonly implemented in separate computer central processing units that are data linked to regulatory control computers.

Another important requirement is procedures for maintaining the model fidelity/accuracy: both process and economic. The real maintenance requirement for such optimizations is continuing effort to ensure that the model representations are current.

Guidelines. The basic guidelines for when to attempt online optimization are to be sure that it will be profitable and that the requirements of Table 4 can be met. If any ingredient is missing the effort will fail. The key things to look for are sufficient independent variables or degrees of freedom for optimization and the sensitivity of these operating variables on plant performance and profitability.

Another thing to look for in process plants is the presence of fixed variables that are held constant for years simply for convenience rather than for business. Examples are operating pressures of distillation columns and steam systems.[5] These are often held at fixed constant values for operator convenience or control system convenience or to reduce problems and complexities. However, they sometimes represent degrees of freedom for business optimization.

Another guideline is to consider process dynamics when implementing steady-state optimization. Strictly speaking, no process is ever exactly at steady-state, and it is difficult to quantify a degree of departure from steady-state such that steady-state models become unsatisfactory. Proper timing and sequencing of optimizer outputs to the process must account for dynamics with proper filtering of moves.[4]

The operator must be able to activate and deactivate the optimizer and various modules or portions of it. If there are some prerequisite variables that must be active to properly implement the optimizer, a check by either the computer or the operator that the prerequisites are met must be included. The operator should be informed of the status and results of the optimization.

Proper and complete documentation of the engineering design and application of the optimizer is a must.

TABLE 7—Questions for optimization applications

Function—What does optimizing computer do? What is its function, job?

Limitations—What does optimizing computer not do? What is done by people or other computers?

Benefit—What is it worth? How do I clearly see where the money comes from?

Cost—What does it cost? What are total costs and commitments?

Procurement—How do I get it? On time, in budget?

Maintenance—How do I keep it going, closed-loop, accurate, profitable?

Profit—How do I minimize my investment risk? How do I ensure computer optimization is a fast payout, low risk endeavor?

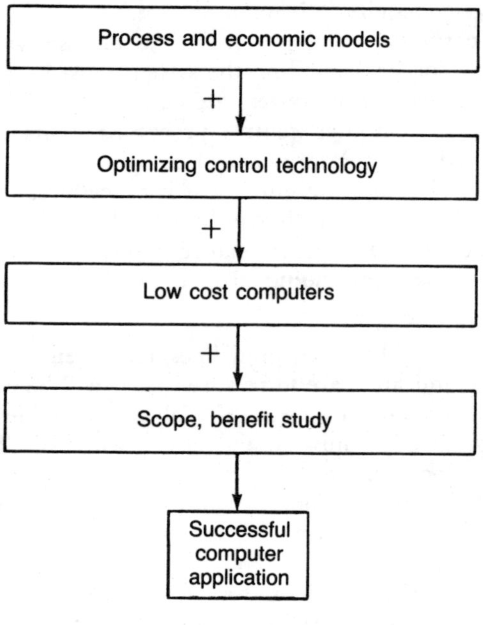

Fig. 2—State of the art for online optimization.

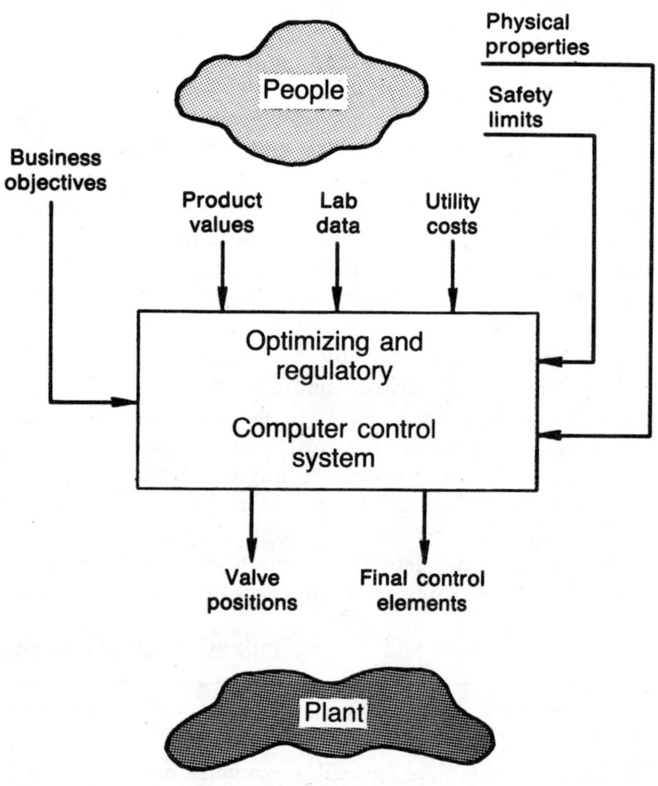

Fig. 3—Complete automation system.

These are for the operators, engineers and business managers.

An effective link between the optimizer and the business groups who have the information required for the optimizer is another guideline for successful implementation. It is obvious that if the business conditions or environment change for the plant, the proper organizational procedures must be followed to be sure the optimizer performance equation reflects these changes. The online optimizer becomes the tool which insures that these groups do influence plant operations.

Online optimization is not a panacea for all problems. Like any other technology it has its role and its limitations. We have listed some things that an online optimizer does not do in Table 5. This list is presented to clarify the role of an optimizer in operating industrial plants. All of these items currently must be performed by people. People must think and do work. These items are already done in one fashion or another for process plants, usually ad hoc. They must be formalized and quantified for the online optimizer. Offline computers for data processing and models from management science may be employed by other groups in the corporation to help people determine some of the items in Table 5.

Optimization in the future. The future outlook for application of online optimization for control of industrial processes is bright. The technical concepts are fundamental to progress.

In general terms, the status of hardware is satisfactory, generalized software is adequate, available sensors are adequate, application software technology is under development and skilled people are in short supply. The perspective for online optimization is summarized in Table 6.

The major effort is in the work for implementation. The principal questions for optimization applications are listed in Table 7. The work of the future will be oriented towards answering these questions for each application. Little fundamental research is required in these areas. This work is mostly engineering design, justification, programming, implementation and enhancement.

In 1977, Ted Williams reported on the state of digital hardware[6]: "There are no remaining technical barriers to achieve any computer control task." This means we can build automated chemical plants or petroleum refineries. There is no question that computerization and automation are on the way; they have already been achieved. The ingredients in the current state-of-the-art are illustrated in Fig. 2.

The overall shape of process control systems is becoming more generalized. Rather than appearing as a complicated set of loops on a process and instrument drawing with hundreds of setpoint input values, it can be conceived as a general functional box illustrated in Fig. 3. From this prespective optimizing control becomes a broad, complex technology. It is an evolving philosophy for bridging the gap between the plant equipment and the desires of people. It is impinging on the work interface of the process operators, instrument people, plant engineers and corporation management. We are entering an era of increasing digital automation and control of our business operations. If there are still any doubters of the digital computer today, their number is in decline.

LITERATURE CITED

1 Latour, P. R., "Process Design Decisions Related to Process Control," Paper 60b, AIChE 74th National Meeting, New Orleans, La., March, 1973.

2 Harrison, T. J., and Stout, T. M., *Minicomputers* in *Industrial Control*, ISA, Pittsburgh, Pa., 1978, p. 208.

3 Latour, P. R., "The Hidden Benefits From Better Process Control," ISA-76 Conference, Houston, Oct. 13, 1976.

4 Latour, P. R., Koppel, L. B., and Coughanowr, D. R., "Time Optimum Control of Chemical Processes for Setpoint Changes," *I&EC Proc. Des. Dev* 6, 4, 452 (1967).

5 Shinsky, F. G., *Energy Conservation Through Control*, Academic Press, New York, 1978, p. 79.

6 Williams, T. J., "Two Decades of Change-A Review of the 20-Year History of Computer Control," *Control Engineering*, *Vol. 24*, No. 9, Sept., 1977, p. 71.

To Optimize the FCC Unit

A model that establishes a patterned search for the optimum is described and then applied to variables of the fluid catalytic cracking process

William J. Fiero, Davison Chemical Div. of W. R. Grace & Co., Baltimore, Md., and **Peter E. Kelly,** The Pace Co., Consultants & Engineers, Inc., Houston

PROCESS OPTIMIZATION of a fluid catalytic cracking process requires sophisticated analysis because the process is among the more complex of all the petroleum refining processes. Total optimization of yield, for example, requires a detailed identification of the effects of feedstocks, mechanical design and the changes in operating variables. This paper illustrates a method for optimizing a process and applies that method to catalytic cracking variables.

Frequently, control and optimization of the cat cracker is delegated to changing only one variable: reactor temperature. The effect of changes in reactor temperature on yields is well known for a particular unit over the normal operating range. For example, if an increase in gasoline yield is needed to meet seasonal demand requirements, the reactor temperature is increased to convert more of the feed into lighter components—particularly gasoline. Seldom are other variables considered primarily because of the difficulty in determining if some other adjustment in operating variables will give a more optimum operation.

Seven independent operating variables can be identified:

- Feed preheat
- Reactor temperature
- Weight hourly space velocity
- Unit charge rate
- Recycle rate
- Catalyst activity
- Feedstock quality.

Although these variables are independent, they do interact, that is, the best setting of one variable depends on the values of the other independent variables. This interaction makes determination of the optimum combination of independent variables for maximizing profits or for obtaining a desired product mix a complex problem.

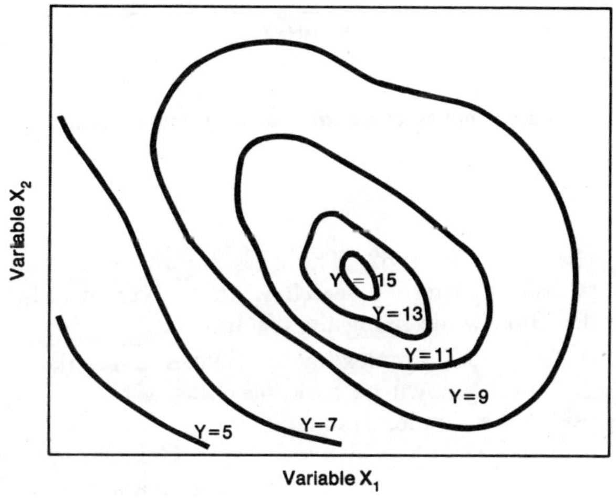

Fig 1.—Constant profitability lines related to variables.

The objective of this paper is to illustrate the effectiveness of a multivariable search technique for finding the set of independent variables which maximize FCCU operating revenues.

Sequential simplex method. There are many multivariable search techniques which could be applied to this type of optimization problem. The sequential simplex technique was chosen since it is one of the easier to apply and is one of the most effective.

Consider the situation illustrated in Fig. 1. Using X_1 and X_2 as the independent variables and the profitability Y contours as shown, the optimum combination of the variables (X_1, X_2) lies in a contour represented by Y equals 15. The selection of the optimum is obvious because the contours of the profit function are available. However, to construct enough contours for an accurate determination of the optimum requires an immense quantity of data. In many situations this is not practical. The sequential simplex method provides a systematic approach to obtain the optimum with a reasonable amount of data.

The simplex method was developed by Spendley, Hext

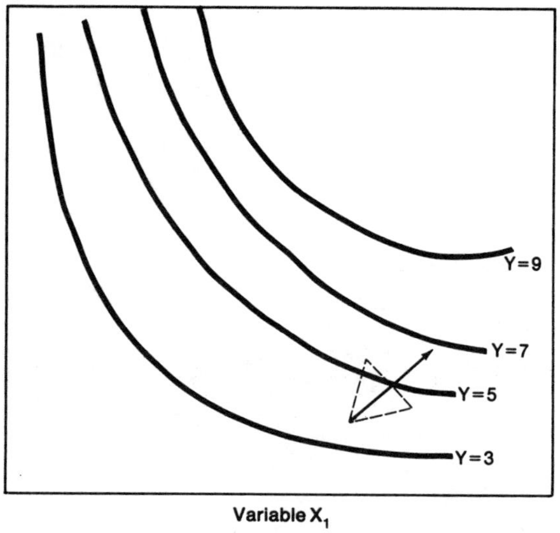

Fig. 2—Search moves away from least profitable point.

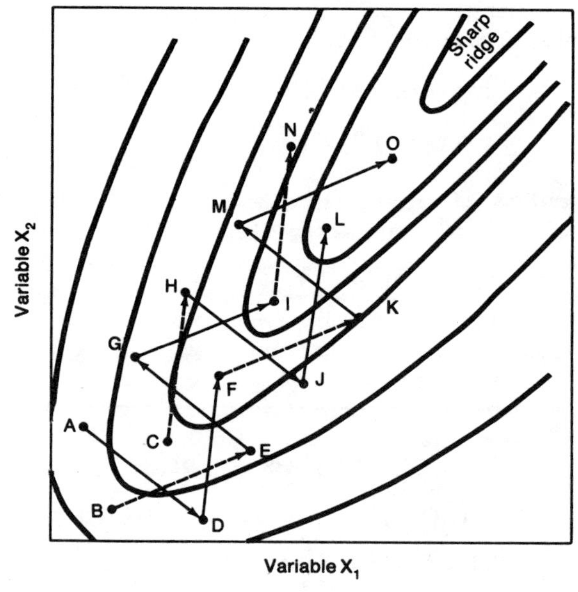

Fig. 3—Rule 2 prevents the search from being stranded.

and Hinsworth in 1962. The basis for the procedure is to establish a patterned search which moves in a favorable direction while fixing the distance of the move.

The sequential simplex method takes a regular geometric figure (known as a simplex) as a basis. In two dimensions the simplex is an equilateral triangle; in three dimensions an equilateral tetrahedron. Cases are located such that the net revenue or profit is evaluated at the points formed by the vertices of the geometric figure. The vertex with the least value is rejected. The general direction of the search is then taken away from the worst point such that the new point preserves the geometric shape of the figure (see Fig. 2). The method proceeds by the process of vertex rejection and regeneration until the figure straddles the optimum. At this point no subsequent moves would lead to further improvement since the last few geometric figures are repeated. Closer approach to the optimum is achieved by reducing the step size of the simplex.

Three rules govern the application of this technique:

Rule 1. The point rejected is the one with the lowest value of the objective function and is replaced by a reflected point in the opposite side of the figure.

Rule 2. No return can be made to points which have just been left.

Rule 3. If the best-valued vertex remains unchanged for more than M iterations, then the simplex size is reduced, for example, by halving the distance of all other vertices from that vertex. Spendley suggested that:

$$M = 1.65\,n + 0.05\,n^2$$

Where n = number of variables.

The second rule applies when the new point generated has a lower value than the previous point. In this case, the point with the second lowest value should be rejected. The effect of this on the behavior of the process in the neighborhood of a rising ridge is shown in Fig. 3. Here the arrows drawn as continuous lines represent steps in which the lowest values of Y (profitability) are rejected in turn, while the arrows drawn as dotted lines represent the application of rule 2 in which the second smallest value of the profitability function is rejected. Application of rule 2 allows the search to move up the ridge instead of becoming stranded.

Development of the simplex. An initial figure must be obtained in order to begin the search. Once this has been established, the successive simplex figures will be determined by changing only one vertex in each cycle. It is not important which simplex is used to start the procedure as long as it conforms to the desired type of figure.

In generalized terms, Table 1 gives the initial simplex for a problem with variables starting at the origin.

It can be seen that apart from the first row and a diagonal of p terms, the table is composed of only q terms. The value of p and q are determined from the length of

TABLE 1—Coordinates for a set of simplex vertices

Vertex	Coordinates of each vertex					
j	$A_{1,j}$	$A_{2,j}$	$A_{3,j}$	...	$A_{n-1,j}$	$A_{n,j}$
1	0	0	0	...	0	0
2	p	q	q	...	q	q
3	q	p	q	...	q	q
4	q	q	p	...	q	q
.	.	.	.	...	.	.
.	.	.	.	...	.	.
n	q	q	q	...	p	q
n + 1	q	q	q	...	q	p

TABLE 2—Product values

Products	$/bbl
$H_2 - C_2$	11.40 (Fuel oil equivalent)
$C_3^=$	14.15
C_3	8.73
$C_4^=$	11.76
iC_4	10.21
nC_4	9.79
C_5—430° F	14.20
Light cycle oil	13.06
Clarified slurry	10.97

the edge of the simplex, a, (desired magnitude of change in the variables for each iteration) and the number of variables, n, as expressed below:

$$p = (a/n\sqrt{2})\ (\sqrt{n+1} + n - 1)$$

$$q = (a/n\sqrt{2})\ (\sqrt{n+1} - 1)$$

Generation of a new simplex. Following establishment of the starting simplex, the remainder of the sequential simplex method involves rejecting one vertex and generating a new one. The new simplex in each cycle is therefore made up of the remaining vertices of the previous simplex together with the new vertex. The new vertex is derived as a mirror image of the rejected vertex reflected through the center of gravity formed by the remaining vertices of the original simplex. The coordinates of the new vertex, N, for an n-dimensional (variable) case can be generalized by:

$$X_{i,N} = [2/n]\ [(\Sigma_{j=1}^{j=n+1} X_{i,j}) - X_{i,R}] - X_{i,R}$$

where $i = 1, 2 \ldots, n$
N = new point
R = rejected point.

Applying the equations for development of the starting simplex, generation of new vertices and the three simple rules of rejection, regeneration and cessation, the sequential simplex method can be used to provide an optimum solution for any number of variables. The ability of the sequential simplex method to optimize a multivariable problem is well suited for use in catalytic cracking. To illustrate the use of this approach for optimization of an FCCU, a sample two variable problem has been worked through. A two variable problem was selected to both simplify the calculations and to enable the reader to visualize the procedure.

THE FCCU PROBLEM

The primary variables in catalytic cracking which have the greatest effect on yields are reactor temperature and the catalytic activity of the catalyst. The catalytic activity is normally measured by a microactivity (MA) test. As noted previously, the reactor temperature is often changed to make yield adjustments. In general, variation of the catalytic activity is made only when a major change in yields is required such as when changing operations from maximum gasoline to maximum furnace oil. Seldom are both of these operating parameters adjusted on a frequent basis to obtain an optimum operation mainly due to the difficulty in determining the optimum value for these variables on a timely basis.

The example considered is based on a 30,000 bpd riser cracker with conventional regeneration. It was assumed that the unit is presently operating at a 970° F reactor temperature. Fresh catalyst is added at a rate of 1 tpd to maintain a 69 microactivity. Fig. 4 shows the relationship between catalyst additions, as a function of the percent inventory replaced, and the equilibrium microactivity. In this case increased activity is achieved by increased fresh catalyst additions and not by a change to a more active catalyst.

The product yields resulting from each combination of temperature and activity were calculated using the Pace

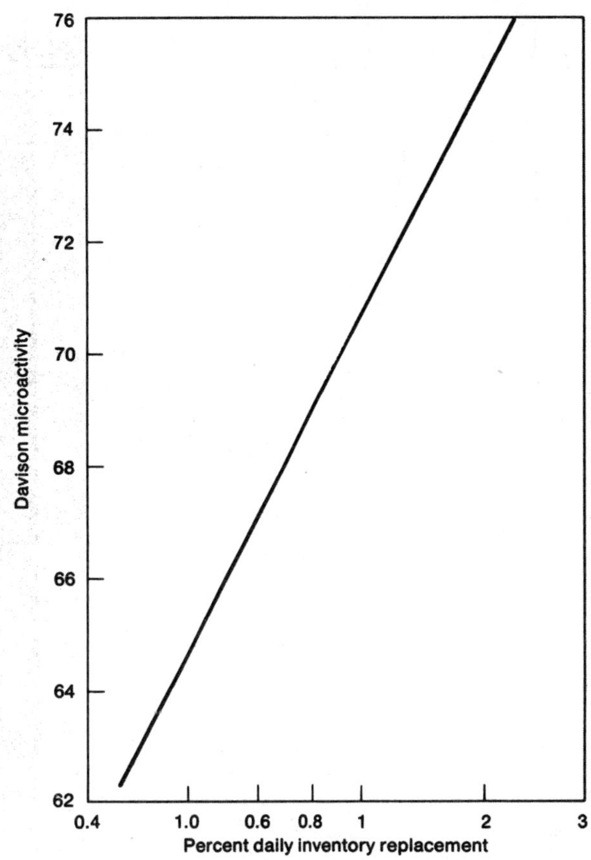

Fig. 4—Microactivity controlled by adding fresh catalyst.

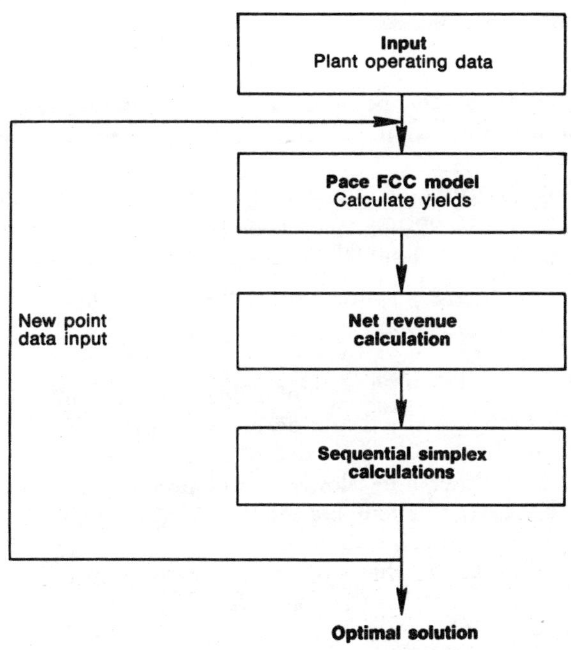

Fig. 5—Schematic diagram of program flow.

FCCU model. The product values shown in Table 2 were used to determine the gross revenue. Catalyst cost ($765/ton) was subtracted from the gross revenue to obtain a new revenue. The simplex method was then used to locate the next point. The optimum was con-

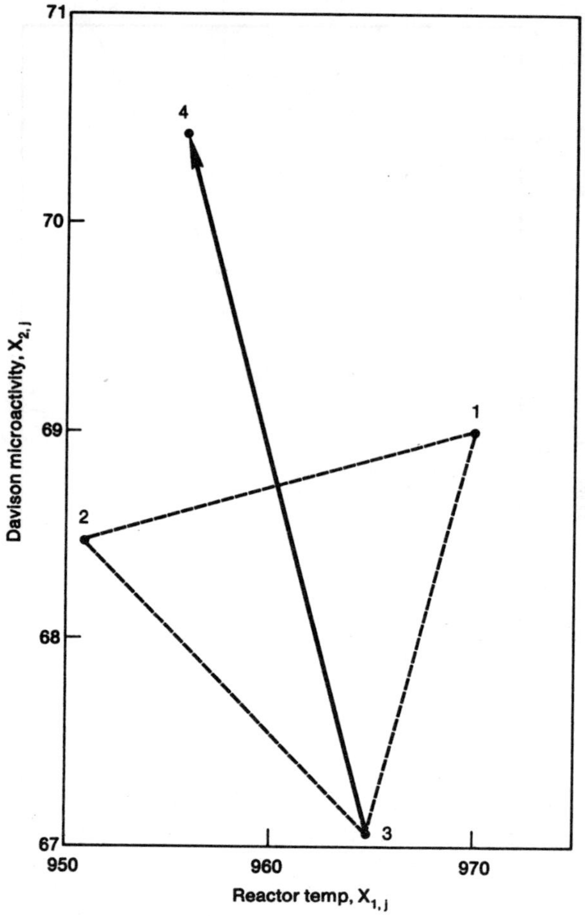

Fig. 6—Rejecting low profit point leads to better point.

sidered to be the point of maximum net revenue. The overall optimization procedure is represented by Fig. 5.

Initial simplex. A starting simplex must be generated to begin the optimization technique. The optimization is for a two dimensional case where $n = 2$ so that:

$$p = (a/2\sqrt{2})\ (\sqrt{2+1} + 2 - 1) = 0.9657\ a$$

$$q = (a/2\sqrt{2})\ (\sqrt{2+1} - 1) = 0.2587\ a$$

The size of *a*, the length of a side of the triangle, determines both the speed and accuracy of the initial search. That is, the smaller the value of *a* the more accurate the search but the more slowly it will progress. A value of 20 (20° F reactor temperature, 2.0 MA) was chosen as a reasonable step size to obtain both a rapid and fairly close approach to the optimum. A smaller step size would require additional costly computer runs whereas a larger step size would result in a larger area to search when a reduction in step is required to obtain a closer approach to the optimum.

Using a step size of 20 and the starting values for the equilibrium microactivity of 69 and the reactor temperature of 970° F, the coordinates of starting simplex are shown in Table 3 and were calculated as follows:

Vertex 1: $X_{1,1} = 970$, reactor temp;
$X_{2,1} = \ 69$, microactivity.

TABLE 3—Searching for maximum revenue (a = 20° F or 2.0 MA).

Points, j	Reactor temp., $X_{1,j}$	Micro-activity, $X_{2,j}$	Marginal revenue $/day	Point rejected, R	Rule used	Points in simplex
1	970	69.0	0	..	..	Starting simplex
2	950.68	68.48	+627	..	..	
3	964.83	67.06	−136	..	..	
4	955.85	70.42	+1008	3	1	1, 2, 4
5	936.53	69.9	+1303	1	1	2, 4, 5
6	941.7	71.84	+1383	2	1	4, 5, 6
7	922.38	71.32	+1500	4	1	5, 6, 7
8	927.55	73.26	+1612	5	1	6, 7, 8
9	908.23	72.74	+1619	6	1	7, 8, 9
10	913.4	74.68	+1674	7	1	8, 9, 10
11	894.08	74.16	+1653	8	1	9, 10, 11
12	899.25	76.10	+1562	9	1	10, 11, 12
13	918.57	76.62	+1561	11	2	10, 12, 13
14	932.72	75.20	+1587	12	2	10, 13, 14

TABLE 4—Reducing the step size (a = 5° F or 0.5 MA).

Points, j	Reactor temp., $X_{1,j}$	Micro-activity, $X_{2,j}$	Marginal revenue $/day	Point rejected, R	Rule used	Points in simplex
1A	913.4	74.68	0	..	..	Starting simplex
2A	912.1	74.20	− 2	..	..	
3A	908.6	74.55	−23	..	..	
4A	916.9	74.33	−40	3A	1	1A, 2A, 4A
5A	918.2	74.81	− 2	2A	2	1A, 4A, 5A
6A	914.7	75.16	− 6	4A	1	1A, 5A, 6A
7A	909.9	75.03	− 1	5A	2	1A, 6A, 7A

Vertex 2: $X_{1,2} = 970 - pa = 970 - 0.9657(20) = 950.68$
$X_{2,2} = \ 69 - qa = \ 69 - 0.2587(2.0) = \ 68.48$
Vertex 3: $X_{1,3} = 970 - qa = 970 - 0.2587(20) = 964.8$
$X_{2,3} = \ 69 - pa = \ 69 - 0.9657(2.0) = \ 67.06$

Using the Pace FCCU simulation program, the gross revenue of each point was calculated. Catalyst costs were subtracted to obtain a net revenue. The first three points of Table 3 summarize the marginal revenue in dollars per day of the starting simplex using the current operation as a basis.

New point calculation. In all subsequent steps, the remainder of the sequential simplex method involved rejection of one vertex and the generation of a new one such that the geometric shape of the simplex was preserved. In this case vertex 3 was rejected as having the lowest net revenue and a new vertex, 4, was generated. Using the formula for vertex generation, the coordinates (value of the variables for the next simulation) of the new vertex are:

$N = 4$, vertex number
$n = 2$, number of variables
$R = 3$, rejected vertex number.

For the reactor temperature:

$$X_{1,4} = [2/2]\ [(\sum_{j=1}^{j=3} X_{1,j}) - X_{1,3}] - X_{1,3}$$
$$= (970 + 950.68 + 964.83 - 964.83) - 964.83$$
$$= 955.85$$

For the microactivity:

$$X_{2,4} = [2/2]\ [(\sum_{j=1}^{j=3} X_{2,j}) - X_{2,3}] - X_{2,3}$$
$$= (69.0 + 68.48 + 67.06 - 67.06) - 67.06$$
$$= 70.42$$

A diagram of the result of rejecting vertex 3 and generation of the new vertex, 4, is shown on Fig. 6. By taking a mirror image of the worst point, together with the two other points, a new equilateral triangle was formed, which was used as the basis for the next move.

Using the new values for the variables, a simulation was run and a net revenue was obtained. On the basis of the net revenue for the three points in the simplex and the application of rule 1 or 2, the procedure of vertex rejection and regeneration continued until the figure straddled the optimum. Table 3 and Fig. 7 show the progression of the sequential simplex method. As can be seen, a cycling has occurred around vertex 10. At this point, the optimum is within an area bounded by microactivities 72.74 to 76.62 and reactor temperatures 894.08 to 932.72. A closer approach to the optimum can only be achieved by reducing the step size.

Step size reduction. Using vertex 10 as as basis and recalling rule 3 regarding a step change, a new simplex was generated reducing the step size from 20 to 5 (5.0° F, 0.5 MA). The starting simplex for the new step size is given by the first three points in Table 4. Again applying the rules for vertex rejection and the procedure of vertex regeneration, Table 4 summarizes the progression of the sequential simplex at the reduced step size. Coincidentally, cycling around the starting point occurred even at reduced step size. Although a closer approach to the optimum could be achieved by a further reduction in step size, the differential between the maximum and minimum net revenue being only $19/day indicates that the optimum lies in a fairly flat plateau and as such the present optimum can be assumed to be a good approximation of the true optimum.

The optimum for this unit represents an approximate $500,000/year gain in net revenue over the initial operation. The gain in revenue was achieved by reducing the reactor temperature and increasing catalyst additions resulting in approximately $775/day in additional catalyst cost. The additional catalyst costs were more than offset by both the improved activity and selectivity of the equilibrium catalyst through increased fresh catalyst additions.

RANGE OF APPLICATION

The simplex method can be applied to a unit design optimization or to optimization of an existing unit. In the case of a design, the profit function would account for equipment investment as well as product value and operating costs. The cases then would be compared on a discounted cash flow basis.

In the case of operating units, the model could be replaced by obtaining operating data on the unit instead

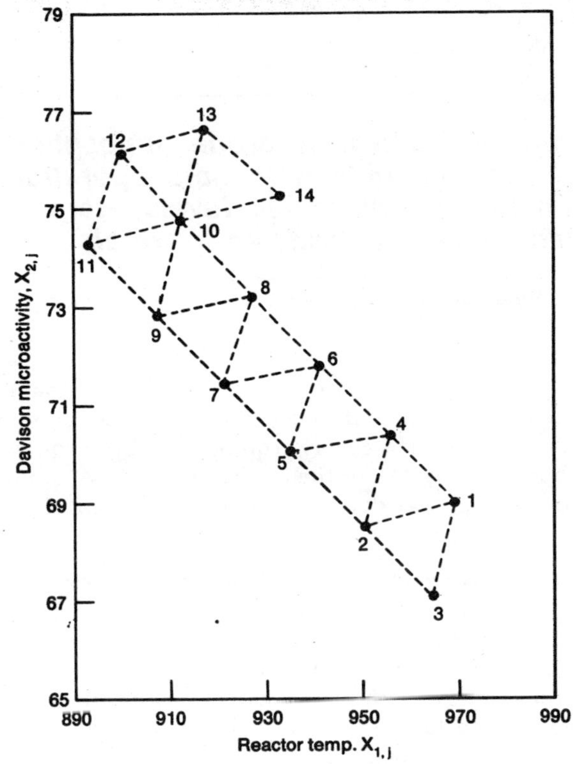

Fig. 7—Progression of the search sequence.

of simulating the operation with the model. However, this technique can be difficult and expensive to apply. Process "noise" often masks the true response of a variable change. and it may be necessary to repeat an experiment several times and average the results. Also, experimentation may require operating at conditions that are suboptimal resulting in severe revenue losses. Thus, the model greatly facilitates the optimization procedure.

Summary. Although only a two variable problem was considered here, the sequential simplex method is easily extended to include any number of FCCU operating variables. Also, although constraints such as gas compressor or blower limitations were not considered, the method can be easily modified to take these constraints into account. The procedure provides not only a method to optimize a multivariable problem but is easy to program, conceptually simple and requires no statistical analysis. The overall optimization technique is an effective tool which can be used on a timely basis to maximize FCCU operating revenue.

Computer control of crude atmospheric and vacuum units help hold operation at optimum energy saving levels. Here are some operating variables to watch

C. J. Ryskamp, H. L. Wade and **R. B. Britton,**
The Foxboro Co., Foxboro, Mass.

ONE INCENTIVE FOR CLOSE CONTROL of crude fractionation is to save some of the energy used to get crude separation. The separation energy required usually is equivalent to approximately 1% of the crude oil charge. A computer control system was developed and is in successful operation to hold energy requirements to a minimum.

Energy savings may come about in a number of ways. One of the most obvious savings may come from improved control of the furnace for more efficient fuel utilization. There are, however, many other places where improved control can lead either to a direct reduction of energy usage, or else to an increase in product value at no increase in utilities cost.

A typical atmospheric crude tower is shown in Fig. 1. The places where potential opportunities exist for saving energy are numbered in this figure and are listed in Table 1. The control techniques for accomplishing energy savings are discussed in the following sections.

Utility costs vs. product values. A study of utility cost for a crude unit must recognize that as fuel costs have increased, the crude costs and product values have risen also. This creates a need to evaluate product distribution along with the potential energy reduction.

TABLE 1—Location of potential crude unit energy savings

1. Adjust flows through parallel heat exchangers for maximum heat transfer.
2. Allocate and control pumparound heat duties to maximum values, limited by constraints.
3. Control top reflux to an optimum value to minimize energy loss to the atmosphere.
4. Increase the yield of more valuable products and the product value added by improved separation at no increase in utilities cost.
5. If gas-oil is being produced for FCCU feed, maximize its production from atmospheric tower, subject to a color constraint, thereby decreasing load on vacuum tower.
6. Adjust heater firing rate for minimum overflash or to balance heat input requirements to the tower, subject to maximum firing rate limitations.
7. Balance flows through furnace passes for maximum heat transfer.
8. Control furnace air-to-fuel ratio for maximum efficiency and safety.
9. Optimize stripping steam to crude and vacuum tower bottoms.
10. Minimize stripping steam to side strippers as required by product specifications.
11. Reduce slop oil treating and reprocessing by improved desalter control.

The relationship between product composition and utility consumption is similar to that for a two product distillation column. For such a column, Shinskey[1] defines separation as a numerical quantity which is a function of the two product compositions. For each value of separation, there is a value of distillate-to-feed ratio which maxi-

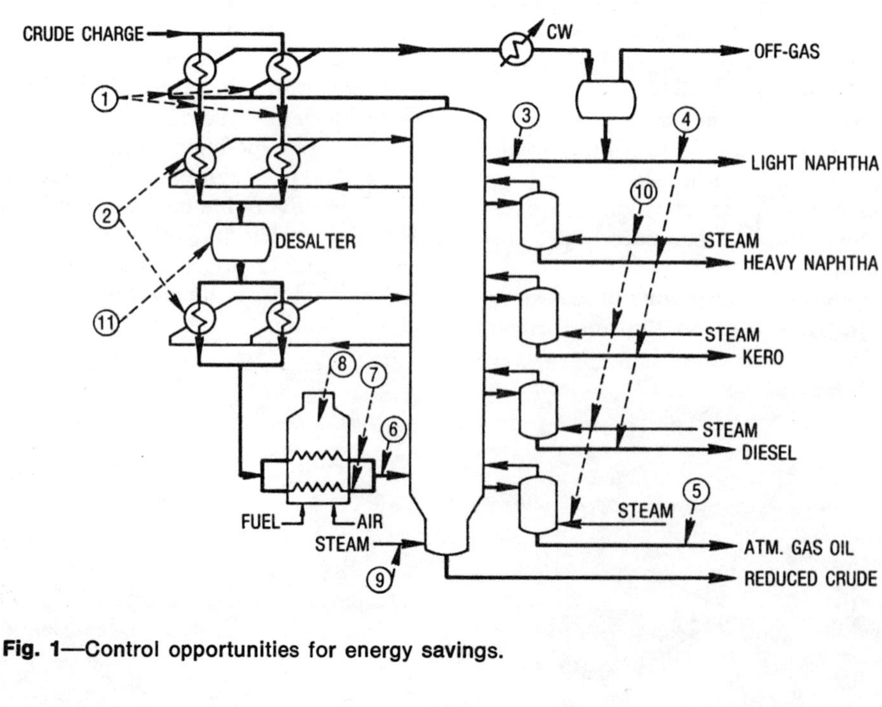

Fig. 1—Control opportunities for energy savings.

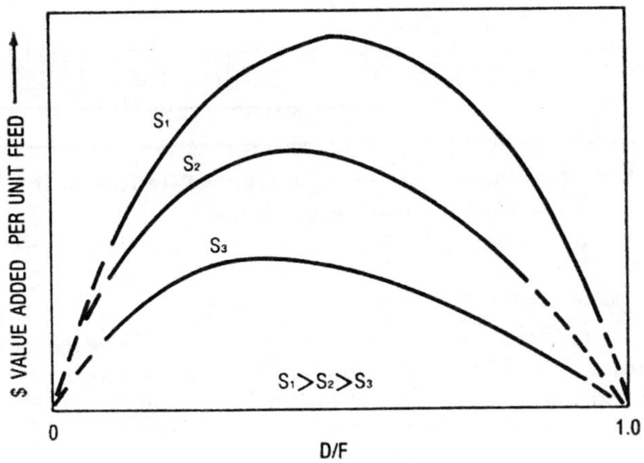

Fig. 2—Product value minus feed cost gives value added.

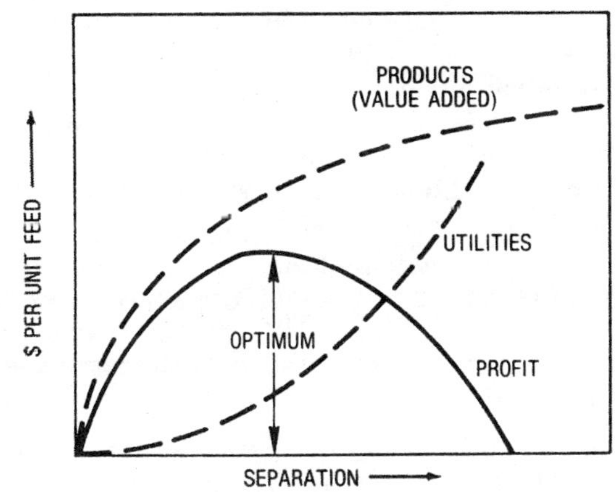

Fig. 3—Value added minus utility cost gives profit line.

mizes the value added per unit of feed. This is shown in Fig. 2. If the maximum value added per unit feed is plotted versus separation, the curve rises rapidly at first as shown by the uppermost dashed line in Fig. 3. However, increasing separation follows the law of diminishing returns, so that the value added becomes asymptotic to a horizontal line.

A plot of the cost of utilities versus separation, shown by the lower dashed line of Fig. 3, is concave in the opposite direction; that is, it becomes increasingly steep as separation is increased. At the origin, where both value added and cost of utilities are zero, and again at the point where these two curves cross, the difference between the two curves, or the net profit shown by the solid line in Fig. 3, is zero. Between these limits there must be a maximum profit. This is the point of optimum separation and should be the operating goal for the column, as long as physical constraints are not violated. If one or more constraints exist which prevent raising the separation to the optimum, then the control system should keep the unit operating at the maximum separation subject to the constraints. If the constraints shift, so will the maximum separation.

This concept also applies to the various sections of a crude tower. Increasing internal reflux flow will improve separation between two neighboring products. This means the overlap will be reduced or the gap increased between the ASTM end point of one product and the initial boiling point of the next heavier product. Because a distillation or flash point specification generally limits the yield of the more valuable product, better separation allows the cut point to be adjusted to increase this yield, increasing the total product value. Obviously, if internal reflux can be increased without any significant change in utility cost, it should be done. However, if increasing internal reflux raises utility costs, the optimum should be determined.

A quantitative method of calculating the optimum separation, or reflux flow, is beyond the scope of this paper. The degree of complexity and the confidence in the results, depends on the time and tools available. A comprehensive off-line simulation of the crude tower with its strippers, etc., could be used in an economic analysis. Alternatively, the process itself may be operated at different levels of constant separation (or reflux-to-feed ratios) and the actual utility costs and product values may be determined and compared.

Parallel exchangers. Parallel heat exchangers become necessary when the crude flow rate is so high that a single bundle for one service would be too expensive or too large. When the crude flow splits into two or more parallel streams, the hot oil product flow will usually be split into the same number of parallel streams. The question then arises whether altering the flow distribution will transfer more heat over-all. It is not sufficient to depend on symmetrical piping to distribute the flows equally; furthermore, equal flows will not necessarily be optimum, especially after some fouling has occurred. Raising the crude preheat temperature 1° F on 100,000 bpd flow rate is worth $10,000 per year in fuel savings, based on $1 per MMBtu and 75% heater efficiency.

The over-all heat transfer rate for parallel exchangers will be maximum if all the crude outlet temperatures are the same and all the hot oil outlet temperatures are also equal to each other. The crude outlet temperatures can be kept equal by adjusting crude flows, and the hot oil flows can be adjusted to keep their outlet temperatures equal. Some dynamic decoupling may be advisable, but not mandatory since relatively slow response is acceptable.

If the control valves used to distribute the flow through parallel heat exchangers are not used to control the total flow, then the system must be designed to always have at lease one valve wide open to minimize the pressure drop due to friction.

If crude vaporization occurs in parallel heat exchangers, then balancing is very important not only for heat recovery, but also to reduce fouling. This system is not self-regulating for if one bundle becomes hotter, the extra vapor generated inhibits crude flow forcing more through the other bundles. The hot one gets hotter and the others cool down further. This problem is more severe with a greater temperature difference between crude and product.

Sidestream energy control. Crude towers have pump-around or recirculating streams which remove heat from various levels. There are normally two or more pumparound streams on a tower. Hot oil is withdrawn from the column and cooled by heat exchange with crude and returned a few trays higher in the column. The pumparound flow rate is generally independent of internal reflux flow entering the pumparound trays. Frequently the lower and hotter pumparound streams may also be used to supply heat for some other service such as a stabilizer reboiler.

The heat capacity is very large for the combined steel and oil in the column, piping, heat exchangers, etc. A specific temperature gradient is required along this heat sink to create the desired tray compositions. Control of the temperature gradient alone, however, is not sufficient to maintain constant tray composition, since the temperatures will change with variations in the partial pressure of the heavier hydrocarbons in the vapor phase. Furthermore, changing heat removal in a pumparound stream may have very little effect on tray temperatures in that part of the column, especially if product flows are adjusted to maintain tray compositions. Basically, the heat removal rates should be used to obtain desired vapor and liquid loading on the trays, with tray compositions controlled by the product withdrawal rates. Because changes in one product draw rate affect the composition of other product draws, a decoupling form of composition control is desirable.

The tray vapor and liquid loading rates are not directly available from on-line measurements. An exception is in the case of total draw trays, where the tray liquid loading rate may be available from external measurement. Generally, however, both the liquid and vapor rates at critical trays may be inferred, using available direct measurements in conjunction with mass and energy balance equations about sections of the tower. This approach has been successfully used for on-line manipulation of pumparound heat removal rate.

The heat transfer rate from a pumparound stream can be calculated from the flow rate, temperature drop and specific heat of the oil. This calculated heat transfer rate can then be controlled by manipulating pumparound flow to keep the heat transfer rate at the set point. Controlled heat transfer rates provide more stable column operation than flow control of the pumparound streams. Pumparound flow variations of up to 20%, over a two hour period, have been observed with pumparound loops under constant heat duty control.

For multiple pumparound loops, the method of allocating the heat removal rate to each loop must be considered. At the time a crude unit is being designed, there are economic incentives to remove heat low in the column. However, once the unit is built and the capital investment made, these incentives no longer exist, and any heat that is transferred to crude by the pumparound streams should be removed as high in the column as possible. This will create higher liquid and vapor flows and increase separation on all the trays between the top and bottom pumparounds. As long as the heat goes to crude and is not lost to atmosphere, there is no significant increase in utility cost for this increase in separation.

The Typical constraints which limit the top pumparound heat removal are:

- Control valve wide open
- Desalter temperature at high limit (incipient vaporization)
- Flooding conditions on or between trays of the top pumparound and the next sidestream drawoff.

Middle pumparound heat removal is normally limited by the first and last of the above constraints, and should normally be kept at maximum also.

If the furnace is firing at a maximum rate, limited by fire box temperature constraint, outlet temperature up to incipient cracking or flame patterns, then the bottom pumparound heat removal rate should not be at its maximum value. The heat removal from the bottom pumparound may be the dependent variable which is manipulated to keep top reflux at a set value. The following section discusses this more fully, where in the long term the reflux flow is controlled at an optimum rate.

If the furnace outlet temperature can be raised, then it is generally most profitable to increase the heat removal in the bottom pumparound to its maximum. The normal constraints are:

- Control valve wide open
- Furnace inlet temperature at maximum (vaporization upstream causing high pressure drop)
- Flooding conditions approached on or below the bottom pumparound trays.

This maximizes crude preheat and is a major factor in saving utilities. Under these conditions reflux flow may be kept at optimum by manipulating heat input with the furnace, by slowly changing the set point of the outlet temperature controller.

The strategy just described of maximum heat removal by the crude tower pumparound streams may require alteration depending upon the heat removal needs of the vacuum tower. If the vacuum tower depends upon heat transfer to crude for a major part of its heat removal, then the temperature of the crude into these exchangers must be low enough to provide sufficient heat transfer. Process economics and constraints must be considered in designing a control strategy which allocates the heat removal in an optimal manner.

If the vacuum tower is a lube oil fractionator, it can be treated similar to the crude tower. If the vacuum gas oils go to catalytic cracking feed, there is no need for separation between the light vacuum gas oil (LVGO) and heavy vacuum gas oil (HVGO). Then the heat should be removed where it can do the most good. This is normally the HVGO pumparound. The LVGO pumparound normally has atmospheric cooling and should be kept at minimum unless extra heat removal is needed to keep tower top temperature down. Generally HVGO pumparound should be manipulated to control tower top temperature, but this can be a slow response loop, so LVGO may need to compensate for transients.

Optimum reflux. Reflux flow to the top tray is normally manipulated to control top temperature, which provides faster response composition control. The set point of the top temperature controller may be adjusted to meet a composition specification. However, the heat loss to atmosphere through the overhead condenser will vary

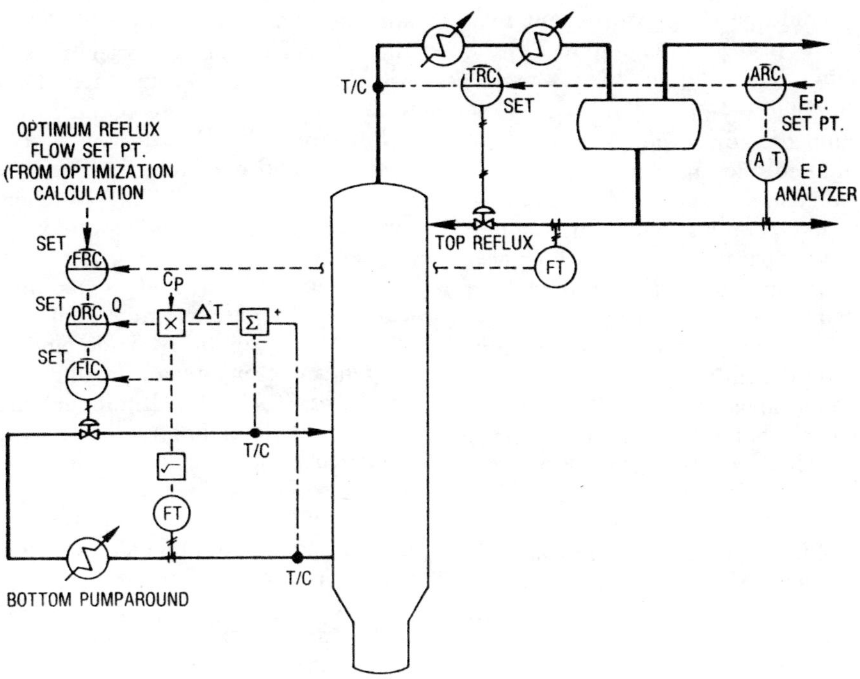

Fig. 4—System to optimize reflux flow.

with reflux flow. When reflux flow is too high, heat may be needlessly wasted, and when too low, the process may not be as profitable as it could otherwise be, due to poorer separation.

The optimum reflux flow may be determined by comparing the cost of heat loss to atmosphere with the benefits of improved separation in the top part of the column as has been discussed previously. The reflux flow should then be controlled to this optimum value in the steady state to assure that utilities are being profitably used. A standard PID control algorithm (or controller) may be used for reflux flow. If the optimum reflux is calculated as a ratio-to-crude charge, or a ratio-to-top product, the flow control set point may be ratioed to the proper measured flow rate. Since in the short term, reflux flow is controlled to maintain overhead temperature (which, in turn is adjusted for composition control) and only in the long term is reflux flow controlled to an optimum value, then the flow controller must be tuned to react slowly. Fig. 4 depicts a scheme for manipulation of reflux for top temperature control plus long-term (steady state) control of reflux at an optimum rate.

Though partially discussed previously, the following principles are of sufficient importance to repeat for clarity and completeness. When reflux is fixed, overhead heat removal is fixed. It is the sum of the heat removed to condense the reflux, top product and steam. Since overhead heat removal is fixed, some other heat flow variable must be manipulated to maintain the heat balance. The furnace is best for over-all utility conservation, or the bottom pumparound if the furnace is limited.

Directly manipulating heat input or removal near the bottom of the column to control top temperature would be too slow. The heat sink is too massive and the temperature gradient too large from the bottom to the top of the column. When the furnace outlet temperature is increased slightly, the extra vapor generated would be primarily gas oil vapor at about 650°F. This would initially be totally condensed by the large mass of oil and steel at the lower temperatures, and would not affect overhead vapor flow or top temperature for several minutes, when enough heat has been absorbed to raise the temperatures of all trays.

Controlling heat input by the furnace or heat removal with the bottom pumparound to keep top reflux at optimum is a major economic benefit. This also tends to stabilize liquid and vapor flows through the whole column because upsets in heat input are smoothed out in the lower part of the column, before compositions are significantly upset, and reflux flow is more steady.

Often cold crude is heated by the overhead vapor before the vapor passes to an air or water cooled condenser. This complicates the economic calculations. Heat loss for top product (light naptha) and steam condensation cannot be avoided, but heat loss to atmosphere for reflux flow can still be optimized. The utility cost for reflux is only a part of what it would be without crude heat exchange. Therefore, higher separation with higher reflux flow is justified. When the fraction of heat transferred to crude increases, an increase in separation is justified and reflux should be raised.

The total overhead condensing duty can be calculated from measured flows of reflux, product and receiver water. The fraction of the total duty transferred to crude can be calculated —from the crude side temperature and flow. The remaining fraction is the heat lost to atmosphere. It can be assumed that the heat removed to condense reflux would be split —into the same fractions. Instead of reflux flow, this duty (the fraction of the heat loss to atmosphere for reflux condensation) could be the measurement to a controller which manipulates heat input or removal in the lower section of the column. Then the heat loss for separation needs can be held at optimum.

If all the pumparound heat duties are at maximum, but the furnace is not limited and there are no flooding problems, then more crude can be vaporized if reflex flow is increased. Under these conditions, optimum reflux would be higher because separation will be improved through the whole column, and the yield of distilled products will be increased, with less reduced crude. The vacuum unit must be included in this part of the analysis to minimize utilities. This optimum crude tower top reflux flow depends on the change in yields, the disposition of gas oils and the useful heat that can be removed from the vacuum tower, as well as spare capacity on the vacuum furnace.

Overflash. The lowest side draw product, atmospheric gas oil (AGO), usually joins the vacuum gas oils and becomes catalytic cracking feedstock. There is no end point specification, but the product must not contain too much black oil because this cokes the cracking catalyst. Therefore, a color specification normally limits the AGO draw rate. The AGO draw rate should be kept at maximum to keep overflash flow to a minimum. This minimizes the utility requirements on the vacuum unit.

Overflash flow is the internal reflux that continues down through the trays below the AGO draw tray. The minimum overflash flow required to keep an acceptable AGO color is very small, only a couple percent of the crude flow. It also may be quite variable because the problem is one of mechanical entrainment of black oil. This depends on liquid surface tension and foaming tendency as well as vapor velocity. The actual overflash flow can be calculated on-line, but the accumulated errors become very large because many measurements are used. An online color analyzer setting AGO draw rate is probably the easiest method to keep minimum overflash.

If a vacuum unit which produces lube stocks is to receive the reduced crude, then more overflash is justified in the crude tower bottom. The extra separation is needed to keep the higher viscosity oils out of the AGO and the diesel fuel out of the vacuum unit. A control system which holds separation constant may be required for this type of operation.

Another example where constant separation would be required (i.e., the gap or overlap should remain fixed) is if diesel fuel were the lowest side draw and most of the diesel fuel should be removed from the reduced crude. To keep separation constant, at optimum, the overflash to crude ratio or overflash to diesel fuel product ratio must be held constant. This means the percent vaporization of the feed is determined. The diesel fuel and all lighter products must be vaporized, and enough additional vapor must be generated so that the extra heat can be removed to provide the required optimum overflash. The furnace firing rate in this case cannot be the dependent heat balance variable for the whole column. Instead, the furnace must be fired to provide the optimum overflash flow.

Furnaces. Parallel passes on the furnaces should be balanced so that the corresponding outlet temperatures are all equal. This forces higher mass flow through the coils with higher heat transfer rates. It not only increases furnace efficiency, but also reduces tube coking tendency.

Furnace air/fuel ratio can be controlled if air flow is measured and controllable. The ratio can be adjusted to obtain the best efficiency. However some safeguards are needed to avoid a reducing atmosphere. In order to guarantee excess air during transients, the controls must force the air to lead the fuel on an increase and the fuel to lead on a decrease in load. A combustibles analyzer on the stack can alarm or override and increase air flow. An oxygen analyzer can alarm or be used in closed loop to set the air/fuel ratio. Limits are needed to avoid serious upset on analyzer failure. The normal problems of sample point location, air leakage, fire box pressure control, etc., must also be considered. The operators should also shut off the air flow to any burner out of service when they are not needed, or are plugged, in order to keep a meaningful value for air/fuel ratio.

Stripping steam. The costs associated with stripping steam are:

- Initial cost for generation
- Condensing cost
- Water treating cost
- Increase in column vapor load may limit pumparound rates
- Decrease in hydrocarbon partial pressure lowers temperatures and heat transfer potential.

Stripping steam should be kept to the minimum required for a product to meet specifications, or used only to the extent where the increase in value of product recovery justifies the cost. Stripping steam does improve separation between products and the concept of comparing utility costs with product values can be applied.

The primary action of stripping steam is to lower the hydrocarbon partial pressure which vaporizes more oil. Most of the heat for vaporization comes from the oil itself. As steam flow is increased, there is a diminishing return since each additional pound of steam has less ability to vaporize oil. Generally, the cost for steam in the bottom of the crude or vacuum tower per barrel of oil vaporized should not be allowed to be as high as the cost of vaporization in the heaters because of the other detrimental affects listed above.

Product flow can be used to set steam flow and the ratio lowered to minimum to conserve utilities. When steam flow is changed, it should be done slowly unless a critical specification must be met, such as flash point. Because of the large heat sink, slow changes over about an hour are necessary for column bottom steam to avoid unwanted transients in tray temperatures and compositions. For side products it is conceivable that the ratio of stripping steam to oil flow could be adjusted slighly when the ratio of internal reflux to product flow changes and affects separation. However this has not yet seemed to be justified.

Steam to the vacuum tower and ejectors should not be so high that it overloads the condensers. Hot condenser outlets tend to break the vacuum which creates the need for more fuel in the furnace. This wastes steam as well as increases fuel.

Desalter. The common buoyancy float level transmitter used for indicating level of the interface between oil and water is affected by changes in density of the crude around the float. If the crude is heavier than the nominal gravity used for transmitter calibration, then the transmitter output will be in error on the high side, that is the indicated interface level will be higher than the true interface. A similar statement is applicable if the crude is lighter than the nominal gravity used for calibration. If the level controller which receives the transmitted signal is operated at constant set point, then the true interface varies with crude gravity and temperature, even though the controller maintains a constant indicated interface level.

From the crude gravity and the desalter inlet temperature, a correction factor may be calculated and applied to the transmitter output, producing a true interface signal. Controlling true interface level can reduce water treating and reprocessing costs. An added benefit is the reduction of carryover of water with the oil, which tends to upset the column.

Control implementation. Many of the control concepts presented here could be implemented with analog instrumentation. For other concepts, involving complex calculations such as the computation of pumparound heat removal rate from temperature difference and flow rate, or the desalter level transmitter compensation, the cost of analog calculation hardware may be prohibitive. In still other cases, such as the calculation of tray loading from mass and energy balance equations, the use of analog computational elements is simply not practicable.

All of these functions may be implemented with computer control. It is unlikely that all of the concepts discussed would be applicable at any one installation. The most significant of these concepts, including on-line calculation and closed loop control of tray loading by manipulation of pumparound heat removal, optimum reflux control, decoupling control of crude tower product compositions and desalter interface level compensation have been field proven.

Careful attention must be given to the operability of a computer control scheme of this complexity. Experience has shown that a well designed, highly modular control scheme, composed of computer set analog controls, software "blocks" performing calculation and control algorithms, and Fortran programs performing the highest level calculations, can be readily assimilated by the operators. A good operator training program is also highly desirable. The separate control functions should be easily understood by the operator, and should be accessible for intervention if necessary. For example, it should be possible to manually enter a set point to the heat duty controls in place of the output from the higher level tray loading controller. In this manner, operator confidence and acceptance will be gained so as to obtain maximum utilization of the computer control system.

Summary. The utility savings by computer control of a crude and vacuum unit can be several hundred thousand dollars per year. Often the benefits begin to accrue soon after a process analysis is made, due to alterations in operating philosophy or implementation of minor process changes. Significant benefits are realized, however, after implementation of a well engineered control system. With computer hardware costs falling and utilities costs rising, computer control will be justified on smaller units as time progresses.

Original presentation was before the API Division of Refining Meeting, Los Angeles, May 12, 1976.

LITERATURE CITED

[1] Shinskey, F. G., *Process Control Systems*, McGraw-Hill, 1967.

Improve Steam Cracker Operation

With applications and examples, this new pyrolysis simulation program based on fundamental reaction kinetics offers the olefins producer a stepping stone to pyrolysis know-how

A. G. Goossens, M. Dente and **E. Ranzi,** PYROTEC, Zoetermeer, The Netherlands

A KEY FACTOR in modern olefin plant design and operation for maximum flexibility and feedstock utilization is accurate and detailed yield/furnace performance prediction, including onstream time assessment, for ethane through gas oil feedstocks or mixtures (co-cracking).

The SPYRO program is a unique coincidence of many years of fundamental development work of a highly qualified scientific team, together with the indispensable practical experience and know-how of a company specializing in the design of steam crackers.

It has been extensively tested and demonstrated to represent a stepping stone in pyrolysis know-how, providing excellent precision and versatility.

Above all, SPYRO is a practical and ready-to-use tool for the olefin plant manager or his R & D department in feedstock and energy conservation programs, outperforming regression models or even sophisticated feedstock test units, enabling enhanced savings of at least 1-4% on ethylene production costs.

The design and operation of a modern olefins facility must be optimized for many complex variables to ensure flexibility and efficiency in processing a variety of feedstocks.

The key to combat soaring feedstock and production costs—feedstock costs represent about 70% for a naphtha based olefin plant—is enhanced optimization and control of the steam crackers.

This can be attained through fundamental under-

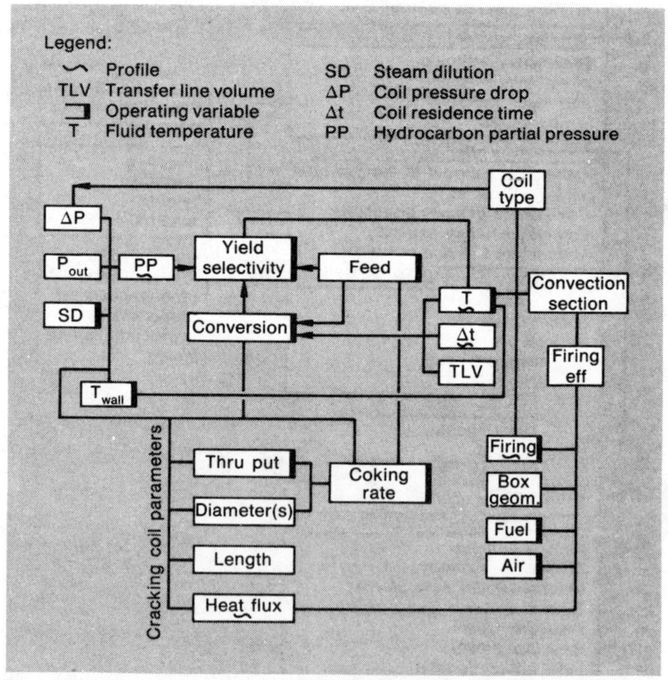

Fig. 1—Multi-parameter interrelationship of pyrolysis furnace.

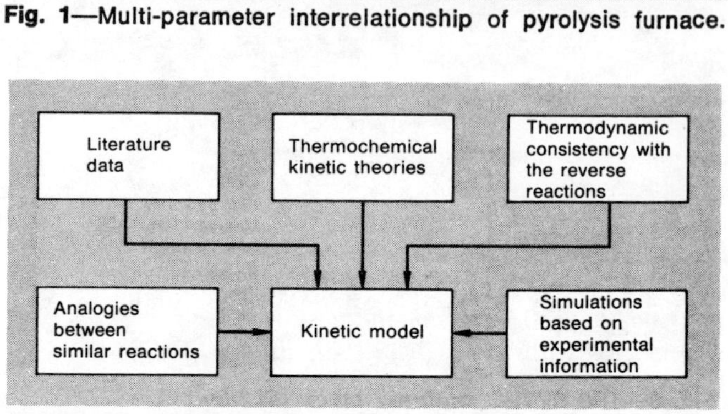

Fig. 2—Sources of kinetic data used in the SPYRO program.

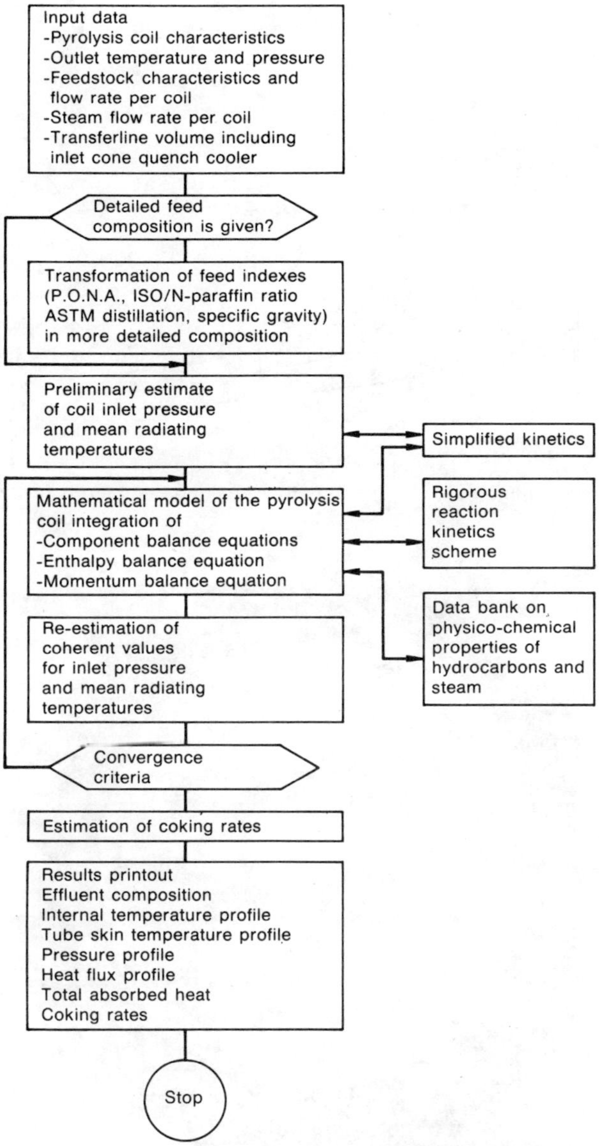

Computer system	CDC IBM 370/115 Univac 1106, 1108 or equivalent
Computer language	Fortran IV
Memory allocation	40 k (decimal) words
Run time	60-120 system seconds

Fig. 3—The SPYRO program block diagram.

standing and mathematical modeling of the actual pyrolysis kinetics and subsequent optimization of the complex furnace multi-parameter interrelationship (Fig. 1), both in design and operation, prior to linking with total plant optimization schemes.

Even marginal yield improvements—beyond analytical detection—have a major impact on incremental profitability. Typically, for a 500,000 metric tons/yr. (mt/yr.) ethylene plant, 0.8 wt. % increase in ethylene selectivity corresponds with a net profit gain of $1 MM U.S.

Essential to this is the highest accuracy and most detailed prediction of yields, furnace performance and coking rates, for any feedstock or mixture (co-cracking).

Our pyrolysis furnace simulation program (SPYRO) is based on a fully mechanistic reaction kinetics model and has been thoroughly tested. It combines nine years of fundamental development work and 15 years of accumulated field data, and is now offered to the olefins manufacturing industry. This program is expected to enable production savings or gains of 1-4% of ethylene transfer price, through better furnace operating management, feedstock selection and production optimization.

The program is unique in that it employs one and the same kinetic model throughout for all feedstocks. It represents a stepping stone in pyrolysis know-how, lifting this from the level of typically semi-theoretical/empirical methods to applied science. Therefore, it may be used with confidence in areas beyond the range of practical experience in the search for new pyrolysis methods as well.

Using only the widely applied computer control may achieve a modest saving of 0.8-1.0% on ethylene transfer price.[2] This is because reliable full furnace effluent analysis techniques are lacking and models based on regression or overall kinetics (despite the input of vast amount of data and effort) fail in the required marginal accuracy. Also, they may not be extrapolated or used for mixture cracking.

Pilot plant tests are time consuming and costly, and are often of insufficient reliability for co-cracking.

PYROLYSIS SIMULATION MODELS

Although well established that thermal cracking proceeds essentially through radical reactions,[6] the prevailing trend followed by most authors was to adopt molecular models.[3,4,5]

Kinetic models can be categorized (in order of complexity and reliability) as:

- Empirical regression of operating parameters;
- Stochiometric modeling or global molecular (apparent) reaction kinetics; and
- Strictly mechanistic models, based on detailed description of main radicals and molecular reactions paths.

This latter approach has generally been confined to single hydrocarbons or simple mixtures and incomplete kinetic schemes and is still a far cry from industrial applicability.

The practical "state of the art" (for light HC only) is well described in a recent article[3] but at the same time it is admitted that the "extent to which the model will predict in regimes outside those of the data base is dictated by the ability of its reaction network to reflect the actual kinetics."

Thus, mechanistic models apply better for mixture cracking and extrapolations beyond range of experimental evidence. In addition, they allow making use of all reliable literature and experimental data available, concerning the significant elementary reactions involved, independently from the specific case considered, thus contributing enormously to reliability of the model.

Finally, they are the only practical way to account properly for "point effects," such as intermediate injections of hydrocarbons or the influence of the (adiabatic) transferline volume.

On the other hand, the general consensus appears that for practical reasons of stiffness of coupled differential equations involved, mainly caused by free radical reaction kinetics and complexity of the reaction network, the rigorous mechanistic modeling of pyrolysis reactions is an almost impossible, if not an impractical goal.

Nevertheless, this last approach was followed in preparing a large kinetic model to simulate accurately the cracking of single hydrocarbons or mixtures from ethane through naphthas and gas oils.[7]

Finally, it is imperative for correct simulations of industrial steam cracking furnaces and prediction of yields, as operations are essentially non-isothermal and non-isobaric, that the kinetic model is properly coupled with the systems' enthalpy and momentum balances.

This requires in addition an intimate knowledge of theoretical and practical aspects of design and operation of pyrolysis furnaces.

GENERAL STRUCTURE SPYRO PROGRAM

The essential elements constituting the Simulation of Pyrolysis Operations program are presented in the simplified block diagram (Fig. 2).

It is obvious that a detailed description of the program would go far beyond the objectives of this article. Several blocks, however, need further clarification for a proper appreciation of the unique qualities of the program. They are: the kinetic model, the mathematical model, the coking model and the input/output features of the program.

Kinetic model. Seven classes of elementary (radicalistic or molecular) reactions are considered[7] as summarized in Table 1.

The criteria on which the development of a consistent kinetic model were based are illustrated by the scheme of Fig. 3.

The kinetic constants have either been drawn directly from the literature[9,10,11,12] or estimated. This was done by utilizing typical methods of thermochemical kinetics[8] and fitting some reliable and suitable experimental data, or utilizing obvious mechanistic analogies, or, finally, checking for thermodynamic consistency between forward and reverse elementary reactions.

The estimation of kinetic parameters could have been performed through a non-linear optimization program. However, excessive computer time cost and the difficulty to define in advance a satisfactory object function precluded this approach. Thus, judicious adjustment of parameters was made manually.

It is noticeable also that in the pyrolysis of small saturated molecules (ethane) true molecular elementary reactions can play a non-negligible role (as observed first by Benson[8]). Obviously, these molecular reactions are of greater importance in determining the fate of certain cyclic unsaturated molecules (i.e., cyclopentenes or cyclohexadienes, etc.) On the other hand, these reactions are negligible with increasing C-atoms number of the saturated molecules (due to the strong increase of the relative contribution of free radicals to the decomposition mechanism).

Another significant point, quite important for purposes of extrapolation and prediction of rate constants, concerns the analogies among similar reactions pertaining to the same class.

For instance, it is convenient to assume the hypothesis,

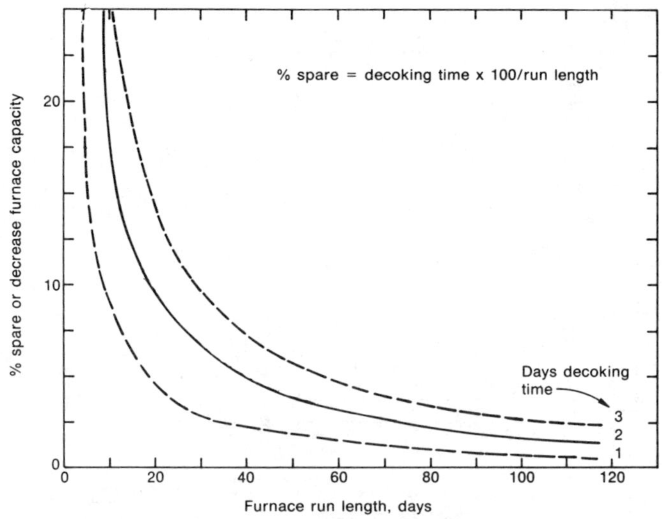

Fig. 4—Influence of run length/decoking time on furnace capacity.

Fig. 5—Section of coked pyrolysis tube.

TABLE 1—Elementary reaction classes

1. Chain initiation reactions
   $R-R' \longrightarrow R^\bullet + R'^\bullet$ (ex. $C_2H_6 \longrightarrow 2\ CH_3^\bullet$)
2. Hydrogen abstraction reactions (metathetical reactions)
   $R^\bullet + R'H \longrightarrow RH + R'^\bullet$ (ex. $CH_3^\bullet + C_2H_6 \longrightarrow CH_4 + C_2H_5^\bullet$)
3. Radical decomposition reactions
   $R^\bullet \longrightarrow RH + R'^\bullet$ (ex. $C_2H_5^\bullet \longrightarrow C_2H_4 + H^\bullet$)
4. Radical addition reactions to unsaturated molecules
   $RH + R'^\bullet \longrightarrow R''^\bullet$ (ex. $C_2H_4 + CH_3^\bullet \longrightarrow C_3H_7^\bullet$)
5. Chain termination reactions
   $R^\bullet + R'^\bullet \longrightarrow R-R'$ (ex. $CH_3^\bullet + C_2H_5^\bullet \longrightarrow C_3H_8$)
6. Molecular reactions
   $RH + R'H \longrightarrow R''H + R'''H$ (ex. $C_2H_4 + H_2 \longrightarrow C_2H_6$)
   and molecular dimerisations of Diels-Alder type:
   (ex. $C_4H_6 + C_2H_4 \longrightarrow$ cyclohexene)
7. Radical isomerisation reactions
   $R'^\bullet \longrightarrow R''^\bullet$ (ex. $CH_3CH_2CH_2CH_2^\bullet \longrightarrow CH_3\text{-}CH_2\text{-}\dot{C}H\text{-}CH_3$)

of fairly general acceptance, that the Arrhenius parameters of any single H-abstraction reaction depend only from the H-type (primary, secondary, tertiary, allylic and so on) and from the attacking radical ($H^\bullet$, $CH_3^\bullet$, $C_2H_5^\bullet$, etc.) and are **practically** independent from the attached molecule. The acceptance of this assumption (with rare special exceptions) permits strongly reducing the amount of the elementary kinetic data. With this in mind, the adjustment of the reference values of the Arrhenius parameters was started essentially from the literature information; they have undergone subsequent modifications to account for often quite different operating conditions and, above all, the experimental evidence (in terms of product distribution obtained from several pure hydrocarbons and mixtures). This is the only simple way to obtain a reliable pre-estimation method of the total H-abstraction reactions rates. For the various radicals here considered, the kinetic parameters generally obtained are in the range of the published data.

With the obvious exceptions, this analogy-approach was used also for other classes of elementary reactions.

In this way the total number of truly independent sets of kinetic parameters necessary to develop a satisfactory kinetic model (for ethane through gas oils) has been substantially reduced from more than 2,000 possible reactions to about 500, involving 18 chain propagating radicals and 85 molecular species and pseudo-compounds (from H_2 to C_{22}). Most radicals containing more than 4 carbon atoms decompose (almost) instantaneously into smaller species.

The results obtained through numerous simulations made over a large number of pure hydrocarbons or mixtures, cracked in quite different conditions, ranging from bench scale experiments to industrial crackers, were very encouraging.

Mathematical model. As a complicating factor, not only compositions but also temperature and pressure are varying along the cracking coil.

Local enthalpy and momentum balances are to be solved simultaneously with the continuity equations of molecular and radical species.

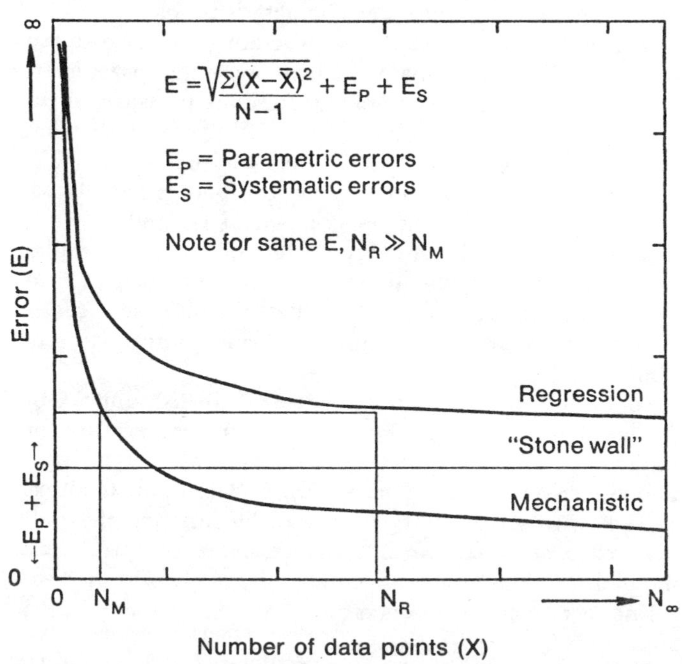

Fig. 6—Attainable accuracy of models.

Proper boundary conditions are to be given. Due to the values of the various reaction rates, covering several orders of magnitude, the stiffness of the resulting very large system of differential equations is unavoidable.

As a result, the ratio between the concentration of a molecular reactant and the concentration of a very reactive radical (as $H^\bullet$) may be 10^8 or more.

Several general purpose numerical techniques have been published to deal with large stiff systems. Nevertheless, to gain computation efficiency and stability without sacrificing accuracy, a specifically conceived technique was developed.

Any numerical method of integration should be judged on the basis of:

1. The non-violation of physical bounds (e.g., non-negative values for the component concentrations).

2. The sensitivity for the step-size of integration, concerning stability and speed to reach convergence.

A satisfactory method, based on the use of a computation criterion intrinsically stable and giving positive values for radical concentrations, has been published.[7]

To by-pass the ill-conditioning of the systems of differential equations that could call for very small integration steps, it is convenient to assume steady radicals concentrations (total rate of formation = total rate of disappearance of radicals).

In fact, it was demonstrated by Blakemore and Corcoran[13] that the time interval required to obtain a practical steady concentration is a very small fraction of the time necessary to obtain a sensible evolution of the remaining system.

Clearly, the pseudo-steady radicals concentration is continuously varying along the coil and has to be re-estimated at any new integration step.

The pseudo-steady state approach is not strictly necessary, but the use of this procedure has ensured an acceptably fast and stable convergence to the (positive) values for the radical concentrations, without any appreciable difference with the non-steady state procedure, up to very high temperatures (1000° C).

In general, the necessary total number of increments is in the range of 40-100, and is specifically imposed by the non-isothermicity of the system, rather than the continuity equations integration method used.

Coking model. Of particular interest for design and operation of pyrolysis furnaces is the correct estimation of the coking rates in the coils, which is an important constraint. As a consequence:

- Run-time/thruput is limited by maximum permissible metal skin temperature or coil pressure drop.

- Increased downtime (Fig. 4). Furnace run-time of less than 50-60 days influences olefin production costs sensibly by increased maintenance and reduced equipment lifetime and capacity utilization. Combined coil/quench cooler cleaning time requires typically 2-3 days.

- Increased utilities consumption. Decoking consumes steam and fuel. Increase of tubeskin temperatures decreases firebox efficiency, or necessitates reducing thruput. Typically, during one onstream cycle, tubeskin temperature may rise 100° C, resulting (for same thruput) in a drop of 1-2% in firebox efficiency, thus increasing fuel consumption about 5%.

- Increased coil pressure drop decreases selectivity of cracking (less ethylene).

Thus, if coking can better be accounted for both in design and operation, substantial savings may be obtained. The SPYRO kinetic model also provides a mechanistic approach towards the carbon deposition rate.

For commercial conditions, it is essentially based on the following presumptions, based on ample theoretical and experimental evidence.

- The coke deposit is covered by a continuous layer of polymeric material, transforming gradually into coke. It can be proven that for these conditions steam is practically inactive as a decoking agent.

- The limiting step in the carbon deposition is the addition of unsaturated radicals or molecules to the surface of the polymer layer. Typical coke-precursors are vinylic or phenylic radicals, whereas unsaturated compounds, particularly acetylenics, may combine directly with the growing (polymeric) surface through Diels-Alder type reactions.

- Based on radical lifetime, the effective radicals are those produced (by the gaseous reactions) in the direct vicinity of the growing surface, thus at or near the deposit surface temperature.

The "coating" effect of the polymeric layer precludes the assumption of increased surface activity by porosity and substantially inhibits catalytic effects of metal surface or metal included in the coke deposit. Such effects only play a role in the initial coking rate and are usually suppressed by sulfur compounds.

The literature [14,15,16] reports poorly conceived coking mechanisms as being mass transfer controlled. Instead, the boundary film, due to its higher temperature regime, would act as a source rather than as a barrier for coke precursors.[17,18]

However, no molecular but only radicalistic species have concentrations low enough to justify observed coking rates, but their lifetime is much shorter than time needed to diffuse through the total boundary layer.

Nevertheless, small bore tubes, having lower Reynolds numbers, are subject to more relative contribution of film cracking, leading to higher concentrations of coke-precursors near the coke-surface.

A definite proof of coking being kinetically controlled is given in Fig. 5. The deposited coke layer of a section of cracking tube clearly demonstrates the sensitivity of coke deposition to the circumferential temperature distribution. In fact, maximum coke thickness is here about 1.6 times the minimum coke thickness located at 120° angle from maximum, in accordance with theoretical heat distribution for radiation for staggered tube configurations, corresponding with a temperature difference of about 25° C, and indicating a local overall apparent activation energy for coking of about 25,000 Kcal/Kmol.

Thus, an increase of 10° C of max. fluid film surface temperature may increase coking rate about 18%. This is an important consideration with respect to both design (single vs. staggered tube configuration, maximum design clean inside surface temperature) and operation

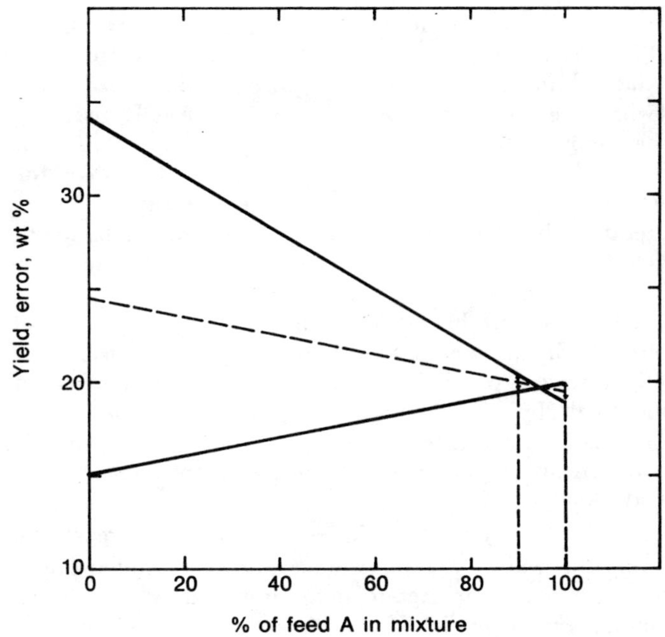

Fig. 7—Error magnification for extrapolated co-cracking yields.

(don't overshoot exit temperature set point), as coke lay down increases exponentially with temperature and is irreversible.

Both (circumferential) average and maximum coking rate predictions of the SPYRO model have been found to be quite satisfactory, taking into account the considerable difficulty to obtain consistent data from commercial plants over an extended period of time.

Not so easily accounted for are spot effects (i.e., burner impingement or mal-operation), but at least on a relative basis (trends), predictions are reliable. In contrast, test furnaces cannot furnish such information for obvious reasons.

PROGRAM INPUT AND OUTPUT

As is indicated in Fig. 2, input and output are basically simple and self-explanatory, i.e., for simulation of an operating furnace, once the characteristics of a cracking coil have been specified to the program (it can be stored input), only the normally known operating parameters must be given as inlet temperature, exit pressure, thruputs, feedstock composition (if variable), days on stream or estimated extent of tube fouling, and object variable.

Full effluent break-down up to C_{10} plus is calculated, including CO, CO_2 and break-down of C_4 and C_5 cuts.

The program is also very versatile, enabling simulation of more complicated problems as well, such as split coils, variable tube diameters, flat tubes, coked tubes, multiple feed injections, different firing pattern or sections.

Input/output units can be metric/SI/British units. Object variables to be specified are coil exit temperature, or required cracking depth, expressed as (either) key-component conversion or, alternatively, certain product ratios.

These are solved by the program by resetting the firebox temperature(s) until convergence is reached. This method guarantees good accuracy, definitely within measuring accuracy of temperatures, as may be concluded from Fig. 9.

Hydrocarbon feed is specified by components. Naph-

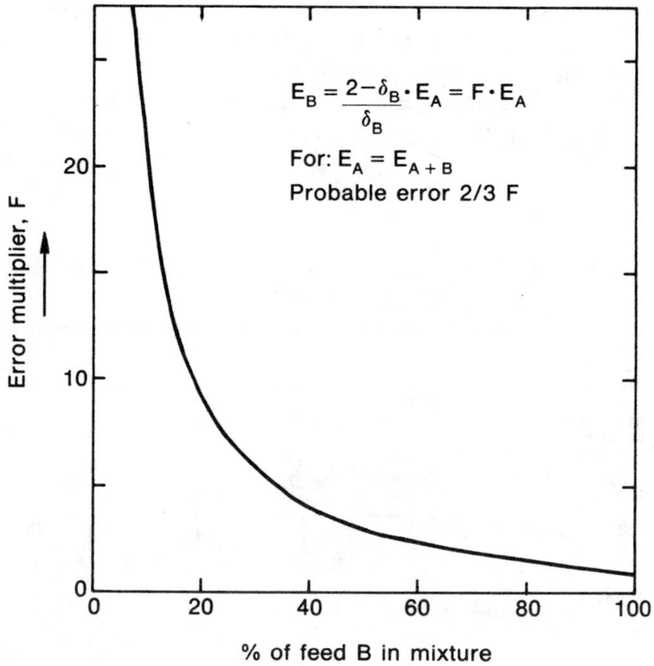

Fig. 8—Error multiplier for extrapolation.

thas and gas oils may be specified by density and boiling range (and PONA for naphtha), but greater precision can be achieved if specified through about 30 components and pseudo-components.

The SPYRO program can easily be linked with short-cut simulation models of quench cooler, firebox and convection section to enable total furnace heat and material balance computations.

TABLE 2—Regression model versus SPYRO

Model	Experimental data needed	Furnace types	Feed types	Extrapolation	Marginal calculations	Flexibility	Accuracy
SPYRO	none	all	all	yes	yes	high	high
Regression	numerous & continuous	one or few	few	no	no	limited	limited

ACCURACY AND RANGE

This aspect is of particular importance, as it is the key factor to justify the development of a mechanistic model to permit "marginal" computations.

The model is in principle very universal. Geometrically, all known furnace configuration can be simulated. The true limitations are not known yet, but the model was successfully applied in the range of residence times of 10 to several thousands of milliseconds, temperatures between 400-1000°C and pressures from sub-atmospheric to several atmospheres, and for a variety of feedslates.

In testing the SPYRO program, a real difficulty was to find experimental data of sufficient accuracy. Data from commercial plants (direct effluent analyses or plant balances) may and often do suffer from large relative errors, thus these may serve merely for a global check.

Only carefully monitored pilot plant runs may provide data of sufficient accuracy and detail useful to check the SPYRO program both on absolute yield prediction as well as on trends (regarding sensitivity to change operating parameters).

Such evaluations were most successfully performed for light hydrocarbons and mixtures, of which numerous data were collected during many years.[19, 20, 21] Also for naphtha, of which experimental data originated from a customer's pilot plant, all SPYRO yield predictions fall within the experimental scatter band.

Generally, the accuracy of a model is determined by the following factors:

- The extent to which the model takes into account all significant parameters; and
- The quantity and quality of independent experimental data used for its evaluation.

Most models, by their very origin, have been fitted against data from a single unit, and in the best case overall cross-checked against other plant data.

The random error may be reduced with the square

TABLE 3—Commercial-plant performance for light HC-cracking versus predictions[3]

	1	2	3	4
Feed, wt. %				
Methane	1.08	0.01	....	0.29
Ethylene	0.02	....	....	0.02
Ethane	97.20	1.54	....	69.31
Propylene	1.35	0.70	....	0.46
Propane	0.35	97.12	1.57	28.96
n-Butane	....	0.63	95.33	0.96
i-Butane	....	....	3.10	....
Operating conditions				
Res. time above 650° C, sec.	1.39	1.35	1.07	0.85
HC partial pressure out, ATM	1.30	1.35	1.51	1.32
Obs./pred/SPYRO outlet temp. °C	829/830/829	834/834/834	793/793/795	849/846/845

	Obs.	Pred	SPYRO	Obs.	Pred	SPYRO	Obs.	Pred	SPYRO	Obs.	Pred	SPYRO
Yields, wt. % on feed												
Hydrogen	3.51	3.61	3.52	1.50	1.45	1.30	0.63	0.61	0.74	2.90	3.05	2.85
Methane	6.43	6.10	6.75	23.74	23.35	24.60	18.94	19.05	18.98	11.92	11.20	11.90
Acetylene	0.19	0.28	0.26	0.50	0.63	0.40	0.19	0.26	0.14	0.53	0.41	.40
Ethylene	46.59	46.32	46.13	33.49	33.36	33.33	27.02	26.18	27.30	46.21	45.50	46.16
Ethane	37.50	37.53	37.41	4.94	5.02	5.46	6.92	7.02	6.04	27.45	27.43	27.80
MA & PD } Propylene	1.70	1.50	1.60	0.20	0.54	0.18	0.30	0.34	0.13	0.14	0.20	0.05
Propylene				15.05	14.86	14.85	21.96	23.64	21.36	4.09	4.03	3.78
Propane	0.09	0.18	0.16	9.72	9.71	9.72	0.91	1.41	1.03	1.74	1.85	2.06
Butadiene	1.25	1.30	1.23	1.92	2.64	2.21	2.19	2.35	2.28	2.10	1.72	1.59
Butenes	0.32	0.20	0.26	0.90	1.04	0.85	2.54	2.87	2.92	0.59	0.38	0.34
Butanes	0.29	0.21	0.37	0.11	0.29	0.05	11.18	11.41	12.30	0.06	0.24	0.19
C_5 plus	2.13	2.77	2.31	7.93	7.11	7.05	7.22	4.86	6.78	2.27	3.99	2.88
Total	100.00	100.00	100.00	100.00	100.00	100.00	100.00	100.00	100.00	100.00	100.00	100.00

root of the number (minus one) of experimental points (see Fig. 6).

However, any systematic and parametric errors remain, resulting in a stone-wall effect regarding attainable accuracies. Only a true fundamental and universal model can use data from any source for further verification or improvement of kinetic constants.

In this way, the systematic errors of data sources become random errors to SPYRO, and its accuracy may increase again with the square root of number of independent sources.

For this reason, the SPYRO model can be regarded to be several times more accurate then the best possible, but less mechanistic, regression or pseudo-kinetic model, and will even outperform the most sophisticated pilot plant.

Table 2 gives a survey of features of SPYRO model vs. regression type models.

Particularly interesting is the case of mixture cracking, if one would attempt to investigate in a pilot plant the effect of the cracking of, say, a 90/10 mixture.

One can easily conclude from Figs. 7 and 8 that any error is magnified almost 20 fold if one attempts to quantify the effect of co-cracking, leading to relative errors far too large to allow economic calculations.

Extrapolating the yield from one mixture composition to the other is not permitted as selectivities are influenced by kinetic interactions and depend on mixture composition.

The SPYRO model does not suffer from these problems; due to its mechanistic nature, it is ideally suited to reliably enable such "marginal" computations (i.e., co-cracking of ethane or raffinates).

As an example, it was recently concluded[22] that co-cracking of ethane and propane would adversely influence (ultimate) yields on ethylene, apparently based on misleading pilot plant data as mentioned before.

However, SPYRO can prove the opposite is true, also confirmed in the literature.[3]

As a simplified explanation, one may say that the enrichment of the reaction system with hydrogen (from ethane) leads to the overall effect that higher olefins (i.e., from propane pyrolysis) will be more hydrogenated to yield lower olefins (particularly ethylene) and methane. In addition, pyrolysis tar formation decreases.

Similar effects are observed with co-cracking of ethane with naphtha, or even admixture of free hydrogen to the reaction system.

EXAMPLES

To demonstrate SPYRO's qualities, examples have been selected from published information, as far as possible, covering a broad range of feedstocks and operating conditions.

Unfortunately, most sources are quite incomplete in reported operating conditions, etc. Nevertheless, for the cases selected, this missing information was assumed on the basis of practical knowledge of analogous (industrial) cases.

For **light hydrocarbons,** SPYRO predictions are compared with observed and predicted yields of a recent publication,[3] as demonstrated in Table 3. The agreement must be regarded as excellent, being at least within range of experimental uncertainty.

Experimental C_5 plus yields are usually rather inaccurate, but SPYRO seems to be in better agreement with the observed data—especially for heavier compounds—than the published stochiometric model.

For **naphthas,** both absolute yields and trends with feedstock type and exit temperature, as predicted by SPYRO are within experimental error with published

TABLE 4—Commercial naphtha cracking versus SPYRO predictions[23]

	1		2		3		4	
Feed, wt. %	Light naphtha				Heavy naphtha			
Normal paraffins	43.3				31.8			
Iso-paraffins	37.9				32.0			
Naphthenes	14.7				26.9			
Aromatics	4.1				9.3			
Density	0.671				0.715			
IBP/FBP, °C	39-107				52-163			
Coil Exit temp. °C	820		840		820		840	
	Pred.	Obs.	Pred.	Obs.	Pred.	Obs.	Pred.	Obs.
Yields, wt. % on feed								
Hydrogen plus methane	15.9	16.0	17.4	17.8	14.9	15.0	16.4	16.4
Acetylene	.3	.3	.5	.4	.3	.3	.5	.4
Ethylene	26.9	27.7	28.9	29.3	24.6	24.7	26.1	26.4
Ethane	5.0	4.1	4.6	4.0	4.1	3.8	3.9	3.8
Sum C_2	32.2	32.1	34.0	33.7	29.0	28.8	30.5	30.6
Propylene	17.5	17.6	16.1	16.4	15.3	15.2	13.9	14.2
Sum C_3	18.7	18.9	17.6	17.7	16.4	16.3	15.1	15.2
Butadiene	4.6	4.6	4.7	4.4	4.7	4.6	4.6	4.6
Sum C_4	11.5	11.4	10.0	10.0	10.1	10.4	8.7	9.2
C_5 plus	21.7	21.6	20.9	20.8	29.5	29.5	29.4	28.6
Benzene	6.0	5.3	7.0	6.2	6.8	6.2	8.1	7.1
Toluene	2.5	2.1	2.7	2.5	4.7	4.0	4.9	4.3
Propylene/Ethylene ratio	.65	.64	.56	.56	.62	.62	.53	.54

NOTES: 1. For simulations, full naphtha analyses were used as published.
2. Published yield data have been normalized to 100 wt. %, assuming "losses" originated from accumulated measuring errors.
3. Other operating conditions: HC partial pressure out = 0.80 ATM
Res. time above 650° C = 0.48 sec.

TABLE 5—Trend analysis commercial naphtha cracking, pred/obs

	Trend LN 820/840		Trend HN 820/840		Trend L/H 820/820		Trend L/H 840/840	
	Pred	Obs	Pred	Obs	Pred	Obs	Pred	Obs
Hydrogen plus methane	+1.5	+1.8	+1.5	+1.4	−1.0	−1.0	−1.0	−1.2
Acetylene	+0.2	+0.1	+0.2	+0.1	±0	±0	±0	±0
Ethylene	+2.0	+1.5	+1.5	+1.7	−2.3	−3.0	−2.8	−2.9
Ethane	−0.4	−0.1	−0.2	±0	−0.9	−0.3	−0.7	−0.2
Sum C_2	+1.8	+1.6	+1.5	+1.8	−3.2	−3.3	−3.5	−3.1
Propylene	−1.4	−1.2	−1.4	−1.0	−2.2	−2.4	−2.2	−2.2
Sum C_3	−1.1	−1.2	−1.3	−1.1	−2.3	−2.6	−2.5	−2.5
Butadiene	+0.1	−0.2	−0.1	±0	+0.1	±0	−0.1	+0.2
Sum C_4	−1.5	−1.4	−1.4	−1.2	−1.4	−1.0	−1.3	−0.8
C_5 plus	−0.8	−0.8	−0.1	−0.9	+7.8	+7.9	+8.5	+7.8
Benzene	+1.0	+0.9	+1.3	+0.9	+0.8	+0.9	+1.1	+0.9
Toluene	+0.2	+0.4	+0.2	+0.3	+2.2	+1.9	+2.2	+1.8
Propylene/Ethylene ratio	−0.09	−0.08	−0.09	−0.08	−0.03	−0.02	−0.03	−0.02

TABLE 6—Wide range contact time naphtha cracking[24]

Specific gravity: 0.70
ASTM, °C IBP: 37
50%: 82
FBP: 196
Molecular weight: 93
H/C Atomic Ratio: 2.16

Paraffins: 73.5% Vol.
Naphthenes: 21.2% Vol.
Aromatics: 5.3% Vol.

Furnace type	Conventional (.3 sec.)		Millisecond (0.1 sec.)	
Yield, wt. %	EXP	SPYRO	EXP	SPYRO
H_2	17.6 (H₂ + CH₄)	.9	16.7 (H₂ + CH₄)	1.2
CH_4		17.6		16.8
C_2H_2	0.6	.7	1.0	1.8
C_2H_4	28.2	27.0	31.0	31.0
C_2H_6	3.9	4.4	3.3	2.9
C_3H_4	0.6	1.0	1.3	1.7
C_3H_6	12.5	12.6	12.7	11.2
C_3H_8	0.5	.4	0.2	.2
C_4H_6	3.6	4.1	4.9	4.9
C_4H_8	2.7	2.8	2.8	1.9
C_4H_{10}	0.2	.1	0.5	.1
C_5^+	29.6	28.4	25.6	26.4

TABLE 7—SPYRO prediction versus pilot plant data

	1			2		3		4		
COIL EXIT TEMP. °C FEED*	814 Naphtha			822 Naphtha		827 90% N + 10% R		Extrapolated 100% recycle stream yield		
	EXP	SPYRO	Δ, %	EXP	SPYRO	EXP	SPYRO	EXP	SPYRO	Δ, %
Yields, wt. % on feed										
Hydrogen	0.79	0.74	− 6.3	0.82	0.81	0.79	0.75	0.79	0.84	+ 6.3
Methane	13.42	13.47	+ 0.4	14.43	14.42	14.25	13.98	21.72	18.57	− 14.5
Ethane	3.49	3.59	+ 2.9	3.47	3.48	3.07	3.15	−0.71	−0.81	− 14.1
Ethylene	25.88	25.80	− 0.3	26.89	26.81	25.45	25.30	21.58	20.80	− 3.6
Acetylene	0.31	0.28	−10.0	0.37	0.39	0.37	0.34	0.91	0.88	− 3.3
Propane	0.42	0.41	− 2.4	0.39	0.38	0.35	0.35	−0.28	−0.19	+ 32.1
Propylene	15.24	15.09	− 1.0	14.24	14.12	14 81	14 88	10 94	12.99	+ 18 7
MA & PD	0 62	0.51	−18.0	0.69	0.64	0.87	0.79	3.12	3.31	+ 6.1
Normal butane	0.11	0.11	± 0	0.08	0.09	0.17	0.20	0.71	1.01	+ 42.3
ISO-butane	0.06	0.02	−66.7	0.05	0.02	0.08	0.07	0.26	0.52	+100
Butene-1	1.46	1.29	−11.6	1.07	0.95	1.20	1.07	−1.14	−0.91	+ 20.2
ISO-butene	2.21	2.42	+ 9.5	1.92	2.09	3.37	3.43	13.81	12.52	− 9.3
Butenes-2	0.80	0.76	− 4.7	0.61	0.62	0.76	0.78	0.40	0.96	+140
Butadiene (+ C_4H_4)	4.89	4.79	− 2.0	4.83	4.72	4.89	4.90	4.89	5.89	+ 20.5
C_5^+	30.30	30.72	+ 1.4	30.13	30.47	29.57	30.01	23.00	23 62	+ 2.7
	100.00	100.00		100.00	100.00	100.00	100.00	100.00	100.00	
Sum C_4	9.53	9.38	− 1.6	8.56	8.48	10.47	10.44	18.93	19.98	+ 5.6
Benzene	6.73	7.07	+ 5.2	7.44	7.92	7.42	7.62	13.63	12.57	− 7.8
Toluene	4.47	4.38	− 2.0	4.32	4.46	4.67	4.35	6.47	4.08	− 37
Xylenes	2.18	2.18	± 0	2.71	2.22	2.04	2.23	0.78	2.68	+244
Propylene/Ethylene ratio	0.589	0.585	− 0.7	0.530	0.527	0.582	0.588	0.507	0.625	+ 23.2

* For SPYRO simulation, detailed composition was used; Naphtha density = 0.724; R = Recycle stream.
NOTE: Relative deviations for **trace** components may become very high (Column 1), whereas those of Column 4 should be divided by 13.5 if compared with those of Column 1 (see also text)

TABLE 8—Commercial gas oil cracking data versus prediction

	Plant 1		Plant 2	
Gas oil feed				
Specific gravity	0.83		0.81	
Boiling range, °C	190-345		210-330	
H/C atom ratio	1.88		1.99	
Operating conditions				
HC-partial pressure out, ATM	0.99		1.17	
Contact time above 600° C, sec	.87		0.95	
Outlet temperature, °C	786		780	
	Obs.	Pred	Obs.	Pred
Yields, wt. % feed				
Hydrogen	0.6	0.7	0.5	0.6
Methane	10.6	9.6	10.0	10.0
Acetylene	0.1	0.1	.1	.1
Ethylene	19.3	19.5	25.0	23.9
Ethane	5.2	3.9	4.7	5.4
PD & MA	0.1	.2	.2	.2
Propylene	13.8	14.3	16.8	17 1
Propane	0.8	.4	.5	.5
Butadiene	3.5	3.5	4.8	4.2
Other C_4's	5.2	5.4	5.9	5.9
C_5-C_{10}	18.2	18.4	18.5	18.0
C_{10}^+	22.6	24.0	13.0	14.1
	100.0	100.0	100.0	100.0
Benzene	3.0	3.0	3.3	3.8
Toluene	3.1	1.9	2.8	1.9
Xylenes & EB	....	1.2	1.2	0.8

TABLE 9—Typical SPYRO applications

In operation:

- Precise and detailed yield assessment on any hydrocarbon feedslate for: optimum feed selection; assessment of possibility/economics of any recycle stream to crackers, and non-conventional mixtures.
- Control/monitoring of pyrolysis furnaces operation.
- Obviating pilot plant tests and direct full effluent analyses.
- Check on plant material balance.
- Optimization of pyrolysis furnace operation (including run-time constraint). This is specific for any combination of furnace type, production target and feedstock.

In design:

- Design/revamp studies of pyrolysis sections.
- Cracking coil design and analysis.
- Planning/design of new olefins facilities.

data[23] as per Tables 4 and 5. No ethane recycle was practiced in this plant, which may have influenced accuracy of measurements.

Another example for naphtha cracking was chosen for its wide range of residence time, as per Table 6.[24]

Fig. 10 gives excellent agreement between SPYRO-ethylene yield prediction for a full range naphtha for a commercial cracker and similar data as generated (and correlated) from a plant-site test unit.

The slight difference both in absolute yields and in trends can quantitatively be explained from the fact that the pilot plant pyrolysis coil had a negligible pressure drop, in contrast to the commercial coil simulation. Similar satisfactory trends were obtained for all other important components as well.

Table 7 gives a typical example of a practical case one may encounter in commercial plants, if certain low-value byproducts are considered for recycle to the cracking furnaces. The extrapolated recycle stream yields (4th column) from co-cracking with a full range naphtha include, of course, kinetic interactions, and serve merely the purpose to enable material balance computations.

Following conclusions may be drawn from this comparison of SPYRO predictions versus experimental (pilot plant) data:

1. Columns (1) and (2) demonstrate that SPYRO accurately follows trend of experimental data.

2. Column (1) indicates also relative deviations, which are within range of experimental scatter—minor components obviously suffer from more scatter as they are close to analytical detection limits.

3. Column (4) gives remarkable agreement between extrapolated SPYRO and experimental data, which are the average of about 40 pilot plant runs.

One should realize from Fig. 7 that the relative deviations indicated should be divided by at least a factor between 10 and 20 if compared with those of column (1).

TABLE 10—Typical influence N/I-paraffin ratio of naphtha on yields and profitability

N/I-paraffin ratio	1.19	0.84
H_2	.90	0.89
CH_4	14.73	14.94
C_2H_2	.43	.43
C_2H_4	27.34	26.49
C_2H_6	4.29	4.29
C_3H_6	14.27	14.31
1, 3-C_4H_6	4.54	4.64
BTX	13.38	13.57
C_5^+	28.59	28.84
M/E ratio	0.539	0.564
P/E ratio	0.522	0.540
Gross margin, $/MT naphtha*	79.3	77.6
Δ GM, $/MT naphtha	.....	−1.7

Basis: Constant feedstock characteristics and operating parameters
* Same basis as per Table 11.

On the other hand, despite the effort, these experimental results are obviously too uncertain to allow marginal economic calculations in commercial practice.

In contrast, on mechanistic grounds, one may attach more confidence to the SPYRO predictions.

For **gas oils,** Table 8 gives for two commercial plants SPYRO predicted yields versus observed data with remarkable agreement.

From the above examples, these conclusions may be drawn:

1. The SPYRO model is consistent for all feedstocks and mixtures, ranging from ethane through gas oils.

2. Trends regarding changes in feedstock, temperature, steam dilution, feedrate, pressure, residence time are accurately followed.

3. The agreement found between detailed SPYRO predictions and observed data is the best possible one may expect in the light of prevailing experimental uncertainties.

APPLICATIONS

A summary on typical application possibilities of the SPYRO program is given in Table 9.

As a matter of course, maximum benefit of the SPYRO program can only be attained if it is judiciously used, either in a stand-alone version—as an ideal "test furnace"—or coupled with general plant control or optimization programs, and with a good general understanding of interrelationship of pyrolysis operating parameters.

Several olefin manufacturers operate a sophisticated pilot plant (bench scale units are too inaccurate) for feedstock testing and production optimization. Others are contemplating doing so. Present costs for materials, engineering and construction of such a unit is estimated to be about $1 MM. In addition, qualified staff, utilities and maintenance will lead to yearly operating costs of about $0.5MM. Engineering, construction and running-in problems (training, analyses) may easily give a delay of four years, before some meaningful (reliable) results may be obtained from the pilot plant. Instead, the SPYRO program offers a faster, more economic, more versatile and reliable alternative to the above.

Optimum feedstock selection promises to be a most attractive application for SPYRO. A simple illustration of the impact of feedstock quality on profitability is

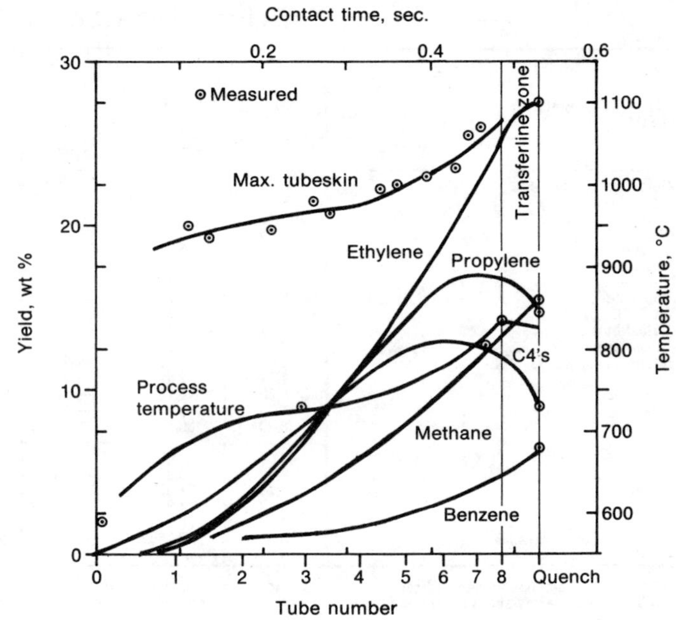

Note: effect of transferline cracking

Fig. 9—Prediction versus measurements for commercial naphtha cracking, medium fouled coil.

given in Table 10, but actually an almost infinite number of variations (including recycle stream or co-cracking) on this theme is possible.

The importance of proper control of pyrolysis furnace exit conditions is illustrated in Table 11. Exit temperature control alone (although a most sensitive parameter) is not adequate for optimum operation, as attainable temperature measurement accuracy (± 10° C) is obviously insufficient.

Furnace control through key-component conversion or other cracking depth parameter alone is not necessarily more accurate (equivalent to ΔT exit of 5-10° C).

In addition, for lowest fuel consumption and maximum run-time, exit temperature should be kept as stable and as low as possible. The temperature stability requirement

TABLE 11—Typical influence coil exit temperature on runtime, yield and profitability for full range naphtha

Coil exit temperature, °C	840	850	860
H_2	0.89	0.92	.95
CH_4	14.19	14.64	15.09
C_2H_2	.39	.47	.57
C_2H_4	27.26	28.15	29.0
C_2H_6	4.23	4.20	4.15
C_3H_6	14 73	14.09	13.33
C_4H_6	4.61	4.57	4.47
BTX	12.97	13.27	13.56
C_5^+	28.38	28.29	28.39
On-stream time (days)	80	58	38
N-pentane conversion (%)	91.2	93.6	95.7
C_3H_6/C_2H_4 ratio	0.54	0.50	.46
Gross margin, $/MT naphtha*	79.13	80.9	82.22
% Furnace runtime**	97.5	96.5	94.5
Gross margin, MM $/year	123.44	124.91	124.32
Net Δ, MM $/year***	−1.47	0	−0.59

Basis: Furnace naphtha capacity = 1.6 MTA
Ethylene value (E) = $340/MT
Naphtha value = $140/MT
C_3H_6 + BTX value = 0.7 E
Butadiene value = 1.1 E
Acetylene value = E
Ethane value = 0.8 E
Rest value = 0.25 E

** Basis: 48 hours furnace/TLX cleaning time
*** Note that average exit temperature setpoint deviations will result in loss of profitability of approx. U.S. $ 1.40/MT ethylene.

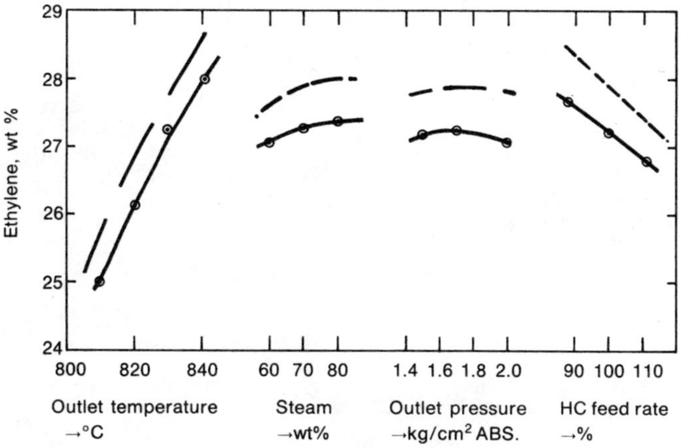

Fig. 10—Trend analysis from SPYRO (solid lines) versus correlated commercial test unit data (dotted lines) for full range naphtha.

implies an addition of feed forward control of fuel/combustion air fluctuations to furnace heat input.

The best pyrolysis furnace set point control will be obtained by giving both the exit temperature and measured cracking depth as input (object functions) to the SPYRO program or derived (and regularly "up-dated") regression model.

This combined input (independent error ranges should approach or overlap as much as possible) enables the program to predict/simulate true setpoint with better probability, at any furnace condition.

At the same time, full effluent composition and other relevant furnace performance data within measuring accuracy are obtained, enabling accurate computations of optimum setpoint value. Attainable gains in profitability may amount $1-2/metric ton of ethylene produced.

Another typical application of SPYRO, for plants operating below capacity, could be for optimizing (higher) steam dilution (within plant constraints) for increased selectivities.[25]

The examples given by no means imply that most plants are already operated relatively close to optimum conditions. In the contrary, due to lack of proper understanding of inter-relationship of cracking parameters, or because of "keep the plant running" philosophies, often overriding optimization considerations, the majority of olefin plants are operated much more remotely from optimum setpoint than indicated. Hence, potential gains could be much higher.

Also, ethylene plants have become today utility plants and might be compared with banking operations: because of large thruput (feedstock or money), marginal improvements in yields (interest rates) give major net gains.

Presently, KTI B.V. takes advantage of the SPYRO program both for design as well as for improving (revamping) older coil configurations to boost capacity and selectivity.

With SPYRO it is now possible to objectively and meticulously compare various current pyrolysis coil designs on the same basis. Again, marginal differences in selectivities may give quick pay-out!

The program is further used for the development of new (advanced) cracking coil designs, and to study various feedstock/recycling alternatives, whereas application of the model is anticipated for adiabatic heavy feedstock cracking.

Recently, after careful evaluation, a major European olefin manufacturer decided to make use of SPYRO for enhanced production optimization. Several other companies are in various stages of evaluating SPYRO.

LITERATURE CITED

1 Wett, T. *Oil & Gas Journal*, p. 51, Oct. 17, 1977.
2 Hammett Jr., J. L., L. A. Lindsay, *Chem. Eng.*, Nov. 8, 1976, p. 115.
3 Ross, L. L., Shu, W. R. *O&GJ*, p. 58, Oct. 17, 1977.
4 Sundaram, M., G. F. Froment, *Chem. Eng. Sci.* Vol. 32, p. 601 (1977).
5 Snow, R. H., H. C. Shutt, *Chem. Eng. Prog.* Vol. 53 p. 133 (1957).
6 Rice, F. O., *J.A.Ch.S.* Vol. 53, p. 1959 (1931).
7 Dente, M., E. Ranzi, G. Antolini, F. Losco, "Symposium on the use of computer in the studies preceeding the design of chemical plants," Florence, April 27-30, 1970.
8 Benson, S. W., "*Thermochemical Kinetics*," J. Wiley, New York (1968).
9 Steacy, E.W.R., "*Atomic and Free Radical Reactions*," Reinhold Publ. Corp., New York (1954).
10 Kerr, J. A., M. J. Parsonage, "*Evaluated Kinetic Data on Gas Phase: Addition Reaction*," Butterworths (1972).
11 Kunugi, T. et al., *Ind. Eng. Chem. Fundam.* Vol. 8, p. 374 (1969).
12 Kunugi, T., K. Soma, T. Sakai, *Ind. Eng. Fundam.* Vol 9, p. 319 (1970).
13 Blakemore, J. E., W. H. Corcoran, *I & EC*, Process Design & Dev. Vol. 8, p. 206 (1969).
14 Fernandez-Baujin, J. M., S. M. Solomon, *Oil & Gas Journal* August 2, 1976, p. 84.
15 Chen, J., M. J. Maddock, Hydrocarbon Processing, May 1973, p. 147.
16 Chambers, L. E., N. S. Potter, Hydrocarbon Processing, January 1974, p. 121.
17 de Blieck, J. L., A. G. Goossens, Hydrocarbon Processing, March 1971, p. 76.
18 Mol, A., Hydrocarbon Processing, July 1974, p. 115.
19 van Damme, P. S., S. Narayanan, G. F. Froment, *AIChE J.* Vol. 21, No. 6, November 1975, p. 1065.
20 Froment, G. F., B. O. van de Steene, P. S. van Damme, S. Narayanan, A. G. Goossens, *I & EC*, Process Design & Dev., Vol. 15, p. 495, October 1976.
21 Froment, G. F., B. O. van de Steene, P. J. van den Berghe, A. G., Goossens, *AIChE J*, Vol. 23 No. 1, p. 93, January 1977.
22 Hatch, L. F., Mater, S., Hydrocarbon Processing, January 1978, pp. 135-139.
23 Ullmanns Encyklopadie der technischen Chemie, 4th Edition, 1974, Band 8, pp. 164-165.
24 Leftin. H. P., D. S. Newsome, T. J. Wolff, J. C. Yarze, ACS - Symp. Series (1976) 32, pp. 373-391.
25 Goossens, A. G., R.F. Westerduin, A. Mol, *Oil & Gas Journal*, August 25, 1975, p. 92.
26 ECN, Feb. 24, 1978, Vol. 32, No. 825, p. 6, p. 21.

Tray Flooding Sets Crude Thruput

On-line calculations determine tray load so that crude distillation thruput can be pushed closer to its flooding point

Harold L. Wade* and **Carroll J. Ryskamp,**
The Foxboro Co., Foxboro, Mass.

An increase in the thruput of a crude distillation tower requires the tower to be operated nearer its flooding conditions. Unless the tower's ultimate capacity is greatly in excess of its required operation, the flooding conditions should be determined by on-line calculations in order to get the most thruput without going into flooding conditions.

Control methods used for binary separation are not necessarily applicable to crude towers. For example, flooding inferred by measuring the differential pressure across a binary tower is not a suitable guide for crude towers because in crude distillation there can be a wide variation in both physical properties and velocity profile along the tower. Even under stable operating conditions, trays are not equally loaded in a crude tower. Therefore, a method is outlined here for making on-line calculations of tray loading at several points in the crude tower.

There are several ways to use the calculated set of flooding values. If the pumparounds are duty-controlled, rather than merely on flow control, the flooding values can be used in closed loop control to allocate the heat removal rate to each pumparound.[1,2] At the very least, the flooding values can be used as operator guides, allowing a more knowledgeable setting of operating variables.

General requirements. On-line control calculations differ from off-line design calculations in the following respects:

- Must be executed in real time and must take advantage of simplifying assumptions and simplified mathematics.
- Must supplement the available measurements with operator-entered data in cases where process values are obtained only from laboratory sampling and analysis, where a process value may change so slowly that installation of a sensor is not justified, and where substitutions must be made for measurement values of on-line sensors which are temporarily out of service.
- Should make validity checks on both sensed data and intermediate calculation results, and may switch to alternative modes of calculating according to process conditions—as in the case of dividing by a flow rate which may occasionally be shut off.
- Can afford to be less accurate since the process behavior itself can be used to update the calculations—although feedback control often takes care of minor model inaccuracies.

A computer program used for the calculation of flooding conditions will have access to a number of inputs from on-line sensors. These include: flow, temperature, pressure and specific gravity. In addition, there must be a set of subroutines for calculating the physical properties of hydrocarbon fractions (such as enthalpy, molecular weight and density) from the input measurements. Finally, the method of flooding calculation applicable to a particular tower, relative to actual tray construction and tower geometry, must be provided.

It is unlikely that all of the desired measurements will be present for any one installation. For example, the specific gravity of the pumparound stream may not be known or the vapor return from a stripper may not be

*Present address for Wade is Biles & Assoc., Houston

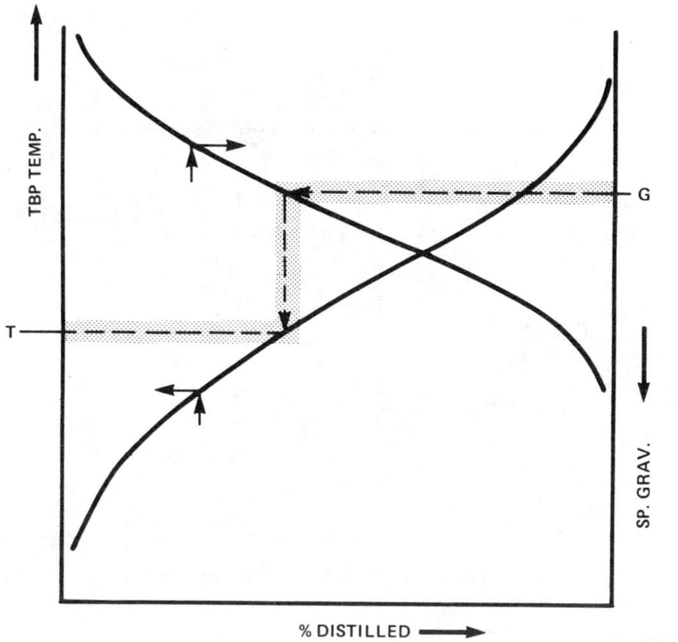

Fig. 1—Graphical representation of the equations used to relate a crude fraction's specific gravity to its TBP. Then Watson-K = $(TBP + 460)^{1/3}/Sp.\ Gr.$

measured. Thus, the calculations must be organized, along with assumptions and approximations, to make maximum utilization of available measurements.

One form of approximation which is widely accepted for calculating enthalpy and other physical properties is to define the composition of a petroleum fraction by its relation to two parameters: the specific (or API) gravity and the Watson-K characterization factor.[3] Gravity is usually measured on-line; the Watson-K factor, however, is an empirical quantity and is not directly available. An expedient, used to arrive at enthalpy data from on-line measurements, is to determine the Watson-K factor for a fraction of a particular crude from the gravity measurement of the fraction, plus the TBP (true boiling point) temperature and the specific gravity curves of the crude as a whole. This method is depicted in Fig. 1. It is assumed throughout this article that this method is used to determine enthalpy and other properties of a crude fraction from temperature and gravity measurements alone.

TABLE 1—Program steps required for on-line calculation of crude tower tray loading

1. Read on-line sensor values: flow, temperature, pressure, specific gravity. Calculate mass flows and enthalpies.
2. (Optional) Normalize mass flows so that algebraic sum = 0. May be applied only to oil streams, or to all streams.
3. Calculate quantity stripped, draw stream specific gravity and Watson-K factor for each stripper.
4. Calculate heat removal rates for overhead condenser system and for each pumparound stream.
5. Calculate internal vapor and liquid mass flow rates and densities for each envelope. Add external draw rates to internal liquid flows to obtain total liquid flow across trays.
6. Calculate percent flooding for each envelope. Output to closed-loop control system or display results.

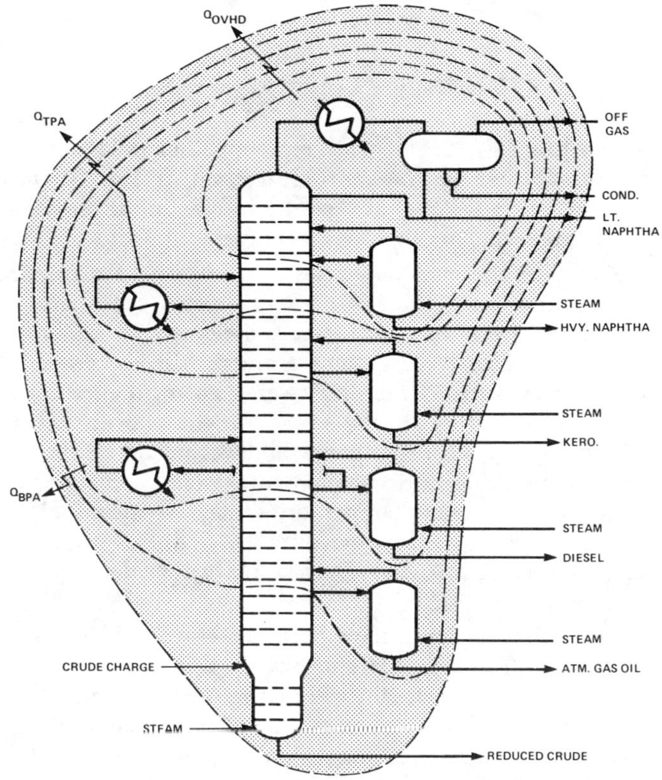

Fig. 2—Envelopes separate the crude distillation tower into sections for calculating vapor and liquid flow rates.

THE CALCULATION METHOD

The following paragraphs provide the theoretical development of the calculation method. A step-by-step summary of the calculation procedure is given in Table 1.

Internal tower flows. Regardless of the tray construction, a starting point for the determination of tray loading is the calculation of internal liquid and vapor velocities. These velocities can be easily determined if the mass flow rates and densities of the vapor and liquid traffic are known. The calculation of the liquid and vapor flow rates is based on mass and energy balances made around envelopes which include sections of the tower plus external condensers, strippers, and pumparounds.

A typical tower and the chosen envelopes are shown in Fig. 2. The main criterion for the choice of the envelopes is that, except for the internal streams, the mass and energy flow rates of all streams crossing the envelope boundary must be known from either on-line measurements, prior computation or operator-entered data. This dictates that envelopes be chosen to cut the tower just below an external draw tray, either a pumparound or a stripper draw. Since the most heavily loaded trays are likely to be those within a pumparound circuit, this choice of envelopes assures that flooding calculations are made at critical points in the tower.

A typical envelope cutting the tower beneath the Nth tray is shown in Fig. 3. Mass and energy balance equations then are:

$$V_{N+1} - L_N = \Sigma F_i \quad (1)$$

$$V_{N+1} H_{N+1} - L_N h_N = \Sigma F_i h_i + \Sigma Q_i \quad (2)$$

The terms ΣF_i, $\Sigma F_i h_i$, ΣQ_i represent algebraic sums of the mass or energy flow rates of external streams. Streams leaving the envelope are regarded as positive flows; streams entering are regarded as negative flows.

Assuming the mass flow rates of all external streams are known and that external energy removal rates can be calculated from on-line measurements, then the entire right-hand sides of Equations 1 and 2 will be known. If the internal stream enthalpies were known, then the two equations could be readily solved for the two unknowns, L_N and V_{N+1}. These enthalpies, however, are dependent on their respective stream temperatures and compositions. Whereas the stream temperatures may be measured, or estimated from nearby measurements, the compositions are, in general, unknown. Therefore, further development is necessary.

Since the envelopes have been chosen to cut the tower below draw streams, the temperature and composition of the internal liquid stream will be the same as those of the draw stream. However, the gravity of a pumparound or a stripper draw stream is not customarily measured. Rather, the draw stream gravity must be calculated or estimated by other means.

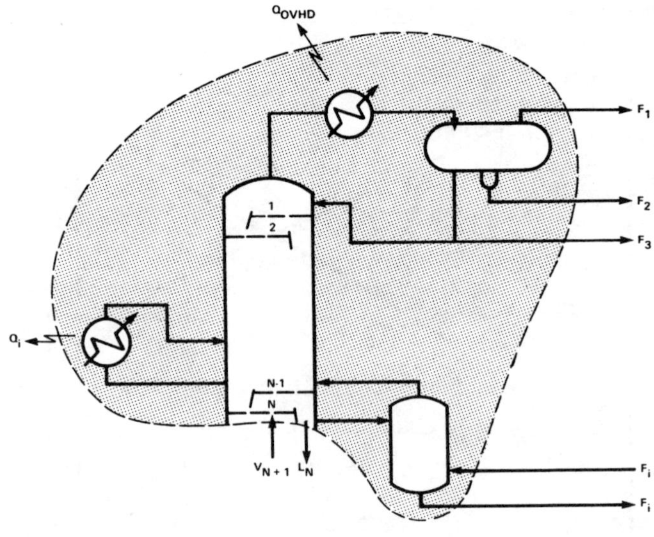

Fig. 3—Typical envelope around a tower section.

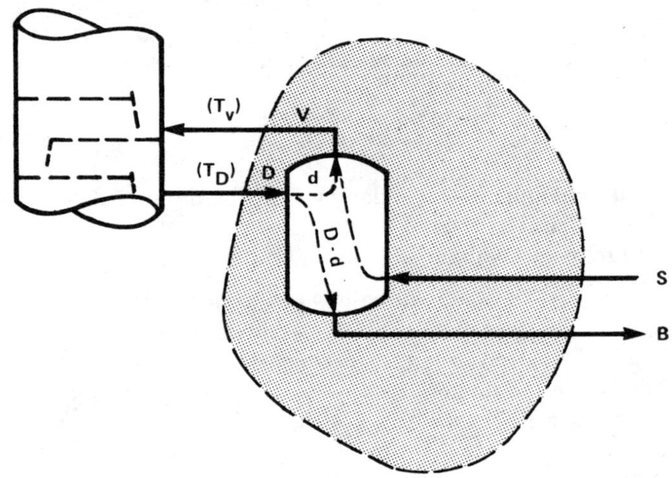

Fig. 4—Envelope used for stripper calculations.

Stripper draw stream. An envelope drawn around a side stripper is shown in Fig. 4. It is assumed that mass flow, temperature and gravity of the stripper product are known, and the mass flow, pressure and temperature (or quality), of the stripping stream is known. The stripper draw stream temperature must be measured; if the vapor return temperature is not measured, it may reasonably be assumed to be the same as the draw temperature.

The first step of the stripper calculations is to determine the fraction stripped. Mass and energy balance equations about the stripper are:

$$D + S = B + V \quad (3)$$

$$Dh_D + SH_S = Bh_B + VH_V \quad (4)$$

Therefore, disregarding any difference between the water dissolved in the stripper product and the draw stream:

$$B = D - d \quad (5)$$

$$V = S + d \quad (6)$$

Express enthalpies h_D and H_V as weighted sums of the enthalpies of their constituent streams:

$$h_D = [B/D]\,[h_B\,(T_D)] + [d/D]\,[h_d\,(T_D)]. \quad (7)$$

$$H_V = [S/V]\,[H_S\,(T_V)] + [d/V]\,[H_d\,(T_V)]. \quad (8)$$

Combine Equations 7 and 8 into 4 and solve for the quantity stripped:

$$d = \frac{B\,[h_B\,(T_D) - h_B] - S\,[H_S\,(T_V) - H_S]}{H_d\,(T_V) - h_d\,(T_D)} \quad (9)$$

The gravity of the stripped quantity, *d,* is unknown. However, a reasonable assumption is that the stripped quantity is of the same gravity as the next lighter product stream. If the gravity of this next lighter product is known, then the specific gravity of the draw stream can be computed by combining the pseudostream "quantity stripped" and the product stream.

$$D/g_D = [B/g_B] + [d/g_d]$$

$$g_D = \frac{[D]\,[g_B]\,[g_D]}{[B]\,[g_d] + [d]\,[g_B]} \quad (10)$$

Pumparound stream. If a pumparound draw is from the same tray as a stripper draw, then the method described above for calculating the gravity of the stripper draw automatically provides a specific gravity value for the pumparound stream. If, however, the pumparound draw is independent of the other draw streams, then its gravity must be estimated.

If there are draw streams (for example, stripper draws) of known or calculated gravity both above and below the pumparound draw, then an expedient is to interpolate the gravity of the pumparound draw between the known gravities of the higher and lower draw streams. The interpolation may be linear and made on the basis of the effective number of trays between the upper and lower draws, with the trays within the pumparound counted as one effective tray. Alternatively, a linear interpolation may be made on the basis of temperature of the known draw streams and the temperature of the pumparound draw. In specific situations, other feasible methods of estimating the unknown specific gravity of the pumparound streams may suggest themselves.

Liquid and vapor flow. Now to return to the problem of solving Equations 1 and 2 for V_{N+1} and L_N. Methods were shown by which the specific gravity of an external draw stream may be determined. These will be the same as for the internal liquid stream. With the specific gravity and temperature known, the liquid enthalpy, h_N, may be determined.

From Equation 1, the vapor stream V_{N+1}, is composed of a number of constituent streams:

$$V_{N+1} - L_N = \Sigma F_i$$

Assuming negligible heat of mixing, its enthalpy is the weighted sum of the enthalpies of each of its constituent streams:

$$H_{N+1} = [L_N/V_{N+1}]\,[H_{L,N}\,(T_{V,N+1})] + \Sigma\,[F_i/V_{N+1}]\,[H_{F,i}\,(T_{V,N+1})] \quad (11)$$

Here, $[L_N/V_{N+1}]$ and $[F_i/V_{N+1}]$ are the weighting factors, and $[H_{L,N}\,(T_{V,N+1})]$ and $[H_{F,i}\,(T_{V,N+1})]$ represent the vapor state enthalpy at the conditions, specifically temperature, of the vapor stream. The vapor temperature $T_{V,N+1}$ may be either measured or estimated as a constant differential above the liquid temperature.

Multiplying Equation 11 by V_{N+1} and substituting into Equation 2 gives the following results:

$$L_N = \frac{\Sigma Q_i - \Sigma F_i\,[H_{F,i}\,(T_{V,N+1}) - h_i]}{H_{L,N}\,(T_{V,N+1}) - h_N} \quad (12)$$

Once Equation 12 is used to calculate L_N, then Equation 1 may be used directly to calculate V_{N+1}.

$$V_{N+1} = L_N + \Sigma F_i \quad (13)$$

Density. In order to compute the liquid and vapor velocities, the densities at operating conditions as well as the mass flow rates must be known. Calculation of liquid density is straightforward. Since the specific gravity (60° F/60° F) has already been calculated, it need only be adjusted to the actual liquid temperature and then multiplied by a constant to obtain the liquid density.

The over-all density of the vapor stream may be determined by combining the specific volumes of each of the vapor stream constituents. The equation is similar to Equation 11, which was written to combine enthalpies of each vapor stream constituent. For the densities, the equation is:

$$1/\rho_{V,N+1} = L_N/V_{N+1}\,\rho_{L,N} + \Sigma F_i/V_{N+1}\,\rho_i \quad (14)$$

Equation 14 requires determination of the vapor state density of each constituent of the vapor stream. Since the crude tower operates at a relatively low pressure, the vapor density of each constituent may be determined from the ideal gas laws using the vapor temperature, column pressure and molecular weight of the constituent stream.

Molecular weight determination depends on the type of the constituent stream:

- For water vapor constituents, the molecular weight is known.

- For gas constituents (for example, non-condensables from the overhead receiver), the gas specific gravity may be multiplied by a constant to determine molecular weight. If the gas gravity is not measured, it may be inferred from correlation with the overhead receiver temperature; alternatively, a constant approximate value accessible for manual updating for different operating conditions may be used. Since the mass flow rate of this constituent is small in comparison to the other streams, errors in the specific gravity of the gas do not significantly affect the final results.

- For oil streams, including the internal liquid when considered as a constituent of the vapor, knowledge of the specific gravity of the liquid state and the Watson-K factor is sufficient to determine the pseudo-molecular weight, using, for instance, Figure 2B2.1 of Reference 4.

Heat duty. Equation 12 requires the energy removal rates for the overhead condenser system and for pumparounds. For the overhead condenser, an energy balance may be made around the envelope shown in Fig. 5. The vapor stream leaving the tower is composed of a number of constituent streams, with each constituent representing one of the other streams leaving the envelope. Proceeding in a manner similar to the computation for tower envelopes, the user can determine the enthalpy of each constituent stream, both at its own flowing condition and in the vapor state at the conditions of the tower overhead vapor. The overhead heat removal rate is then computed from:

$$Q_{ovhd} = \Sigma F_i\,[H_i\,(T_{V,ovhd}) - h_i] \quad (15)$$

where the summation includes those streams, except for the overhead vapor itself, leaving the envelope shown in Fig. 5.

For the pumparounds, the specific heat may be determined from the gravity. From this, with the mass flow rate and the measured temperature difference between draw and return, the heat duty calculation is straightforward.

Flooding. Having calculated the liquid and vapor flow rates and densities at each of several points in the tower, the user may now calculate the percent flooding at these points. The term "flooding" is used to describe a number of phenomena which result from excessive vapor velocity and cause inoperability of the tower. Most important of these conditions are:[5]

- **Hydraulic flooding**—where the pressure drop between adjacent vapor spaces is sufficient to cause the height of the liquid in the downcomer to reach the height of the liquid over the weir of the tray above, thus preventing additional liquid from passing down the tower.

- **Priming**—where the foam height on the plate becomes sufficient to cause backmixing on the tray above, thus greatly reducing the separation effectiveness.

In determining the loading condition, or "percent flooding," of a tray, the user may consider separately the hydraulic flooding and priming conditions, then take the worst of the two cases. The calculations are based on experimental correlations for a particular type of tray, and require tower geometry factors such as the type of tray, tray spacing and tower diameter. Correlations for flooding calculations are given in the literature for bubble cap,[6] perforated trays[7] and ballast trays.[8]

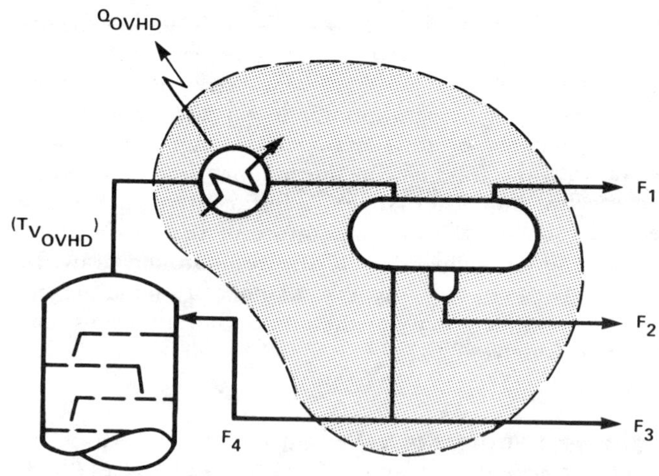

Fig. 5—Envelope used for overhead heat removal rate.

APPLICATIONS

For complex calculations of the type described here, it is advisable for the programs to be organized in a modular fashion. This permits easier understanding, troubleshooting and modification in the event of process alterations. Each functional step may be programmed as a separate software element or module. These modules may communicate with each other through data files, rather than maintaining the computational results within the program area. Furthermore, each module may be driven by its own data file, which identifies the type of computation, source of input data and other parameters, and the destination of computational results. In this manner, a minimum amount of parameters, equipment identification, etc., is written into the programs, permitting reuse of the programs on similar items of process equipment, or even at different installations.

Operating guides usually are specified to operate a tower at some fixed percent of flood; for example, 90 or 95 percent. However, characteristics of different crude stocks vary, for instance in foaming tendency, so that different crudes may be processed at different calculated values of percent flood. For some crudes the allowable rate may exceed 100 percent of the calculated rate. With frequent on-line calculation of percent flood, experience will soon show the limits to which the tower may be operated for each type of crude stock processed.

The calculation method described here is used successfully on closed loop control of a crude tower where the maximum calculated value of flooding was used in allocating pumparound heat duty. Calculations are performed every 5 minutes. Elapsed time for execution, at each calculation cycle, is approximately 15 seconds; this allows for data read-in, bulk memory transfer, processor time plus execution of higher priority tasks. The control action keeps at least one section of the tower fully loaded. In addition, the system has proven its ability to provide advanced warning of conditions which would have led to a major upset under conventional control.

NOMENCLATURE

- B Mass flow rate, stripper product
- d Mass flow rate, lights stripped in stripper
- D Mass flow rate, draw to stripper
- F_i Mass flow rate of external stream
- g_i Specific gravity, stream i
- h_i Liquid enthalpy of stream i at its own temperature
- $h_i(T_k)$ Liquid enthalpy of stream i at the temperature of stream k
- H_i Vapor enthalpy of stream i at its own flowing conditions
- $H_i(T_k)$ Vapor enthalpy of stream i at the temperature of stream k
- L_N Mass internal flow rate of liquid from tray N
- Q_i Heat removal rate at point i
- S Mass flow rate, stream to stripper
- T_k Temperature, stream k
- V_{N+1} Mass flow rate of vapor from tray $N + 1$
- ρ Vapor recovery

LITERATURE CITED

1 Ryskamp, C. J., Wade, H. L., and Britton, R., "Computer Control of Crude-Vacuum Units for Energy Conservation," presented at 41st Midyear Meeting, *American Petroleum Institute, Refining Division*, Los Angeles, May 10-13, 1976.
2 Ryskamp, C. J., Wade, H. L., and Britton, R., "Improved Crude Unit Operation," *Hydrocarbon Processing*, May 1976, p 81.
3 Nelson, W. L., *Refinery Engineering*, 4th ed., p. 169, McGraw-Hill Book Co., Inc., New York, 1958.
4 Technical Data Book—Petroleum Refining, American Petroleum Institute, Washington, D C., 1970.
5 Van Winkle, M., *Distillation*, McGraw-Hill Book Co., Inc., New York, 1967.
6 Fair, J. R., and Matthews, R. L., *Petrol Refiner*, April 1958, p. 153.
7 Fair, J. R., *Petro/Chem Engineering*, September 1961, p. 45.
8 Bulletin 4900, *Ballast Tray Design Manual*, F. W. Glitsch and Sons, Dallas, Texas, 1961.

Save Utilities with Better Control Systems

As HPI complexes and energy costs grow, so does the need for better utility control systems. Energy savings and operating flexibility are the justifications. Here's how to plan a centralized utility control system to maximize the benefits.

P. N. Berkowitz Tex-A-Mation Engineering, Inc.
Laporte, Texas

As REFINERY and petro chemical complexes have expanded over the years, localized control points have developed for the generation and distribution of utilities. Under normal operating conditions, independent control points present no problem. However, in normal operation units come on and offstream plus other contingencies such as extreme weather conditions facilitate the need for close coordination in the generation and use of all utilities in the entire complex. The purpose of an integrated utility control system (UCS) is to provide the necessary control and coordination of utility functions from a single point. To aid these functions, an extensive system of instrumentation, computer systems and control consoles will provide the operator with rapid and accurate information of all energy supplies, reserves and uses. To achieve this coordination, a Utilities Control System becomes the basic requirement for tying together the control of steam, electric, air, gas and water utilities. Considerations for a properly designed UCS are presented.

The average profit earned by energy companies in 1978 was 3.9 percent. Based on the savings in Table 1, the eight UCS projects saved nearly $23 million annually, therefore, the combined sales necessary to generate the additional profit would have been approximately $600 million. To further emphasize the significant gains derived from a UCS project it may be easier to state the investment in the financial terms of ROI. If a UCS project were evaluated on "return on invested capital" the summarized table would have had an ROI of 48.6 percent versus 11.9 percent for the 1978 energy industry average.

The economics of incorporating such a system is more than justified when considered in terms of today's fuel costs. Payout for energy savings can result from the effective management of utilities and the elimination of uncertain operation. Project costs can be recovered in early operation if the UCS is prudently planned.

DEVELOPING UTILITY CONTROL SYSTEMS

Good control systems are developed when the user can separate the utility objectives from the prescribed order of system design and implementation. A proposed order in development of a Utilities Control System is: utilities objectives; utilities operations profile; assimilate as control problem; information plan; utilities control plan, and system design.

It is generally understood that the objectives are divided into the categories of operations and the costs to perform that operation. Thus, it is necessary to first understand the operating profile before assigning costs to utility savings. The UCS objectives are as follows:

► To enable the coordinating task to be performed from a single location, namely the Utilities Control Center (UCC).

► To make available to the Utilities Control Center operator rapid and accurate information on all energy supplies and their demands.

► To provide computational and visual aids for easy and quick evaluation of systems status and to provide a basis for sound operating decisions by the operators.

► To anticipate possible contingencies and thus provide remote operation of the control loops from the Utilities Control Center.

Operating objectives. In utilities control it is important to note that although certain loads can be switched, demand is uncontrollable. Load can be shed, but not really controlled. The nature of control is in the supply or production. In regulating the supply, there are five operating guidelines. In order of importance these are:

1) Keep the process plants running (maximize return on products)

2) Minimize the cost of process operation

3) Maximize reserve or surplus (insurance against loss)

4) Keep the system stable (keep steam and electric production smooth and in a controllable range)

5) Safe operations.

TABLE 1—Cost vs savings for utilities control projects

Type complex	Investment	First year savings
Petro-chemical	$ 250,000	$ 450,000
Refinery	1,000,000	950,000
Chemical	1,750,000	1,500,000
Refinery	10,000,000	1,500,000
Petro-chemical	3,900,000	2,812,000
Petro-chemical	10,000,000	5,150,000
Chemical	18,000,000	10,000,000
NGL	2,083,000	460,000

To achieve these operating objectives, the individual power plants must be controlled along these guidelines. The principal role of the UCS is not to control generation facilities alone, but to achieve the overall goals by coordinating and balancing production between manufactured and purchased utilities to meet the complex's demands.

Operation and maintenance of individual boilers requires that each be out of service one to two weeks per year (assuming no waiver). To maintain schedules and operations and still meet emergencies on a complex-wide basis, careful coordination of all sources and loads in the system is required. The use of steam and electricity dictate how utilities are expended. When purchased power is used an added objective, contract utility cost is imposed on the operations.

The first objective of controlling "paid for" utilities is defined as *maintaining load factor*. The second part of minimizing the cost of electric energy is known as *demand peaking control*. Exercising demand peaking control will require some override procedures on normal events. This last requirement is the most critical because it implies that management may need to alter operations to endorse that policy.

The minor utilities of air, water and fuel have objectives different from steam and electricity. Since control is not an objective for these utilities it is more appropriate to think of their needs in terms of maintaining supply. The objective can be maintained by assessing as quickly as possible the situations and advising users of the impending need to curtail operations for safety reasons. The result of monitoring these utilities will be an assessment of the overall requirements, the necessity to reduce consumption, availability of alternate supplies and the shut down procedures.

Operations profile. Understanding how utilities are manufactured, used and distributed is essential to the operations profile. Steam is produced from boilers, waste heat recovery from process boilers, and process heat exchangers. Transduced sources result from letdown stations and prime movers. In addition to the process uses, there are "prime mover" applications which include the following: turbogenerators; gas turbines; mechanical drivers; switchable loads, and vents and letdowns.

Primary electrical power may be manufactured, purchased or supplied as a combination of both. The function of the electric transmission system is to provide a network or grid of power at transformed levels. This network is made up of a series of transformer groups or substations. These substations and their interconnections are collectively referred to as the transmission system. The management of the transmission system is the responsibility of the UCC.

Air may be produced in several ways: station, local and portable compressors. The principal uses for water in a petrochemical complex are for cooling and boiler feeds. In addition to these prime uses, the water supply can support the fire system and domestic service. Multiple sources are usually available.

Finally, the operations profile can determine how the objectives may be met. Table 2 represents a summary of the system operations as they relate to objectives.

TABLE 2—System operation

Objectives	Actions
Labor reductions	Consolidate control points Coordinate dispatch Eliminate manual data collection Automate reporting
Defer capital outlay	Assess current utilization Improve efficiency
Maintenance	Schedule maintenance Continual performance assessment Test and measurement control
Improve reporting	Develop standard costs Allocate costs Performance evaluation

TABLE 3—Power

Inputs	Monitor
Calendar Divisions of time Peak hours, days Contract schedule Power factor Desired reserve	Time kva kvars
Outputs	**Alarms**
Summaries Contract kwh Usable kwh Available kwh Recommended kva, kwh Forecast	High rate Low load factor Low pressure Recovery profile Anticipated peak Peak penalty assessment

Control problems. There are a number of possible control actions available to the UCC for anticipating, avoiding or recovering from problems. Some actions are listed in Table 4. All of these actions may be implemented indirectly by voice request, the UCC operators or through control systems.

Table 4 also lists specific actions possible at the UCC to help anticipate, understand, avoid or minimize steam problems and certain jobs that can be done to anticipate and avoid electric load, demand, or failure problems.

Table 5 represents the electric steam interaction in terms of a control problem.

Air control actions are more indirect and are as follows:

- Direct the start-up of local or portable compressors
- Ask plants to preserve system pressure by using the compressors in their own plants
- Dispatch men to operate valves to cut plant air into the instrument system
- Dispatch men to operate isolation valves to block failed segments.

Water control actions are also indirect in that a voice request is usually issued to someone in the field. Under these circumstances, the actions open to the operator are as follows:

- Reduce domestic consumption
- Reduce tank and tower re-supply selectivity
- Connect in well supplements
- Proceed with orderly plant shut-down.

In addition to these actions, the UCS must monitor pressures and predict need for auxiliary pumps, monitor water quality and need for treatment and plan pump maintenance to ensure continuity of water service.

TABLE 4—Possible UCC Control Actions

Change or add boilers
Switch from electric to steam or vice-versa on switchable loads
Shed loads
Adjust turbines
Reduce steam usage at turbines
Control reducing stations
Control venting station
Control isolation and diversion valves.
Block in plants
Bring generator(s) on-line
Reduce/increase turbine power generation
Adjust generator excitation
Close breakers
Open breakers
Activate emergency power supplies

Possible steam problem actions
Balance steam at the operating levels
Determine points of imbalance
Predict trends in production, capacity and surplus
Audit performance of boilers and turbines
Schedule maintenance

Possible electrical problem actions
Calculate electric surplus
Predict consumption trends
Perform fault isolation
Audit performance of generators
Schedule against demand period

TABLE 5—System optimization

Objectives	Problems	Actions
Electric	Low load	Breaker control
Maintain load factor	High peak	Shed loads, switch
Peak control		Turbine adjustment
Minimize power costs	Trips	Emergency power
Steam		
Minimize production	Steam storage	Add boilers
Costs	Deficient Surplus	Switch, shed loads
Maximum ready surplus	High/low pressure	Adjust turbines
Maintain stability	High/low	Vent/letdowns
Maintain pressure	temperature	Isolate, divert

For gas, the primary objectives are to maintain pressure and heating valves. The gas distribution system must be monitored for pressure flow and mixture. The following actions can be taken to correct problems or minimize their effect until the problem is solved: operate isolation valves to block failure; put compressor on line; bypass faulty regulators; manually switch to oil, pitch or other gas; line up alternate fuels; request supplier to raise pressures; improve quality at blend tanks, and flare refinery gases.

Information plan. With the objectives and a knowledge of existing operations it is possible to identify information available and information needed for the UCS. Defining what information will be monitored, used in computational analysis and/or control actions will result in a data flow plan and a set of display requirements. The information plan must support the way operations are to be performed and to what extent the operator can input change data or a computer system can compute and control the desired change.

Production data is available for boiler capacity. Data on each boiler may be manually inputted to the UCS and that data must contain the facility for change. Modification of boiler feed water temperatures will result from analog control or computational systems. Waste heat production will be available for waste heat boilers. Analog data is available on steam flows in headers and turbines. Alarm limits for low flows should be modifiable by the UCS operators. Master pressures and temperatures should be available for the steam headers. There may be vents which are monitored for a flow/no flow condition and vents where flow is measured in analog form. Some letdown stations are monitored for flow/no flow indications and some for actual flow. For engineering and accounting purposes, flows are monitored for a variety of fans, pumps, compressors, heaters and drivers.

When a decision is made to shed loads, a voice instruction should go to the pre-agreed control points to start the shedding procedure. After an acknowledgment from the plants involved, the control signals may be transmitted by the UCS.

In many complexes there are loads driven by steam or électric power. Electric motors are usually equipped with contacts to provide indication of whether or not the motor is in operation. If control operations are to be changed it should be UCS initiated. Also, motor-operated valves may be controlled from the UCC for isolation and diversion procedures.

Certain steam data is important to the electric service for power generation control. Steam inflows to turbo-generators and helper turbines are needed to calculate the material balance for steam and electric.

Data on the transmission and distribution system is in five forms, namely, status of circuit breakers, bus feeder voltages, currents and power factor. Purchase power measurements include kva, kwh, kvars, and time.

Air data includes the station air produced, flow and header pressures. To have some indication of overall system status of plant and instrument air, central pressures are to be measured.

Flow data on clarified and unclarified river water needs to be monitored at or near the river water station. Header pressures in deep and shallow well headers, as well as pump discharge and manifold pressures must also be monitored.

Data on purchased gas should be available from flow rate transducers. The gas system status refers to pressure monitored at points throughout the system. The specific gravity of fuel gas should be measured at each blend tank.

Finally, the data information presentation and record keeping can be expressed by data handling, computation, consoles and mimics.

Equipment must be provided to collect, process and route the available data to appropriate displays or devices. Data collection involves scanning all the input signals, both analog and digital. The incoming data will be checked and conditioned as required before routing to the computer or hard wired displays. Data handling equipment will also have to be provided to route control signals from the console to the appropriate valves, motors, or breakers. Amplifiers, drivers, encoding, decoding and data distribution equipment all come under the data handling requirement. Supervisory functions are also in this equipment.

Control plan. Control may be from the UCC console position, local unit operator or computer initiated. Some circuit breaker control and motor operated valves may be operator controlled but the trend is to have them computer controlled by the UCS.

The computer will be one of the key elements in the coordination of utility production and distribution. The computational tasks will include at least the following: alarm or limit checking; capacity calculation for boilers; steam surplus calculation; electric surplus calculation; peak demand prediction; flow integrations; steam balances; report generation; alarm logging and display, and display formatting.

The system must be designed to operate twenty-four hours per day in the real-time, on-line mode. A processing capability that permits the central processor to schedule tasks according to their relative importance is necessary if software maintenance is to be run while real-time data is processed for monitoring and control.

On many signals, a limit will be set in the computer by the UCS operator. When the limit is exceeded the UCS would prepare a display message, drive an alarm and the occurrence would be logged on a typewriter. Changes in state for two level signals would also be interpreted as alarms. Hence, the state of bi-levels must be stored and compared.

One of the limits set will be on boiler house production. These limits are entered by the UCS operator for a standard feedwater temperature. The computer may then sample the actual temperature and adjust capacity according to a given algorithm. The computer can continually subtract production data from adjusted capacity to obtain surplus by boiler and pressure level. To determine electric surplus the computer can assess the steam capacity and loads and determine the potential kw production. This potential, the demand kw, and load status will be combined in the surplus algorithm to determine electric surplus. Cost payout comes from not setting new peaks when purchase power is used. The computer can do demand extrapolation on kw demand to provide 2, 5, and 10-minute warnings of an excess demand. Table 3 contains the elements in making electrical decisions.

All flow data can be integrated by the computer to provide cumulative flows on demand or for regular reports. The system will be periodically balanced by the computer for an evaluation of stability. Inflows and outflows are compared for trending net steam. Balance can be by area or system wide.

Every time a calculation, prediction or signal exceeds a limit set in the computer, the event can be logged and an alarm initiated. If an event changes state (vent open, breaker trip, etc.) the change can be logged with a time tag and an alarm issued. The computer has the capability to discriminate between hard and soft alarms, thus the UCS can be designed to modify alarm limits.

One of the main jobs of the computer will be retrieving and formatting displays for console presentation. This refers to dynamic displays of incoming data or calculated quantities. Summary reports and special reports on all utilities provide operation managers with meaningful data. This is another reason to bring data into the computer.

UCS operation usually requires an integrated console so that all primary monitor and control functions can be accomplished normally by the operator. Primary utilities (steam and electric systems) can be managed from the same console. Normal UCS functions should be accomplished at the console position.

To supplement the console requirement and provide a means of status indication and alarm on all systems, wall displays may be required. These displays might include a mimic of the electric and steam systems. In addition, monitoring equipment for minor utilities (air, gas, water) might be provided on the wall or as a graphic on the display console. Hardwired recorders are also required for permanent recordings and control backup when the computer is inoperative.

System design. System design includes all of the aforementioned data and the facilities and support systems to operate the utilities complex. To establish requirements for implementing the UCS, it is necessary to define the subsystems required to support operations:

- Data acquisition subsystem
- Computer subsystem
- Display/control subsystem
- Communications subsystem
- Facilities
- Support subsystems

The data acquisition subsystem is the collection of instrumentation and cabling required to sense, transduce, transmit, receive, process and route all utility parameters to the central computer and appropriate display devices.

Data from various plant utility systems would be routed to the UCS computer system. The function of the computer is to process the data for cathode ray tube displays, output reports, and provide sequential events logging. The display equipment will enable an operator to rapidly assess the status of the utilities systems. The computer and displays provide immediate data so the operator can determine the best course of action to meet the objectives of integrated utilities control. Equally important is the system software. System software is that required to schedule and run real-time programs, system input and output, convert and compute data, and the application software that measures the utility system performance. The system software also includes programming and debugging tools, such as utility programs, dumps, diagnostics, loaders, table builders and language processors.

The Communications Subsystem includes both data and voice communication equipment to provide the following:

- A highly reliable communications network with alternate routes, in case of a primary link failure
- To allow the operator and foreman, from control panels, to access individual stations, groups of stations, or all stations
- To allow certain utility stations to communicate with other stations, independent of the UCC.

The Utilities Control Center should be sized to include support systems needed for the UCS operations. The control room and equipment rooms should be as close as possible. Distinct operational advantages, such as efficient controller traffic flow, effective system coordination, centralized viewing of operational areas by shift foreman and multiple utilization of displays support this concept. The computer area of the facility should be able to accommodate an expanded data processing system and its I/O systems. This area will be needed for future additions. The equipment area should house the analog and digital I/O equipment, terminations, miscellaneous equipment, other control measurement equipment, communications and telephone equipment. The control area should house the alarm and setpoint controllers, console, video terminals and mimic equipment. Operators' consoles, filing and storage areas should be positioned so all personnel can see the wall displays.

An uninterruptible power system should be employed for UCS power requirements. The system should supply the instruments, video displays, computer, critical wall and console displays on a continuous basis with power from an inverter(s). Lead-calcium station-type batteries should be used to maintain supply to the inverter in the event of electric power failure. Alternate power should also be supplied to the UCC. Standby air conditioning needs to be provided for backup of the control room and equipment room environments.

Cost justification. Approximately 6-10 percent of the energy value of a barrel of crude is consumed within the refinery for production of utilities. An average barrel of oil will cost companies between \$16-\$30 at today's prices. Knowing the operating status of the utilities system can reduce demand while increasing production. Leaking steam traps alone can cost more than \$1000/year per trap. Large petrochemical complexes may have more than 10,000 steam traps, thus the annual savings can be quite significant if this is properly monitored.

Installing a UCS for reducing the need for excess surplus and electrical peak demand control can save 2 percent or more of the annual utilities cost. Table 1 is an illustration of some typical results from existing UCSs. Although different returns are experienced by different operating companies, all agree that energy savings have been achieved. According to Houston Lighting and Power Co.'s 1979 electric rates, the penalty for setting a new 5,000 kva peak will cost a customer on 138 kv service almost \$900,000. It is evident that a small error in utilities judgment can represent a significant cost to operations.

A typical example of an integrated UCS that resulted in \$4.5 million annual savings was as follows:

Given: Conversion rate = 1,600 Btu/lb.
Oil weight = 350 lbs./barrel
Daily use = 640,000 bpd
1978 Avg. cost of energy = \$1.40/MM Btu

Annual cost = wt. of oil processed (1600 Btu/lb.) (Cost/MM Btu)

Annual cost = \$183,142,400

Percent utilities savings = 4,500,000/183,142,400 = 2.46 percent

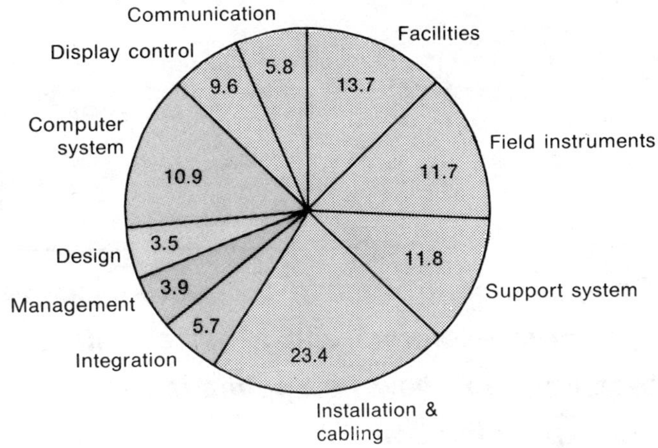

Fig. 1—System cost percentages.

Of course, the cost to implement a UCS is related to the size and needs of the complex versus the extent of the proposed monitoring and control. However, a general guide to system cost percentages is shown in Fig. 1.

Utility control systems are more than cost saving investments, they provide the basis for safer operation with the benefit of improved operations. Plant managers and operating supervisors will have access to current information and can make decisions based on factual data. Engineers will have historical data to analyze plant upsets and normal operations. Operators will be in command of the situation and for the first time will be in a position to act rather than react.

A word of caution must be applied since an integrated utility philosophy from a central location is not uniformly accepted. The system must be designed with the objectives in mind and the implementation must be complete if the objectives are to be met. The term instrumentation needs to be expanded from *measurement* to include *control* and *optimization* of the UCS. New operating philosophies will be needed to cope with increased complexities in petrochemical complexes, as well as installing and operating an integrated Utilities Control System. Plans for implementation and transition from the old system to the new must be developed long in advance of actual operation. It is best to install field instrumentation during scheduled downtime, otherwise project costs will be uncontrolled. Early training of operators will serve two purposes; one, gaining their support for the new system and two, getting feedback on procedures that are not documented in the existing system. This also allows time for incorporating other ideas and the correction of misunderstood data.

ACKNOWLEDGMENT

Based on material originally presented at Texas A&M University 24th Annual Symposium on Instrumentation for the Process Industries, Jan. 18, 1979.

Double Cascade Controller Tested

Computer control using a double-cascade system is explored with a pulse testing technique to show its application in an industrial situation

F. Joseph Schork, E. I. du Pont de Nemours & Co., Louisville, Ky., and **Pradeep B. Deshpande,** University of Louisville, Louisville, Ky.

A DOUBLE-CASCADE CONTROLLER was successful in a control system to isolate a polymer for a continuous, industrial process. A pulse testing technique identified the dynamics of the open loop process and helped determine the settings for each level of the closed-loop control. The system included an inner control loop that is analog and two outer loops that are under computer control.

Cascade control is an effective means to eliminate the effects of load disturbances upon the dynamic response of feedback control systems. To review, a cascade control loop uses two feedback controllers. Only one of these controllers, referred to as the "slave" or "secondary" controller, has an output to the process. The other controller, called the "master" or "primary" controller, is used to adjust the setpoint of the slave controller. The process output is fed back to the primary controller while an intermediate process variable is fed back to the secondary controller. This arrangement for cascade control is shown in Fig. 1.

If the load disturbance enters near the beginning of the process, the secondary controller starts corrective action before the process output shows any deviation. Thus, the performance of the control loop is greatly improved. On the other hand, if the load disturbance enters near the end of the process, the performance of the cascade arrangement is still better than that of the feedback system because the former has a higher natural frequency and higher controller settings are permitted on the primary controller.[1]

The industrial application described here uses a double-cascade system. The purpose of the double-cascade system is to maintain the controlled variable, C, at setpoint despite changes in loads L_2 and L_3. The innermost control loop is analog and the two outer loops are under computer control. A block diagram of this system is shown in Fig. 2.

The purpose of this project is to identify the dynamics of the open-loop process and that of each successive level of the closed-loop system. The resulting dynamic data are used to design the innermost analog controller

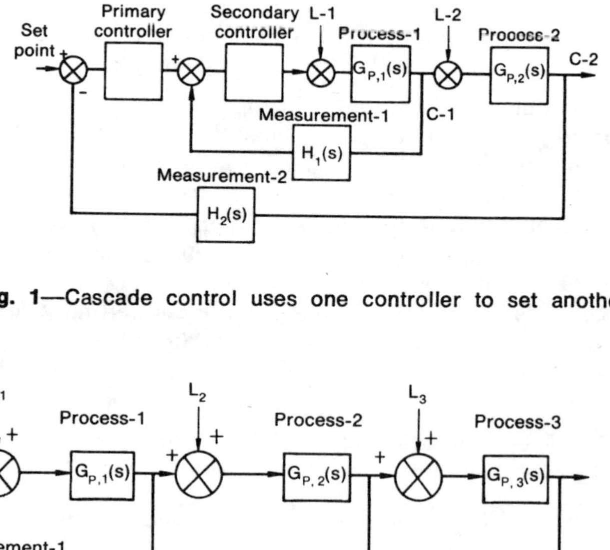

Fig. 1—Cascade control uses one controller to set another.

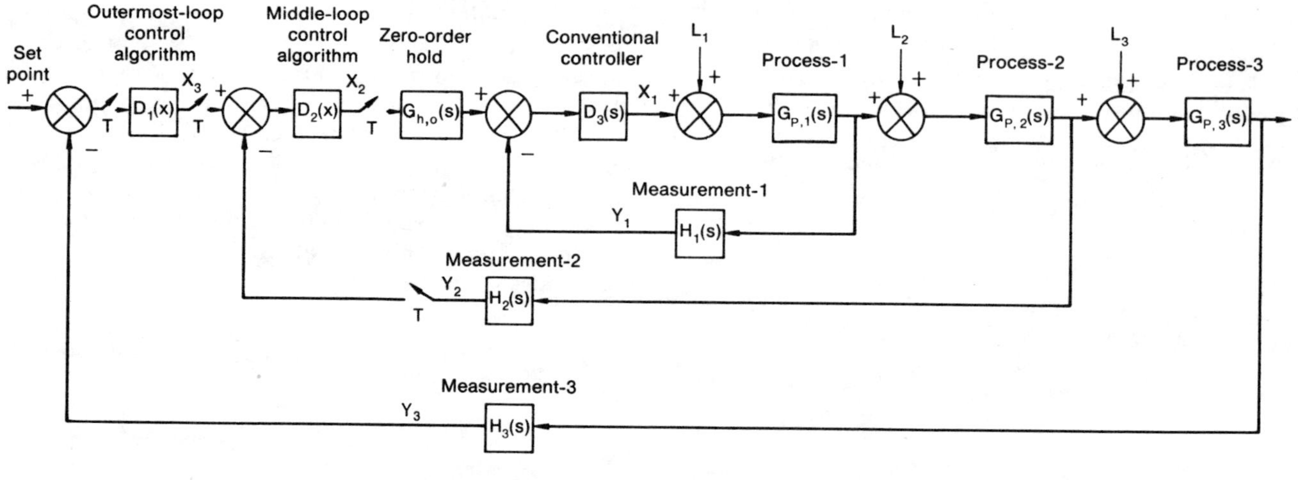

Fig. 2—The test system used a double-cascade control.

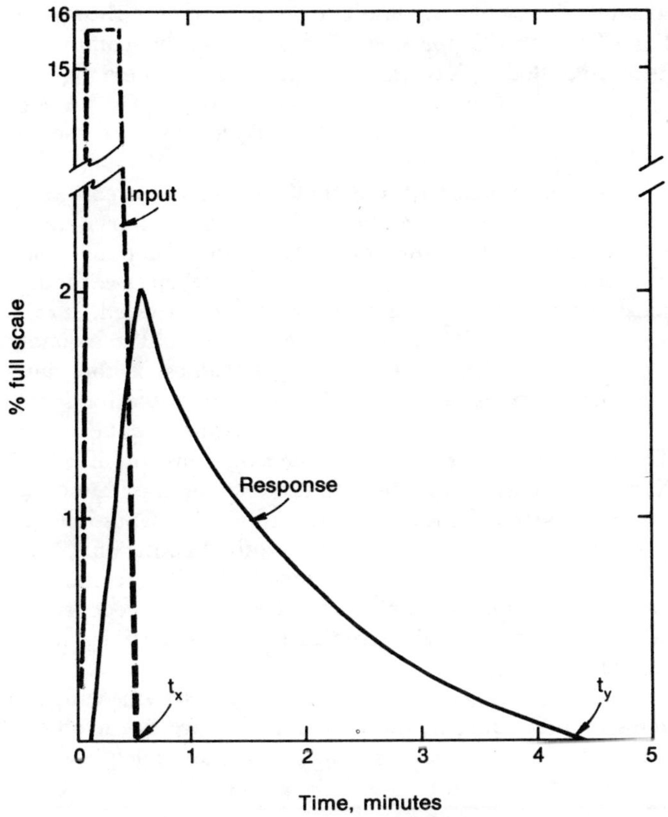

Fig. 3—The input and response for the inner loop.

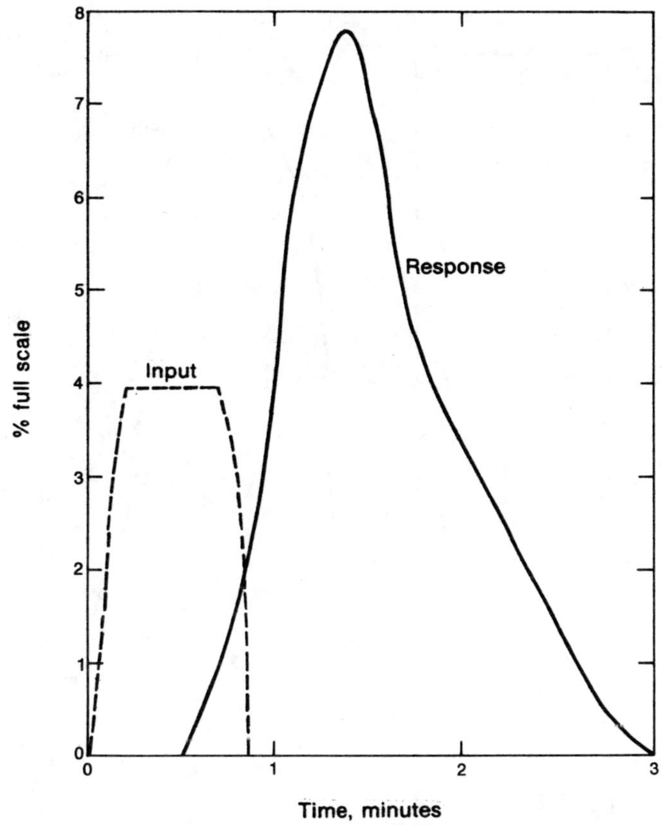

Fig. 4—The input and response for the middle loop.

and the two outer digital control algorithms. Then controller settings are applied to the process operating under closed-loop control, in the presence of load disturbances, to evaluate the adequacy of their design.

Dynamic information from the process control system can be obtained by any of several methods: The information can be based on a dynamic mathematical model of the process or it can be based on experimental tests on the actual plant. In the latter approach, a suitable disturbance is introduced into the process, operating in manual control, and the transient open-loop response is recorded. The analysis of the input/output data yields frequency response from which controllers can be designed.

Several forms of disturbance inputs have been successfully applied. Among them are step input, pulse input and sinusoidal input. The pulse input is especially convenient because it upsets the process only for a brief period of time and the process returns to the original steady-state upon the conclusion of the test. The present application is based on the pulse testing method.

The pulse testing technique, for a single-loop, feedback control system, applies a pulse of arbitrary shape to the input of the process that is operating in manual under steady-state conditions. The transient response of the process, Y, as well as the input pulse, X, are recorded. These records shown as Bode plots help to develop a frequency response for the process and its controllers. The derivation of the control functions is detailed later.

COMMERCIAL TESTS

To prepare for the plant tests, the input, X_1, and the output, Y_1, (see Fig. 2) were connected to separate pens of a two-pen strip-chart recorder. When the desired steady-state was achieved and the process was free from load disturbances, all control loops were switched to manual. A pulse of desired magnitude and duration was introduced into the innermost loop by rotating the manual knob on the innermost controller. By observing the time-record of the input pulse, it was possible to return the input to the initial steady-state operation. The test was repeated for different pulse widths and heights (above and below the initial steady-state values). For each run, numerous values of the input and output were read from their respective time-records into an off-line computer program which solved Equations 7 and 8 (see equations derived in a later section) so as to determine frequency response. From these data, the tuning constants for the innermost *PI* (Proportional and Integral) controller were determined by the Ziegler-Nichols method. These settings were implemented on the innermost controller and that loop was switched to automatic. The transients were allowed to disappear and when the steady-state was reached, the second pulse test was conducted.

The innermost controller is a Moore Products Synchro-III computer set control station. The controller is equipped with a stepping motor which accepts a pulse train from the process control computer at a rate of up to 30 pulses per second. The stepping motor moves the setpoint needle of the controller. Each pulse into the stepping motor moves the setpoint needle one thousandth of full-scale. The stepping motor responds to discontinui-

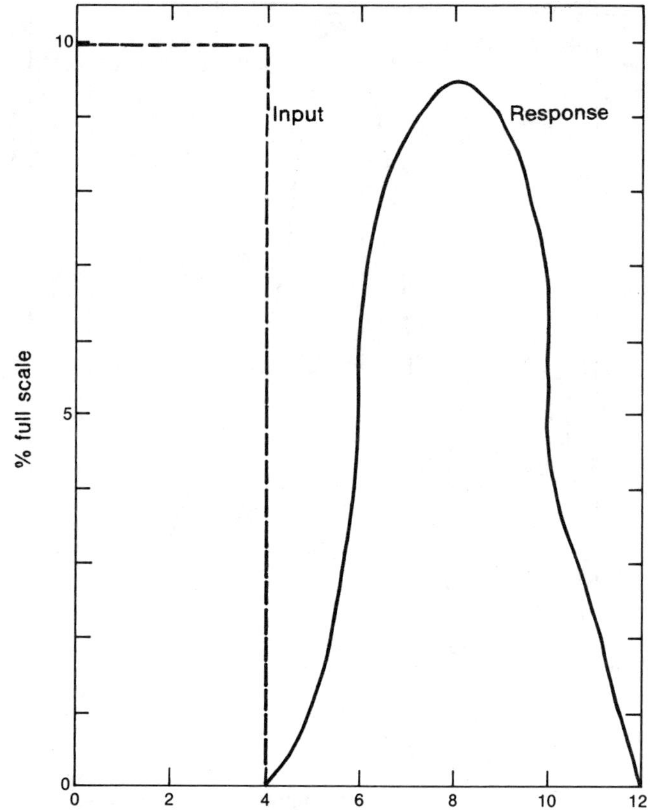

Fig. 5—The input and response for the outer loop.

ties in voltage between "closure" and "common" terminals. Thus, by disconnecting the computer from the control station and by connecting two leads to the "closure" and "common" terminals, the setpoint can be pulsed any desired amount by repeatedly contacting the two leads.

In this manner, several pulses of desired magnitude and direction were introduced into the middle loop at X_2 and the transient response, Y_2, was recorded. As before, time records for X_2 and Y_2 were processed on the offline computer so as to determine the Ziegler-Nichols settings for the middle-loop *PI* algorithm. These settings were divided by two (the reason for this will be explained in the next section) and then read into the on-line control computer as tuning parameters for the middle-loop algorithm. The middle-loop was then switched to automatic by reconnecting the pulse contacts from the computer to the control station.

With the two inner loops in automatic, the outermost loop was switched to manual. A pulse of desired magnitude and duration was generated on the on-line control computer and introduced into the input to the outermost loop at X_3 and the transient response, Y_3, was recorded. The time-records for Y_3 and X_3 assisted in the determination of the *PI* control algorithm constants of the outermost loop. The test was repeated using a pulse of different magnitude.

These settings were halved and then read into the control-computer program as the tuning parameters of the outermost algorithm. All three loops were switched to automatic and the transient response of the process, in the presence of load disturbances, was obtained and analyzed.

Sample time records of input pulse and the associated transient response for each loop are shown in Figs. 3, 4 and 5 for one of the tests. The computer-generated Bode plots from the analysis of these figures are shown in Figs. 6, 7 and 8, respectively. The tuning parameters determined from these Bode plots are shown in Table 1.

If all three loops of the double-cascade system were analog, these settings would have been satisfactory. However, since the two outer loops are under computer control, their performance depends not only on the tuning constants but also on the sampling period. Indeed, it can be shown by the Z-transform techniques, that a second-order system, which is stable for all values of the gain for a conventional system, can become unstable for some values of the sampling period in a sampled-data loop.[2] The performance of the two outer loops with the Ziegler-Nichols settings, found from pulse testing, was found to exhibit unstable behavior. Since it was not possible to alter the sampling period due to other constraints, the Z-N settings had to be reduced. Digital computer simulations, using Z-transforms, showed that good transient response could be obtained if Z-N settings were halved.

Finally, the closed-loop response of the system, with all loops in automatic, is shown in Fig. 9. Subsequent observations have shown these settings to be quite adequate.

TABLE 1—Tuning parameters for pulse data

Loop	Inner	Middle	Outer
Natural period, minutes	0.50	1.88	15.4
Maximum controller gain	14.00	1.92	8.60
Ziegler-Nichols gain	6.30	0.86	3.87
Actual gain implemented	6.30	0.43	1.94
Reset time, minutes	0.42	1.57	12.83

EQUATIONS DERIVED

This section is intended for the reader who wants to follow the derivation of the equations relating the transient response of the process, *Y*, to the input pulse, *X*. Refer to Fig. 10 to define the symbols used here for a single-loop, feedback control system.

The input and output variables as functions of time, *t*, are related to functions of *s* by using Laplace transforms. Then $X(t)$ and $Y(t)$ are related by the open-loop transfer function, $G(s)$, which is the product of $G_u(s)$, $G_p(s)$ and $G_m(s)$. The relationship is as follows:

$$G(s) = \frac{Y(s)}{X(s)} \tag{1}$$

Laplace transforms substituted for Equation 1 permits the equation to be written as follows:

$$G(s) = \frac{\int_0^\infty Y(t)\, e^{-st}\, dt}{\int_0^\infty X(t)\, e^{-st}\, dt} \tag{2}$$

The upper limits on the two integrals can be replaced by t_y and t_x, which are the time periods during which changes in *Y* and *X* occur. Also *s* is replaced by ω, the frequency, multiplied by *j*, the square root of minus one. Then Equation 2 becomes:

$$G(j\omega) = \frac{\int_0^{t_y} Y(t)\, e^{-j\omega t}\, dt}{\int_0^{t_x} X(t)\, e^{-j\omega t}\, dt} \tag{3}$$

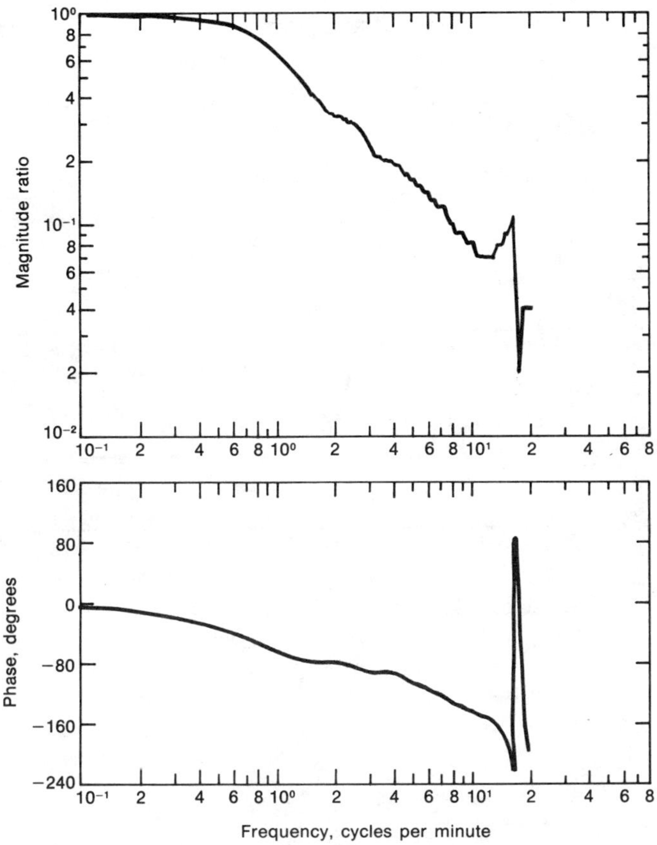

Fig. 6—Frequency response for the inner loop.

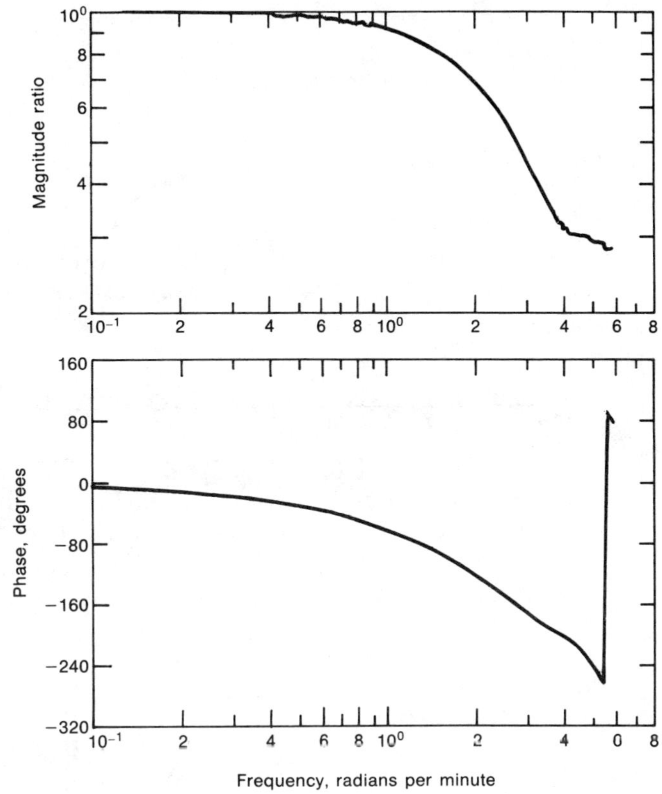

Fig. 7—Frequency response for the middle loop.

The numerator in Equation 3 is the Fourier transformation of the time function $Y(t)$ and the denominator is the Fourier transformation of the time function $X(t)$. Since $e^{-j\omega t}$ can be replaced by $\cos(\omega t) - j \sin(\omega t)$, Equation 3 can be written as:

$$G(j\omega) = \frac{\int_o^{t_y} Y(t) \cos(\omega t)\, dt - j \int_o^{t_y} Y(t) \sin(\omega t)\, dt}{\int_o^{t_x} X(t) \cos(\omega t)\, dt - j \int_o^{t_x} X(t) \sin(\omega t)\, dt} \quad (4)$$

To simplify Equation 4, the four integrals will be represented by A, B, C and D to give:

$$G(j\omega) = \frac{A - jB}{C - jD} \quad (5)$$

Then the right-hand side of Equation 5 is multiplied by the complex conjugate of the denominator to give:

$$G(j\omega) = \frac{A - jB}{C - jD} \bullet \frac{C + jD}{C + jD}$$

$$= \frac{AC + BD + j(AD - BC)}{C^2 + D^2}$$

$$= \left(\frac{AC + BD}{C^2 + D^2}\right) + j \left(\frac{AD - BC}{C^2 + D^2}\right)$$

$$= Re\, G(j\omega) + j\, Im\, G(j\omega) \quad (6)$$

where Re symbolizes the real portion of the complex number and Im symbolizes the imaginary portion.

The amplitude ratio, AR, and the phase angle, ϕ, are determined by taking the magnitude and the angle of the complex number, respectively. Thus,

$$AR = |G(j\omega)|$$

$$= \sqrt{[Re\, G(j\omega)]^2 + [Im\, G(j\omega)]^2}$$

$$= \sqrt{\left(\frac{AC + BD}{C^2 + D^2}\right)^2 + \left(\frac{AD - BC}{C^2 + D^2}\right)^2} \quad (7)$$

$$\phi = \arctan\left(\frac{AD - BC}{AC + BD}\right) \quad (8)$$

To use these equations, a specific value of frequency, ω, is chosen. The integrations are performed which yield a single value of A, B, C and D. Substituting these values in Equations 7 and 8 then gives a point on the frequency response diagram, corresponding to the chosen frequency. The procedure is repeated for numerous values of ω and the complete frequency response diagram is prepared.

Since the time functions would be available in graphical form, the integrals in Equation 4 are most conveniently evaluated using numerical integration techniques on the digital computer[3]. Once the frequency response diagram is completed, the controller tuning parameters can be determined by Ziegler-Nichols method.

The commercial test described earlier used the pulse testing method on the double-cascade system in the following general procedure: First, the technique is employed to identify the dynamics of the innermost open-loop process. From its Bode plot, the inner-most controller is designed. Next, these settings are implemented on the inner-most controller and that loop is switched to automatic. Then, the pulse testing method is applied to the middle loop so as to identify its dynamics and to design the middle-loop control algorithm. These parameters are read into the computer memory and the middle

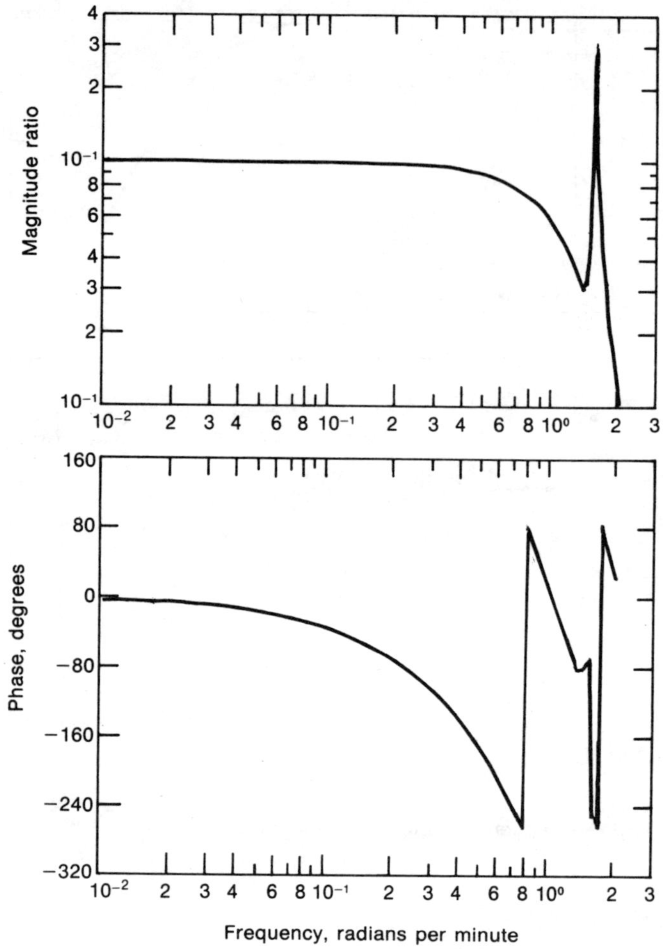

Fig. 8—Frequency response for the outer loop.

loop is placed on digital control. The procedure is repeated so as to design the outermost control algorithm.

In order to determine the adequacy of the parameters and tuning constants found through pulse testing, all three control-loops are switched to automatic. The process is operated under closed-loop control, in the presence of load disturbances, and the transient response is obtained and analyzed.

CONCLUSIONS

The commercial test was a successful and interesting application of the pulse testing method for the control system design of a double-cascade loop. With slight additional work, it should be possible to automate the entire procedure described in this paper. Then the computer can, from time to time, update the controller settings to compensate for any changing process characteristics.

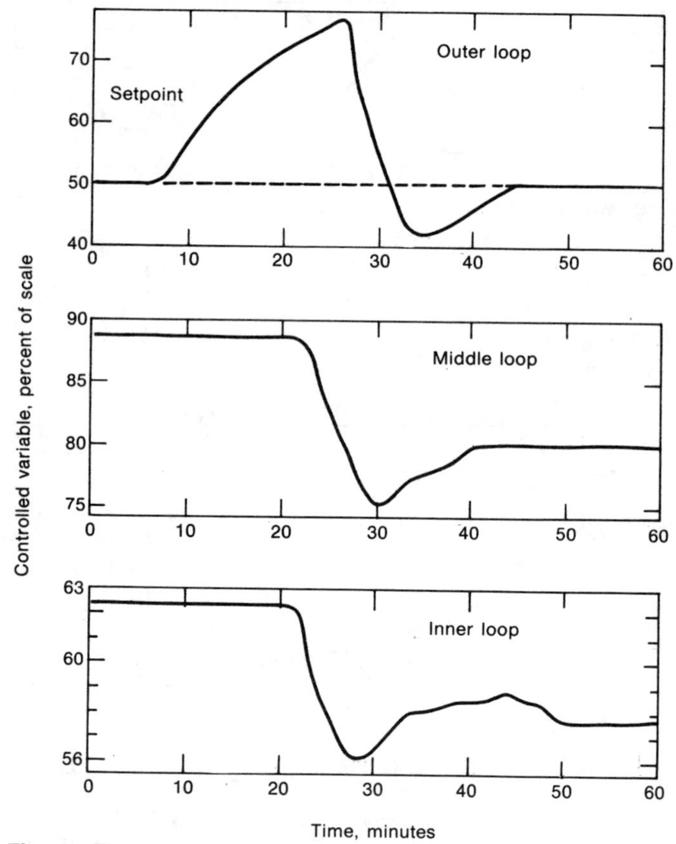

Fig. 9—The system response to a load change.

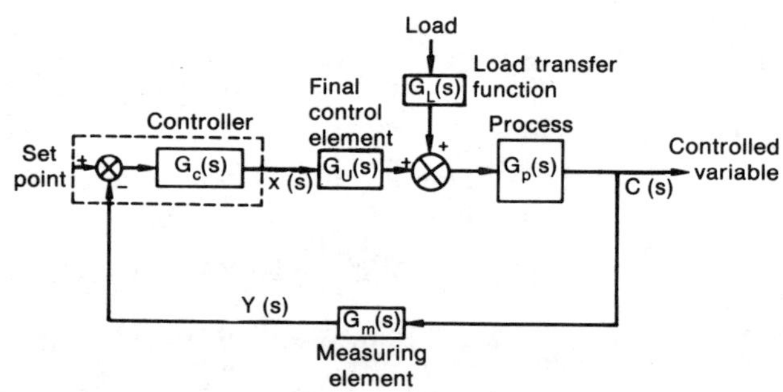

Fig. 10—Deriving the equations for a feedback controller.

ACKNOWLEDGMENT

The authors wish to thank the du Pont Co. for their permission to publish this work. Original presentation was before the American Institute of Chemical Engineers meeting, New York, Nov. 13-17, 1977.

LITERATURE CITED

[1] Harriott, P., *Process Control,* McGraw-Hill Book Co., New York, N. Y., 1964.

[2] Tou, J. T., *Digital and Sampled-Data Control Systems,* McGraw-Hill Book Co., New York, N.Y., 1959, pp. 242-44.

[3] Luyben, W. L., *Process Modeling, Simulation, and Control for Chemical Engineers,* McGraw-Hill Book Co., New York, N. Y., 1973, pp. 282-92.

Computer Control . . . How Will It Pay?

These program methods are useful for project management, determination of economic process operations and potentially for closed loop process optimization control on a larger scale than is currently practiced

W. E. Cox,
Exxon Chemical Co. U.S.A., Baton Rouge, La.

A COMPUTER process control project can normally be expected to yield economic credits resulting from increased product recoveries, reduced energy consumption, fewer process upsets and other factors. Before the investment decision, a study is made to estimate savings which can be achieved by computer process control. From this study, a schedule may be prepared showing the anticipated rate of credit generation during the life of the project (Fig. 1). During the implementation period, actual progress may be monitored and compared to expected results. As capital expenditure increases, this process becomes increasingly important to ensure adequate return on the investment and to provide additional information leading to even greater savings. Following is a description of a real-time computer program to calculate credits generated by computer control. The greatest potential for such credit calculations is in conjunction with a linear program to provide automatic process optimization control by closing the loop and allowing the computer to change process variables.

These programs, which are intended to be run on a supervisory process control computer, compare current operating data with previously determined base conditions for each major unit operation. These data are used in conjunction with an economic model of the process to calculate the dollar effect of differences between current operating conditions and the pre-computer control basis. Economic data generated by the programs are useful for project management, determination of economic process operating guidelines, and potentially for closed loop process optimization control on a larger scale than is currently practiced.

Economic data base. Care must be taken in developing a sound, realistic economic data basis; since as in all computer applications, the quality of the results is dependent on the quality of the input data. The economic data base consists of utilities and raw materials costs, product values and intermediate costs. Utilities, product values and raw materials have well defined costs. These should be manually input to the computer and updated as often as conditions warrant. However, there are other costs which must be given more careful consideration.

These costs include such items as refrigeration systems, steam boilers, intermediate products and recycle costs. For example, if a unit is running at full capacity, recycling in the process backs out fresh feed, reducing effective capacity. Therefore, recycle streams should show a debit

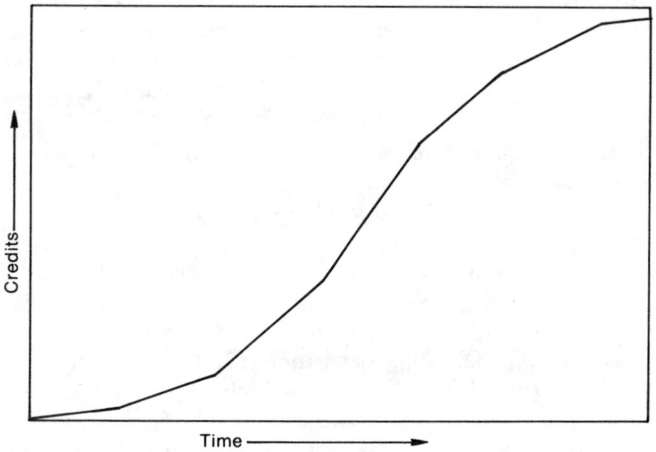

Fig. 1—Computer control credits vs. implementation time period.

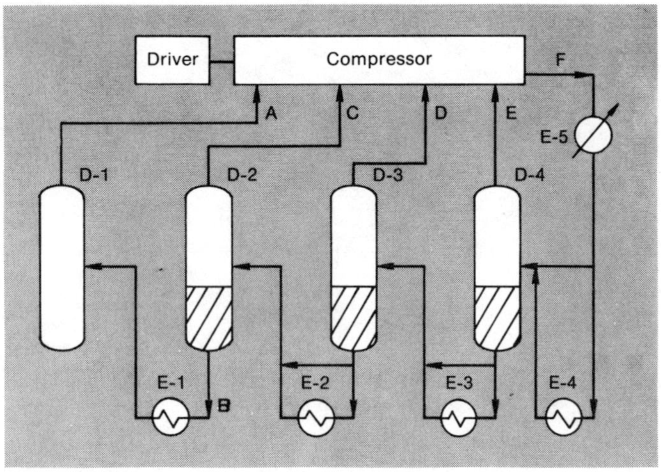

Fig. 2—Typical four stage refrigeration system.

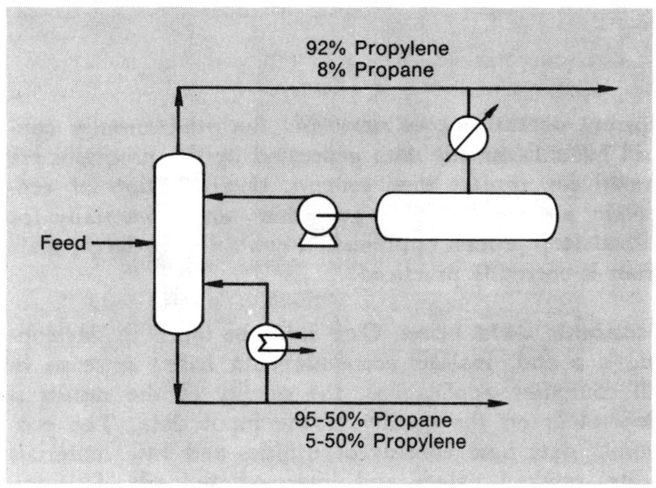

Fig. 3—Propylene/propane fractionator.

under certain conditions for unit capacity reductions. This cost would be in addition to energy costs associated with the recycle stream, such as pumping, heating and cooling.

Detailed costs can also be developed for secondary utilities generated within the plant, such as steam and refrigerant. Real-time programs can be developed to calculate current costs for steam and refrigerant based on the more fundamental costs of fuel and electricity. In this manner, the calculated costs more accurately reflect changes in operating conditions and basic utility costs.

As an example of a cost calculation program, consider the 4-stage refrigeration system in Fig. 2. Intuitively, lower pressure levels of refrigerant should cost more per Btu than higher levels, due to the greater amount of compression required. Total cost of operating the driver is a relatively straightforward calculation. However, a detailed heat and material balance must be calculated to determine what fraction of the total cost should be attributed to each stage of refrigerant users. This is necessary, because the lowest level users will cause some vapor to be flashed in each of the higher pressure levels, and the cost of recompressing this vapor must be added to the cost for compressing the direct flow through the lowest pressure suction. For example, if 100 M lb./hr. of refrigerant is required at point B (Fig. 2) some amount of vapor will be flashed in D-2 upstream. This increases the load to the compressor at point C. Similarly, additional vapor will be flashed in D-3 and D-4, increasing the load to each stage of the compressor. Thus the total cost charged to E-1 is the vapor flow to each stage times the cost of compression from suction pressure to discharge pressure. A sample cost calculation is shown in Table 1. This procedure must be repeated for each exchanger in the refrigeration system. Note that the cost for E-2 will be lower, due to lower compression cost for stages two, three and four. This type of calculation should at least be done manually, but an on-line calculation more accurately reflects changes in operating conditions. Experience has shown these costs can vary as much as 20 percent over a 24-hour period, depending on changes in unit feed rates and process upsets. A similar program can be developed to calculate steam cost based on fuel cost for firing the boiler.

Pre-computer control data base. To determine the basis for plant operation before computer control is implemented, data should be collected for a minimum of one month (more if feasible). This defines an average set of plant conditions over a wide range of operations. Specific data required depend on the economic model of the process, which will be discussed later. Data acquisition and analysis are greatly simplified if one has access to historical data stored automatically by the computer. Note however, this type of data is subject to wide variations, and a very large number of data points is required to ensure that a true mean has been established.

Process economic model. A credit calculation program should be created by the process control engineer for each major unit operation, such as distillation towers, compressors, furnaces and reactors. Emphasis will be given here to distillation towers, although the same principles can be used on any operation if the required process data are available.

On a generalized distillation tower (Fig. 3), there will be at least four types of credits to be considered. These are:

(1) Reduced overhead condenser duty
(2) Reduced reboiler duty
(3) Increased product recovery
(4) Reduced product losses.

Since each of these parameters can be expected to vary significantly with feed rate and composition, each should be expressed as some function of feed rate. This relationship must be determined by the base data previously collected. For example, base operations may indicate that reflux rate is always held constant, regardless of feed rate. At low feed rates, this will result in overfractionation, thus wasting energy, and giving away the amount of product which exceeds the purity specification. Computer control will seek to minimize this type of inefficiency, and the process economic model will be used to calculate resulting credits. In most instances, however, the operator does manipulate reflux to some extent to maintain product purity as feed rate and composition varies. Therefore, a simple ratio is often adequate. Fig. 4 represents an example of base data which indicate that reflux is held to an approximate ratio of feed rate. If feed composition varies significantly, reflux should be determined as a function of the rate of the heavy key component in the feed.

Reboiler duty can be treated similarly, but should be related to the rate of light key component in the feed. In the absence of feed analyzers, product analyzers may be used to back-calculate feed composition. But for simplicity, feed rate itself will usually suffice.

Product recovery and losses can best be discussed by means of a specific example. Consider again the propylene/propane tower in Fig. 3. Propylene product is taken overhead at a desired purity of 92%. This means it is desirable for precisely 8% of overhead product to be propane. If any less propane is taken overhead, the propylene above the amount required for 92% purity is being "given away" at the bargain price of 92% propylene. Similarly, any propylene which is allowed to go bottoms can be considered as lost potential product, and will be sold for a much lower price than the 92% propylene product. These two indices are highly interactive; but are really separate, controllable quantities, which can be reduced at the expense of higher energy consumption. This is the economic trade-off which computer control should seek to optimize, after relative credits have been specified. Disposition of off-specification product must also be considered in the economic model, otherwise, credits will increase steadily as propane content goes above the 8% limit. Off-spec product should be valued according to its disposition, such as fuel, recycle or flare. Obviously, if off-spec product must be flared, it will have a value of zero.

With this economic model in mind, it is relatively simple to develop equations necessary to make credit calculations. For the reflux condenser (see Nomenclature):

$$CRED_R = \left[(R/F)_B F_c - R_c \right] \$_R \frac{8760 \text{ hr./yr.}}{1000} \quad (1)$$

The product, $(R/F)_B F_c$, represents the amount of reflux expected under base operating conditions for current feed rates. For example, if the base ratio of reflux to feed was 0.425 M Btu/lb. feed, and the current feed rate is 92 M lb./hr., a reflux condenser duty of 39.1 MMBtu/hr. would have been expected under base conditions. For a current reflux duty of 38.0 MMBtu/hr., the difference of 1.1 MMBtu/hr. is the energy savings. This savings is multiplied by the cost and annualized to obtain credits in M$/yr. Thus, if refrigerated condenser duty costs $7.65/MMBtu as in the previous example, the credit in this case would be 73.7 M$/yr. The equations for reboiler duty, product giveaway and losses are similar:

$$CRED_Q = [(Q/F)_B F_c - Q_c] \$_Q\, 8760/1000 \quad (2)$$

$$CRED_G = [(G/F)_B F_c - G_c] \$_G\, 8760/1000 \quad (3)$$

$$CRED_L = [(L/F)_B F_c - L_c] \$_L\, 8760/1000 \quad (4)$$

The total credit for current operations over base conditions is simply the sum of the four components.

$$CRED_T = CRED_R + CRED_Q + CRED_G + CRED_L \quad (5)$$

Note that current credits have been extrapolated to annual rates, so that the results will be in the desirable units of M$/yr. Meaning if current instantaneous conditions were maintained precisely for one year, the credits indicated would be obtained.

Although only a simplified distillation model is presented here, similar economic models can be developed for compressors, furnaces and reactors. The concept is identical but different indicators are used, such as horsepower, fuel Btu's and conversion rates all as functions of feed rate.

These programs have been implemented on each major unit operation of a large computer control project. Resulting data have been found to be a powerful tool in guiding efforts of the process control engineer into areas where the greatest potential exists for improvement to plant operations. It can also be helpful in determining plant operating guidelines and reporting progress of computer control implementation to management. Potentially, this can be the most powerful indicator for closed-loop process optimization control since it reflects all the factors which affect efficiency of plant operation.

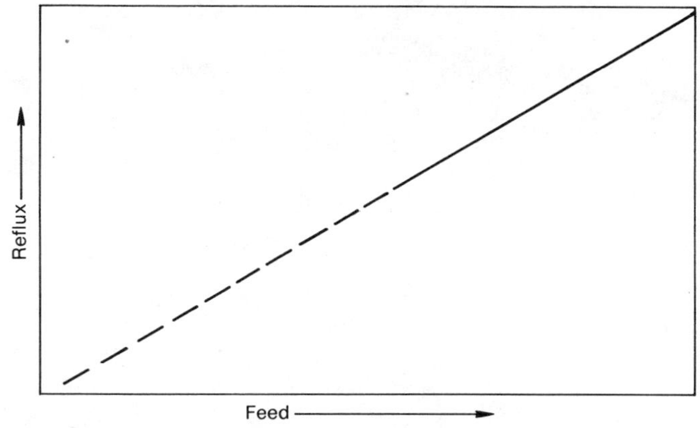

Fig. 4—Base data for distillation tower.

TABLE 1—Sample refrigerant cost calculation

Flow through E-1 100 M lb./hr.	×	Compression cost from pressure A to F $5/M lb.	=	First stage cost $500.0/hr.
Additional vapor flashed in D-2 due to E-1 flow 15 Mlb./hr.	×	Compression cost from pressure C to F $4.20/Mlb.	=	Second stage cost $63.0/hr.
Additional vapor flashed in D-3 due to E-1 flow 11 Mlb./hr.	×	Compression cost from pressure D to F $3.60/Mlb.	=	Third stage cost $39.6/hr.
Additional vapor flashed in D-4 due to E-1 flow 8 Mlb./hr.	×	Compression cost from pressure E to F $3.05/Mlb.	=	Fourth stage cost $24.4/hr.
		Total Cost	=	$627.0/hr.

Total cost $627.0/hr. ÷ E-1 duty 82.0 MMBtu/hr. = Cost per Btu for E-1 $7.65/MMBtu

NOMENCLATURE

CRED	Credits, M$/year
B	Base conditions
C	Current conditions
F	Feed rate, M lb./hr.
G	Product giveaway due to exceeding the product purity specification, lb./hr.
L	Lost product, such as propylene in bottoms in example, lb./hr.
Q	Reboiler duty, MMBtu/hr.
R	Reflux condenser duty, MMBtu/hr.
T	Total
$\$_G$	Value of on-spec overhead product, $/lb.
$\$_L$	Cost of downgrade from overhead product to bottoms product, $/lb.
$\$_Q$	Cost of reboiler heating medium, $/MMBtu
$\$_R$	Cost of condensing medium, $/MMBtu

ACKNOWLEDGMENTS

Based on a paper presented at the Texas A&M University Instrumentation Symposium, Jan. 18-20, 1978. The author extends his gratitude to co-workers who have been instrumental in developing the calculation techniques presented herein: Messrs. R. A. Ours, C. P. Senyard, N. I. Silverman, R. D. White, Dr. D. E. Haskins and Dr. D. L. Ritter.

Interface Computers to Existing Plants

Existing plants can often benefit from the addition of a digital control computer. Many factors such as existing control systems, computer interface hardware requirements and operator acceptance must be considered. These considerations are presented here.

F. P. Bader, Moore Products Co., Spring House, Pa.

A DIGITAL CONTROL computer can often improve the operation of an existing plant. But there is a tendency to think that such an addition by itself guarantees successful results. As opposed to original system design concepts, upgrading an existing plant has restrictions which must be considered as part of the design. For example, present equipment (sensors, valves and controllers), transmission means (pneumatic or electronic), operator procedures and maintenance capability play a role in defining the nature of the system. Prudent design utilizes the existing situation as best as possible, without jeopardizing the ultimate control computer objective of greater profitability. The ongoing addition of computers to existing plants has witnessed both pronounced and subtle technological changes in interface schemes that are a blend of modern and time-proven concepts.

Supervisory or DDC computer control strategy, the collection of various electronic and pneumatic analog inputs, the integration of the computer output into the control system, and simplicity of back-up schemes which gain operator acceptance of the "other" operator, are worthy of more detailed examination.

SUPERVISORY OR DDC CONTROL STRATEGY?

The bulk of computer control utilizes the supervisory concept (Fig. 1). By definition, the computer adjusts the setpoint of a separate instrument controller. The controller, in turn, adjusts the valve to keep the process variable at the setpoint. In DDC, the computer bypasses the automatic controller to reset the valve directly with the duplicated control function of the instrument now resident in the computer (Fig. 2).

Several factors assure ready acceptance of supervisory control. The strategy gives the appearance of cascade control with the computer acting as a complex primary controller manipulating the setpoint of a secondary flow or level controller. A lower load demand is placed on the computer, since the fast response secondary control is in the instrument. This chain of control ensures the use of the automatic controller in the separate instrument thus guaranteeing its availability in the back-up mode. When the local control configuration tracks the computer-set mode, transfer to back-up is transparent to the process. Of course, absence of the computer from the control

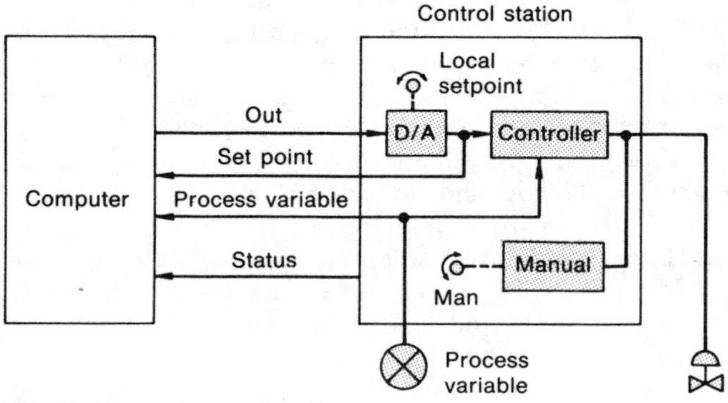

Fig. 1—Supervisory control.

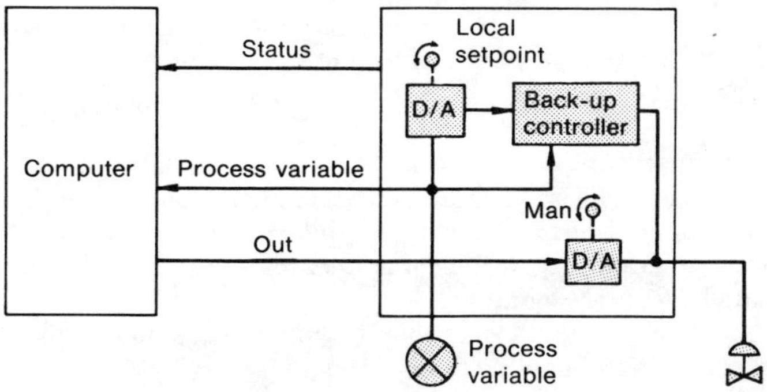

Fig. 2—Direct digital control.

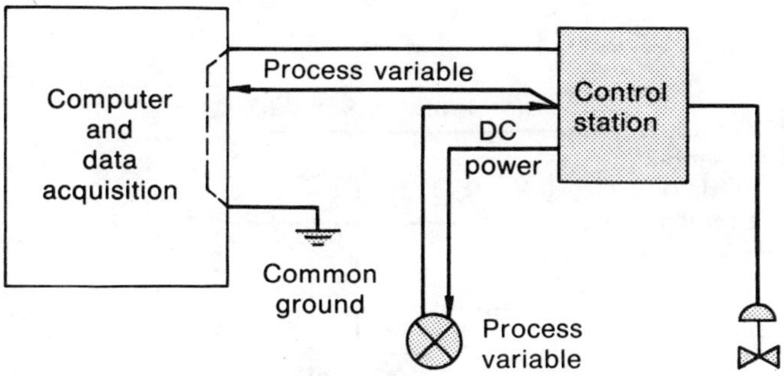

Fig. 3—Adjacent computer/instrument systems.

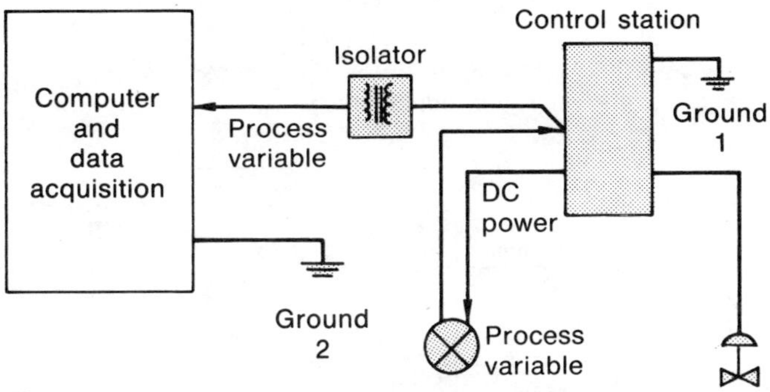

Fig. 4—Remote computer/instrument systems.

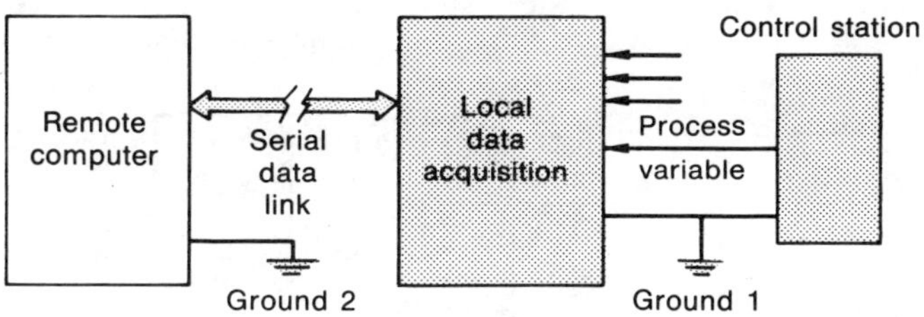

Fig. 5—Remote computer/data acquisition system.

scheme will likely have a long-term effect on process performance. The amount of back-up should be confined to the necessary schemes required by the operator to maintain reasonable control over the process. A particular aspect of back-up capability is discussed later.

Since DDC bypasses the analog controller, there is always a question of the controller's availability, current tuning and calibration accuracy when the back-up mode is selected. Transfer to back-up analog from DDC is more complex than supervisory to analog back-up, and a fully tracking back-up mode is almost imperative to assure a smooth transition. Many systems incorporate a combination of supervisory and DDC control, utilizing the advantages of both computer concepts.

DATA ACQUISITION AND INPUT CONSIDERATIONS

The configuration of the real-time data acquisition subsystem is defined by the nature of the system inputs. Since both electronic and pneumatic control systems exist in plants, both types of systems are handled readily by proven hardware concepts. Location of inputs with respect to the computer also plays a role in the selection of data acquisition components.

Electronic systems. Data acquisition of electronic analog and on-off signals is a well-defined subject by the many suppliers of this type equipment. Integration of the computer and instrument system, both of which use the input data, is an important aspect of a successful system (Fig. 3). For two-wire sensors, such as differential pressure transmitters, the instrument furnishes DC voltage to power the sensor. This is a logical arrangement, since integrity of the back-up system must be maintained before the computer can perform its task through the back-up system. What if the computer and its input data acquisition system are located remotely from the instrument panel? Here, the possibility of multiple system grounds is confronted.

Since more than one ground in a system generates noise and errors in the signal transmission, multiple grounds are to be avoided at all costs. Two possible solutions exist. The most obvious solution employs isolation modules to physically detach the DC common return between the two ground points (Fig. 4). Both transformer and opto-coupling type isolators are available to eliminate the ground problem on a single-loop basis. If many signals are involved in the grounding problem, isolation at each loop is an expensive proposition. A second approach locates the computer data acquisition in the area of the instrumentation (Fig. 5). Now, the signals share a common system ground. The analog data is converted to its digital counterpart and transmitted over a digital link to the computer mainframe. (The data link employs optocouplers to separate the data-acquisition system ground from the computer ground). Not only is the ground problem resolved, but the number of wires between the control room and the computer room is reduced for lower installed costs. "Smart" remote data acquisition systems have extensive capabilities, including outputs as well as inputs and sophisticated signal conditioning such as thermocouple linearization, square root extraction and engineering unit conversions.

The major disadvantage of such remote systems is the obvious risk involved in the dense signal load of a shared link. Ultimate reliability is essential. Factors such as noise, communications protocols, speed, hardware and software compatibility affects this desired reliability. For this reason, elaborate schemes have been devised to provide the necessary performance. Dual redundant data links and signal validity checking are used to avoid errors, but each devised scheme can lead to a lack of uniformity or standardization between pieces of the system.

The use of multiple-grounded, remote data acquisition systems with data links, versus single system grounds with dedicated connections, must be considered for each application. The greater the distance and quantity of input/outputs, the more attractive remote systems become. There is no specific distance or quantity of input/outputs that establishes the choice, since both types of systems can be installed for satisfactory performance.

While determining the type of data acquistion equipment to utilize in the system, bear in mind that the back-up system usually requires a high-level process variable input (Fig. 6). Thus, signal conditioning equipment for low-level inputs (thermocouples, RTD, etc.), for pulse-inputs (turbine meters, speed transmitters) and a large variety of analyzers and special sensors is required before connecting to both the back-up system and the computer data acquisition. Sensor inputs not required by the back-up system for minimal operation without the computer, may be connected directly to data acquisition input channels.

Electronic systems, installed according to intrinsic safety requirements, add an extra dimension of multiple ground considerations to a system. Since the system barriers must conform to the intrinsic safety grounding procedures, the required computer ground must reference to the same safety ground to avoid the multiple ground problem.

Pneumatic systems. Pneumatic systems are as compatible with computers as electronic systems—despite the fact that no digital computers exist with pneumatic inputs or outputs. In many respects, it is accurate to state that the input side of the computer control system is incompatible with pneumatics, while the output side of the same system is incompatible with electronics. This latter statement is attributed to the widespread use of pneumatic valves as the final control element. When adding a computer to a pneumatic system, the transmitters and transmission system are almost entirely pneumatic, and it is difficult to justify eliminating this equipment in favor of its electronic counterpart when no performance improvement is likely from the switch. Even on new systems, some users prefer pneumatic control systems because of the inherent electrical safety, separate power (air) source, ruggedness, noise immunity and existing maintenance capability. Others prefer pneumatic transmitters in the field for electronic control systems because of harsh environmental factors. Transducing the pneumatic signals into their electronic counterparts for the computer then becomes an integral part of a computer/pneumatic system.

Individual pressure-to-electric (P/E or P/I) converters may be used on each pneumatic variable. Recall from Fig. 1 that each supervisory loop usually requires a process and setpoint feedback to close the control path with the computer. Thus, each pneumatic loop requires two transducers. When considering large numbers of signals, the concept of a wall covered with transducers produces the reality of considerable expense from initial investment to installation and maintenance. Fortunately, there are alternate transducer techniques that save space and expense. One such packaging concept utilizes conventional 19 in. relay rack mounting with high-density transducer arrangements. Using a common DC power supply bus, the racks provide individual transducing at lower initial investment.

A transducer in a class by itself is the Scanivalve multiplexer (Fig. 7). Sixty-four pneumatic inputs to a multiported valve are converted by a single high-accuracy transducer. A common motor drives the valve and a digital encoder. The computer reads the synchronized encoder output with the common transducer output in a continuous scan mode or handshake routine. The device yields a very attractive cost/point, and requires minimal space requirements. Furthermore, each 64 port "deck" can be stacked in a 19 in. relay rack and driven by a common motor. The self-contained system can handle 768 pneumatic signals. Recent addition of a micro-computer to this cabinet digitizes and stores the ports and

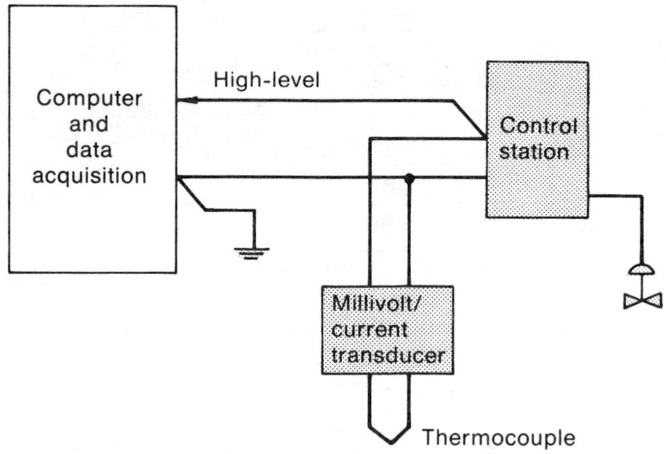

Fig. 6—Low level signal conditioning.

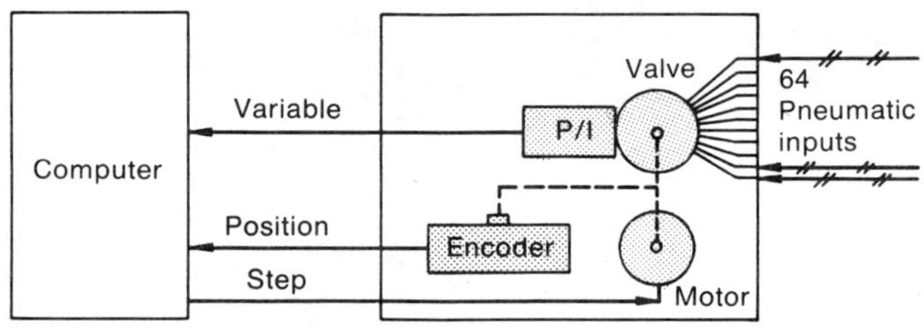

Fig. 7—Pneumatic muliplexer.

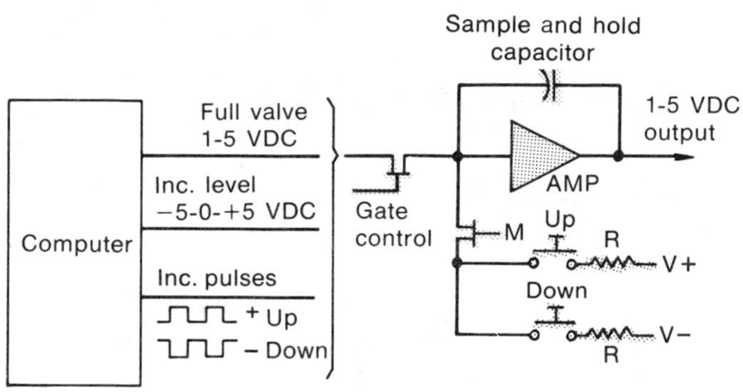

Fig. 8—Sample and hold interface.

associated signals, thus relieving the control computer of the task. Block transfers of data can be made via a serial data link or a parallel interface to the computer. Scanivalve also includes thermocouple and electronic analog data acquisition in the same cabinet.

Maximum operating speed for the Scanivalve multiplexer in continuous duty is 6 ports/sec. This is entirely adequate for supervisory control but not multi-loop DDC. The Scanivalve Co. has a high speed scanner, operating on a different principle, which operates at 64 ports/sec.

Status update. An important piece of information for the computer is the operating status of the control system. The typical feedback signal is an on-off contact or solid-state switch which the computer reads in its scan sequence (Figs. 1 & 2). This contact tells the computer if the controller will accept or ignore the computer command. For many computers, the information becomes vital to the software strategy, since ignored commands would otherwise imply to the computer that a failure has occurred in the system.

COMPUTER CONTROL OUTPUTS

It is desirable to have a control system assimilate com-

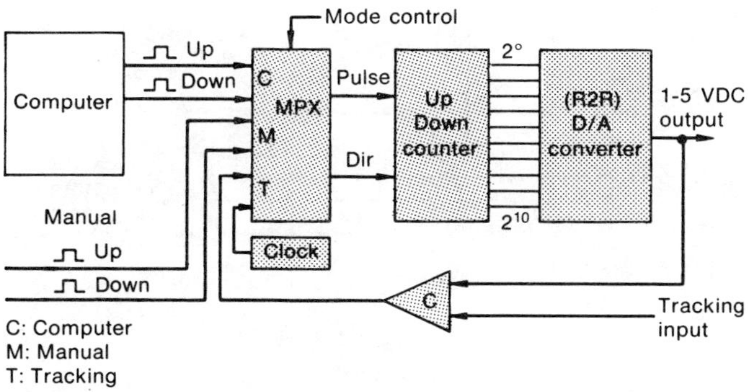

Fig. 9—Electronic digital to analog converter.

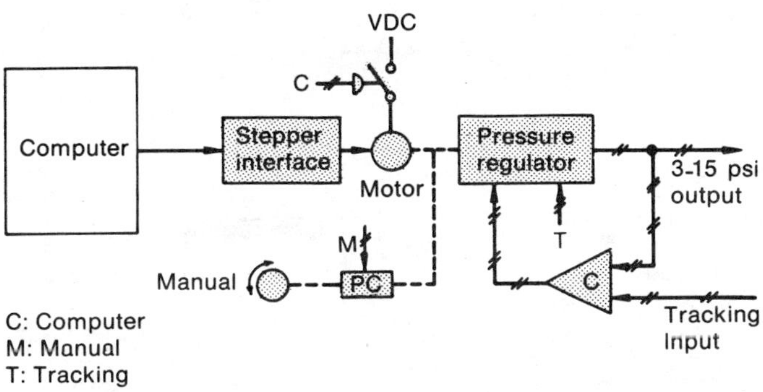

Fig. 10—Pneumatic digital to analog converter (Motorized self synchronizing regulator).

mands from the computer in a manner which allows simple control transfer on and off the computer by the operator. Preferably, the transfer shall involve no procedures, and there will be no shift in the process after these transfers take pace. This continuity between the computer and the operator gains quick operator acceptance of the computer system. In short, the computer should not make the operator's job more difficult.

Referring ahead to Figs. 9 and 10, the instrument interface device has three inputs and a common output to achieve this operating simplicity. A computer input, manual input and tracking input all drive a common output. The computer input is active in computer control, the manual input is active in the back-up mode requiring direct operator setting, and the tracking mode is active when the interface device is aligning itself for bumpless, procedureless transfer to manual or computer-set. If, for example, a separate non-tracking manual device is used to deliver the back-up output, desirable, procedureless-switching simplicity is violated.

With the computer acting as another operator, the computer mode and local mode should be isolated from each other. Only one operator should manipulate the process control loop at any instant, with mode selection usually given to the operator. Remember that status feedback always informs the computer of the current mode so the computer can monitor and store the operator's activity. But, who will keep the operator from negating the computer commands while the computer mode is active? It should not be possible for the operator to override the computer, unless he physically switches the loop from the computer mode. Obviously, this capability must be included in the computer-set instrument.

Several techniques have evolved to cover the needs of the ideal instrument interface.

Electronic sample and hold. This early interface scheme utilized existing state-of-the-art electronics to receive commands from the computer (Fig. 8). A high impedance circuit literally traps a charge on a capacitor which is indicative of the analog output (setpoint or valve position) desired by the computer. The charge is varied by several means. The most straightforward method simply replaces the charge with a new value through a solid-state high-impedance switch. This is known as full-value implementation. The capacitor charge may be incremented up or down by DC level changes, by integration of opposite polarity pulses, by integration of a set of up-down pulses, or by integration of a single pulse with an auxiliary direction logic. All have been used at one time or another.

The advantage of the sample and hold technique is its rapid response to an input change. Full scale change can occur in a matter of milliseconds. A set of normally open pushbuttons furnishes the operator's back-up adjustment of the same circuit, thus insuring bumpless transfer between computer and local modes. By their very nature, sample and hold techniques are not drift-free, since the capacitor voltage eventually discharges. Careful consideration to circuit impedances, capacitor quality and moisture protection must be used to minimize this effect.

Electronic D/A converter. D/A converters overcome the drift problem of sample and hold analog techniques by virtue of a digital register or counter used to hold the computer command (Fig. 9). The up-down counter responds to an incremental input, changing its binary output by the number of counted input pulses from the computer. One thousand pulses full scale is a generally used range. A resistive network (R2R) converts the counter's binary output into an equivalent analog value to be used in the control system. A solid-state switching matrix (MPX) adds the back-up adjustment to the D/A converter to assure procedureless transfer. This device is essentially a switching circuit which selects computer, manual or tracking inputs in accordance with the selected control mode of the instrument.

Typical D/A converters accept pulse rates to 1 kHz, allowing a full scale change in one second if desired. By modifying the input gate circuitry, the D/A converter can respond to modified versions of pulse input techniques. Although dual pulse gate inputs are preferred, a single pulse gate and direction gate is sometimes used. This technique is not failsafe, since a failure of the direction gate alone could send the D/A converter output in the wrong direction. The tracking circuit may be used to accept an analog input, as used in sample and hold circuits. This has a disadvantage similar to pulse/direction techniques, since the D/A converter will track any analog input, even if it should inadvertently go to zero or full scale.

Pneumatic D/A converters. In pneumatic controllers, a motorized regulator converts digital commands from the computer into the standard 3-15 psi signal (Fig. 10). These devices allow the use of small, low power (typically less than 10 W) motors and produce accurate, repetitive response equivalent to the electronic D/A converter. Two types of motors are used. Bi-phase synchronous motors are simple to interface to the computer and come in a host of full scale timing ranges from approximately 10 sec. to several thousand seconds. A mercury-wetted contact at the computer gates 115 VAC line voltage to up or down terminals in a pulse duration concept, with the pneumatic output changing by a percentage of full scale time. Synchronous motors sacrifice low increment resolution at fast speeds, so the preferred motor in most interface applications is the stepper. This motor, powered by DC voltage, responds in fixed output shaft increments for each applied input pulse. The motor has a peculiar logic of operation, thus requiring an electronic interface card to translate the serial pulse information from the computer into correct motor response. This interface is included in the instrument with the motorized regulator. The typical input range is 1,000 pulses full scale, producing 0.1 percent resolution at the pneumatic output.

The pneumatic regulator also includes three inputs with one output for the various operating modes. A knob for manual output adjustment, and a tracking mode complete the integration of the motorized regulator into the instrument. A piston-like coupling (PC) isolates the knob in the computer mode, preventing the operator from interfering with the computer adjustment.

SYSTEM IMPLEMENTATION

Successful systems employ several noteworthy concepts, which enhance operator acceptance of the computer.

Emergency switching: Computer-set controllers equipped with remote switching connections allow the transfer to backup in the event of a computer malfunction, without operator assistance. This multi-loop transfer must follow the switching continuity requirements already discussed to be effective. It is not desirable, for example, to have a number of controllers shifting to arbitrary non-tracking local setpoints when a common transfer occurs.

Multi-variable backup: Systems with cascade, ratio, remote-set, etc. require careful analysis to provide smooth transition to backup. Ordinarily, the complex control schemes become part of the internal computer control strategy and the only backup required are single-loop controllers. On occasion, it is necessary to install complex backup when the computer is not available.

Fig. 11 shows a cascade backup scheme. The computer adjusts the secondary (fast) loop through the DDC mode of the primary. The backup modes include manual, local secondary-set and local cascade-set. To assure complete

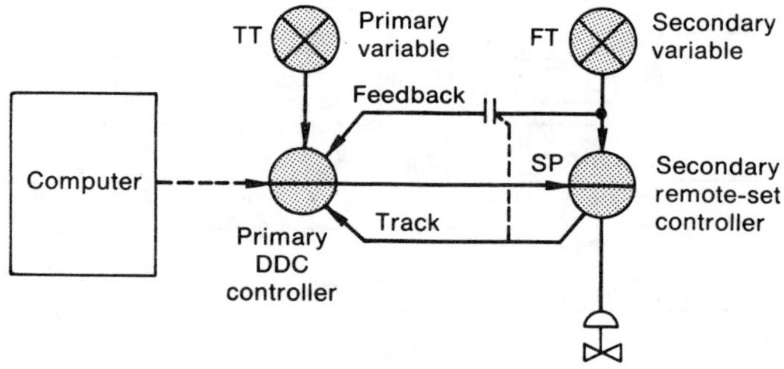

Fig. 11—Cascade back-up for supervisory control.

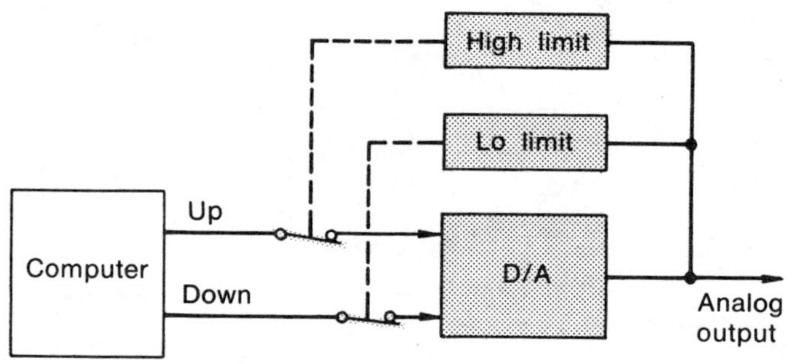

Fig. 12—Adjustable high-low limits.

backup synchronization, the primary controller tracks the secondary variable through a feedback connection whenever the cascade mode is de-activated. This prevents primary reset windup, and gives complete confidence to the operator for procedureless, non-shift switching on and off computer control. While not required on every loop, such a scheme on very important loops can make or break an entire system.

Combination instruments: Some users have found application for instruments that include both supervisory and DDC in a single unit. The application deals with occasional process dis-continuities, or non-linear operating levels that may be enhanced by a special DDC control strategy.

Hardware limits: Adjustable limits built into the instrument are used as a safety envelope on the computer commands (Fig. 12). Whenever the computer attempts to drive the interface output above or below the limit setting, the input signal is disconnected in that direction. While software limits are usually part of the computer, some users prefer the extra assurance of dedicated, independent hardware limits located in the instrument.

LITERATURE CITED

[1] Bader, F. P., "Linking Pneumatic Control Systems to the Computer," ISA Conference, Philadelphia, Pa. 1970.
[2] Mamzic, C. L., "Pneumatic Controls Interface with Computers," Instruments and Control Systems, October, 1975.

How Feed Affects a Stabilizer

The influence of feed composition on the butane-pentane split from a gasoline stabilizer helps determine the best way to control the column

C. R. Clark, The University of Leeds, Leeds, England

THE FEED to a gasoline stabilizer can vary to influence changes in the pentane content of the bottoms product and the butane content of the overhead material. This is espcially true when the stabilizer feed comes from a refinery gas recovery system handling many independent gas streams. A computer model is formulated and the results help show which factors most influence the stabilizer's products.

The stabilizer. The basis for this article was obtained from Continental Oil (U.K.) Ltd. in the consideration of one of their gasoline stabilizers. At the time of writing, the company had recently completed building a refinery at South Killingholme, near Grimsby. The gasoline stabilizer being considered is on this site and handles unstabilized light gasoline which is condensed in the gas recovery plant.

There will be approximately 12 independent streams fed into the gas recovery plant and this will affect the composition of the feed to the tower. Main variations to be considered will be in the butane/pentane content of the feed as these are the two key components for the top and bottom products.

The feed to the tower is de-ethanized in the preceding tower. Bottom product in the form of stabilized gasoline from the column passes to a gasoline splitter, while the top product or overheads are fed to a propane/butane splitter.

The control instrumentation for the tower is shown in a simplified form in Fig. 1. The design material balance for the tower is given in Table 1. Temperatures and design, reboiler and condenser loads are also given, although there is no compatible heat balance available for this tower. Some general equipment data are given in Table 2.

For normal operation, it is reasonable to assume that the feed/flow quality will be subject to very frequent, but possibly relatively small changes. With the feed originating from a wide number of sources, almost any change in refinery operation will have some effect on the system. Taking the design material balance to give the base composition profile, various changes were made in the feed composition for an equivalent computer model. These included interchanging, by up to 10 percent, the

TABLE 1—Design material balance

Component	Stream rate, lb. mol./hr. Feed	Bottom prod.	Reflux	O.H. prod.
Ethane	6.74	......	8.12	6.74
Propane	621.55	......	748.97	621.55
Isobutane	230.06	......	277.22	230.06
n-Butane	490.90	2.56	588.45	488.34
Isopentane	73.98	70.85	3.77	3.13
n-Pentane	84.91	84.91	......	......
Hexane	453.55	453.55	......	......
Total	1,961.69	611.87	1,626.53	1,349.82

TABLE 2—Equipment design data

Tower	Tray 1	Tray 34
Tower i. d., inches	84	84
Tower area, sq. ft.	38.48	38.48
Tray spacing, inches	24	24
Downcomer area, sq. ft. (approx.)	7.5	7.5
No. of liquid flow paths	2	2
Hole area, % bubbling area	9.0	9.0
Hole area, sq. ft. (approx.)	2.4	2.4
Vapor to tray, lbs./hr.	159,000	147,000
Liquid from tray, lbs./hr.	90,000	202,000
Vapor density, lbs./cu. ft.	1.42	1.84
Liquid density, lbs./cu. ft.	31.6	33.2
Liquid viscosity, c.p.	0.12	0.20
Liquid surface tension, dynes/cm	6.87	4.95
Overhead condenser, MM Btu	22.421	
Reboiler, MM Btu	18.875	

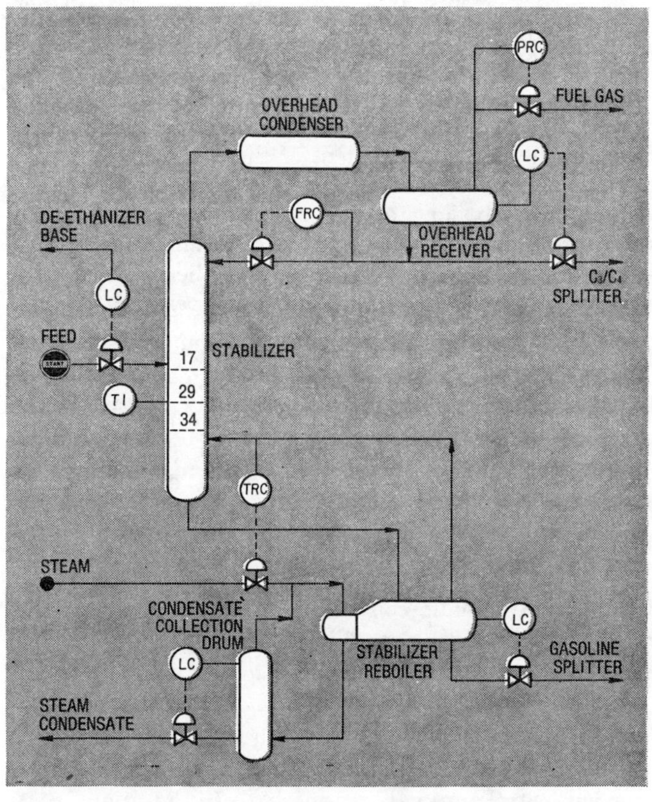

Fig. 1—Computer model studies stabilizer.

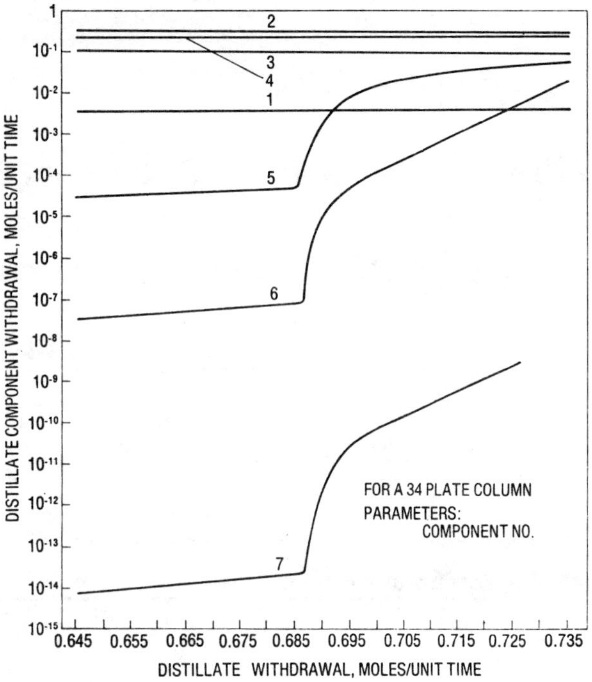

Fig. 2—Distillate composition as distillate rate increases.

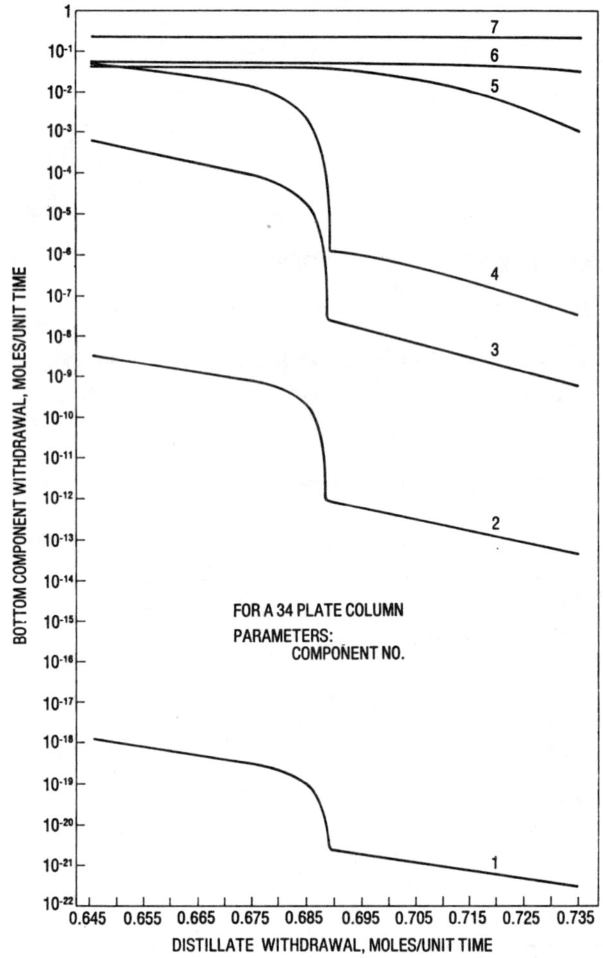

Fig. 3—Bottoms composition as distillate rate increases.

light and heavy key components of the feed. This would test the effect of comparatively small changes of the most sensitive components. It was also suggested that further practical conditions might be simulated by reducing the lighter components in the feed, such as ethane and propane.

When changes are made in the feed composition the temperature and vapor content of the feed will also vary. This is caused by flashing from the pressure of the pre-column reboiler to the pressure of the gasoline stabilizer. From this variation, a different temperature profile for the tower will be found and hence some control can be obtained by noting this temperature change at some point in the column.

The control system is based on controlling the flow of reflux, controlling the top product receiver and reboiler levels by regulating the top and bottom product flows. Indirect control of the bottom product composition is achieved by controlling the temperature of a plate in the stripping section. From observing the temperature indicator on plate 29 the steam flow to the reboiler can be regulated. Automatic control of the steam flow to the reboiler is given by controlling the temperature of the returning vapor from the reboiler to the column.

Economics. As mentioned previously, the overheads from the tower pass on to a propane/butane splitter for liquefied petroleum gas production. In addition, part or the whole of this stream can pass to refinery fuel directly from the tower or at other downstream points. The gasoline recovered from the base of the tower has a relatively low octane number and is reprocessed for quality improvement.

If fractionation is poor in the tower it will possibly result in an increase of pentane in the top product and also possibly an increase of the propane or butane in the bottom. In order to give the top product some specification it was decided that to meet the liquefied petroleum gas specification, the pentane content of the pentane plus butane moles passing overhead should not exceed two percent.

For economic considerations only 70 percent of both the propane and butane in the feed to the tower are required for light petroleum gas. The remaining top product from the tower is used for refinery fuel.

The bottom poduct is to be treated here as a product gasoline blending component, which has a true vapor pressure specification of 16 psia maximum.

For any economic evaluation the following generalized costs and values based on the North European prices will be used:

- Ethane, propane—
  £10.8/metric ton, or £5.4/metric ton as refinery fuel
- Isobutane, *n*-butane, isopentane—
  £10.8/metric ton, or £5.3/metric ton as refinery fuel

• Bottom product—
£8.5/ metric ton as gasoline blending component.

Numerical experimentation. Investigations have only been carried out on a conventional column, in an ideal steady state. As the main tool, a computer program was written and published[1] based on the theory of Holland's "Theta Method".[2] The computer used was the University's I.C.L. 1905E with the language being Fortran IV.

The computational basis is the Thiele and Geddes method for finding a solution to a steady-state problem. From this basis a theta multiplier was incorporated for the correction of the calculated top and bottom component molar withdrawal rates until convergence is attained. At this point the corrected and calculated component molar withdrawal rates being equal gives theta a value of unity.

The tower being considered has 34 plates, condenser, reboiler and the feed entering at plate 17. A pressure for the column is given so that expressions of the equilibrium constants can be found for the varying temperature of each of the seven components in the feed. This being partially vaporized at approximately 63 percent liquid and at an approximate temperature of 180° F. The distillate withdrawal rate is 69 percent of the total feed in moles with a reflux ratio of 1.2.

The multicomponent distillation column program was used with this basic data while the total top product withdrawal rate was varied. Starting with 64.5% in moles of the feed being taken off at the top, the value was increased by 0.5% until 73.5% was reached. These figures were chosen to give a representative spread of results either side of the suggested value by Continental Oil of 69%.

The plot of the amount of each component in moles in the top product against the varying total top product in moles, taking the feed as unity is given in Fig. 2. Similarly, in Fig. 3 is given the plot of each component in moles in the bottom product against the same varying total top product in moles, again assuming a feed of unity.

It will be noticed that for components 5 through 7 in Fig. 2 and for components 1 through 4 in Fig. 3 there are apparent discontinuities or jumps in the curves. To explain this jump consider the components of the feed are in the following mole fractions

	Component	Mole fractions
1	Ethane	0.003436
2	Propane	0.316844
3	Isobutane	0.1172765
4	*n*-Butane	0.2502435
5	Isopentane	0.037712
6	*n*-Pentane	0.043284
7	Hexane	0.231204

then the possible total of the first three light components is 0.4375565. But, if these are all taken off at the top of the column when the top product withdrawal rate is 0.688 moles (the feed being assumed as one mole), then the amount of the other components in this stream must be 0.2504435 moles. This should be compared with a value of 0.2499435 moles with a top product withdrawal rate of 0.6875 moles. Since the maximum amount of component 4 is 0.2502435, a certain jump must occur in the amount of the heavy components taken off at the top. The possible maximum of the first four components is 0.6878. So, if the amount of top product exceeds this value, the heavier components 5, 6 and 7 will increase considerably in the top.

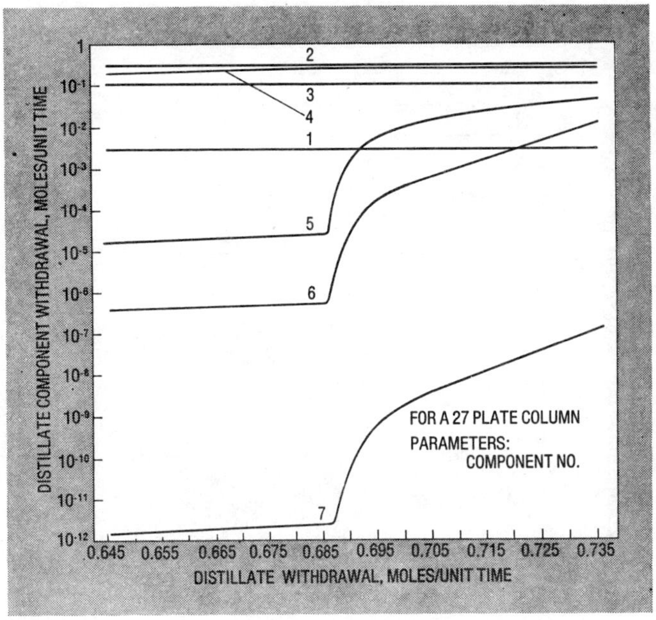

Fig. 4—Distillate composition for 27-plate column.

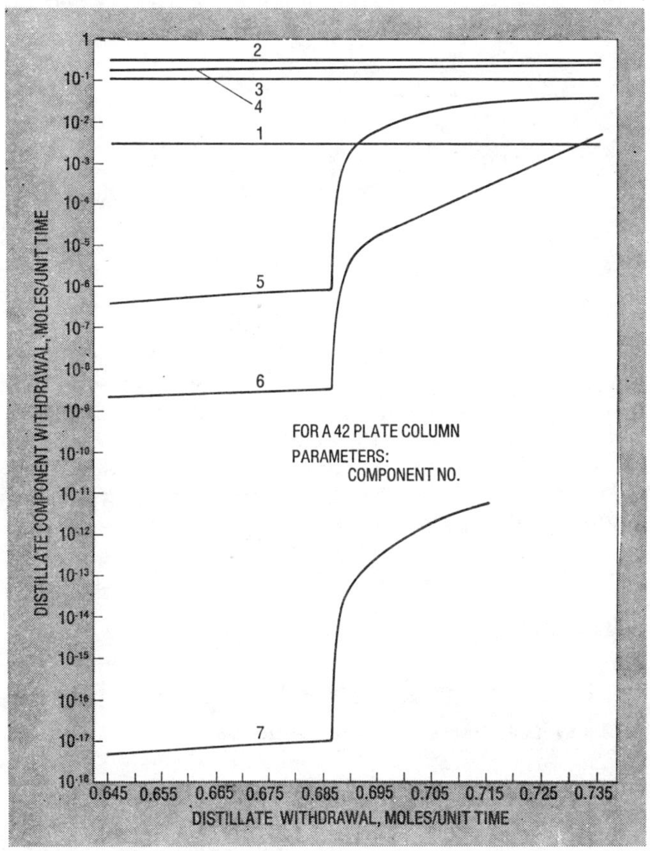

Fig. 5—Distillate composition for 42-plate column.

Plate efficiency. For comparison, plots of the amount

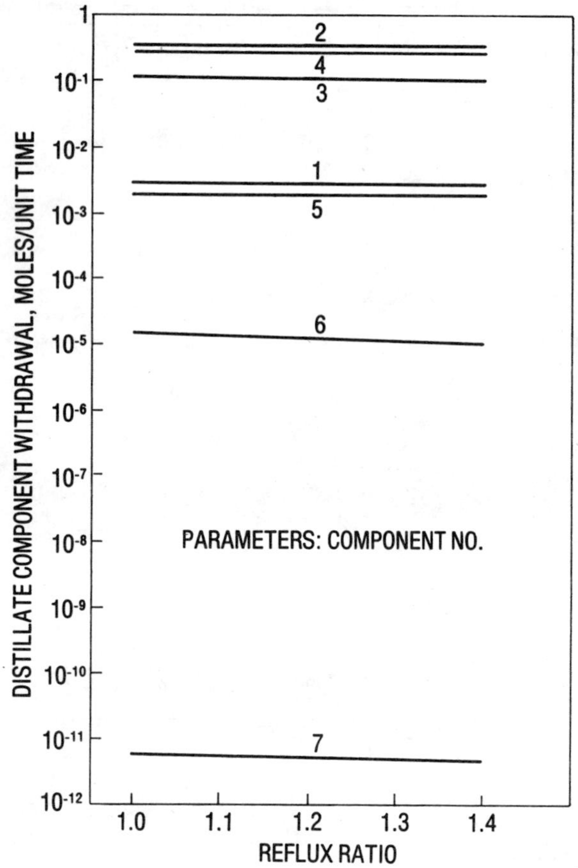

Fig. 6—Reflux ratio changes have slight influence.

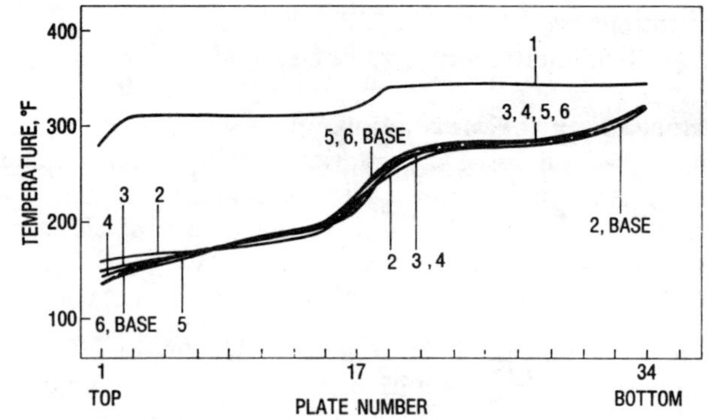

Fig. 7—Case 1 gave off-specification distillate.

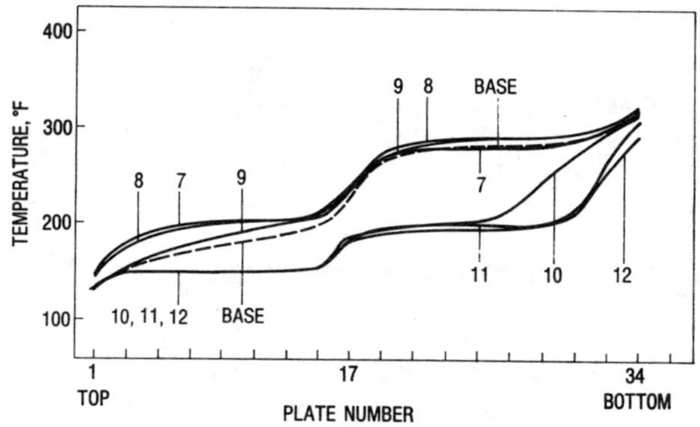

Fig. 8—Cases 7 and 8 also gave off-specifications.

of each component in moles in the top product are given in Figs. 4 and 5 for columns with 27 plates and 42 plates, respectively. They are of a similar form to the previous 34 plate model, although with the increasing of the number of plates a finer or better separation of the lights and heavies will occur. These two extra models were considered in an attempt to simulate plate efficiencies of approximately 80 percent and 125 percent, respectively.

TABLE 3—Variations in feed composition

Case	Concentration of given component*				Temp. °F	Liq. frac.
	1 Ethane	2 Propane	3 Isobutane	7 Hexane		
1	0	0		Inc.	250	1.0000
2	0	0	Inc.		203	0.7530
3	0	−50%	Inc.		191	0.6795
4		−50%	Inc.		191	0.6695
5	0		Inc.		181	0.6124
6	0	Inc.			180	0.6187
Base	0.003436	0.316844	0.1172765	0.231204	180	0.6150
	4 *n*-Butane	5 Isopentane				
7	Dec.	+100%			182	0.6303
8	−10%	Inc.			183	0.6240
9	Dec.	+10%			180	0.6184
10	Inc.	−10%			181	0.6166
11	Inc.	−50%			180	0.6123
12	+10%	Dec.			179	0.6013
Base	0.2502435	0.037712			180	0.6150

* Relative to base case concentrations as shown. Increase (Inc.) and decrease (Dec.) in other component made to compensate.

By dealing with these two, the various extremes in the literature are covered for the Continental Oil distillation column.

From the AIChE Manual[3], a publication by Hydronyl, and our own method, the higher value is attained; while Continental Oil and O'Connell[4] would point towards the lower value.

Reflux ratio. The multicomponent distillation column program was re-run once again keeping the basic data specified by Continental Oil but this time varying the reflux ratio. From a reflux ratio of 1:4, reduction was made in steps of 0.05 until 1 was reached. This gave coverage about the suggested value of 1.2. The plots of varying reflux ratio against the amount of component top product take off in lb. moles are given in Fig. 6.

Since there is little variation in the amount of each component taken off as distillate, it is not surprising that the top and bottom products stay within their specifications.

Feed composition. With the basic data, different feed compositions were used in the model. The variations followed two main ideas. Firstly, that the actual feed would vary about the key components, *n*-butane and isopentane. Secondly, that the lighter components, ethane and propane, would be reduced. An infinite number of changes could be made to the feed but 12 were chosen as shown in Table 3. [Editor's note: The case numbers in Table 3 and the following text were rearranged and renumbered for better comparison of results rather than in the order of the author's experimentation.] The base case feed composition is the same as for the example given earlier.

Temperature variations on all the 34 plates of the column are shown in Figs. 7 and 8. The temperature and fractions in the liquid state of the feed for each of these feeds are given in Table 3. Any variation seems to be shown in the temperature profile at the top of the column more than in the bottom for the cases chosen. In Case 1, the original feed of 31% heavies and 69% lights is changed to 63% heavies and 37% lights, a more drastic temperature change occurs than for any of the others. Although in feeds 10, 11 and 12 where the amount of lights is increased, there also is a change in the whole of the temperature profile.

The specification for the vapor pressure of the bottom product and the composition of the liquefied petroleum gas in the top product were included in the multicomponent theta program, and then rerun once again with the varying feed compositions stated previously. This time the temperature of the feed and its fraction liquid as stated were included for each variation.

In the variations, the specification for the bottom product seemed very insensitive to change. The upper bound for the specification being unity, the values obtained throughout the variations were all between 0.3 and 0.67 of this amount.

The top product specification went greater than its upper limit of unity on three of the 12 variations that were considered. This happened when the heavy components were considerably increased. The feed changes which were sufficient to send the top product outside its specifications were: Cases 1, 7 and 8.

These are all an increase of at least 8 percent in the total amount of components 5, 6 and 7 in the feed and so a decrease in the light components. This results in less lights being able to be taken off as top product and so a reduction in the total amount of distillate required would bring the liquefied petroleum gas back to within the desired specifications. The actual amount of distillate to be taken off would vary with each case.

Distillate rate. If the distillate withdrawal rate, D, is more than the sum of the light or first four components, then caution is needed to assure liquefied petroleum gas specification for the top product would be satisfied.

For example, define the liquefied petroleum gas specification in terms of the amount of individual components d as follows:

$$(d_5 + d_6) / (d_3 + d_4 + d_5 + d_6) \leqslant 2/100 \qquad (1)$$

or

$$d_5 + d_6 \leqslant (1/49)\ (d_3 + d_4) \qquad (2)$$

But in the converged state,

$$D = d_1 + d_2 + d_3 + d_4 + d_5 + d_6 + d_7. \qquad (3)$$

Substitution of Equation 3 into Equation 2 gives

$$D \leqslant d_1 + d_2 + (50/49)\ (d_3 + d_4) + d_7.$$

Since the sum of the light components in the feed can be represented by

$$F\ (X_1 + X_2 + X_3 + X_4)$$

then if

$$D > F\ (X_1 + X_2 + X_3 + X_4)$$

it can be assumed that most of FX_1, FX_2, FX_3 and FX_4 are contained in the top product. Then

$$d_1 \cong FX_1,\ d_2 \cong FX_2,\ d_3 \cong FX_3,\ d_4 \cong FX_4.$$

In addition, it is hoped that only a negligible amount of the heaviest component would be in the top product, so assume $d_7 \cong$ O. This then gives

$$D \leqslant \mathrm{F}\ [X_1 + X_2 + (50/49)\ (X_3 + X_4)] \qquad (5)$$

if the liquid petroleum gas specification is to be satisfied.

Bottoms rate. There is analogous case of the bottom product withdrawal rate, B, being greater than the sum of the heavy components:

This case was considered with an increase in the first two light components, although this does not simulate the expected changes in the feed components. The feed considered is:

	Component	Mole fraction
1	Ethane	0.006872
2	Propane	0.475266
3	Isobutane	0.1172765
4	*n*-Butane	0.2502435
5	Isopentane	0.034276
6	*n*-Pentane	0.043284
7	Hexane	0.072782

The temperature and fraction liquid of the feed were found to be 144° F and 0.4491.

With this as the feed composition the bottom product went outside the vapor pressure specification. To bring it back to within the stated specification it was necessary to reduce the bottom product withdrawal rate from 0.31 lb. moles per unit time to between 0.16 and 0.17 lb. moles per unit time. If this variation in the feed composition was likely to occur, an expression similar to Equation 5 could be formed for the value of B.

LITERATURE CITED

[1] Clark, C. R., *Brit. Chem. Eng. and Proc. Tech.*, Vol. 17, No. 1, 1972, pp. 44-46 and 48.
[2] Holland, C. D., *Multicomponent Distillation*, Prentice-Hall, Inc. 1963.
[3] AIChE, *Bubble Tray Design Manual.*
[4] O'Connell, H. E., *AIChE Trans.*, *Vol.* 42, 1946, p. 741.

Part II.

Microprocessors and Programmable Controllers

Microcomputers . . . Don't Get Left Behind

Process control instrumentation . . . its past, its future and a basic discussion of microcomputers

F. J. Berto, Standard Oil Co. of California,
San Francisco, Calif.

HPI CONTROL INSTRUMENTATION is changing at an accelerating rate. In three decades we have gone from local field control to central control houses to direct digital control with cathode ray tube operator consoles. The cost of keeping current becomes higher each year, both in money and in demand for technical manpower: engineers, programmers, analyzer mechanics, etc. If we don't keep current we risk becoming non-competitive with the industry leaders in this environment of change.

Industry is also faced with similar accelerating change; pollution regulations, a 10-fold increase in energy costs, new products with tighter specifications and increased federal regulation. Long-range planning has become very uncertain.

Modern instrumentation can be an effective solution to many of these changes.

Changes in control. Instrument engineers have often been the engines that accelerate this change. Let's review the major changes in refinery control (Table 1). From the turn of the century to the 1930s, refineries were run by stillmen, firemen and levelmen in the field reading indicators and recorders. Precise control was "half a spoke" of a globe valve.

TABLE 1

Changes in HPI Control		
Manual control to automatic control	1900 - 1940	40 years
Local control to central control	1940 - 1960	20 years
Analog control to computer control	1962 - 1972	10 years
Changes in Electronic Components		
Vacuum tubes to transistors	1930 - 1955	25 years
Transistors to integrated circuits	1955 - 1970	15 years
Integrated circuits to microprocessors	1970 - 1978	8 years

Direct-connected, pneumatic, automatic controllers and spring-diaphragm control valves came into general use in the middle '30s. The first major change from manual control to automatic control took about 40 years.

Primarily to save manpower, but also to better operate the processes, we began to consolidate automatic controllers into local control houses. The development of pneumatic transmitters, particularly the dP cell for flow, allowed the local control houses to be consolidated into central control houses for an entire plant, and then for groups of plants.

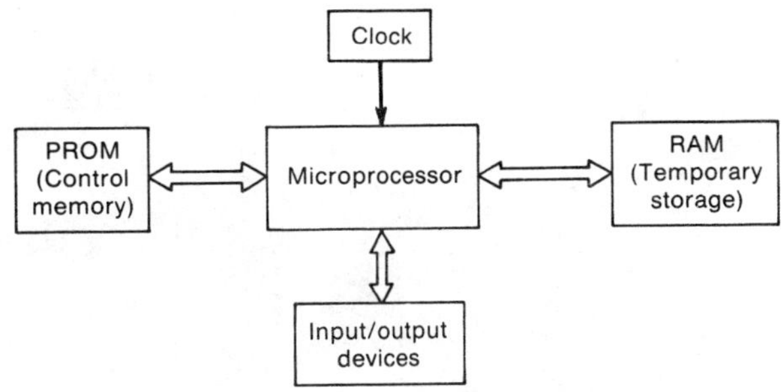

Fig. 1—Microcomputer block diagram.

With more and more controllers in one location, the control panels grew so long the operators needed roller skates. In the early 1950s, we went from large-case to small-case instruments. To help the operators comprehend the volume of information, we also developed full-graphic panels. This proved unnecessary and costly and was replaced by the three-high or four-high semi-graphic panels. Multipoint thermocouple instruments and annunciators kept pace with the central control concept. The second major evolution, from local automatic control to central automatic control, took 20 years.

Electronic controllers appeared in the 1950s and used vacuum tubes. They increased transmission distance but just about everything else was wrong. By 1960, we could buy reliable solid-state electronic controllers from several vendors and electronic semi-graphic panels became common.

Computer control of processes was the next major change, starting in about 1960. Control came in two methods: Digital Set Point Control (DSPC) and Direct Digital Control (DDC). In DSPC, the computer adjusted the set point of an analog controller which made the process comparison and sent the signal to the valve. In DDC the computer compares the set point and the process variable and generates a control valve signal.

Computers weren't very reliable, so there were always backup electronic controllers. Computers were expensive and sensitive to their surroundings so they were installed in air-conditioned central control rooms. This accelerated the trend to electronic instruments and to central control houses for entire refineries.

Field multiplexers (mostly for thermocouples) were used to reduce the high cost of miles of wires from plants in the field to the central control house.

By the early 1970s, entire refineries, on-plots and off-plots, were being run by very large DDC computers with

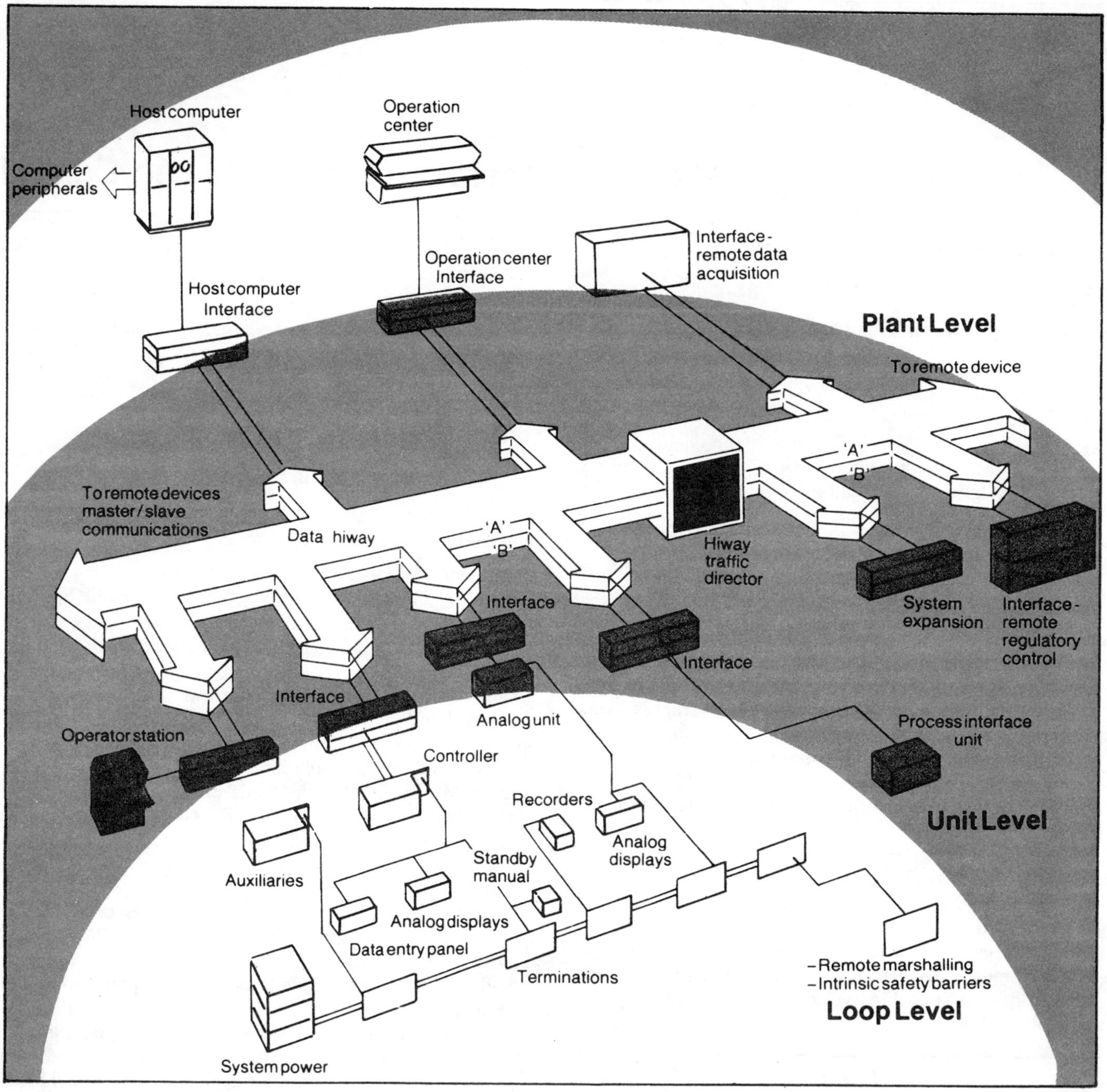

Fig. 2—Data highway concept.

three or four console operators reviewing and modifying the computer's actions.

Computer control has been a more traumatic change to the operator than the original switch to automatic control. He is another step removed from seeing the direct result of his actions. Now he can tell the computer to change the 90 percent point of a side cut and as many as 10 interacting variables will be manipulated.

When something goes wrong, the operator has to determine if he has a process problem (change of feed) or an instrument problem (transmitter drift) or a computer hardware problem (loss of parity check) or a software misfit.

New varieties of black box wizards, systems applications engineers, digital process engineers and process programmers have been added to the instrument engineers and meter men.

The change from analog control to direct digital control took about 10 years and sets the stage for microcomputer-based instruments.

It is obvious that the pace of change has been accelerating, but there's another conclusion that is less obvious. The first changes were gradual enough that they became almost universal in new construction, and older plants have been largely upgraded.

Computer control has been too fast and too controversial. Even now we see non-computer controlled plants or whole refineries being built and the 1970-style computer control is obsolete. Its replacement is on stage.

Computer salesmen, both those working for vendors

and forward thinkers in oil companies, have not been able to convince management to make changes fast enough.

Changes in electronic components. Let's begin our electronic component history (Table 1). In 1957, the hottest name in computers was the IBM-650. It used vacuum tube registers. "On" was a 1 and "off" was a zero. With eight tubes, measuring 10 inches by 8 inches, it could count to 2^8 or 256. My introduction to computers was designing the air-conditioning for an IBM-650. It took three 200-hp freon compressors to keep it cool.

Commercial application of the transistor was the breakthrough that made big computers and reliable electronic controllers possible. Discrete transistors, diodes, resistors and capacitors were soldered into printed circuit boards. Most companies bought their first electronic instruments in the middle 1960s. They were more expensive but installation costs were lower, maybe, and longer transmission distances were possible. We talked about future computer control.

Integrated Circuits (ICs) came into use in the late 1960s. Using photo-reduction and masking techniques, numbers of diodes and transistors were deposited and connected on silicon wafers. The number of components per circuit has been doubling every years. It went from tens to thousands and integrated circuits became Large Scale Integrated Circuits (LSIs). Something in the same fashion that supertankers became Very Large Crude Carriers (VLCCs).

About 1968, you began to hear the word minicomputer. This wasn't rigidly defined, but to most control people it meant a small control computer that had 16 bit or smaller words, sold for less than $25,000, and fitted into a relay rack. As LSI moved into minicomputers, the price came down and the performance went up.

Today you can buy a $50,000 minicomputer that will outperform an IBM-1800 that cost $500,000 in 1968.

LSI didn't have that much impact on industrial instruments because of its peculiar economics. A custom LSI chip might cost $100,000 for the first one. After that, they cost, in volume, say, $20 each. Instruments didn't have the kind of volume to justify custom LSI chips.

Microcomputer description. This set the stage for the microcomputer. In 1971, Intel introduced the first microprocessor, the 4004. The breakthrough idea was to put a computer's "brain," the Central Processing Unit (CPU), on a single chip. By adding programmable memory and input-output chips to the basic microprocessor, it could become a traffic light or elevator controller or a fuel gas Btu calculator by changes in programming.

Intel broke the first cost barrier to custom-designed large scale integrated circuits. By using Read Only Memories (ROMs), these microprocessors could be programmed to substitute for unique LSI or for a large number of individual logic chips. Reducing the number of components reduces cost but it also increases reliability by decreasing dramatically the number of connections.

The 4-bit word Intel 4004 was superseded by the 8-bit Intel 8080. There are probably 100 different microprocessors from 10 different vendors available today.

The Intel 8080 is 2 inches by 0.5 inch. It's that big mainly to take care of the 40 pins that connect the chip to the outside world. The chip itself is about ⅛ inch

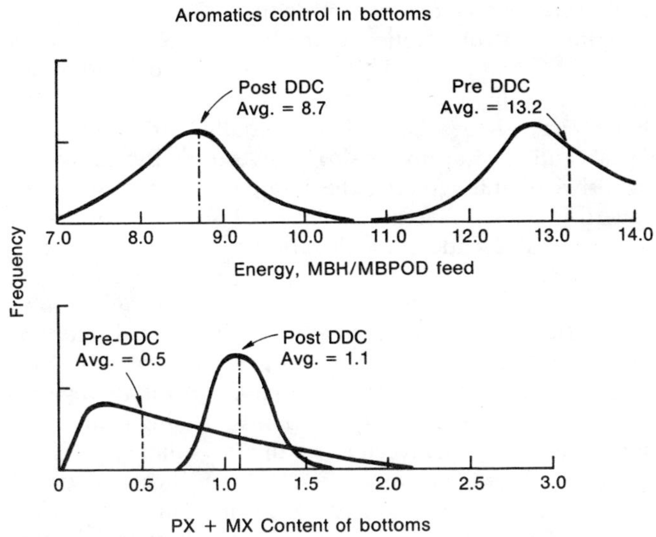

Fig. 3—Savings from advanced control.

square. It contains 10,000 discrete components.

Fig. 1 is a microcomputer. You convert a microprocessor into a microcomputer by adding Programmable Read Only Memory (PROM), Random Access Memory (RAM), Input Output devices and a Clock. This board cost $300 in 1976. Programming the two 1,000-word PROMs determines what kind of task this microcomputer will do. The PROM is programmed by first erasing memory (shining ultraviolet light through its window) and then electronically inputting the new program. Each one of these PROMs contains the same counting ability as 1,000 of the 8-tube registers. They sell for about $5 each.

The 8080 is several years old. It in turn has been superseded by the concept of an entire microcomputer on a chip. Intel calls this the 8098 and there's lots of competition.

Again note the accelerating pace of change and that the technology is changing much faster than the user can adapt to it.

Features of modern instrumentation (Table 2). Let's look at how the process control computer has changed refinery instrumentation in the past decade and how the microcomputer is going to accelerate these changes.

TABLE 2

Features of Modern Digital Instruments
Process operator consoles
Shared digital control
Data highways
Field multiplexing
Hierarchical control
Features That May Not Survive
Central control houses
Electronic analog backup controllers
DDC minicomputers controlling plants and driving operator consoles

The first and most visible item is the **Process Operator's Console.** Until about two years ago, operator consoles were always backed up with analog recorders and controllers, often very densely packaged. This backup instrumentation was used less and less as computers became more and more reliable.

The second feature is **Shared Digital Control.** Shared control means using one microprocessor controller to handle more than one control loop.

Digital control implies that the control equation is solved by comparing digital numbers, zeros and ones, rather than by comparing analog voltages. Computer control has always been done digitally with analog to digital and digital to analog converters. By using inexpensive digital circuits, the latest microprocessor controllers include as standard features such sophisticated functions as cascade control, ratio control, feed-forward, error^2, etc.

The **Data Highway** is the third feature of modern control systems. All of the field information to and from the transmitters, thermocouples, control valves and controllers and all of the set points, tuning constants, alarms, etc., are converted to digital numbers. These digital numbers, and an address or identification, travel down the data highway in serial fashion, one item after another. The data highway connects to all of the devices that generate or require data: operator consoles, digital controllers, field multiplexers, control computers, etc. There are large savings in wiring costs. Fig. 2 shows Honeywell's TDC-2000 data highway. It uses a coaxial cable with a capacity of 150,000 bits of information per second.

Field Multiplexing is the fourth feature. The field multiplexer works with the data highway to convert data to digital form near the transmitters, control valves, etc. The major saving is wiring cost. Properly done, the data are less subject to noise and interference.

The fifth major feature of today's control computers is the concept of **Hierarchical Control.** In trivial form, this could be analog cascade control, level resetting flow. The words have come to mean big management information computers, communicating with medium-sized optimizing computers which in turn communicate with lower level direct digital control computers. Now we are seeing microprocessor-based instrument systems replacing the lowest level DDC computer and analog backup instruments.

There are some other features of most of today's digital control systems that may not be around five years hence. (Table 2).

We may evolve back to smaller local control houses containing shared digital controllers and simple operator displays. This field hardware would communicate over the data highway to higher level computers in a much smaller master control room.

Small microcomputer applications. Many vendors are using microcomputers to upgrade their older instruments or to provide powerful and cost-effective new kinds of instruments.

Here's a quick review of just five of the many microcomputer-based instruments introduced in the last few years.

A chromatograph controller and peak integrator that can control up to four chromatographs. I know of four vendors providing similar packages.

A multiloop process controller which can control up to 8 control loops providing many of the sophisticated modes of control that used to require a DDC process computer. I know of seven similar packages from vendors.

A gas or liquid flow computer which can handle up to six orifice or turbine flowmeters along with temperatures and pressures. It reads out in volume, mass or energy content. I know of three similar devices.

A tank gage readout package that has tank strapping tables in its memory so that it can read out and alarm in feet and inches or in barrels.

A blender controller which can handle up to 24 components.

All of these microprocessor based instruments are lower in cost and higher in performance than the devices they replaced. Installation costs can be markedly lower and the first feedback indicates that maintenance is indeed better.

The proliferation of new instruments from new companies has made the instrument engineer's job much harder. Picking the right instrument for each application is not easy and today's selection will probably not be valid a year from now.

Pressures on profitability. Industry profitability has changed dramatically in the post-OPEC era, as I'm sure you all know. (Table 3).

TABLE 3

Pressures on Profitability
Nearly fixed feed and product prices
Escalating energy costs
Government regulations on product quality
Environmental restraints
Difficulty in building new plants

Advanced control is an inexpensive technique with the ability to decrease energy use, hold tight product specifications, minimize pollution and maximize volumes in an existing refinery.

Advanced control is the effective combination of modern instrumentation, analyzers, computers and the necessary trained technical support people.

Fig. 3 is a real example from a recent DDC/chromatograph installation of an aromatics distillation column. The payout was less than one year after taxes. It is an example of the kind of results that are being realized with today's advanced control.

The microprocessor is an important device to help new instruments meet the needs for increased efficiency.

I expect those companies that can keep up with the accelerating pace of change to be the most profitable.

ACKNOWLEDGMENT

Based on a paper presented at the API 43rd Midyear Refining Meeting, May 9, 1978, Toronto, Canada.

Microprocessor Maximizes Coker Thruput

Economically it is more attractive to charge cokers with as much excess resid as possible. Presented is the development and application of a microcomputer designed to achieve this goal

L. H. McGrievy, Marathon Oil Co., Robinson, Ill.

Marathon's Robinson refinery has two delayed coking units. The larger unit, called "Regular Coker," is charged with vacuum residuum from the crude units to produce a regular grade coke. The smaller unit, called the "Special Coker," is charged with a mixture of vacuum residuum and thermal residuum to produce a premium grade coke. It is economically attractive to maximize the amount of residuum oils charged to the cokers. This article describes development and application of a microprocessor specifically designed to aid the operating people in accomplishing this objective.

Delayed coker capacity limitations. In the operation of a delayed coker, when striving for maximum thruput, it is desirable to have the flexibility to adjust the charge rate so the final coke drum level will be as close as possible to the target drum outage* (17 ft.) at the end of each charging cycle. The coke drum charging cycle is usually 22 or 24 hours long, but for simplicity, it will be assumed to be 24 hours long in this article. The complete coking cycle, including drum charging, cooling, drilling and drum warmup, would be 48 hours long.

The maximum daily average charge rate for any given feed is dependent upon:

- Capacity of the charge heater
- Capacity (volume) of the coke drums
- Capacity of the fractionation and gas recovery sections to handle the liquid as gas products.

* The "outage" is the average gaged distance from the top face of the coke drum head flange to the coke level in the drum. The smaller the outage, the fuller the drum.

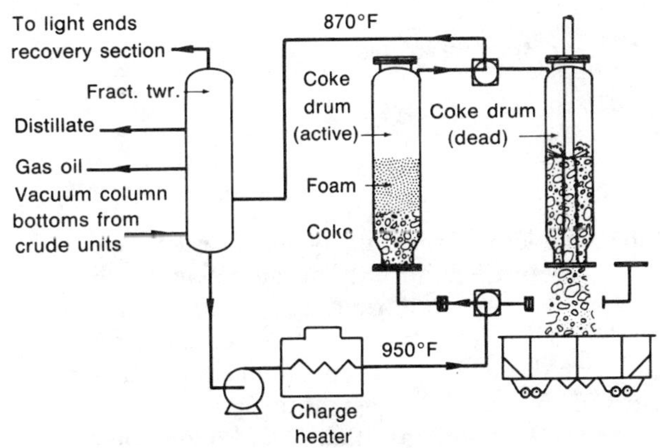

Fig. 1—Simplified flow diagram of Regular Coker.

The Special Coker is limited by the heater and the fractionation and gas recovery sections. Since the drums are not a limiting factor, the unit's capacity can be fully utilized by holding the charge rate high enough to keep the heater and the fractionation and gas recovery sections loaded.

On the other hand, the Regular Coker (Fig. 1) has adequate heater and fractionation and gas recovery sections to permit filling the drums on almost any given charge. Therefore, the limiting factor, as far as total thruput capabilities are concerned, is the volume of the coke drums. Since it is almost impossible to determine where the coke level is in the drum at any given time during the drum charging phase of the coking cycle, it is very difficult to determine a steady charge rate which will cause the drum to be filled to the desired outage level (17 ft.) at precisely drum switch time (4 a.m.). This fact has made the task of keeping the Regular Coker fully loaded quite troublesome.

The problem. The coke-producing characteristics of coker charge stocks vary considerably depending on a

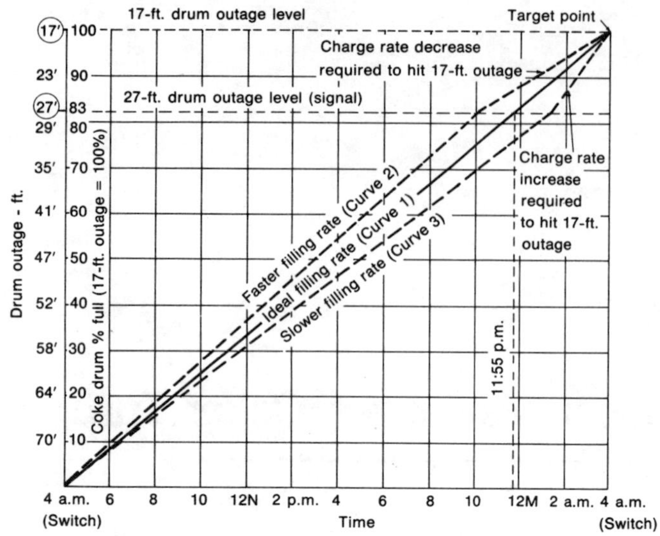

Fig. 2—Coke drum filling rate (Regular Coker).

number of factors including types of crudes from which they are derived, uniformity of operating conditions on the crude unit and uniformity of operating conditions on the coker, i.e., recycle rate, heater outlet temperatures, etc. Because of the wide variation in coke-producing characteristics of charge stocks, there is no easy way of accurately determining a charge rate that would allow the drums to be filled precisely to the desired drum outage during the 24-hour drum charging period. When the drums are overfilled, due to too high a charge rate, there is danger of coke foam going overhead to the fractionator and necessitating a shutdown. On the other hand, when the drums are underfilled, there is an economic loss due to reduced thruput, during periods of long residuum oil supply.

During periods when there is excess coker capacity for the charge available, charge rate can be reduced to a comfortable rate and there need not be any concern about maintaining low outages. Outages of 25 to 30 feet are quite common during these lax periods. However, this article will be oriented toward the problems involved in keeping the unit operating at its maximum capacity during periods of long vacuum residuum supply, which now constitute more than 90 percent of the time.

In the past, efforts have been made to determine the required charge rate to hit the 17 ft. target outage level by 4 a.m. drum switch time, based on the softening point and even the Conradson carbon content of the charge. However, the coker charge stock coke-producing characteristics were so variable due to changing crude types, variations in crude unit operating conditions, etc., that these indicators were of little value. As time went on and as the pressure to run more charge increased, it became more and more frustrating for the operators to try to minimize drum outages with little to go by as a guide. On several occasions the drums were accidentally overfilled to well above the 17 ft. outage level and coke foam was carried over through the vapor line into the fractionator. This necessitated unscheduled shutdowns for equipment cleanout, and resulted in costly production losses. Quite understandably, it took only two or three such incidents to make the operators more conservative. As a result, they developed a tendency to underfill the drums, sometimes by substantial amounts, to be on the safe side. This, in turn, resulted in an economic loss by forcing more residuum to be disposed of by burning as plant fuel oil and No. 6 fuel oil sales.

Method to determine ideal charge rate. A study of the problem was begun with the objective of developing a method which would not have to rely on laboratory-generated data and would be dependent only on currently available unit operating data.

This work resulted in development of a straight forward calculation method to determine the coke production of the charge by integrating the charge rate between switch time (4 a.m.) and the time of day that the coke level passes the 27 ft. outage level detector. In using the method, one would calculate the amount of coke that a given charge stock produced in $ft.^3$ of coke/bbl. of coker charge, knowing drum volume and the number of barrels of vacuum bottoms had been charged to the unit up to the precise moment the coke level passed the 27 ft. outage level detector. Then, using this $ft.^3$ of coke/bbl. of charge figure, he could determine what the charge rate should be to hit the 17 ft. target outage level at precisely 4 a.m. drum switch time.

If the coke-producing capabilities of a given charge stock were known in advance, and if a constant composition could be maintained, the precise charge rate, which would allow the drum to be filled at exactly switch time (4 a.m.) every day, could be calculated. This calculated rate could be called the "ideal charge rate." Curve 1 in Fig. 2 shows that if the active drum were filled at the "ideal charge rate," the coke level would pass the 27 ft. outage level detector at precisely 11:55 p.m. every day and would hit the 17 ft. outage level at precisely 4 a.m. switch time. But since the coke-making characteristics of the charge stock are so variable and almost never precisely known in advance, drum filling is rarely at the ideal rate. In actual practice, the charge rate is either too high or too low and rate adjustments must be made to hit the 17 ft. outage level precisely at the 4 a.m. switch time.

If the drum is filled faster than the ideal rate for a given charge stock (Curve 2, Fig. 2), the 27 ft. outage level is reached before the 11:55 p.m. time. In this instance, the charge rate must be reduced if the target 17 ft. outage level is not to be exceeded before switch time. On the other hand, if the charge rate is lower than the "ideal rate" for a given charge stock (Curve 3, Fig. 2), the 27 ft. level will not be reached until after 11:55 p.m. In this event, the charge rate would have to be increased to hit the 17 ft. outage level by switch time. Such charge rate adjustments must be made as soon as possible after the coke level passes the 27 ft. level detector, which records in the control room.

Microprocessor development. It was decided to develop a microprocessor which would perform all the required monitoring and calculations, assuming the necessary inputs could be provided.

Instrumentation specialists at the Denver Research Center were consulted and requested to design, construct and install a microprocessor which would accomplish this to aid the operators in filling the coke drums to the 17 ft.

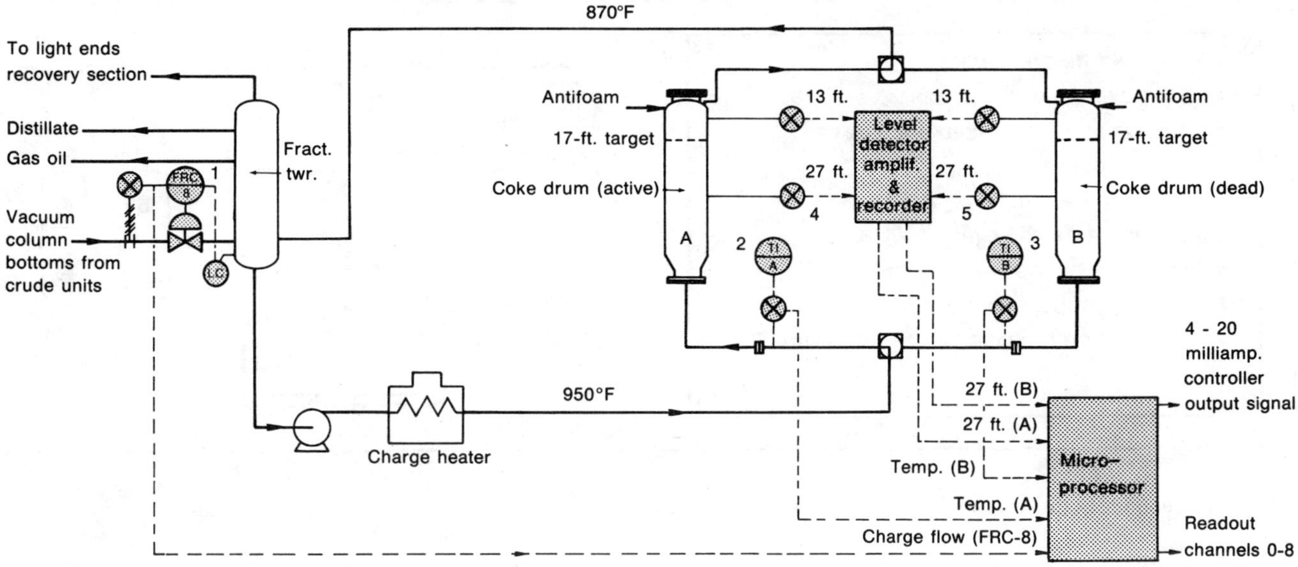

Fig. 3—Regular Coker microprocessor installation.

outage level every coking cycle. The basic calculations, previously described, were programed into the microprocessor.

It should be noted that this instrument, even though referred to as a "microprocessor," is technically a microcomputer since it has input and output capabilities, D/A and A/D converters, memories and a central processing unit. However, in this article it will be referred to simply as a "microprocessor."

Although the possibility of closed loop control of the coker charge rate was considered even before the microprocessor was developed, it was decided to design it initially for open loop operation to be used only as an operating aid. At the time, it was felt that closed loop control of the unit charge could be reconsidered after operator confidence and acceptance had been gained.

The microprocessor can be programed for any coking cycle length; however, only 22 and 24-hour cycles have been used. The microprocessor scans all the inputs on frequent predetermined time cycles and provides much more information than just the charge rate required to hit the 17 ft. outage level at switch time.

► Inputs to microprocessor

Fig. 3 shows graphically how the microprocessor fits into the coker process scheme. There are only five process inputs to the microprocessor. These are:

(1) **Coker charge rate (FRC-8).**

(2 & 3) **Drum inlet temperature to each drum (TI-A and TI-B).** At drum switch time, as soon as the charge is switched into the empty drum, the drum inlet temperature rapidly increases from about 650° to over 900° F. When the thermocouple temperature on the inlet to the active drum reaches a predetermined point between these two temperatures, a signal is sent to the microprocessor indicating that adrum switch has been made and a new operating cycle has begun.

(4 & 5) **The 27 ft. outage radiation level detector outputs from each drum (No. 4 and No. 5 points).** These signals indicate to the microprocessor whether or not the coke level is at the 27 ft. outage level. When the coke level passes the 27 ft. outage level, the computer recalculates the charge rate required to fill the active drum the remaining 10 ft. to the 17 ft. outage level based on the coke accumulation rate between switch time and the time coke level passed the 27 ft. level detector on the active drum.

► Microprocessor digital readouts (channel selector)

The channel selector panel (Fig. 4) is located on the left front of the microprocessor and consists of nine red push buttons with the title of each alongside. When one of these buttons (numbered 0 through 8) is depressed, the channel number and its current numerical value appear in the digital display window located on the right side of the microprocessor front panel.

The following is a list of each of the nine channel readouts on the panel selector, and a brief explanation of how the numerical value of each is determined:

(1) **Channel 0—present time**—The microprocessor has a highly accurate quartz clock incorporated in its circuitry because all of the calculations are based on time. When button 0 is depressed, the current time appears on the digital readout. The time, after being manually set, is displayed as military time, i.e., 01:00 to 24:00 hours.

(2) **Channel 1—last switch time**—When button is depressed, the last drum switch time is displayed in the readout window.

(3) **Channel 2—time 27 ft. last passed**—When button 2 is depressed, the time of day the coke level last passed the 27 ft. outage level appears in the readout window. During the period of time when the lower part of the active drum is being filled, this number will refer to the time of day that the coke level passed the 27 ft. level detector in the previous coke drum

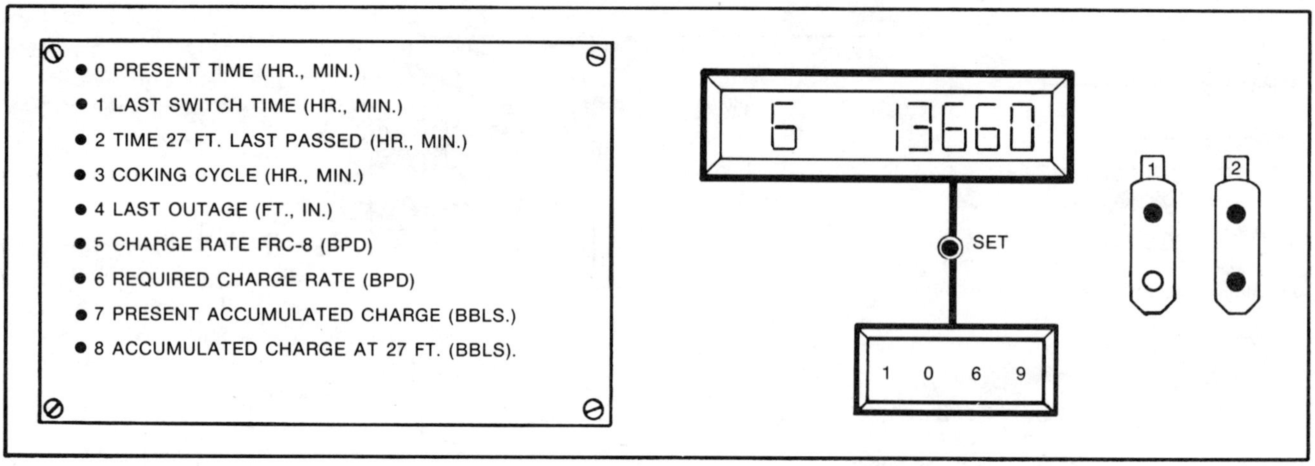

Fig. 4—Front panel of Regular Coker microcomputer.

(the "dead" drum). The red lights in the top and bottom half of each miniature coke drum shown on the panel of the microprocessor indicate whether the coke level is above or below the 27 ft. outage level (Fig. 4). The instant the coke level passes the 27 ft. level detector in the active drum and the signal is received, the Channel 2 readout will change to indicate the new time of passage. At the same time, the red light will go out on the lower part of the miniature coke drum and the red light in the upper part will light up.

(4) **Channel 3—coking cycle**—When button 3 is depressed, the cycle time length (22 hours, 24 hours, etc.) is displayed in the readout window. This figure is fed manually into the microprocessor when it is first put in service. Cycle length significantly affects charge rate corrections calculated by the computer.

(5) **Channel 4—last outage**—When button 4 is depressed, the last calculated drum outage, in feet and inches, appears in the display window. This number refers to the outage of the dead drum until coke level passes the 27 ft. level detector in the active drum.

This figure is determined by calculating the coke production in $ft.^3$ of coke/bbl. of feed between opening switch time (4 a.m.) and the time that the coke level passes the 27 ft. outage level detector approximately 20 hours later. Then, by multiplying this $ft.^3$ of coke/bbl. of feed by the total barrels of feed (by integrating the FRC-8 unit charge meter signal) that was charged to the unit between the time the coke level passed the 27 ft. outage level detector (around midnight ± 1 hour) and the present time, the current coke level is determined. This figure is continually updated every 60 seconds until closing switch time at 4 a.m. when a final outage figure is calculated. This final outage figure then remains in the channel 4 memory until the coke level passes the 27 ft. outage level on the following drum cycle.

It should be pointed out that in calculating the drum level during the final 10 ft. of drum filling (27 ft. outage to 17 ft. outage) it is necessary to extrapolate only for the final 4 hours (± 1 hour) using the $ft.^3$ of coke/bbl. of charge figure obtained during the first 20 hours of operation.

(6) **Channel 5—charge rate (FRC-8)**—When button 5 is depressed, the present charge rate in bpd will be displayed in the readout window. This figure is derived from the signal received from the unit charge meter (FRC-8).

(7) **Channel 6—required charge rate**—When button 6 is depressed, the charge rate required to make the coke level hit the 17 ft. target outage at the scheduled closing drum switch time (4 a.m.) is displayed in the readout window. This figure is computed by using the values stored in:

- Channel 0—(Present time)
- Channel 1—(Last switch time)
- Channel 3—(Coking cycle)
- Channel 7—(Present accumulated charge)
- Channel 8—(Accumulated charge at 27 ft.)

From opening switch time until the coke level passes the 27 ft. level detector, this value depends on the data in channel 8 memory (accumulated charge at 27 ft.) from the previous (dead) drum. However, the instant that the 27 ft. level is passed in the active drum, the channel 6 value is re-computed based on the newly updated values in channels 2 and 8.

If channel 6 is being displayed in the readout window when the 27 ft. level is passed, one would see a significant change in this value if the coke make/bbl. of charge in the present (active) drum is very different from that of the previous drum. The channel 6 figure for the final 10 ft. of drum volume (from 27 ft. to 17 ft.) is based on the channel 2 and channel

8 values of the active drum. Since this final 10 ft. represents only about one sixth of the usable drum volume, even a major charge compostion change during the remaining time would not seriously affect the final calculated outage figure.

Toward the end of a drum-filling cycle (the last hour or so), the channel 6 display number may increase or decrease at an increasing rate if the charge rate has not been changed as recommended by the computer. For example, if the computer had called for a 13,500 bpd charge rate (channel 6) after the coke level passed the 27 ft. level detector and the charge rate FRC-8 (channel 5) was 12,000 bpd but was not changed, the channel 6 figure would increase at an increasing rate until closing switch time. This figure may get ridiculously high, far exceeding the capacity of the unit. Numbers as high as 99,999 bpd might be seen just before drum switch time. This is because the computer does not know that the coker cannot process enough charge to hit the 17 ft. outage level in the remaining time until drum switch. Once this value goes above the heater charge capacity (about 14,500 bpd) it is, theoretically, no longer possible to reach the 17 ft. outage level on the active drum in the time remaining until 4 a.m. switch time.

(8) **Channel 7—present accumulated charge**—When the channel 7 button is depressed, a figure appears in the display window that represents the total number of barrels of vacuum resid charged to the coker since opening switch time. This number is calculated by integrating the FRC-8 signal from opening switch time to the present time.

The display for this channel will read 0 for the first minute or so after opening switch time and then increase at a rate of about 8-10 bpm when the charge rate is in the 12-14,000 bpd range. When the coke level passes the 27 ft. outage level detector, the channel 7 value is registered into the channel 8 memory to replace the value stored from the previous (dead) drum. However, this value also remains in channel 7 and integration continues until closing switch time.

(9) **Channel 8—accumulated charge at 27 ft.**—When the channel 8 button is depressed, the accumulated charge (in barrels), charged to the coker between opening switch time and the time the coke level passes the 27 ft. level detector, is displayed in the readout window. When the charge is switched into the new drum, this value will pertain to the previous (dead) drum until the 27 ft. level is passed in the new or active drum. At that time, a new figure will be transferred in from channel 7, as mentioned previously. This figure will pertain to the active drum being filled; and a new figure for channel 6 (required charge rate) will be computed for use as a guide in filling the remaining 10 ft. (27 ft. to 17 ft.) on the active drum. This same figure will be used in calculating the required charge rate (channel 6) for the next drum until the 27 ft. level is again passed by the coke level.

Problems encountered. When the microprocessor was first installed on the regular coker, problems which seriously hampered its operation were encountered. The most serious problems were caused by the 27 ft. outage level radiation detector transmitter on each coke drum. The transmitted signals were very weak and "noisy." This periodically caused premature triggering of the 27 ft. level signal which would throw the microprocessor off for an entire cycle. In addition, any outside interference, such as stray electromagnetic fields from nearby welding machines, etc., would aggravate the signal noise problem even more. The noise problem was handled on a temporary basis by inserting a software filter in the program.

These and other less serious problems did not do much to develop operator confidence in the microprocessor, but the problems were eventually corrected by installing better radiation level detectors. Present plans call for further improvement of the coke level detecting system with the installation of improved detectors on each drum at the 27 ft. outage level.

Another problem has been the lack of properly trained refinery maintenance people who can troubleshoot and make repairs on the microprocessor. Whenever anything major went wrong, the equipment had to be returned to our Denver Research Laboratory for troubleshooting and repairs. However, this problem is being remedied by "hands on" training of the instrument maintenance personnel.

Nevertheless, the microprocessor has functioned very well for long periods of time and has given very good service in predicting drum outages and in determining charge rates required to hit the target 17 ft. outage level.

Future possibilities. It was earlier mentioned that closed loop control of the coker charge rate was considered prior to development of the microprocessor. To provide for this future possibility, the microprocessor was provided with an analog output for controlling the charge flow control valve (FRC-8). However, for closed loop control, the various fractionator pumparound streams would also have to be controlled to maintain the necessary tray temperatures. The microprocessor has adequate capacity to do all this, and only minor programing changes would be required.

ACKNOWLEDGMENT

Based on material originally presented at the NPRA Computer Conference, Anaheim, Calif., Oct. 29-Nov. 1, 1978.

Microprocessors Control Chemical Addition and Coking Unit Cooling

Here are some guidelines for evaluating microprocessor systems for control applications, and examples of microprocessors applied to add chemicals to cooling water and to automate the cool down cycle of a coking unit

J. W. Mitchell, Amoco Texas Refining Co., Texas City, Texas

AMOCO'S TEXAS CITY REFINERY frequently has needs for small, special purpose controllers. During the past 2-3 years, microprocessor systems have competed with conventional systems for these applications. These systems are justified and two applications are presented here.

When evaluating a microprocessor control system for an application, the following five questions are asked:

► What exactly is the application, and what is the function of the control system? Some applications are naturals for microprocessors where others are more difficult to handle. It is vital that the process personnel describe exactly what they require before proper evaluation can be made.

► Is there a suitable controller available on the market? If none is available, this may have to be an in-house development.

► How much will the controller cost, including programing and development?

► Who will maintain the controller and how much training will be required?

► When is the system needed? Although each step in creating the controller may be short, considerable lead time will be needed for project completion, depending on complexity. Lead times of three months to a year are typical for in-house developments.

After the microprocessor approach has been selected, four basic problems must be encountered in firming up system design. Hardware and software are considered integral and not separate parts of system design.

• Inputs generally consist of measurements called analog inputs and contact closures called discrete inputs. Computer outputs may be variable voltage called analog or D/A outputs, and individual contact closures called discrete outputs. The inputs and outputs are the controller's ears, eyes and fingers. Insufficient sensors partially blind or deafen the microprocessor from the process which it is controlling. Selection of I/O is vital to good control. Put yourself in the role of the controller and visualize which inputs and outputs are necessary to adequately do the job. Be careful only to include those points which are absolutely essential.

• How often have we been asked, "What happens if?" on a new piece of equipment, and not had a satisfactory answer? During system design we investigate each type of failure which might be encountered, and add hardware or software which will reduce or eliminate the adverse consequences which might result. This usually adds considerable inputs and outputs to the controller.

• Although hardware may be developed to run by itself, it is run and monitored by operators. The operator may do no more than flip a switch or monitor a light, but he is still operating the equipment. Good operator interface is essential for a successful installation. The designer must include the necessary Input/Output for operator communications.

• Microprocessor systems can be developed from several different starting points. One can begin with the microprocessor chip, a microprocessor card, or even a microprocessor based computer. The designer should remember that one of the main reasons for choosing a microprocessor is low initial cost. He should take care he does not offset those savings with unnecessary development cost. We have developed systems beginning from each of these levels, and tend to favor the microprocessor card for small applications and the microprocessor based computer for larger applications.

COOLING TOWER CHEMICAL ADDITION

Amoco's Texas City refinery is currently using microprocessors for both monitoring and control. One application is the control of chemical additions to cooling tower basins.

Application. Make-up water is added to a cooling tower basin by local level control. The water either evaporates from the tower or is blown down to the sewer. As water

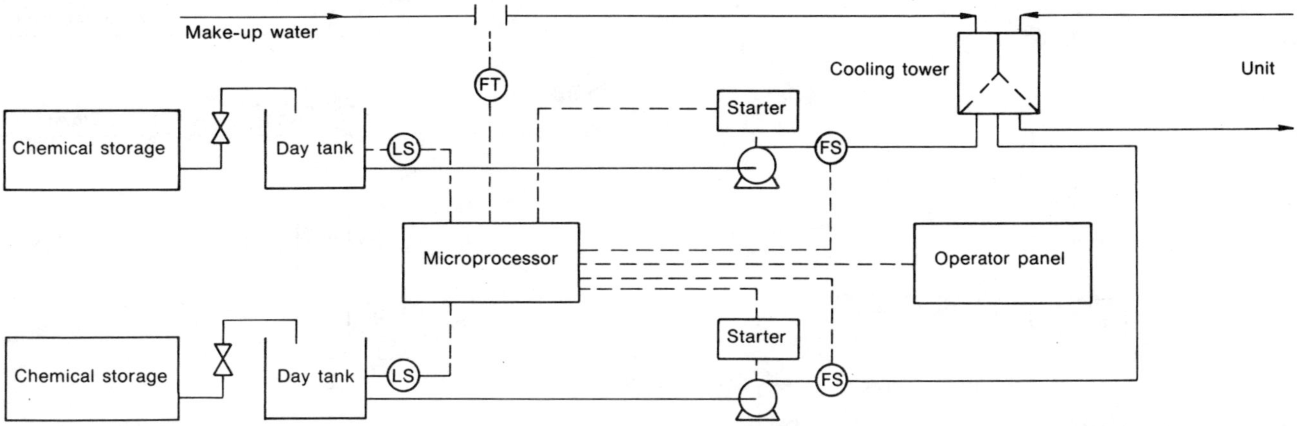

Fig. 1—Cooling water treatment total inputs and outputs.

is blown down, chemicals are lost. Two of the chemicals are generally added by the operator with small pumps. The purpose of this project was to have the microprocessor monitor make-up water rate and add the chemicals proportional to this rate. The chemicals are added intermittently with positive displacement pumps from small day tanks. The quantity of chemicals pumped would be determined by the pumping interval.

Availability. There was not a suitable control system on the market, so in-house development was warranted.

Cost. Total cost of the control system, including development, was estimated between $1,500 and $2,000.

Maintenance. There was sufficient in-house expertise to maintain the equipment.

Timing. The system would be required approximately six months after the project was begun. Three months lead time would be required to develop, construct, and program the control system, which was well within the time frame available.

System design. One analog/digital input and two discrete outputs are the only inputs and outputs necessary for the application. The A/D input is the make-up water flow transmitter; the discrete outputs are the contact closures which would start the chemical pumps.

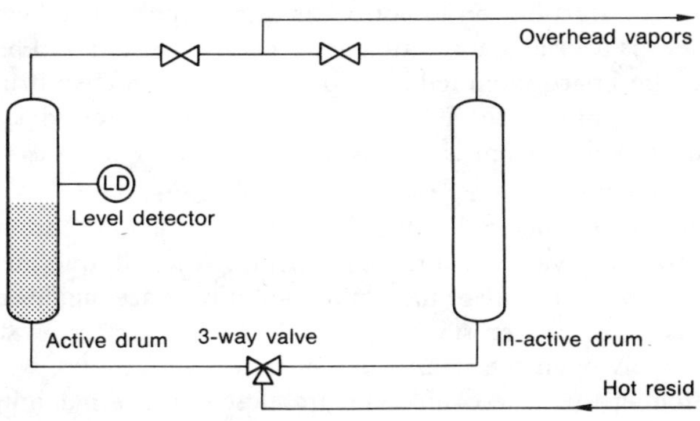

Fig. 2—Delayed coking unit.

Input/Output to prevent failures. Although the basic I/O as described is all that is necessary to run the system, additional I/O is needed to correct for failures of the entire system (Fig. 1). The following system failures were investigated.

- What would happen if the pumps failed to start, or started and failed to pump? Software and hardware support was added to correct this problem. This consisted of a pump run switch or a chemical flow switch plus the necessary alarm output to inform the operator of the malfunction.

- What would happen if the chemical day tanks were empty when pumping was required? Low level switches were added with the appropriate software to shut down the pump on low level and alert the operator.

- What would happen if the microprocessor failed, especially with the pump running? This could empty the day tank and possibly even damage the pumps. We developed a small computer fail module which would alert the operator and disable the pump outputs on computer failure.

- What would happen on power failure? The program is stored in programable read only memory (PROM or EPROM) and is not affected by power failure. The software and hardware were developed to bring the system back up following power failure.

- What would happen if the program added too much or too little chemical? That is, what if the actual chemical required differs from that in the program? In order to correct this problem, a thumbwheel switch was included for each pump. These switches provide the operator with a means of inserting a chemical rate setpoint externally. The predicted chemical rate is the mid-range setting on the thumbwheel switch. Increasing the setting would increase the chemical pump time, and conversely, decreasing the setting would decrease the pump time.

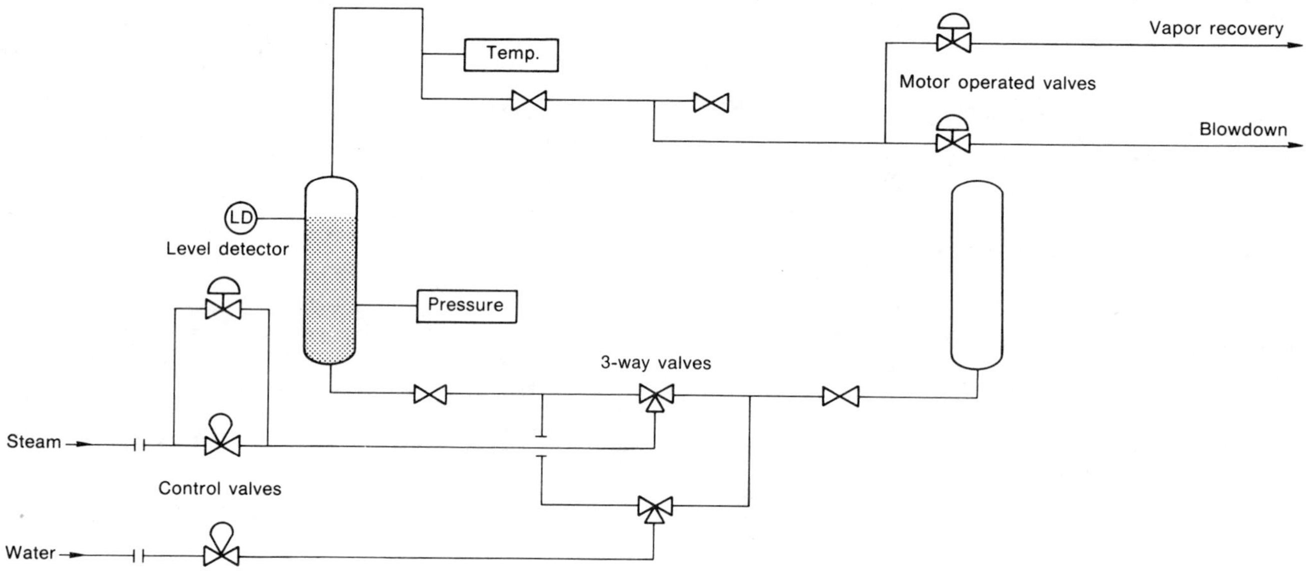

Fig. 3—Detail of each drum.

Operator I/O. The next step was to design the operator interface. Basically we wanted to provide the operator with pump status lights, low level lights, start and stop switches, thumbwheel switches and a general purpose alarm.

Three of these systems have been built and placed in service. The first was built from the chip level using the Motorola 6800 chip. The next two were built from the board level using the Texas Instruments TM990/100m microprocessor card.

AUTOMATIC COOL-DOWN CYCLE—COKING UNIT

There are three coking units at Amoco's Texas City refinery. Each coking unit has two coke drums (Fig. 2). One drum is filled with hot resid which forms coke and lighter hydrocarbons. At the same time the coke is removed from the other (inactive) drum. De-coking of the inactive drum consists of steaming—to remove the hydrocarbons; filling with water—to cool the coke; removing the coke, and finally heating the drum for service. The steam and water additions compose the cool-down cycle. Traditionally these have been done manually by opening a number of block valves. Considerable time was lost during the cool-down step. This is due primarily to the extra long steaming and water times which result from manual time-keeping methods. Thus, a project was initiated to automatically control the cool-down cycles of the three coking units.

Availability. Although many microprocessor systems on the market could be programed to do the job, we did not know of any packaged system designed for this particular application. Thus, we decided to utilize a microprocessor for which we had a development system. This was the Texas Instruments 990/4 microcomputer.

Cost. The application included numerous control valves, solenoid valves, transmitters and major piping changes. It also included a 3-foot control panel, a scanivalve and a microprocessor system with floppy disk. The cost of the control panel with the associated microcomputer equipment was about $60,000.

Maintenance. Because we chose a microcomputer for which we had prior maintenance experience, maintenance should not constitute a problem.

Timing. The timing of this project, about six months, was not any problem. However, as it turned out, the controller was not required for approximately a year.

System design. Before system design could be started, a good understanding of the process was essential (Fig. 3). I will briefly step through the process, visualize the basic I/O required, the preventive I/O required, and the operator I/O required.

Hot resid is pumped into one of a pair of coke drums on a coking unit. The resid cracks, forming hydrocarbons—which pass overhead, and coke—which remains in the drum. Radiation level detectors are used to determine the level of coke in the coke drum. When the drum is full of coke, the operator switches to the second drum, and the cool-down cycle begins.

The operator prepares the drum for steaming, and presses a start button to notify the microprocessor to begin. There is a local start button near each coke drum. There are also board-mounted start buttons for each coke drum. The microprocessor will detect from the depressed start button which drum is inactive.

Immediately upon starting, the computer will open a small steam solenoid valve. The steam will pass through a three-way valve into the coke drum. A small amount of steam is maintained until the operator once again depresses the proper start button. Then the processor will gradually open the steam controller in five minutes, while watching the coke drum pressure closely. Once maximum steaming is achieved, a timer is started. When the timer has expired, the steam controller is nearly closed, and water addition is begun through the three-way valve. As soon as a positive water flow is determined, the steam is cut completely out. Once again, the processor watches the drum pressure closely as it ramps the water control valve open. Overhead vapors from the coke drum continue to pass to the vapor recovery section until the water level reaches the upper level detector. Then the operator must switch two motor-operated valves so that the overhead vapors are routed to the blowdown drum. Failure to do this would cause water to be sent to the vapor recovery section after the coke drum completely fills with water.

Water flow to the blowdown drum is continued until the coke drum overhead temperature drops below 200° F. Then water flow is stopped and the microprocessor notifies the operator that cool-down is completed.

The necessary inputs and outputs per coker are as follows:

► Two D/A outputs for controller setpoints.

► Eight A/D inputs for measurements.

► Two discrete inputs for start buttons.

► Two discrete outputs for the steam solenoid and the termination alarm.

These are the only inputs and outputs necessary to control the coker cool-down process. However, additional inputs and outputs are needed to prevent adverse consequences if something goes wrong.

Input/output to prevent the unexpected.

• The small steam solenoid can be operated manually or with the microprocessor. Digital control of this solenoid is essential for automatic cool-down control to either begin or proceed. Thus, if the steam solenoid is on manual, the microprocessor will sound a panel alarm and will refuse to start a cool-down cycle. This requires one discrete input and one discrete output.

• The microprocessor outputs setpoints to the steam and water analog controllers. Thus, if one of the controllers required for computer control is on manual, the microprocessor must recognize it and take appropriate action. This action involves manual setpoint tracking, as well as sounding a panel alarm. Two more discrete inputs and two discrete outputs are required for this operation.

• The steam and water pass through three-way motor operated valves which divert these flows to the inactive coke drum. These three-way valves can either be operated manually or with the microprocessor. The computer is not concerned if it has control of these valves but rather in their position. Limit switches were installed to monitor the valve position. The microprocessor checks the limit switches every second and will alarm if the valves are in the wrong position in manual, or if the valves cannot be moved into the right position while in automatic. This requires six discrete inputs and four discrete outputs.

• Throughout the steam cycle and during most of the water cycle, coke drum overhead vapors are diverted to a vapor recovery section to recover the hydrocarbons.

During the latter part of the water cycle, the coke drum overhead flow is routed to the blowdown drum. Routing of the overhead flow is accomplished with two motor-operated valves. Limit switches on these valves are sensed by the microprocessor. These limit switches are monitored every second. A panel alarm is sounded if the positions are incorrect. This requires two discrete inputs and one discrete output.

• Two system failures include microprocessor failure and power failure. The hardware and software were modified to minimize the effect of these failures. One additional discrete output was required.

Operator input/output. In addition to the I/O previously mentioned, Input and Output must also be added for operator interface. The following I/O were included for operator interface:

► Status lights were needed to indicate which drum was being cooled, and whether the steam or water cycle was in progress.

► A four-digit display was included to indicate how long the cool-down has been in progress.

► A four-digit display was included to show the temperature of the coke drum overhead in the cool-down cycle.

► Thumbwheel switches were included from which the operator can enter the time interval desired for the steam cycle.

Microprocessor starting point. Because of the complexity of the process, the microprocessor based computer route was chosen. The Texas Instruments 990/4 microcomputer was used. Approximately 5,000 words of 16-bit memory were required. The program is loaded at system startup with a floppy disk. The program was written to handle all three cokers simultaneously with each coker in any mode of cool-down.

Conclusions. The two microprocessor systems are among several monitoring and controlling systems which have been developed and installed at Amoco's Texas City refinery. Microprocessors will undoubtedly continue to be used on small and medium-sized applications.

ACKNOWLEDGMENT

This article is based on material originally presented at the NPRA Computer Conference, Anaheim, Calif. Oct. 29-Nov. 1, 1978.

Microcomputer Systems . . . Where Do You Begin Troubleshooting?

New microprocessor-based process control systems require a different approach to maintenance and trouble-shooting

Martin Zielinski, Honeywell Inc.,
Process Control Div., Fort Washington, Pa.

THE MICROPROCESSOR has brought about dramatic changes in process control instrumentation and control system strategy. These new systems require a different approach to maintenance and trouble-shooting. The conventional techniques used are not usually successful. A microprocessor based system must be trouble-shot based on its *architecture* (configuration) rather than its function. The required technique is compared with the conventional analog method in an example for the control system of a simple feed heater.

Trouble-shooting an analog control loop. Suppose the control loop in question is a feed heater or cracking furnace found in refineries and petrochemical plants. Fig. 1 shows this process. A liquid feed material such as crude oil or naphtha is to be preheated to some desired temperature before being fed to a later stage of the process, typically a distillation column. A transmitter monitoring the temperature at the feed heater outlet, provides the process variable input to a temperature controller. The output of this controller serves as the setpoint to a fuel flow loop. The flow controller maintains the flow required by the temperature controller, taking into account pressure variations in fuel supply pressure. Hence, this cascade control system tries to maintain a desired feed output temperature, minimizing the effect of fuel supply pressure fluctuations.

A technician applying conventional techniques in trouble-shooting a malfunction in this control scheme would probably make the first functional subdivision at the output of the temperature controller. If the output indicated that a larger fuel flow was required in response to a drop in outlet temperature and vice-versa, the repairman would conclude that the temperature controller and transmitter were functioning correctly. However, if the controller did not give the proper response, he would probably assume the flow loop was functioning properly and investigate the temperature controller.

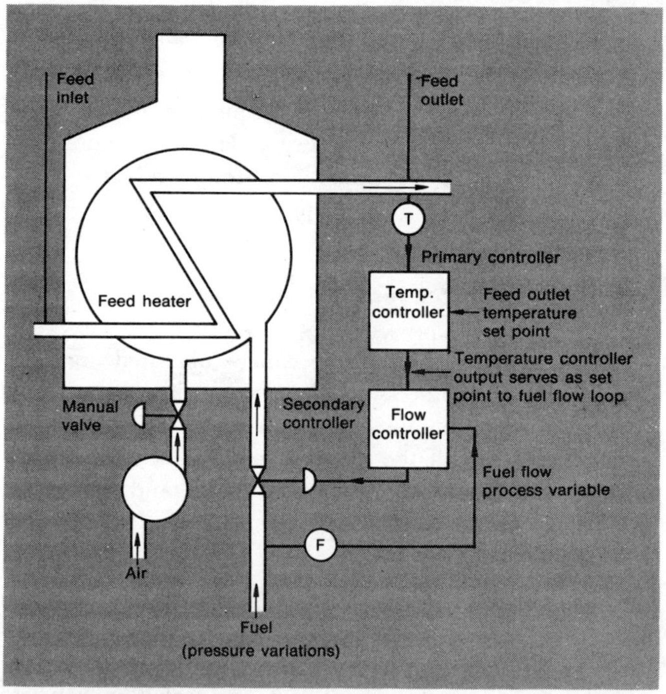

Fig. 1—Cascade temperature control system block diagram.

The next logical subdivision would then be at the output of the temperature transmitter. If its output is representative of the actual feed outlet temperature, the technician would then investigate the actual temperature controller and its related wiring. Eventually, through this process of subdivisions, the malfunction would be located and repaired. This process is successful in control systems consisting of interconnected, individual analog controllers; however, it would have limited success trouble-shooting a multi-loop microprocessor based controller. The assumption of one half of the system being good and the other defective, is not always valid.

In analog control loops each subdivision has its own independent components. In microprocessor based control loops, subdivisions may share common components (e.g. the microprocessor). In addition, since microprocessor based controllers typically control several loops, subdivisions of many process control loops may be common to each other. A technician employing the usual techniques of checking subdivisions might conclude that several subdivisions are malfunctioning instead of just the common component. A service technician must be trained in the architecture of the multi-loop controller

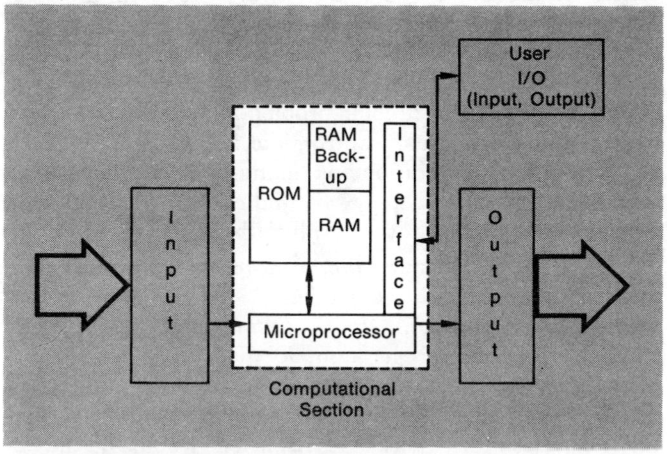

Fig. 2—Microprocessor based controller architecture.

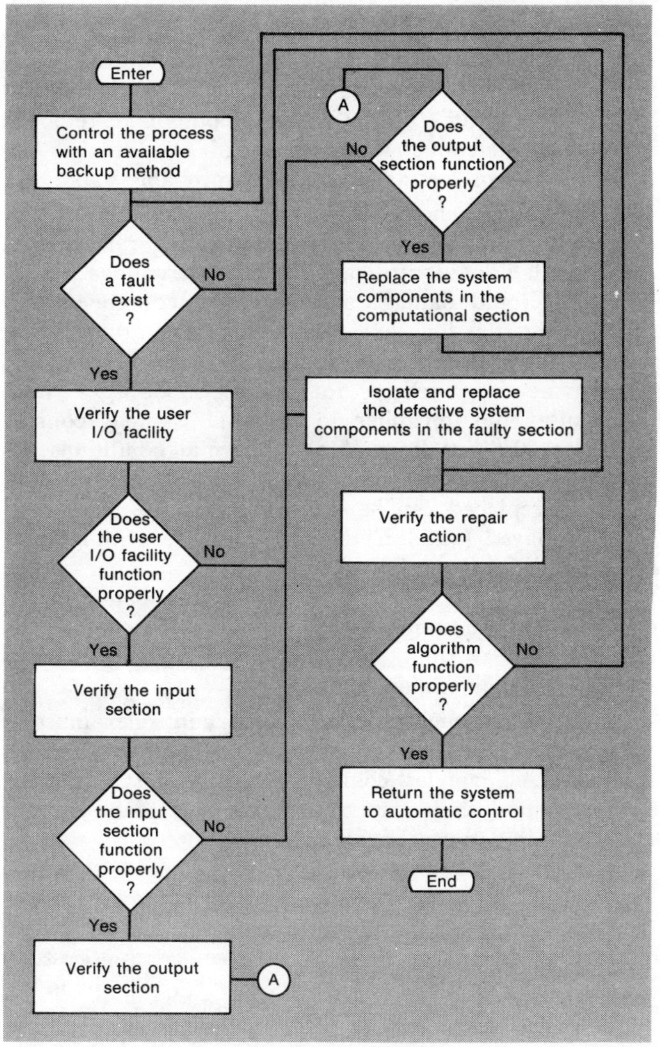

Fig. 3—Microprocessor based controller trouble-shooting flowchart.

to successfully restore the control system to a functional state.

Multi-loop microprocessor based controller architecture. Before discussing a different trouble-shooting technique, the architecture of a typical microprocessor based controller must be described. A majority of the analog hard-wired interconnections are replaced by internal "software" (i.e., programing) connections in a microprocessor.

Fig. 2 shows a functional block diagram. The controller is comprised of three basic parts: *the Input section, the Computational section* and *the Output section.*

The function of the Input section is to convert the analog signals received from the field transmitters to digital information that can be handled by the microprocessor. The Computational section receives these data as inputs to one of the control algorithms (i.e., proportional, integral, derivative) described in Read Only Memory (ROM) and calculates a new output value. The Output section converts this new digital value to an analog signal that can be used for control purposes in the field.

The Computational section must have convenient displays for process monitoring and operation as well as System Configuration (i.e., I/O device). The user must be able to select the control algorithm, enter tuning constants, read the process variables, determine the inputs to the selected control algorithm, etc.

Let us look at how the control scheme in Fig. 1 fits this architecture. For purposes of discussion, let the output of the flow transmitter be Input #1 and the output of the temperature transmitter Input #2. Suppose the computational section is designed such that it has eight separate computational slots. Each slot is capable of doing the calculations for one algorithm. The user selects Slot #2 as the temperature controller, picks the three mode controller algorithm, selects Input #2 as the process variable and determines the setpoint. The user then selects Slot #1 as the flow controller, picks the same algorithm, enters Input #1 as the process variable and selects the output of Slot #2 as the flow loop setpoint.

The selection of the output of the temperature controller as the setpoint for the flow loop, does not require a wired connection. Since all of this information is in the Random Access Memory (RAM), the user configures the microprocessor to read the setpoint for Slot #1 from the memory location where the output of Slot #2 is stored. Likewise, Input #1 could have been selected as the process variable for Slot #2. Its a matter of properly programing the microprocessor.

Trouble-shooting microprocessor multi-loops. Although the actual control scheme shown in Fig. 1 is accomplished, its physical representation is much different than a series of interconnected analog controllers. The trouble-shooting process of functional subdivisions still applies; however, the variations in architecture must be taken into account. In this control system the subdivisions are not the separate controllers, but rather the Input, Output and Computational sections. The user I/O section could also be considered a subdivision. It has no direct control function, but it provides the means for monitoring the process. The trouble-shooting technique is not new, but rather a different approach that accounts for the structure of the microprocessor based controller.

In trouble-shooting a control loop, the technician should first determine that the user I/O facilities are functioning properly. This is accomplished by a check-out procedure peculiar to the display device, but will direct the technician to check the display functions that are independent of the suspect control system. If the I/O device has the capability of being shared among several multi-loop controllers, it can be checked out with a known good controller.

Next, the Input section should be investigated. If all process variables displayed at the user I/O facilities are representative of the actual conditions being measured, this section is functioning properly.

Likewise, the Output section is functioning properly if it transmits the valve signal to the field calculated by the computational section. Notice that the computational section could have calculated the value incorrectly, but the Output section is doing the job it was designed to do.

Assuming the technician found no malfunctions in the user I/O, Input or Output sections, the only remaining area is the Computational section. As shown in Fig. 2, this section is usually made up of the microprocessor working in conjunction with some peripheral memory. The technician could begin to investigate this section; however, he would need detailed knowledge and information to isolate the fault to one of these parts. The recommended repair action is to replace one or both of these system components. The repair can be verified by checking for proper controller action in response to a change in process variable.

It is assumed that some sort of standby manual control is available for both the analog controller example as well as for the microprocessor controller. The process must be controlled while system components are replaced.

The recommended method for trouble-shooting a microprocessor based controller is summarized by the flow chart in Fig. 3. It is important to recognize that this flow chart is designed to isolate an equipment fault. If the process is out of control because the controller is improperly configured (for example, incorrect tuning constants), the technique will not directly help to solve the problem. Hence, the first step in the procedure is to determine if a problem exists. Once the problem is established, the remaining steps of the flow chart will isolate it to the defective section. Being general, this flow chart cannot isolate the fault to a subdivision of one of the sections described in Fig. 2. That requires knowledge of the specific microprocessor based controller under consideration. The knowledge can be provided in the form of detailed flow charts for each section and/or through the experience of a qualified service technician.

For the flow chart in Fig. 3 to be effective a significant portion of the microprocessor controller must be working. Usually, manufacturers of this type of controller design in checks on the instruction set of the microprocessor and its peripherals. If a failure of any of these checks occurs, the system aborts to manual control and displays a message.

Self-diagnostic failures. Since self-diagnostics of the controller are not a 100% test, two types of failures can occur. The first kind has already been discussed. The controller does not fail the internal diagnostics, but it cannot adequately control the process. The second type of failure poses a particular problem. Since the controller has aborted to manual control, the failure mode cannot be analyzed. The equipment manufacturer may design the diagnostics to display an indication of the failed system component, but not all failures.

The typical internal communications structure of a microprocessor based controller is a bus, made up of several lines, common to all system components. Any component could fail in a mode that would load down one or more of the bus lines, invalidating all communications between the microprocessor and its peripherals. Any error message displayed would be meaningless. Recourse for a technician in this situation is to arbitrarily replace the system components individually.

A better technique, would be to monitor the bus lines in a static condition. If an abnormal condition is detected, the technician could remove each component until the fault clears. The last system component removed is probably defective. If no abnormal condition is detected, the error message displayed is probably valid.

A secondary benefit of clearing the bus lines of any abnormal conditions, is that of protecting replacement components. The fault could be an overvoltage condition that would damage a replacement component. Insuring that the bus lines are within operating limits, reduces the possibility of damaging spares. Hence the recommended procedure for trouble-shooting a fault that causes the controller to abort to manual control, is to clear the bus lines of all abnormal conditions before replacing system components. The components that are to be replaced can be determined by the error messages displayed (valid if the bus lines are within operating conditions) or by determining which ones caused the abnormality on the bus lines. Bus lines can be more conveniently monitored by a special maintenance device rather than doing it by hand. Some manufacturers offer such a maintenance tool.

A multi-loop microprocessor based controller must be trouble-shot based on its architecture rather than its function. A manager contemplating the installation of such a system should recognize this and plan to train maintenance personnel accordingly.

ACKNOWLEDGMENT

Presented at the Instrument Society of America's 1977 Spring Industry Oriented Conference, Anaheim, Calif.

How to Apply Programmable Controllers

New programable controllers are designed to do more than just replace relays. Here's how to maximize their capabilities by proper process definition, hardware and software selection and safety and startup considerations

V. J. Maggioli, E. I. de Pont de Nemours & Company, Inc., Newark, Del.

CAPABILITY OF THE special-purpose computer called the programable controller has been increased to the point where it has to be considered a viable alternate when selecting a controller to perform operational logic functions in chemical control applications.

This article describes guidelines used in the application of programable controllers (P.C.) in process control.

Du Pont presently utilizes the P.C. in process applications for interlock, sequencer and computer interface functions. The ability and desire of the P.C. industry to upgrade their equipment to accommodate new technology such as microprocessors, fiber optics, digital data communications, and direct digital control systems, will determine the place the P.C. takes in future process applications.

It is interesting to note that while the P.C. industry basic marketing approach is to consider the P.C. as a relay replacement device, process applications highlight it for what it is, a special-purpose computer.

This article addresses today's fundamental design approach to P.C. process applications at Du Pont by the Engineering Department, Design Division, Electrical Control Group. This group has designed over 20 "stand-alone" process applications involving approximately 50 P.C.'s of approximately seven types and five different manufacturers as well as providing consulting services on additional process applications within Du Pont. The chemical processes were generally batch type (60 percent) with the remaining involving supervisory management, pneumatic conveying, automation enhancement and computer interface of a process system.

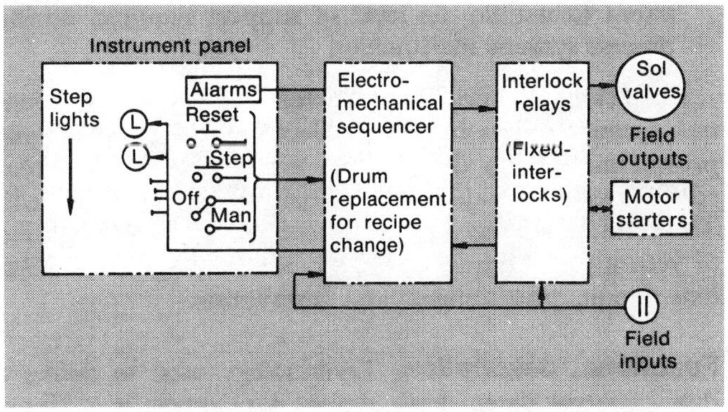

Fig. 1—Process control block diagram of existing system.

System basis. Basic data prepared by the Operating Department (end user), defines the desired performance capability of the control system. It is the communication and cost control "cornerstone" upon which a sound design approach is developed.

Computer technology with its capability to automate processes to a level of sophistication previously impractical, appears in many instances to be the panacea for all problems. Generalizations used to define control system requirements may have been acceptable prior to the arrival of computer-type equipment as a control tool; however, these generalizations today result in control system selection and application which may not address the true needs of the end-user. This potential is compounded by vague terminology (e.g., automatic, computer interface, reliable, easily maintained, self-diagnostic, etc.) in communications between the design engineer and end-user throughout the evaluation, design and installation phases. The result can be a high level of confidence by the design engineer that he is providing the proper system, while the end-user may also feel very comfortable his needs are being addressed.

However, startup and operation of the process may clearly indicate the real control system needs have been missed, such as:

- "Automatic" mode is inadequate without short-term assist modes (e.g., "manual", "maintenance", "by-pass").
- System availability is more important than equipment reliability.
- Process equipment diagnostics is a greater factor in reducing system shutdown than control equipment diagnostics.
- Plant technical and maintenance forces are not prepared to handle the level of support required during process systems malfunction.

Preparation of proper basic design data prior to control system evaluation and selection provide the only proven method to develop the economics assuring that control system capability will satisfy process system needs. The result is management can proceed with an evaluation of return-on-investment prior to authorizing control system design, procurement and installation.

Functional description. Terminology used to define a detail control system basic design data report is a "functional description."

The unique aspect of a functional description is that it reflects the needs of all Operating Department sections (including production, maintenance, and technical) and provides discrete definition of many terms through an individually developed glossary and dedicated mathematical formulas.

Before an existing system can be automated a detail understanding of the basic process operation has to be documented. Engineers are assigned to work in conjunction with the operator allowing documentation of correct operator action, reaction, and nuances employed to make product; no detail is too small for review when developing a functional description. If the process is new, pilot plant or R&D participation may be required to initiate development of the functional description.

The items defined in the development of a functional description varies for each process system and results in a list too expansive to include in this article. Instead an abbreviated list to illustrate the required depth of content is provided, as follows:

- Availability formula (with each term defined)
- Thruput formula (with each term defined)
- Operating procedure (existing and proposed)
- Modes of operation (include production, diagnostic, definitions)
- Inter-Mode transfer requirements (most systems do not allow indiscriminate transfer to a higher-level mode of operation—must be defined)
- Piping diagram
- Instrumentation diagram
- Equipment arrangement
- Operator Interface (local pushbutton, instrument panel, keyboard, C.R.T.)
- Plot plan defining locations of process, central control room, tank farm, services, electrical control room
- Safety considerations (personnel, ecology, equipment)
- Interlock requirements (logic diagrams, description of operations, Boolean equations, etc.)
- Analog system requirements
- Process data collection system (computer)
- Peripheral equipment needs
- Communication-Highway needs for the above
- Planned stage of turnover.

The experienced control engineer may expand this list tenfold prior to having a complete functional description; note the need for a "team effort."

However, the preparation of a functional description provides major technical and implementation problems which should be recognized.

The most serious technical problem is the preliminary definition of the scheme for which the functional description is to be written around. Defining the existing control scheme may be time consuming. Fig. 1 shows this definition in the form of a block diagram of an existing control system.

Writing a description of a new control scheme without a guideline of what and how this is to be achieved is virtually impossible. This guideline generally takes the form of a block diagram (See Fig. 2) defined early in functional description development and reflects agreement reached by all team members after a number of review meetings.

As control system definition proceeds, the original control scheme concept (Fig. 2) will generally be altered as required to illustrate up-to-date concepts. Fig. 3 is included to illustrate the extent of change a control concept may undergo.

The major implementation problem in developing a functional description is its need to be developed early in the project schedule. By its nature it is not a parallel ongoing function, but instead an in-line series action which may delay design and startup of the facility. Also the manpower resources (especially in the operating department) required to develop a functional description may adversely affect production on existing operating lines.

Management approval to assign "key" personnel for functional description preparation must be established prior to proceeding with functional description development.

Hardware selection. With a functional description in hand, the engineer proceeds with the hardware selection process, as follows:

Determine if process system warrants electro-mechanical or solid state controls.

Comments: Reams of articles have been written on electro-mechanical control (E.M.C.) versus solid state control comparisons and will not be reviewed herein. Today's process control applications using P.C.'s involve the following considerations:

- Higher level of automation achievable with P.C. than with E.M.C. (Are automation benefits desirable?)

- Initial investment is higher with P.C. than with E.M.C., especially in single stand-alone system. (Exception: use of remote input/output (I/O) devices reduce the cost gap.)

- System availability higher with P.C. than with E.M.C. P.C. diagnostic features reduce mean time to repair (M.T.T.R.)

- Present experience with P.C. indicates a mean time between failures (M.T.B.F.) in excess of 8,000 hours.

Percent Availability = (MTBF/(MTBF + MTTR)) 100. Using the above formula and including process factors (e.g., raw material availability, storage space availability, preventive maintenance shutdown, shift changes, etc.) provides a forecast system availability which highlights superior reliability of solid state over E.M.C.

Assume solid state selected; determine which members of the solid state control family (e.g., logic, microcomputer, P.C., mini's.) are legitimate candidates.
Comment: Four major classes of solid state controls are generally considered for process interlock applications.

- Solid state logic (with supervisory control features).

- Fixed program P.C. (with PROM memories and no C.R.T. loader).

- P.C. with C.R.T. type programming capabilities (core or semi-conductor memories).

- Microcomputer systems.

To date, the first and third items have shared the majority of the solid state process interlock application. Engineering costs should be broken down into five categories when using P.C., as follows:

► Hardware design costs

► Installation assistance to construction (installer)

► Software (program development, programing, program documentation)

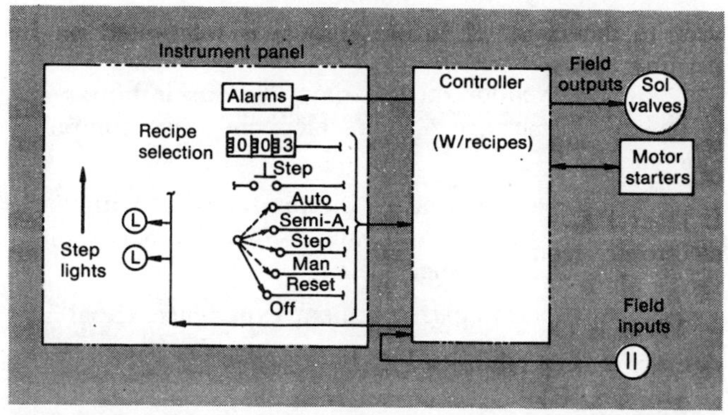

Fig. 2—New system—functional description initial guideline.

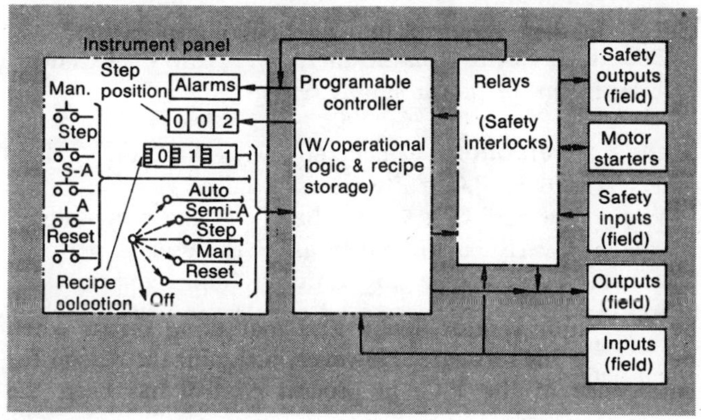

Fig. 3—New system—defined after functional description development.

► Simulation

► Checkout and run-in.

Evaluations establish engineering and hardware dollar comparison. Engineering dollar estimates require cost analysis of software; experienced software engineer (use two software pricing sources) is required to insure realistic dollars are included.

Of the solid state computer family excluding solid state logic the P.C. with C.R.T. has had the largest share of process logic applications to date.

Assume P.C. selected; select which type is required to fit application.

Comment: Matching the job needs [e.g., Input/Output (I/O) count, memory size, programing requirements (e.g., relay interlock, add, subtract, count, latch, matrix handler)] to the P.C. capability is a must. However, this only slightly reduces the choices of vendors. The key factors in P.C. selection include:

a. Cost and delivery

b. How P.C. capability matches application safety requirements?

c. How P.C. architecture and size match desired utilization to minimize impact on production if castrophic loss of P.C. occurs?

d. Does P.C. vendor update user on maintenance report and E.C.O. (Engineering Charge Order); this allows user to determine if future change is warranted on his existing system?

e. Does P.C. vendor offer 7 day, 24-hour instantaneous technical support (e.g., via telephone, via computer, other)?

f. Does P.C. vendor update his product line with latest electronic technological advances; what are his short range plans — long range plans?

g. What is Operating Department experience, capability, desires, as they relate to P.C.?

h. What is P.C. vendor's capability and commitment to provide a nonstandard feature (e.g., Isolated I/O, R.T.L. (resistor-transistor-logic) input, Reed relay output, C.R.T. loader) required in a particular application?

i. Does P.C. utilize volatile memory? How does vendor handle this in his power supply design?

Software selection. Concurrent with hardware selection a choice of P.C. software is required.

The staple of P.C. software has been the relay-ladder language cleverly configured to allow the user to perform "fill-in-the-blank" type programing. This continues to be the major vendor design and marketing thrust when developing his software. However, a significant reason for emergence of the P.C. in process control has been the recent generation of P.C.'s which provide analog I/O, arithmetic capability, matrix handling type programs, and computer type software features (e.g., skip, indirect addressing, P.I.D. [proportional, integral and derivative control algorithm] compare).

This software capability is transferable into process control functions which include the following:

- Charging materials into batch tanks
- Metering these materials as they are charged
- Adjusting set points of analog controllers
- Ramping set points up or down
- Holding set points for a length of time
- Transferring materials at the completion of the process operation

Availability of this more sophisticated language has required greater participation by the engineers in software approach selection and design. The advantages to programming in other than relay format generally includes:

► Better memory utilization

► Ability to handle arithmetic and analog functions of the process generally assigned to other hardware.

Note: This is not to suggest that the P.C. is recommended as an analog controller — it is not. It can handle a small number of analog functions thus helping reduce cost on process systems with limited analog requirements (analog handling in P.C. generally results in poor I/O and memory utilization at present). A typical PID Control Program in an 16 BIT P.C. requires 500 words of memory with each additional control loop beyond the first requiring 65 additional words.

► Higher level of process system diagnostic can be achieved to reduce MTTR.

The disadvantages to programing in other than relay-ladder language includes:

► Greater engineer involvement in software preparation

► Reduce the range of user personnel than can communicate with the P.C. program

► Increase training requirements of programers

► Worsen real-time response P.C. performance in some applications.

The advent of the microprocessor has played a major role in allowing the P.C. vendor to offer more programing features while maintaining an improved cost profile.

Peripherals. P.C.'s selected for process applications generally offer the following peripherals:

- C.R.T. type program loader
- Computer interface capability (serial or parallel port)
- TTY (teletype) type interface
- Magnetic tape cassette interface
- Acoustical coupler interface

Two common considerations are:

- How will the recipes be loaded?
- How will the program documentation be handled?

Recipes may be computer loaded through the data communication port or loaded via magnetic tape cassette. If the first approach is to be used, computer personnel familiar with their equipment communication features should be consulted prior to P.C. selection. Documentation should provide hard-copy in program format (e.g., relay-ladder diagram, recipe) with mneumonics to facilitate legibility. Vendors generally have this capability but require support equipment (e.g., computers) which may not be available at plant. A significant amount of investment dollar can be expended for peripherals; include considerations of this aspect when evaluating P.C. usage.

Safety. Implementation of P.C. equipment in fail-safe applications presents a challenge to the application engineer. Fail-safe as utilized herein is defined as "the failure of any single P.C. item will not result in injury to personnel, equipment, or the ecology." The first priority is to identify the process safety problem areas.

Review with the engineering team is required to understand how the piping, equipment, instrumentation, and such have been selected and designed to adhere to safety needs. The safety needs of the P.C. can then be identified and safely integrated into the overall control scheme.

Each process application results in its own unique method of solving safety problems; however, a basic list of approaches is included, as follows:

- Critical input circuit hardwired jumper to line side of critical output
- Hard-wired bypass in critical output circuit
- External watchdog timer
- Redundant I/O
- Redundant P.C.s
- Redundant processors in one (1) P.C.
- Compare (exclusive or) redundant feature, with alarm or shutdown interlock
- Dedicated I/O assignment to monitor P.C. multiplexing security
- Use of P.C. vendor supplied safety features
- I/O circuit monitoring with P.C. inputs
- Service load output (Solenoid valves, motors) through emergency stop contactors while maintaining inputs and diagnostic outputs (lights, alarms)
- Selected I/O fusing with analysis of impact on system once fuse is blown
- Design of alternate mode of operations (ex. manual, field bypass) to provide orderly process shutdown upon catastrophic loss of P.C.
- Use of read-only-memory (R.O.M.) to reduce programability

The P.C. with its capability to store information can be used as a tool to diagnose problems before they occur as well as when they occur. This coupled with the superior reliability of solid state over electro-mechanical devices offers the engineer the tools for achieving improved safety levels in process control system design.

Startup. P.C.'s because of their inherent features (e.g., small size, programability, C.R.T.); offer the engineer improved opportunities to meet start-up on schedule. However, a well thought-out startup plan involving Design, Construction and Plant forces is required.

This plan should include consideration of the following needs:

- ► Simulation
- ► Installation assistance to Construction
- ► Checkout and run-in
- ► Training
- ► Safety

Simulation of P.C. program for Design engineering team and Operating Department approval is performed on all process control systems. The cost incurred in this procedure is justified as follows:

- Improves program checking prior to startup
- Performed as a parallel function in the project flow chart, hence it is not in the critical path
- May be used as operator training prior to startup
- Facilitates designer/user communications
- Engineering problems found during simulation are in an environment that can bring greater engineering talent to bear on the problem, than may be available during startup.

Use of solid state controls in industrial process environments may expose the solid state components to destructive voltages (e.g., 120V, 460V), and requires installation practices different from those found in electromechanical control installations. This is especially significant in plant-wide systems like those found in chemical process.

The depth of design provided and the level of expertise of the installer in P.C. system installations dictate the level of communication required between engineer and installer. A general practice in P.C. installation requires a visual inspection by the engineer prior to turnover for checkout. Checkout of P.C. systems is used to:

- ► Energize main-frame and insure it is operational
- ► Checkout I/O circuitry and energize
- ► Confirm motor directions correct
- ► Confirm support system (e.g., pneumatic, hydraulic) operative and interface satisfactory with I/O
- ► P.C. peripherals are operative.

This phase should be implemented with Engineering Department and plant forces. This will allow the plant forces to become totally familiar with the P.C. hardware and control system configuration. Some form of acceptance test may be established to obtain approval that checkout is complete.

Run-in of P.C. system requires entry of program tape and operation of process under a strictly supervised mode. Generally, the process is checked in its lower level mode of operation (e.g., Manual Mode) prior to proceeding to more automated modes of operation.

Use of plant forces is imperative at this point as well since the control system operator training can be achieved concurrent with run-in.

Final acceptance test for plant approval is performed during run-in; upon satisfactory acceptance test completion the process system is ready for startup.

Note that each of the phases outlined herein may result in new process related safety questions. Identification, resolution and documentation of safety items prior to their reaching the problem stage is required. There-

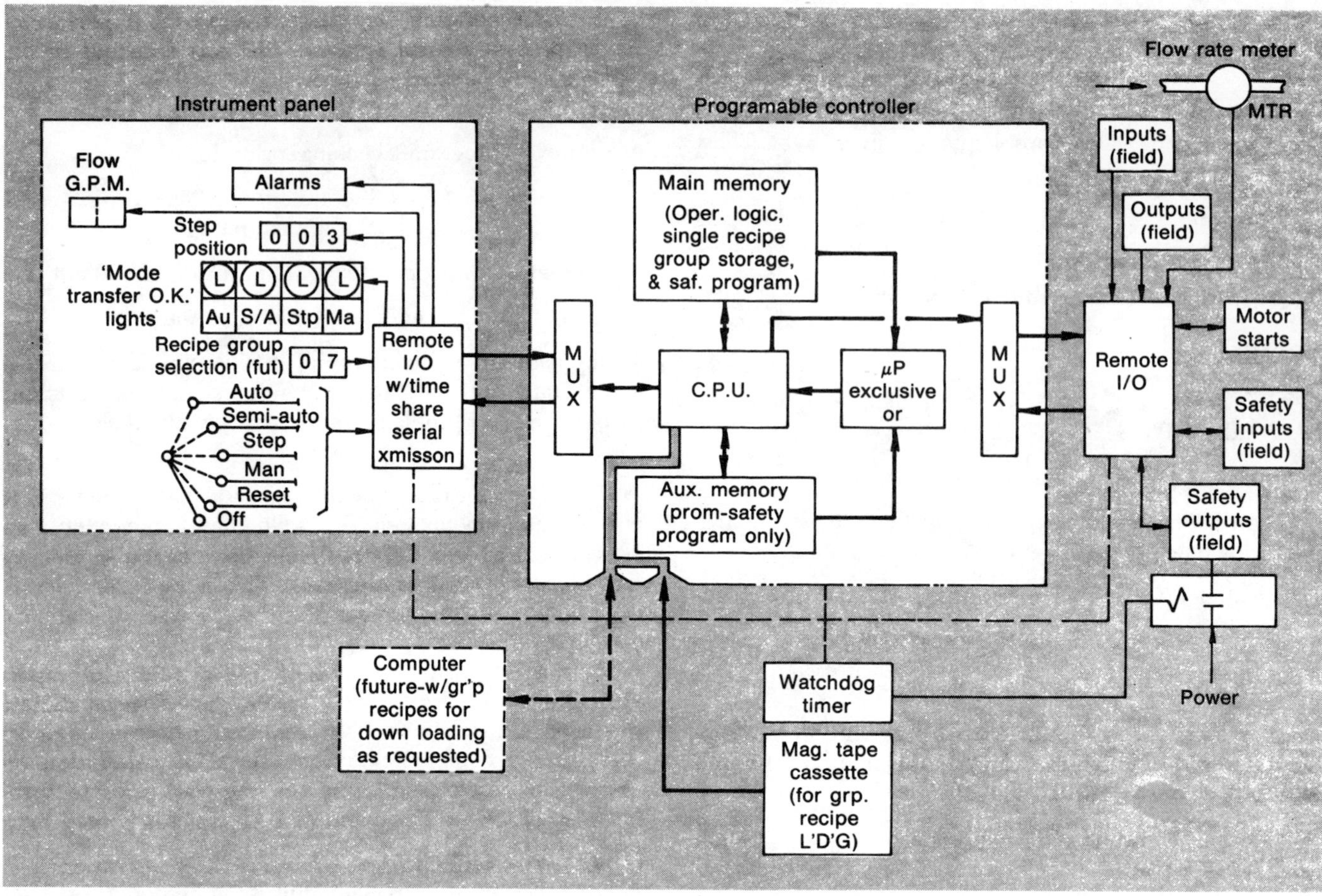

Fig. 4—New system—final arrangement.

fore, an in-depth safety procedure should be prepared and made available prior to start-up.

Fig. 4 is included to illustrate "as built" configuration of the final control system.

Conclusion. If the P.C. is to remain a viable process control tool it will have to incorporate features that are presently difficult to obtain, as follows:

- Ability to handle redundant power supplies
- Ability to communicate between P.C.'s without high I/O and software overhead
- Ability to monitor and isolate faults between communicating P.C.'s
- Small P.C. with dedicated functions that can be "ganged" to provide an overall unitized system
- Self-failure diagnostic I/O (e.g. fail-safe)
- More uniformity in hardware, software, peripherals and communications between P.C.s of various types and vendors
- Fail-safe processor
- Fiber-optic compatible I/O
- Higher level software language
- P.C. configurated to handle analog functions more efficiently
- P.C. configured to interface with packaged direct digital control systems
- Addressable I/O that can be required from more than one P.C.

The future use of P.C. for analog control appears limited except where conversion to higher level languages are implemented. If this feature is incorporated into the P.C., the engineer will have difficulty in differentiating the P.C. from the microcomputer.

BIBLIOGRAPHY

IEEE Standard Dictionary of Electrical & Electronics. Terms - IEEE Std. 100-1977.

Reliability Evaluation - Work Study, August '77.

Robert N. Stauffer, "The P.C.: Carving and Niche in Machine/Process", September '77/Manufacturing Engineering.

Robert T. Fuller, Phillip S. Besler and William B. Crawford, "P.C.'s - Provide Control options that Rival Minicomputers", November '77, Instrument & Control Systems.

Davd Heimann - "Availability - Concept & Definitions". - 140276RM086 - Proceedings 1976 Annual Reliability & Maintainability Symposium.

"Three-Mode PID Analog Control" Allen Bradley Application Data Bulletin 1774.

ACKNOWLEDGMENT

This article is based on material copyrighted by IEEE and originally presented at their 25th annual Petroleum and Chemical Industry Conference, Tulsa, Okla.; Sept. 11-13, 1978.

Optimize Centrifugal Compressor Performance

This microprocessor-based control system provides flexible, reliable compressor control for wide variations in inlet conditions. Energy consumption is also minimized by using a complex algorithm to control within 5 percent of surge line

B. P. Gupta and **M. F. Jeffrey,** Fluor Engineers and Constructors, Houston

CENTRIFUGAL COMPRESSORS can be easily controlled if the thermodynamic factors affecting compressor performance are considered. A control system can be designed to fit the process in which the compressor functions. Wide variations in the thermodynamic variables at the compressor inlet require special strategies for optimal control. Analog controls are not only cumbersome, but also costly. Sometimes analog hardware is not available to implement the required control scheme.

The problem becomes more complex with compressors operating in parallel. It is very difficult, if not impossible, to find two compressors having identical performance characteristics. Slight flow unbalance between parallel compressors can cause "stonewall' in one compressor and useless recycle in another. The problem worsens when the shutdown of one compressor causes a sudden inrush of flow to its parallel compressor causing overload and possible tripping of its motor. Presented is a microprocessor-based control system that alleviates these problems.

WHY USE A MICROPROCESSOR?

A dedicated microprocessor is more flexible than the analog equivalent and has a lower initial cost. Also, a

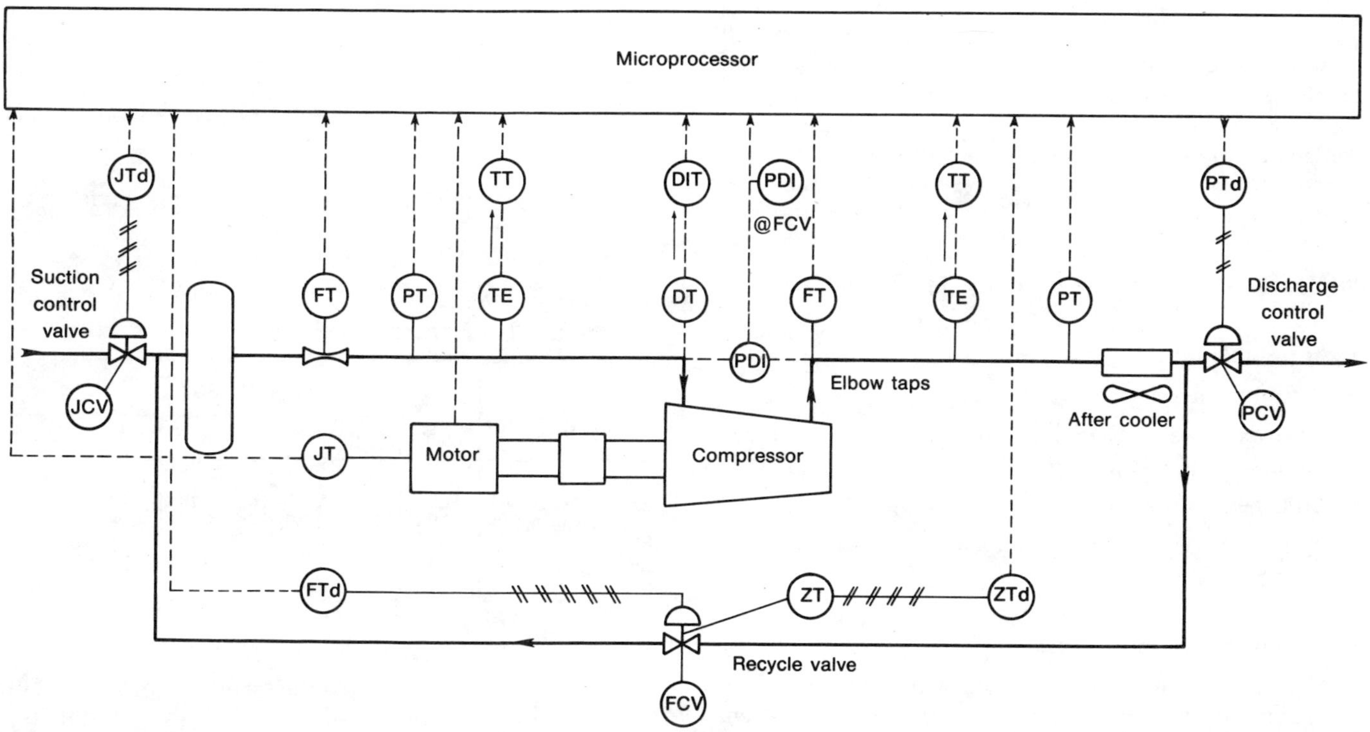

Fig. 1—Compressor layout for microprocessor controls.

new control algorithm can be implemented or an existing control scheme improved with only software changes. Fail-safe strategies can be included in software to handle selected input failures. While startup routines are used to bring up the compressor, shutdown routines are incorporated to decouple the effect of shutdown on its parallel compressor. Strategies are included to accurately predict surge phenomena, even under widely varying thermodynamic conditions at the compressor inlet. This is accomplished by a complex algorithm that allows operation of the compressor close to the surge line to save energy. Dynamic decoupling between the recycle loop and the other loops is easily incorporated. While control actions are important, proper choice of valves and transmitters and their response times cannot be underestimated for optimal compressor control.

General features. The microprocessor converts all inputs into engineering units. Before using a particular input for control, it is checked against its preset upper and lower limits. If the input falls above the upper limit or below the lower limit, an alarm is initiated. Immediately, the microprocessor either substitutes a safe value to replace the input or reverts to its fall-back features. These fall-back features are discussed in detail later. Other variables needed for control are computed internally and compensated for pressure and temperature if necessary. All inputs and variables are made easily accessible on the front panel.

The microprocessor communicates with the operator through a front panel, with the process through valves and with other microprocessors through a communication channel.

A typical layout of the microprocessor for centrifugal compressor control is shown in Fig. 1. The microprocessor implements the following controls on the compressor to make it energy efficient:

Anti-surge controls. The type of control used depends on the driver. For steam and gas turbine driven compressors, speed control is the most commonly used surge control method. For electric motor driven compressors, control becomes more intricate especially for the usual constant speed motors. For this type of driver, the most common controls are:

- Hydraulic coupling between the motor and the compressor
- Adjustable inlet guide vanes
- Butterfly valve for suction throttling
- A power wheel at the inlet of the compressor, upstream of the first stage impeller
- Adjustable diffuser vanes
- Wound rotor induction motors to control speed by varying rotor resistance
- A DC motor to control speed by varying the field current by means of a rheostat.

All types of controls mentioned are either expensive, not energy efficient or require heavy maintenance in corrosive atmosphere service. The most common method varies spillback flow to maintain the compressor inlet flow above the surge point (Fig. 2). The problems of surge control are two-fold. The first is accurate prediction of the surge point. The second is the algorithm to implement the control scheme. Under fairly constant suction conditions of pressure, temperature and molecular weight, surge can be easily predicted and controlled by commercially available analog systems. However, under widely varying conditions of temperature, pressure and molecular weight, surge prediction error can be as high as 48 percent, as shown in Table 1 for the compressor in Fig. 3. While an inaccurate surge control point can put the com-

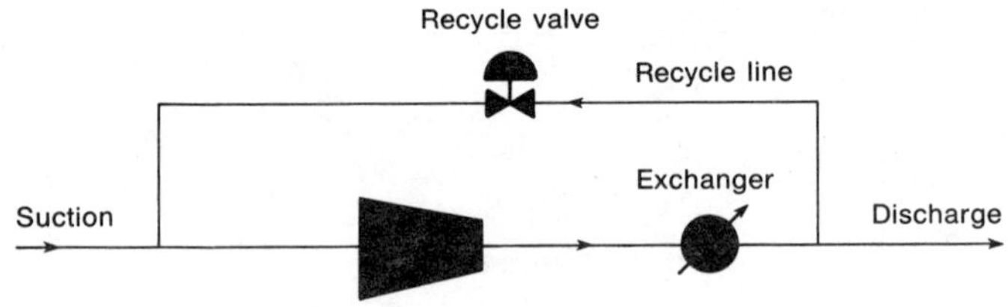

Fig. 2—Typical compressor flow scheme.

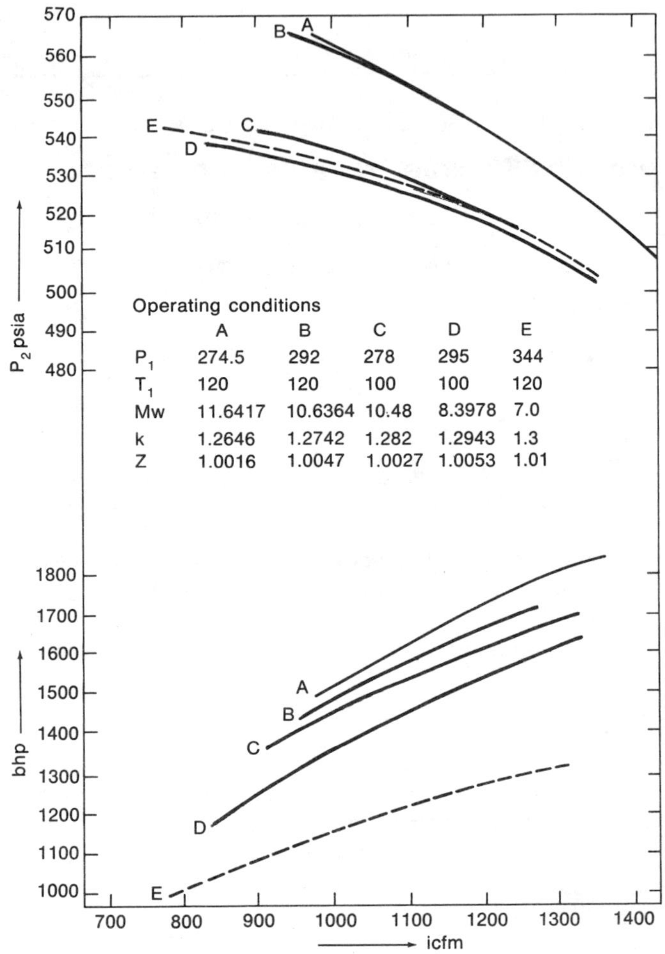

Fig. 3—Typical compressor curve.

TABLE 1—Error analysis in predicting surge

THEORIES	CONDITIONS					
	Summer	Winter	Alt. A	Alt. B	Alt. C	Alt. D
Basic h vs. P........	16.98	15.97	28.31	28.82	0.0	4.17
Temp. compensation.	17.55	13.93	29.01	23.10	0.0	4.05
Density compensation	3.95	4.09	21.28	20.28	0.0	3.87
Dimensional analysis	26.12	23.44	47.63	46.97	0.0	4.74
bhp-wt. flow........	4.52	4.82	0.0	0.563	0.0	0.93

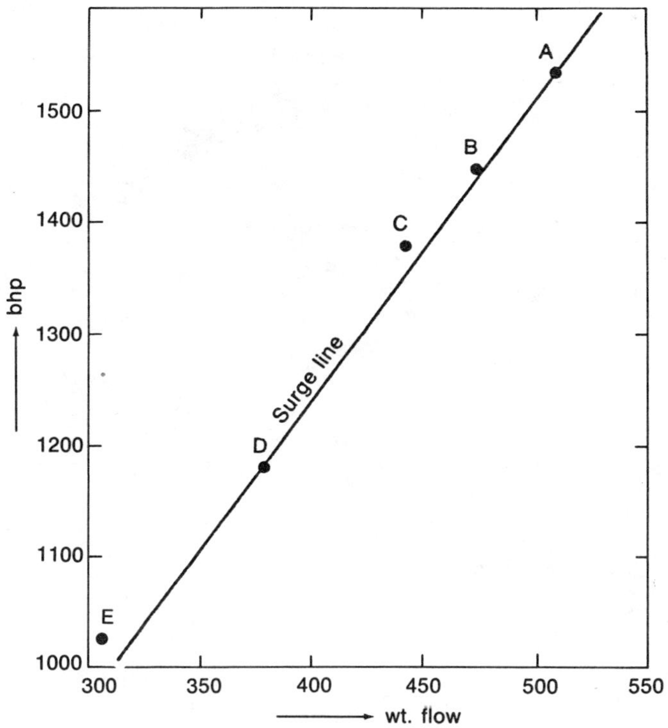

Fig. 4—Brake horsepower vs. weight flow theory.

pressor into deep surge, a conservative surge point results in useless recycle and wasted horsepower. The following scheme takes into account these problems.

Bhp vs. mass flow. The compressor is essentially a head generator (ft.-lbs./lb.). Near surge, the compressor cannot absorb a certain amount of energy input for a certain amount of fluid flowing through it. This hypothesis should not be confused with the basic principle that a compressor is a volume machine. Under actual field conditions and according to various compressor manufacturer's curves, head (H) generated by the compressor remains constant even under widely varying inlet conditions of temperature, pressure and molecular weight. It is also, known that:

$$\text{bhp} = Hm/33{,}000\eta \qquad (1)$$

Efficiency variations near surge are generally less than 2 percent. Moreover a slight increase in compressor head corresponds to a slight increase in compressor efficiency near the surge point only. Therefore, H/η can be considered constant in Equation 1. Equation 1 now can be written as:

$$\text{bhp} \propto \dot{m} \qquad (2)$$

or

$$\text{bhp} = c_1\dot{m} + c_2 \qquad (3)$$

A plot of bhp vs. $\dot{m}$ at surge for the compressor in Fig. 3 is depicted in Fig. 4.

Mass flow can be easily calculated as follows:

$$h = c_3\, Q_1^2\, P_1\, Mw/T_1\, Z_1 \qquad (4)$$

$$\zeta_1 = c_4\, P_1\, Mw/T_1\, Z_1 \qquad (5)$$

so

$$h\zeta_1 = c_3\, c_4\, Q_1^2\, (P_1\, Mw_1/T\, Z_1)^2 \qquad (6)$$

or

$$h\zeta_1 = (c_3/c_4)\, Q_1^2\, \zeta_1^2 \qquad (7)$$

or

$$= (c_3/c_4)\, \dot{m}^2 \qquad (8)$$

$$\dot{m} = c_5\sqrt{h\zeta_1} \qquad (9)$$

For an electric motor driven compressor, bhp can be measured easily by a watt meter. The theory has worked very well in the field for a compressor having 50 percent variations in suction inlet flow, 40 percent variations in suction pressure, 33 percent variations in molecular weight and 20 percent variations in inlet temperature. The compressor was efficiently controlled within 5 percent of the surge curve.

The microprocessor can easily compute the mass flow from flow and density transmitters. The following schemes are programmed into the microprocessor.

- Compressor pressure rise (ΔP) vs. differential across suction flow meter (h)
- Pressure ratio vs. actual volumetric flow
- Break horsepower vs. mass flow.
- Pressure ratio vs. mach number squared
- Incipient surge
- Minimum flow
- Surge spike detection.

Incipient surge can be used with other strategies. ΔP vs. h can be used with combinations of temperature, pressure and molecular weight compensation. Any software strategy can be selected to control compressor surge. The following fall-back and safety features are incorporated.

Minimum flow as first fall-back. The microprocessor operates on any of the strategies listed. If a fault occurs in one of the transmitters used for measurement or surge control (except for flow transmitters), the microprocessor automatically reverts to the minimum flow scheme for surge control. For example, in the bhp vs. mass flow scheme, if the density transmitter becomes faulty, the microprocessor ignores the density input signal and substitutes a predetermined safe value in its place. This safe value is used for control computations. An alarm is initiated to indicate which transmitter is faulty and also indicates that the microprocessor is operating on minimum flow rather than the selected strategy.

This type of control allows the compressor to operate in case of transmitter failures. Even though it is not the most efficient control, the compressor is kept on line.

Incipient surge as second fall-back. If the flow transmitter becomes faulty, the incipient surge control scheme is automatically activated.

Surge spike detection as third fall-back. If the basic control scheme does not work, minimum flow scheme does not work and the incipient surge scheme fails, the microprocessor detects the surge spike built within itself as a last line of defense to protect the compressor. Upon such detection, the recycle valve is made to go wide open and a special alarm is activated to alert the operator that something serious has occurred to the compressor operation.

Safety features on anti-surge controls.

- If the recycle loop is in the manual mode, the microprocessor still computes the output needed for controlling anti-surge. Manual or computed output, whichever is greater, goes to the recycle valve. This is especially useful when the operator believes he needs greater recycle valve opening than calculated by the microprocessor. In this instance, the operator puts the recycle loop in manual to a certain percentage opening of the recycle valve. Now, if all of a sudden a process change dictates more recycle valve opening, the compressor is still protected. Without this feature, the compressor stands a chance of going into deep surge and possibly sustaining damage.

► If the recycle loop is in the manual mode, the recycle valve cannot be closed faster than a predetermined closing rate. This prevents the compressor from going into surge because of operator negligence.

Anti-surge control algorithm. The problem of surge cannot be solved successfully with only accurate surge prediction. A sophisticated control algorithm is needed to implement the accurate surge prediction technique. This technique was simulated on a digital computer and applied very successfully in the field. The microprocessor is programed to differentiate between the operating regions in the control strategy plane, independent of the type of surge control strategy chosen. The microprocessor uses the predictive control algorithm rather than derivative control near the surge control line. This avoids the noise problems of derivative control but still uses the basic principle of derivative control to increase response speed of the control system. The surge control curve is allowed to move closer to the actual surge curve. Details of the control algorithm are shown in Table 2 and Fig. 5.

Valve considerations. The problem of surge protection becomes more complicated with accurate surge prediction and sophisticated control algorithms that allow the surge control line to move closer to the real surge line. There is no advantage in coming closer to the surge line without a recycle valve opening, if at the time of crossing the surge control line, the valve response time is too long.

TABLE 2—Anti-surge control algorithm

MODE OF OPERATION
Normal operation:
A, B = Anti-surge valve partly open
C & D = Anti-surge valve closed
Anti-windup feature shall prevent control action

MOVING TOWARDS ACTUAL SURGE
A = P + I + predictive + set surge flag
(a) Surge valve closed
B & C = Predictive
D = No control
(b) Surge valve open
B (surge flag reset) = P + I
B (surge flag set) = closing ramp
C = closing ramp + predictive
D = closing ramp + predictive + reset surge flag

MOVING AWAY FROM ACTUAL SURGE
A = P + I + predictive
B = closing ramp
C = closing ramp, reset surge flag
D = closing ramp

NOTES
(a) The above modes of operation also apply for minimum control, h.
(b) Dynamic mode transfer from closing ramp output to P + I in region B is bumpless.
(c) P & I constants in region "B" are different than P & I constants in region "A".
(d) Set point for P + I in region "A" is surge control line. Set point for P + I in region "B" is between surge control line and slow control line (set by data entry).
P = Proportional band
I = Integral

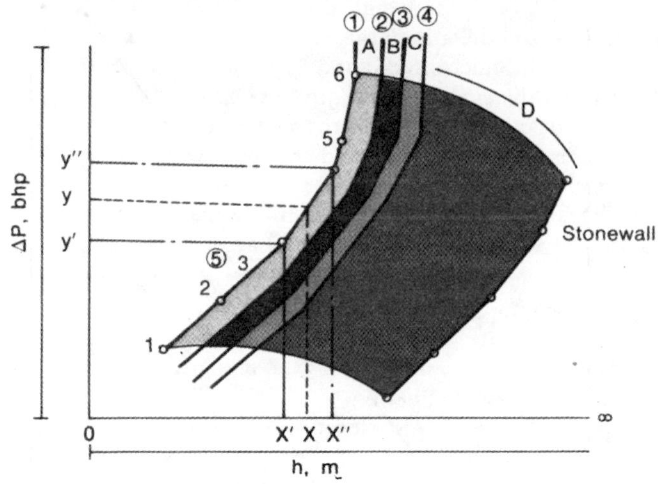

①= Actual surge line
②= Surge control line
③= Slow control line
④= Anticipating C line
⑤= x-y reference values (1 through 6) (set by data entry)
A, B, C, and D = Control regions
To select y' & y'', take signal y and detect the next higher and lower reference point.

$$x = \frac{(y-y')(x''-x')}{y''-y'} + x'$$

Fig. 5—Anti-surge control algorithm.

High response time of the recycle valve can put the compressor into deep surge due to sudden loss of inlet flow. The authors suggest that a 2-second valve response time is adequate to protect the compressor. This can be achieved by using valve boosters or a digital valve. Moreover, the valve characteristic should be linear. An equal percentage characteristic valve takes too long to achieve the required flow thruput and a quick opening valve passes too much flow.

Suction flow control. The microprocessor throttles the suction valve to prevent overload or stonewall when the bhp consumed by the motor exceeds the set point. If suction pressure is above a predetermined level, an alternate fast throttling mode is used to protect the motor.

In parallel compressor operations, a major problem is one compressor running overloaded while the other one recycles to avoid surge. The microprocessor is programed to distribute flow evenly between the two compressors to avoid this problem. This scheme reduces useless recycling and saves horsepower. However, if an overload condition occurs, the load-sharing scheme is automatically deactivated for compressor and motor protection.

The microprocessor is also programed to cause one compressor to recycle early, keeping the other one running without recycle. This scheme is particularly useful if one compressor gives trouble. In any event, either equal sharing or base loading can be easily selected without reprograming.

Discharge pressure control. The microprocessor controls discharge pressure at the set point by proportional and integral modes. If volume between the compressor and the discharge valve is small, fast opening of the recycle valve can cause high discharge pressure and large oscillations on the discharge side of the compressor. To avoid this, the microprocessor is programed to control the discharge valve in a reverse proportional mode to surge control output. The control program automatically returns to the normal control mode when the recycle valve opening slope rate is below a predetermined level and after a predetermined time delay. This minimizes dynamic coupling between the recycle and discharge valves.

If discharge pressure exceeds a predetermined level, adaptive tuning controls the compressor to maintain the discharge pressure. This algorithm is particularly useful if the compressor is operating close to design.

Startup controls. The microprocessor opens the recycle valve at compressor start and closes it at a different ramp rate (different from recycle valve control) until the valve is fully closed. The recycle control valve algorithm can override this ramp to protect the compressor from surge. At this time, either the recycle valve is fully closed or the recycle control valve algorithm is executed, and startup control is automatically deactivated.

Shutdown controls. Shutdown of the compressor is monitored by the motor contact. At the time of shutdown, the microprocessor closes the suction valve and opens the recycle valve.

In the event of a parallel compressor shutdown, the sudden inrush of suction flow can cause the motor to trip. To avoid this shutdown, the microprocessor throttles the suction butterfly valve to a predetermined safe position for a predetermined time. After this predetermined time, the microprocessor reverts to normal control.

TRANSMITTER CONSIDERATIONS

Proper selection of transmitters cannot be overlooked when designing efficient compressor controls. The transmitter ranges greatly affect the tuning constants of the various controllers built in the microprocessor. For example, we may have a 0-100-in. of H_2O range flow transmitter on the inlet of the compressor. Assuming that surge occurs at 50 percent of flow, then the transmitter reading at surge is 25 in. of H_2O. Controlling surge within 5 percent of the surge line means action must be taken when flow is 26.25 in. of H_2O and before the flow drops to 25 in. of H_2O; the valve has to be sufficiently open to avoid surge. Since the flow near surge is already disturbed, at least 30-40 percent valve opening is required in region "A" (Fig. 5) to stabilize the flow phenomenon. After the initial opening of the recycle valve, the valve is closed very slowly at the steady rate required to control surge within 5-10 percent of the surge line. Following is the analysis of the tuning constants of that controller:

- Difference between setpoint and process = (26.5-25)/100 = 1.5 percent
- Proportional band setting = 20 percent
- Valve opening due to proportional band = 7.5 percent
- Reset setting = 5 repeats/minute
- Valve opening due to reset = 37.5 percent
- Total valve opening after one minute = 45 percent

One minute is too long a response time for the required corrective action. Designing optimal controls, sophisticated algorithms and fast opening recycle valves are of no benefit if the transmitter ranges do not allow corrective action in the available time of 1 to 2 seconds. Lowering the transmitter range to 10-40 in. of H_2O will improve the situation immensely. Finally, the authors would like to comment that all compressor installations deserve some attention by control systems engineers even if they are alike. Proper controller tuning cannot be underestimated. All control constants and other variables needed for compressor control are easily accessible on the microprocessor and can be changed very easily. All this data is backed up by nonvolatile memory, which can be retained for one year without reentering at every startup. Since the microprocessor has various levels of backup built-in, backup analog controls are not necessary. Since all process information is easily accessible, future work on improving compressor controls can be done easily and checked in the field.

NOMENCLATURE

acfm	Actual cubic feet/ minute at suction
bhp	Brake horsepower
c_1-c_5	Constants
H	Head generated (ft.-lbs./lb.)
h	Differential pressure across a suction flow meter, inches of water
I	Integral
$\dot{m}$	Mass flow = $c\sqrt{h\zeta}$
Mw	Molecular weight
ΔP	Differential pressure between compressor suction and discharge, psi
P_1	Suction inlet pressure, psia
P	Proportional band
Q	Suction inlet flow (acfm)
T_1	Suction inlet temperature (°R)
η	Efficiency
Z_1	Compressibility factor
ζ	Density

BIBLIOGRAPHY

Gupta, B. P. and Teague, C. C., "Compressor Surge Control," 1977 Symposium on Instrumentation for the Process Industries, Texas A&M University, January 1977, pp. 7-13.

White, M. H., "Surge Control for Centrifugal Compressors," *Chemical Engineering* (Dec. 25, 1972).

Compressor Handbook, Gulf Publishing Co., 1969.

Part III.

Analyzers and Composition Control

Analytical Instrumentation for Process Control

THE INCREASED application of analytical techniques for determining composition of petroleum streams, according to D. D. Zakaib of Gulf Oil Canada Ltd., Canada, has given the refiner a powerful tool for controlling complex processes. These developments have taken place mainly with off-line systems but have been complemented by similar advances in on-line analyzers. Greater reliability and more selective applications of on-line analyzers have led to their increased acceptance. On-line compositional data have been used effectively in closed loop feedback and predictive control schemes.

Here is a review of the application of both on-line and off-line analysis to the study and control of major processes with particular emphasis on fractionation, reforming, catalytic cracking, naphtha cracking, alkylation and sulfur recovery.

FRACTIONATION

The prime use for onstream analytical instrumentation is in the monitoring and control of distillation towers. The onstream gas chromatograph is usually chosen for this application now that improvements in sample handling systems and programmer design have simplified maintenance and increased onstream time. These instruments produce a significant improvement in plant operations even though the majority of chromatographs are used for operator guidance rather than for closed loop control. However, closed loop analyzer control, in most situations, can further improve tower operations.

Fractionators are more sensitive to feed composition and flow changes when operating at minimum reflux. Consequently feedforward control is required to maintain product quality. The composition and flow of the feed are used to predict the set-points required to cope with the changes in load.

More difficult problems require digital computer control. A system incorporating both analog and digital hardware has been developed for the predictive control of an ethylene plant's demethanizer. In this application, chromatographs analyzed the feed and product streams.

NAPHTHA PROCESSING

Naphtha is the feedstock for both ethylene plants and catalytic reformers. The composition of the naphtha is the most important variable in the operation of either unit. Paraffinic naphthas are excellent feedstocks for ethylene production, while naphthas rich in cycloparaffins and aromatics are premium stock for reforming.

Reforming. The primary role of the catalytic reformer is to maintain the octane level of the gasoline pool. Once the target octane for the reformer has been established the operating objective is to produce this constant octane reformate. This is achieved during the operating cycle by raising the inlet temperature of the reactors to compensate for declining catalyst activity. As a result, there is a decline in reformate yield, a decrease in hydrogen make and purity and an increase in light hydrocarbon yield which eventually forces a termination of the cycle.

For semi-regenerative units it is important to monitor catalyst performance during the cycle so that any abnormal changes in activity can be corrected to increase cycle length. Declining catalyst performance is monitored by plotting reformate yield (corrected for changing feed composition, product octane and reactor variables) against the onstream time. Accurate reformate yields can be calculated from the compositions of the feed and reactor effluent. In addition, hydrogen production and light hydrocarbon make can be accurately determined from the analytical data. These data may then be used to calculate the extent of the various reactions such as isomerization, hydrocracking, dehydrocyclization and dehydrogenation which, in turn, point to changes in activity and selectivity of the catalyst during the cycle.

On-line analytical systems have been able to provide a better understanding of the over-all reforming process. Detailed feed and product compositions were obtained by onstream capillary chromatographs coupled to a minicomputer. Four capillary columns were required to obtain a detailed analysis up to C_{12}. The analytical data were then used to calculate on-line, specific component conversions, the octane number, the distillation curve and process parameters. Benefits derived from such data include increased economy through operating mode changes, tighter unit control and improved mathematical models which can be used to evaluate the reformer's role in future processing schemes. A development from this work was the correlation between Research Octane and total aromatics in the product, and this has been applied onstream using a process chromatograph programed to read out Research Octane on a 3-minute cycle.

Naphtha cracking. The pyrolysis of hydrocarbons is the major source of petrochemicals. The most profitable area for analytical control of an ethylene plant lies in the operation of the cracking furnaces. Onstream analyzers are needed to monitor the severity of each furnace to maximize the unit's profitability.

Despite sampling difficulties, complex streams have been designed to obtain representative samples of the furnace effluent for quantitative fractionation into vapor, gasoline and pyrolysis oils. Analysis of these samples gave a detailed yield profile for off-line optimization studies. A simplified approach for on-line monitoring and control used a less complex sampling system to obtain a representative sample of the gas only. This sampling system removed coke, tar and liquid products from the furnace effluent sample and a process chromatograph determined the concentrations of the major light ends in the gas samples. Three of these analytical systems monitored the effluents from six coils. An on-line computer processed the raw analytical data and calculated the ratios of methane, ethane, propane, propylene and total C_4's to ethylene. These ratios measured the effective cracking severity of each coil in order to decide on the changes required to optimize the product mix. The on-line data were used to determine the apparent rate of coil coking and to develop more sophisticated yield models.

CATALYTIC CRACKING

The Fluid Catalytic Cracking Unit (FCCU) is operated to maximize the value of the product mix, the most valuable components being gasoline, light cycle oil and C_3C_4 hydrocarbons. Since the composition of the FCCU feedstock has a large influence on the yield distribution of product, an accurate definition of feedstock quality is required both for feed selection and process modeling. Earlier definitions of feedstock quality based on physical inspection data were found to be inadequate for process modeling. This was particularly evident when evaluating processed stocks for FCCU feed. More accurate feedstock correlations based on mass spectrometric and spectroscopic and chromatographic analyses have been established. Conventional inspection data do not distinguish between mono, di or triaromatics in a sample and therefore cannot provide accurate yield predictions. With zeolite cracking catalysts, it is known that certain feed monoaromatics are gasoline precursors, di and triaromatics yield light cycle oil and decant oil, and the heavier aromatics are coke precursors.

The recognition that polycyclic aromatics are responsible for coke formation has led to the development of a process liquid chromatograph to monitor these components. This instrument may be used as part of a feedforward control scheme for better regenerator operation. By determining variations in the coke precursor concentration in the feedstock it is possible to predict changes in reactor coke make and to allow for this change by adjusting the air rate to the regenerator. This would be useful for units running varying feedstocks with changing coke precursor levels. Feedback control of the regenerator operation has been provided by an on-line flue gas chromatograph. This had proved difficult because of the high temperature and catalyst fines content of the stream, but improved sampling systems have increased the reliability of this application. The chromatograph established the air rate to the regenerator so that the regenerator bed temperature was held at an optimum. Improved regenerator operation will reduce the carbon level on the regenerated catalyst and decrease the catalyst/oil ratio. This will in turn lead to increased liquid volume recovery for the unit.

ALKYLATION

One of the important controlled variables in alkylation is acid strength. The optimum acid strength is a function of reactor temperature and of the hydrocarbon-acid contact time. This strength can be selected using the analytical unit survey data above. An on-line liquid chromatograph system has been developed for the analysis of HF acid. This system determined the concentration of acid soluble oils, hydrocarbons, water and HF in various acid streams. Applications for this analyzer include monitoring the feed and effluent from the HF regenerator as well as the condition of the acid phase in the reactor. This system may be used to establish and control the optimum acid strength.

For sulfuric acid units, pilot plant studies have shown that an improved quality alkylate, as determined from analytical data, is obtained by precise control of the water content of the catalyst. Onstream liquid chromatograph systems can also be applied to sulfuric acid units.

SULFUR RECOVERY

Environmental and conservation demands have given increased importance to the control of Claus sulfur recovery units. There is a great incentive to operate the sulfur recovery plant under best possible control since this may allow legislated levels to be attained without additional recovery units.

The three operational criteria that are used to achieve best control of the Claus process are setting the correct ratio of air to acid gas, selecting optimum catalyst temperatures, and minimizing entrained sulfur losses.

Despite sampling problems the tail gas analysis for hydrogen sulfide and sulfur dioxide has been carried out by onstream gas chromatography and ultra-violet spectroscopy. Both types of instruments were used in closed loop systems where the tail gas ratio adjusted the air rate to the plant.

Onstream gas chromatography has been used to determine the optimum tail gas ratio, to monitor furnace performance and to select the correct temperature for each catalytic converter.

In addition to losses due to incomplete conversion, elemental sulfur losses contribute to the sulfur dioxide emission from the incinerator stack. SO_2 emission has been monitored on a continuous basis with a process chromatograph and the results used for plant optimization. A new approach to continuous measurement of the recovery efficiency is based on an analysis of oxidized samples of feed and tail gas.

Computer control of a sulfur plant has been described in which process gas chromatographs analyzed the acid gas feed and the tail gas, and an ultra-violet instrument monitored the stack sulfur emissions. These analytical data were used by the computer to select the reactor bed temperature, and the air:acid gas ratio for minimum SO_2 emission.

Distill with Composition Control

A computer control system based on product composition from fractional distillation can give energy savings of 5 to 15%

D. E. Smith, W. S. Stewart and **D. E. Griffin,**
Applied Automation, Inc., Bartlesville, Okla.

THE NEED to conserve energy and to use products in an optimum manner provides incentives for control of the composition of fractionator products. The economic benefits of composition control include the following:

- Energy conservation by avoiding overseparation
- Shifting less-profitable components into more-profitable products
- Increasing thruput.

Fractionator control usually is accomplished by controlling physical properties that are related to product composition. However, direct control of product composition permits a more precise separation to be achieved. A high degree of automatic control is necessary if direct control is to give maximum profitability.[1] Thus, a properly designed and implemented control system requires:

- Reliable measurement of at least one product composition
- Control configuration consistent with the operating objectives of the fractionator.

A case against overseparation. Part of the economic benefits of computer control of a single fractionator centers about overseparation of products. With regards to energy conservation, the case against overseparation can be made with reference to Fig. 1. Fig. 1 is for a deisobutanizer and is a plot of internal reflux-to-feed ratio versus isobutane impurity in the bottoms product with four different values of the normal butane impurity in the overhead product.[2] As in all columns, observe that the internal reflux-to-feed ratio increases rapidly as the bottoms product becomes more pure. The shape of the curves would be similar if the graph were constructed by plotting the

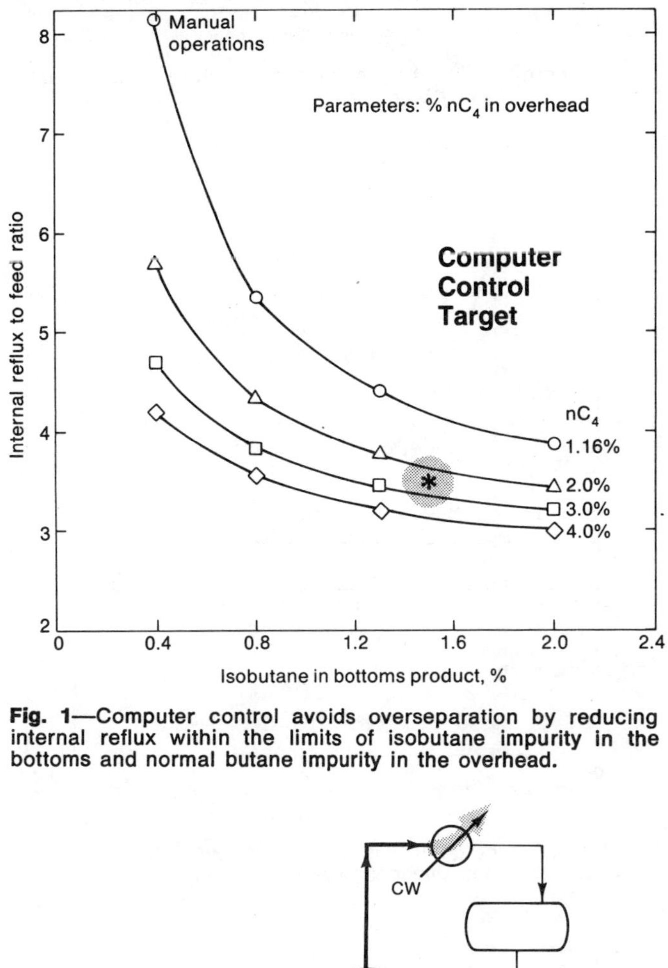

Fig. 1—Computer control avoids overseparation by reducing internal reflux within the limits of isobutane impurity in the bottoms and normal butane impurity in the overhead.

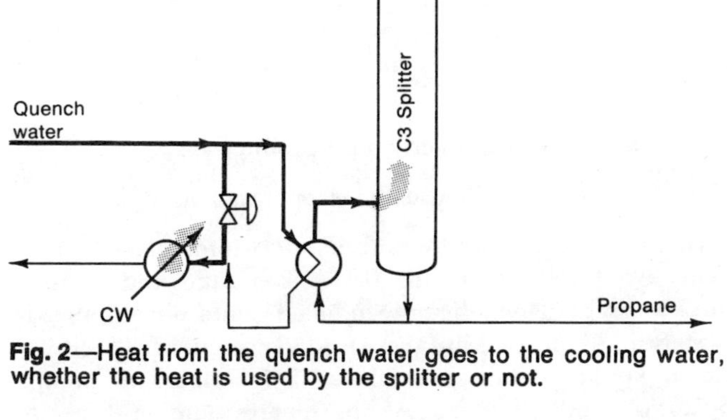

Fig. 2—Heat from the quench water goes to the cooling water, whether the heat is used by the splitter or not.

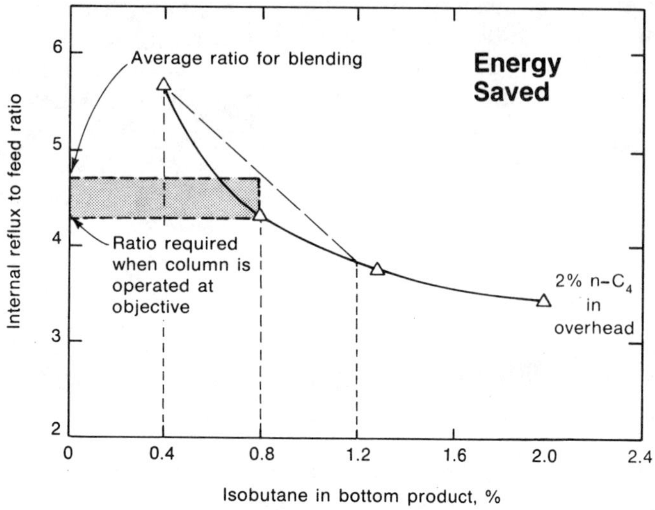

Fig. 3—Energy is saved by holding all product within specifications rather than by blending.

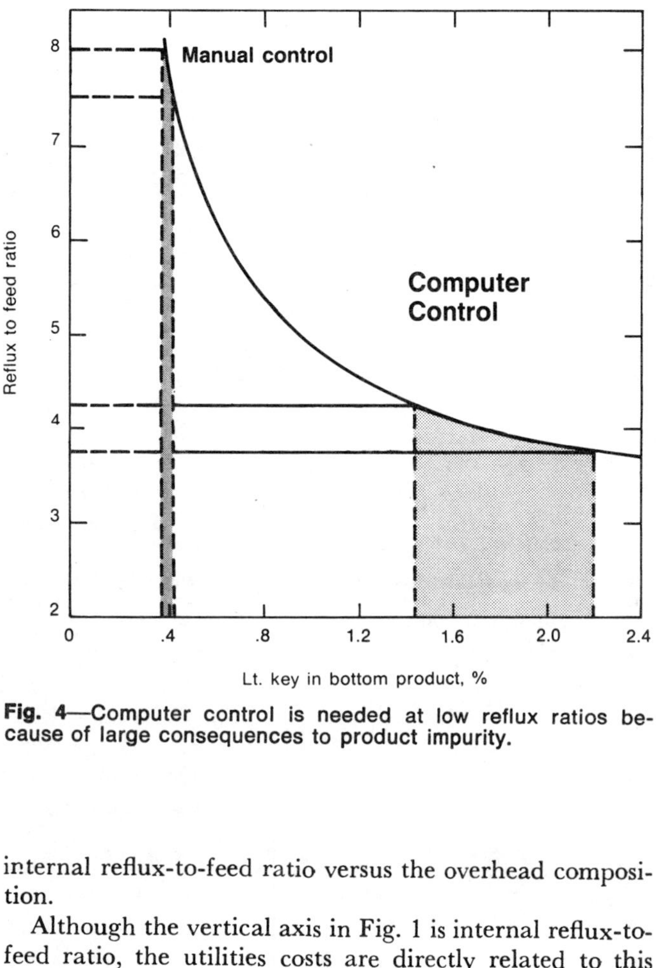

Fig. 4—Computer control is needed at low reflux ratios because of large consequences to product impurity.

internal reflux-to-feed ratio versus the overhead composition.

Although the vertical axis in Fig. 1 is internal reflux-to-feed ratio, the utilities costs are directly related to this ratio. Thus, Fig. 1 is essentially a plot of utilities costs versus product composition. Fig. 1 illustrates two points:

- Penalty paid for overly pure product
- Why the process usually is operated as it is.

In the example deisobutanizer, a bottoms product was being produced containing 0.4% isobutane and an isobutane product was being produced containing approximately 1.1% normal butane. However, contract specifications would permit as high as 2% isobutane in the bottoms product and 2.5% normal butane in the isobutane. This overseparation is clearly costing utilities, and is the source of benefits for computer control of fractionators. For this particular deisobutanizer computer control made the following two changes in composition:

- Isobutane in the bottoms increased from 0.4% to 1.5%
- Normal butane in the overhead increased from 1.16% to 2%.

The result was an energy savings of 10.2MM Btu/hr., or $171,000 annually with energy valued at $2/MM Btu.

Computer control can conservatively obtain a 5 to 15% reduction in utility usage by a fractionator. At the current prices of energy, computer control of fractionators can often be justified by energy savings alone.

The reader is cautioned that there is a distinction between energy saved and energy not used. To illustrate, suppose a reduction in energy consumption achieved on a particular unit was entirely in low pressure steam, while at the same time the plant was venting low pressure steam. In this case, energy has not been saved because the energy inflow to the plant has not been changed.

To further emphasize this point, Fig. 2 illustrates an actual C_3 splitter in which utilities cannot be saved. This column is being reboiled with waste heat, all of which must end up in the cooling water — regardless of how much is used in the C_3 splitter reboiler. Since the waste heat is free in this case, as much reboiler heat as possible should be used to maximize the recovery of polymerization grade propylene, the more valuable product.

One approach to partially offset the extra energy consumed in producing products of excess purity is to blend with products whose purity is below specifications. However, this also results in extra energy consumption as compared to producing all of the product just meeting specification. To illustrate, Fig. 3 reproduces the line from Fig. 1 for an overhead composition of 2% n-butane. Suppose equal amounts of bottoms products with 0.4% isobutane and 1.2% isobutane are produced, and then blended to produce a product of 0.8% isobutane. As illustrated in Fig. 3, the average internal reflux-to-feed ratio required to produce the two individual products is higher than the internal reflux-to-feed ratio required to produce directly a bottoms product with 0.8% isobutane. Thus, extra energy has been consumed in the blending approach.

For minimum consumption of utilities in a fractionator, the following guidelines are recommended:

- Operate the fractionator to produce minimum overseparation
- Implement control to achieve composition control on both ends of the fractionator
- Be certain that the reduction in energy usage is reflected in the energy inflow to the production complex
- Blending of overseparated products results in less efficient use of energy.

Composition sensitivity. Unfortunately, energy conservation in fractionators is not as simple as adjusting set points on existing controllers closer to product impurity constraints or specifications.

It is likely that the operations group responsible for operation of the example deisobutanizer knew that the targeted computer operating point (shown in Fig. 1) was more profitable than the manual operating point (also shown in Fig. 1). The point of manual operation is probably as close as the human operator can get to the targeted computer operating point without producing off-specification product an unacceptable portion of the time. If any sustained improvement over the manual operating point is desired, computer control will be required to obtain it.

A variety of external disturbances occur to a fractionator for which control systems must compensate. These disturbances include:

- Feed rate changes
- Feed composition changes
- Feed enthalpy changes
- Overhead condensate temperature variations
- Reboiler heat variations.

Furthermore, the effects of these disturbances are more pronounced when the degree of overseparation is reduced.

There are at least three reasons why the human operator cannot achieve and sustain the most profitable operating conditions without advanced controls. These reasons include:

- Reliable on-line analytical measurements are usually not provided
- Operators cannot respond as frequently and as precisely as a computer control system
- The fractionator tends to become less stable or more sensitive to disturbances at lower reflux-to-feed ratios.

This tendency toward instability is due to a variation in composition sensitivity, as noted by Rutledge.[3]

Composition sensitivity is the process gain or change in concentration *(ΔC)* of a "key component" impurity resulting for a given change in reflux-to-feed ratio (ΔR). This relationship may be expressed as follows:

$$\text{Composition Sensitivity} = \Delta C/\Delta R.$$

This sensitivity changes with reflux-to-feed ratio. Fig. 4 illustrates the magnitude of change in composition sensitivity at two reflux-to-feed ratios. At a nominal reflux-to-feed ratio of 7.75, the sensitivity is $(0.05/0.5) = 0.1$. However, at a lower reflux-to-feed ratio of 4, the sensitivity is $(0.76/0.5) = 1.52$. Thus, a 15 fold change in composition sensitivity occurs when the reflux is cut in half.

The practical aspect of this increase in sensitivity is that a control system whose performance is entirely satisfactory at high purities (and high energy usage) may give totally inadequate performance when the composition set point is moved closer to the specification or constraint value (lower energy usage). In order to achieve the energy savings possible through reduction of overseparation, the performance of the control system must be significantly improved.

ADVANCED CONTROL SYSTEMS

Almost every fractionator is unique. No "canned" or standard control package will meet the needs of every fractionator. The function of a particular tower within the process, its objectives and its hardware details will vary. Therefore, the control system must be designed to meet the requirements of the particular fractionator. Years of experience in implementing composition control on fractionators has resulted in an orderly engineering approach to control system design.[4,5]

Composition control objectives. A vital first step is a clear statement of the process operating objectives of the particular fractionator. These objectives are used to specify the product composition criteria for that application. Reliable, repeatable measurements of the product compositions are essential to the success of the control project. Process gas chromatographs have been proven in hundreds of closed-loop control applications when proper attention is given to the selection, installation and routine maintenance of the analyzers.

In a two-product column, composition control is always desirable on one of the products. In a majority of the cases, composition control is also desirable on the second product as well, thus producing a requirement for double-ended composition control. But due to limitations on condenser and reboiler capacity or tower vapor loading limits, it may not be possible to control the composition of both ends of a fractionator during periods of peak production. That is, control of the composition of one end of the column must be relinquished in order to maintain control of the composition on the other end. The composition that remains on control is called the *primary composition specification.*

In most cases, the composition specification or set point on one of the products can be derived directly from the product specifications. Normally this product composition will be the primary product specification. The second product specification then specifies the separation or recovery, to be achieved by the fractionator. As will be considered in a subsequent section, the actual value for this composition specification may be the result of an optimization calculation that takes into consideration product values, utility costs and fractionator characteristics. It is this second product specification that is relaxed whenever it is not possible to meet both composition specifications.

The product composition criteria on a fractionator that yields product directly is usually the concentration of an impurity. However, it may be the sum of several impurities even though only one can be controlled on that fractionator. When one or more of the fractionator products are separated further in downstream equipment the product composition criteria may be a ratio of concentration of two components.

To illustrate the manner that column objectives affect

product composition criteria, Table 1 summarizes the objectives and criteria on 10 different deethanizers. The only thing common to all 10 fractionators is that ethane is the major overhead product. The table illustrates the range of objectives that can be encountered for fractionators that may share a name, and how this variety of objectives influences both the form of the composition specification and the choice of primary composition specification. The primary product specification for bottom product was either the ratio of light key to heavy key or light key plus heavier than heavy key. When the primary product specification was based on the overhead, the specification was either heavy key or heavier than heavy key.

Internal product flows. It is necessary to have a clear concept of the important mechanisms occurring within a fractionator in analyzing control configurations. For this purpose, a simplified model relating the internal flows within the fractionator is extremely useful. A very satisfactory model can be derived from heat and material balances on the fractionator, with the simplifying assumption that all components of the fractionator feed have equal heats of vaporization.[4] Most short-cut fractionator calculations are based on this same simplifying assumption, which leads to the often-used assumption of equimolal overflow.

Fig. 5 provides a conceptual steady state internal flow model for a fractionator. This model consists of the following relationships:

The vapor boilup rate V_B equals the heat Q_B added by the reboiler divided by the heat of vaporization (H_V):

$$V_B = Q_B/H_V \quad (1)$$

The vapor rate V above the feed tray equals the vapor boilup rate plus the vapor entering with the feed (feed rate F times vapor fraction V_F, provided the feed is neither sub cooled nor superheated):

$$V = V_B + FV_F \quad (2)$$

The internal reflux rate R_I equals the external reflux rate R plus the vapor that must be condensed in order to raise the reflux to its bubble point:

$$R_I = R[1 + (C_P/H_V)(T_O - T_R)] \quad (3)$$

The liquid rate L below the feed tray equals the internal reflux R_I plus the liquid in the feed.

$$L = R_I + (1 - V_F)F \quad (4)$$

The distillate rate D equals the vapor rate V minus the internal reflux R_I:

$$D = V - R_I \quad (5)$$

The bottoms rate B equals the liquid rate L minus the boilup V_B:

$$B = L - V_B \quad (6)$$

An unsteady state must also be accounted in order that the dynamics of the process can be taken into consideration. This is especially true for fractionators. Fig. 6 extends the steady state internal flow model in Fig. 5 to the unsteady state. This requires four additional considerations:

- Since a change in the reflux rate must work its way down the column, this change will not be seen at the reboiler until the passage of several minutes, with the exact lag depending on column size, number of trays and other parameters. For this reason, two dynamic terms, G_T and G_B, are included in Fig. 6. For most columns, second-order-lag-plus-dead-time dynamics are sufficient.

- Changes in boilup rates will be observed at the condenser in a matter of seconds. Therefore, no dynamic terms are necessary for the vapor streams.

- The liquid inventory in the condenser and/or associated accumulator will change. Equation 5 applies for both the internal overhead product rate D_I and the external product rate D at steady state. However in the unsteady state, the difference $D_I - D$ is the rate of ac-

TABLE 1—Primary composition specifications for various applications

Case—Objective	Product composition spec.	
	On bottoms	On overhead
1. **Propane for sale** by deethanizing feed to a depropanizer	C_2/C_3	C_3
2. **Ethane for pyrolysis furnace** by fractionating furnace feed	C_2/C_3	C_3
3. **NGL vapor pressure** by recovering C_3+ from an expander gas plant	C_2/C_3	—
4. **Ethylene and ethane** separated from feed to sponge oil absorber (Separator not refluxed)	—	$C_3^=$
5. **Polymerization grade propylene** by deethanizing feed to depropanizer and propane splitter	$C_2/C_3^=$	$C_3^=$
6. **Same as Case 5** except propylene is not polymerization grade	$C_2/C_3^=$	$C_3^=$
7. **Propane for sale** by deethanizing overhead from a depropanizer	$C_2 + C_4$	C_3/C_2
8. **Ethane-propane mix** (overhead) **and NGL** (bottoms) from ethylene plant	C_2/C_3 with override on C_4 in O.H.	C_4
9. **Same as Case 8** except there is iC_4 spec. on overhead product	C_3	iC_4
10. **Same as Case 1** except there is C_3/C_2 spec. on overhead product	C_2/C_3	C_3/C_2

Note: Circled components are the ones used for primary composition specification.

cumulation of material in the condenser/accumulator, as would be indicated by a change in accumulator level.

• The liquid inventory in the bottom of the column will also change with time. Consequently, Equation 6 does not hold in the unsteady state. Although the internal bottom product rate B_I is equal to B at steady state, the difference $B_I - B$, in the unsteady state, is the rate of accumulation of material in the bottom of the tower.

Engineers recognized years ago[6] that specifying overhead and bottoms composition also fixed the overhead and bottoms flow rate due to the fractionator material balance at steady state. Changing these product flows will force a composition change.

A highly significant consideration in fractionator control is that the product compositions respond to the internal flows, not to external flows. It is true the internal flows equal the exterior flow when liquid levels in the overhead accumulator and reboiler remain constant. However, accumulators and reboilers are sometimes used as surge tanks for downstream units. Under these conditions, the internal product flows D_I and B_I become extremely important variables for control of a fractionator.

As indicated in Fig. 6, a change in the internal vapor flow affects both of the internal product flows almost immediately. However, there is a significant time lag in the response of the internal bottom product flow B_I to a change in the internal reflux flow due to the liquid flow dynamics of the individual trays. Because of these dynamic characteristics, there is a definite advantage for using internal vapor flow, rather than internal liquid flow, to control the internal product flows of a fractionator. The taller the column and the higher the reflux-to-feed ratio, the more pronounced is this characteristic.

The concepts of internal liquid and vapor flows can be extended to complex fractionators. These include multiple draw towers such as crude fractionators, main fractionators on fluid catalytic cracking units, recycle splitters in hydrocracking units and disobutanizers in alkylation units.

Controlling internal flows. The control of product compositions for a fractionator is primarily a matter of control of the internal flow rates within the fractionator. In considering product separation it is useful to define two terms. These are: the degree of separation and the orientation of separation.

The degree of separation is suggested by the Fenske[7] equation:

$$\text{Degree of separation} = \ln_e(D_L\,B_H/D_H\,B_L) \quad (7)$$

where: D_L = Mol fraction of light key component in the distillate
B_H = Mol fraction of heavy key component in the bottom product
D_H = Mol fraction of heavy key component in the distillate
B_L = Mol fraction of light key component in the bottom product.

The orientation of separation for a given degree of separation is defined as:

$$\text{Orientation of separation} = D_H/B_L \quad (8)$$

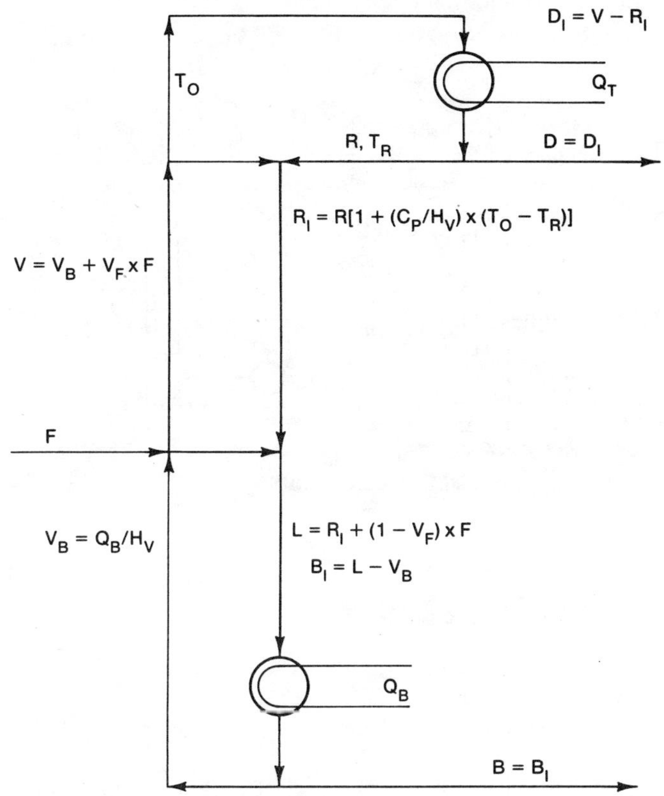

Fig. 5—Steady state internal flow model.

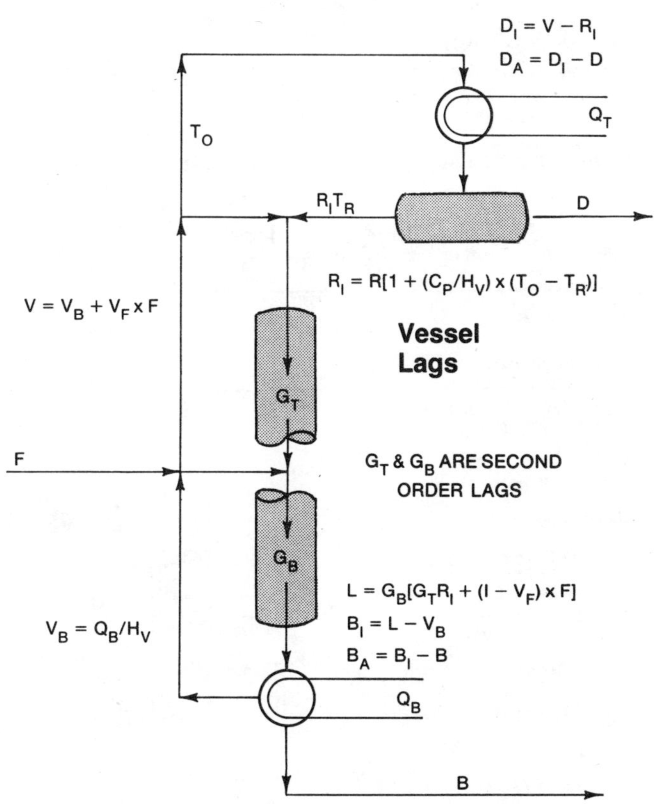

Fig. 6—Dynamic internal flow model includes time lags.

Conceptually, fractionator product composition control can be thought of as a problem of making precise adjustments to the rate of heat addition, Q_B, at the bottom of the fractionator and the rate of heat removal, Q_T, at the top of the column.

Physically, the following happens: the heat removed,

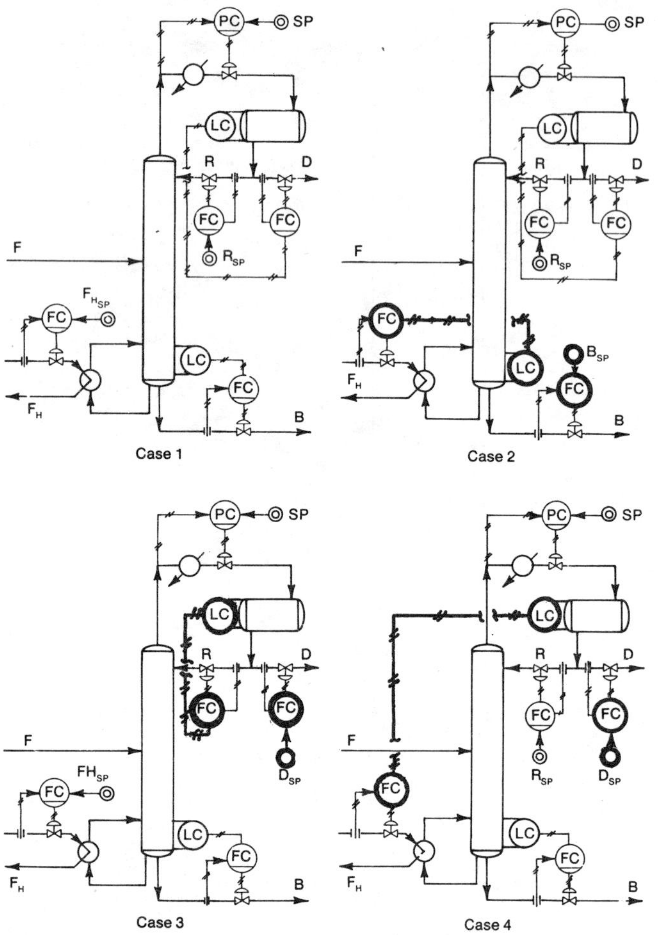

Fig. 7—Ways to use conventional controllers.

TABLE 2—Flow control equations for Figs. 7 and 8

	Eq. used by given case			
	1	2	3	4
Liquid flow rate, feed tray $L=G_B[F(1-V_F)+R_IG_T]$	●	—	—	—
Internal distillate product rate $D_I=F(O_D)$	—	—	●	●
Internal bottoms product rate $B_I=F(O_B)G_B$	●	●	—	—
Vapor boilup rate $V_B=L-B_I$	●	—	—	—
$=R_IG_T-V_FF-D_I$	—	—	●	—
Steam flow $F_S=V_B(H_V/H_S)$	●	—	●	—
Internal reflux $R_I=O_O+A_1F$	●	●	●	●
External reflux $R=R_I/[1+(C_P/H_V)(T_O-T_R)]$	●	●	—	●

Note: O_B, O_D and O_O are operator bias and analyzer controller outputs.

Q_T, from the top of the fractionator determines the internal reflux flow rate; the heat added, Q_B, at the bottom of the column determines the internal vapor rate below the feed tray, and the magnitude of these internal liquid and vapor flow rates determines the circulation rate within the column, which in turn determines the degree of separation between the two key components.

A precise balance must be maintained between the heat added at the bottom of the fractionator and the heat removed at the top of the fractionator. This heat balance determines the difference between the vapor and liquid flow rates in the top section of the column (the internal overhead product flow, D_I). This heat balance also determines the difference between the liquid and vapor flow rates in the bottom section of the column (the internal bottom product flow, B_I). The ratio between the internal liquid and vapor product flows determines the orientation of the separation, and strongly affects the composition of both overhead and bottoms product streams.

Four fractionators with conventional controls for flow and levels are shown in Fig. 7. The equations required for control of the internal flows for each of these configurations are derived from the equations in Fig. 6 and are given in Table 2. These equations are represented as control diagrams in Fig. 8. In each case, two internal flows are being controlled, and each control diagram provides a place for cascaded entry of an analyzer controller output, O_B or O_D. A feedforward signal is used to anticipate effect of feed flow rate changes on internal flows.

Table 2 and Fig. 8 illustrate that, for a given column design, different control equations for controlling internal product flow rates will apply, depending upon the configuration of conventional instruments used. Table 2 and Fig. 8 do not exhaust the possible arrangements of conventional instrumentation, but are merely illustrative.

Both theory and practical considerations are used in designing a product composition control system to assure the greatest return on investment. All of the above control arrangements have been proven and are in use.

Feed enthalpy and column pressure. Feed enthalpy and column pressure are sub-optimization variables. The manner in which these two variables are utilized to achieve two different sub-optimization objectives are illustrated by the following examples:

- With overhead and bottoms product compositions on control, the sub-optimization problem is to adjust feed enthalpy and column pressure to minimize utilities costs and thus increase fractionator profitability.

- With capacity limits over-riding the control of one product composition, the sub-optimization problem is to adjust feed enthalpy and column pressure to maximize fractionator profitability by partially relieving the capacity constraint.

A study of the internal flow model in Fig. 6 indicates that the primary effect of increasing feed enthalpy is to decrease vapor-liquid circulation below the feed tray relative to vapor-liquid circulation above the feed tray. When feed preheat is less expensive than reboiler heat, or when reboiler is limiting, maximum feed preheat is often optimal. When condenser capacity is limiting, or "flooding"

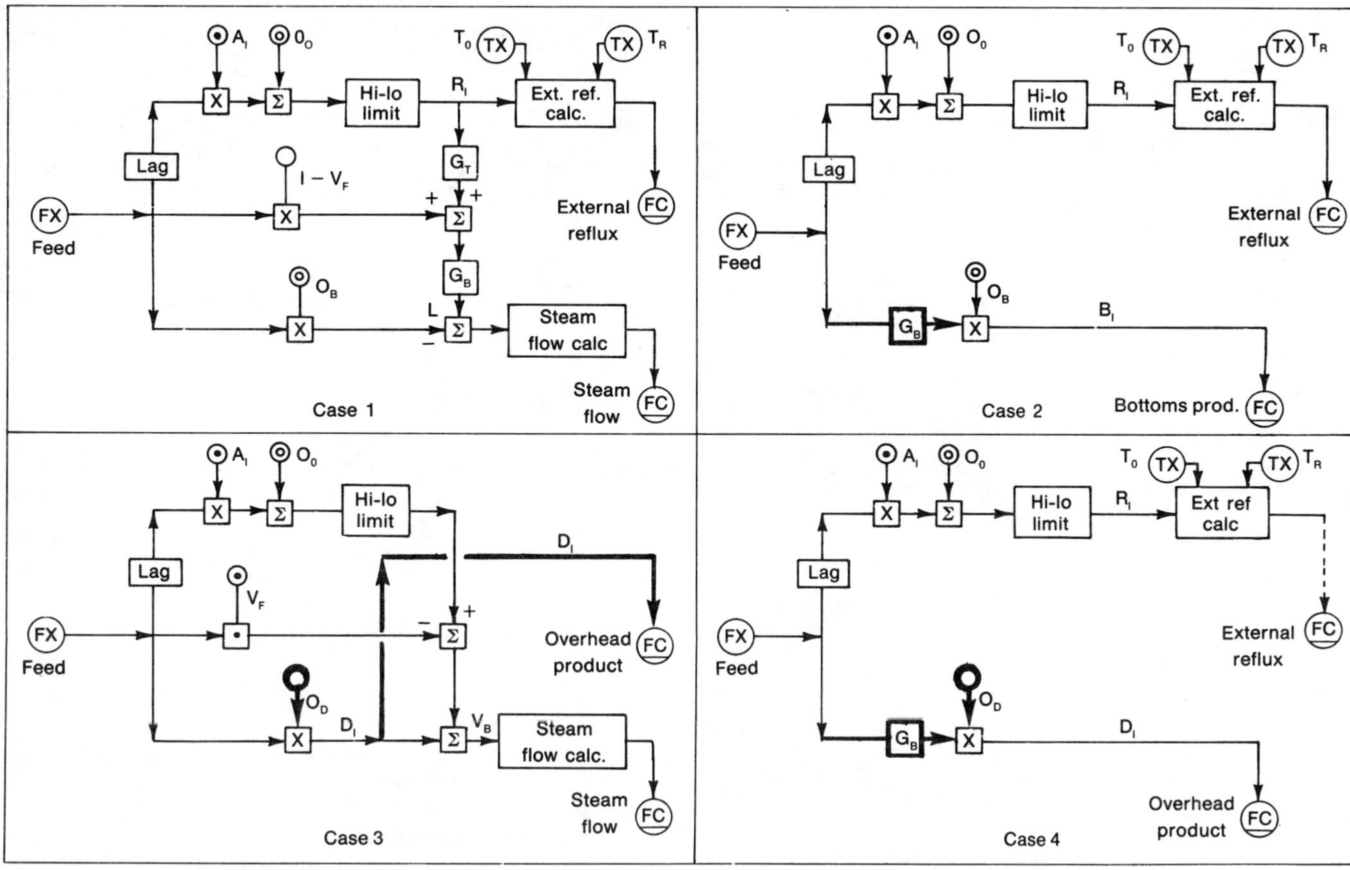

Fig. 8—Control configurations for the four cases.

is encountered above the feed tray, maximum feed cooling is often optimal.

Optimization of column pressure is a complex consideration[8] in which the dependence of the following factors on pressure must be included:

- Relative volatility
- Reboiler and condenser capacity
- Utility costs
- Vapor volume and tower loading
- Tray efficiency (viscosity, surface tension, vapor and liquid densities, etc.)
- Latent heats of vaporization.

Although the optimal column pressure is not always obvious, the effect of reducing column pressure is known to be advantageous for many applications including:

- Expander type demethanizers[9]
- Depropanizers
- Debutanizer
- Butane splitters
- Butane separation.

On the other hand, increased pressure was found to be advantageous for a C_3 splitter with a heat pump.

The C_3 splitter case illustrates the need to investigate carefully the optimum pressure, preferably with the use of simulation. For this purpose, fractionator design routines should be used instead of the simplified fractionator models discussed in this paper. Alternatively, the optimal column pressure can be determined by plant testing, but such testing should be conducted with both overhead and bottom product compositions on automatic control.

Pressure optimization can be applied dynamically using a valve position controller. Normally, one of the control valves in a composition loop is kept "nearly" open by adjusting the column pressure. However, it is important to realize that composition control must take precedent to pressure optimization, i.e., the pressure control loop is tuned to respond slowly in comparison to the composition control loops. Furthermore, potential effects of pressure changes on upstream and downstream units must be considered when dynamic pressure optimization is applied.

PRODUCT COMPOSITION CONTROL

For studying the economics and optimization of product composition control, a simplified steady state composition model of a fractionator has been developed.[4]

For small changes about a typical set of operating conditions, the following equations can be used to summarize the relationships of the product composition model:

$$B_I = F\,[(1 = F_{LL} - F_L - D_H)/(1 - D_H - B_L)] \quad (9)$$

and

$$R_I = (A_1 + A_2F)\,(F_H/D_H)\,(F_L/B_L) \quad (10)$$

where:

F_{LL} = Mol fraction of feed lighter than the light key component
F_L = Mol fraction of light key component in the feed
F_H = Mol fraction of heavy key component in the feed

and A_1 and A_2 are constants in the linearized internal reflux equation. These two equations constitute the feed-

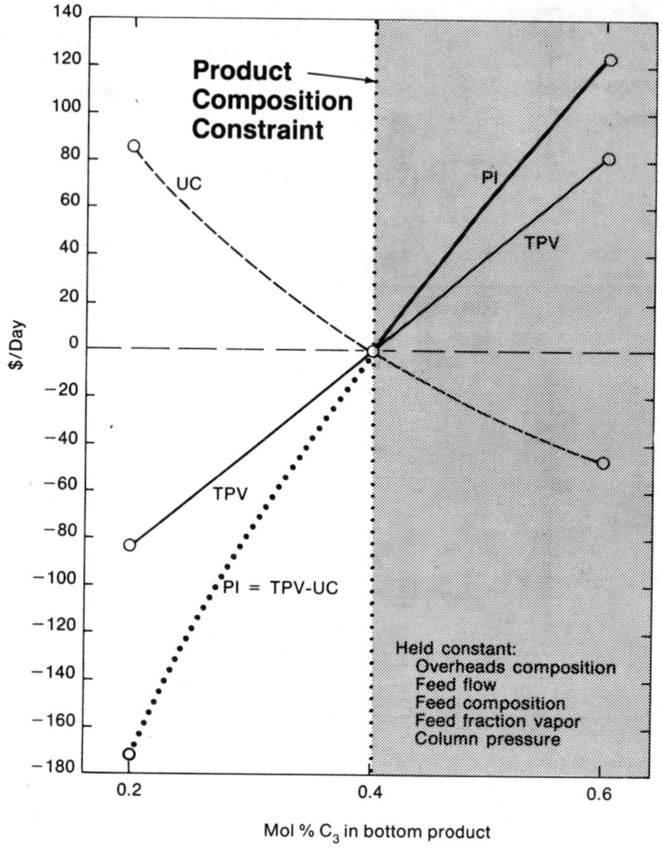

Fig. 9—A greater profit comes from having as much propane in the bottoms as is permitted.

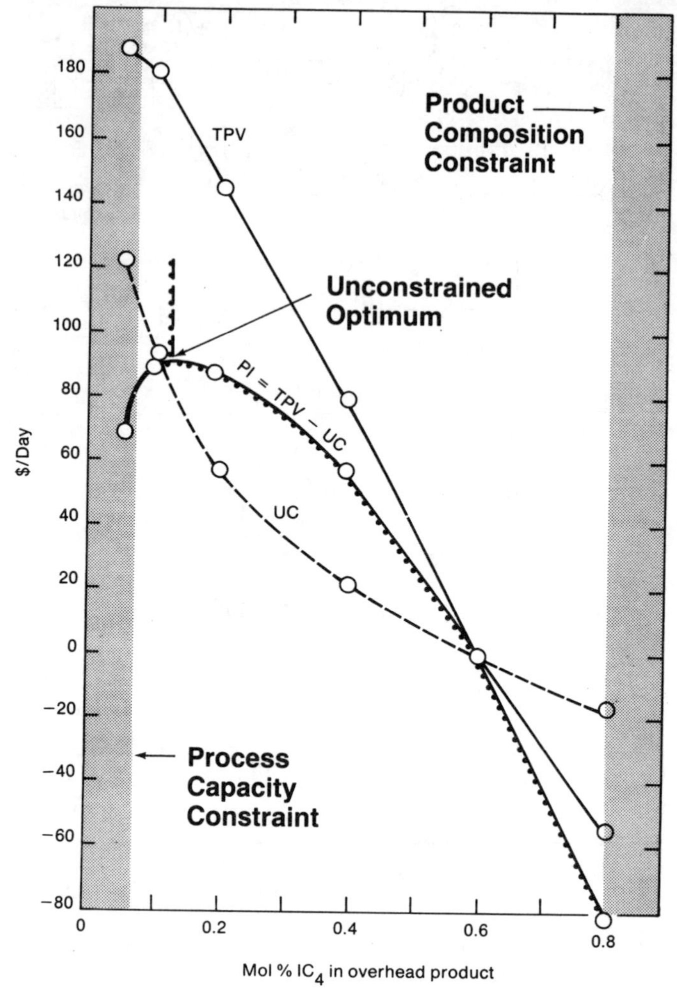

Fig. 10—The optimum profit is established by the amount of isobutane permitted in the depropanizer overhead.

forward form of the composition model. The equation for the internal bottoms product flow, B_I, is rigorous and is derived from a material balance. The equation for internal reflux R_I, is a linearized approximation, but will yield results which are reasonable in terms of magnitude and will always be accurate for predicting the direction-of-change from a base operating point.

If a more accurate estimate of internal reflux is required, a short-cut or tray-by-tray fractionator routine may be used. Reference (10) recommends two short cut methods: the minimum reflux-minimum trays correlation method of Fenske, Underwood, Erbar and Maddox; and the more rigorous method of Smith and Brinkley.[11] The Fenske-Underwood-Erbar-Maddox method is a feedforward method in which feed rate, feed composition, and product composition are specified and internal reflux is calculated. The Smith-Brinkley method is a process-model method in which feed rate, feed composition and internal flow rates are specified and product compositions are calculated.

The response of product compositions to changes in the independent process variables can be determined by solving the feedforward Smith-Brinkley model by trial and error for D_H and B_L as functions of B_I, R_I F, F_{LL}, F_L, F_H, D_H and B_L. However, the use of a trial and error solution does not lead to a better understanding of the fractionation process. The feedforward form can be linearized to a process model form which shows explicitly the effects of the independent process variables on the product composition:

$$B_L = b_1 + b_2F_L + b_3F_H + b_4F - b_5R_I + b_6B \quad (11)$$

$$D_H = c_1 + c_2F_L + c_3F_H + c_4F - c_5R_I - c_6B \quad (12)$$

$$B = (B_I/F) - [(1 - F_{LL} - F_L - D_{HO}/(1 - D_{HO} - B_{LO})] \quad (13)$$

In Equation 13, the subscript O denotes base operating conditions. The process model form indicates that increasing bottom product flow with respect to overhead product flow causes an increase in the light impurity concentration in the bottom product and a decrease in the heavy impurity concentration in the overhead product. Increasing vapor-liquid circulation rates causes a decrease in both impurity concentrations.

Value of control. Economic calculations for fractionator operations must be tailored to individual applications because the calculations depend not only on the physical model of the fractionator but also on other important factors. The economic calculations must include the effect of product composition on product value and thus they depend on the terms of the product sales agreement. If the fractionator products are not final sales products, or if the fractionator product compositions do not determine the quality of final sales products, the economic calculations may need to include the economic effects of fractionator product compositions on downstream processes.

The economic calculations for a very simple application are shown in Fig. 9 in which the fractionator products are final sales products, the product sales agreements are based on maximum impurity specification and the predominant utility cost is the cost of reboiler heat. For such an application, the utilities cost U_C is computed by

$$U_C = A_3 Q_B, \tag{14}$$

the total product value *TPV* is computed by

$$TPV = A_4 B + A_5 D, \tag{15}$$

and the profit index P_I is computed by

$$P_I = TPV - U_C \tag{16}$$

where A_3 is the unit cost of reboiler heat, A_4 and A_5 are unit product values, Q_B is the reboiler heat flow and *B* and *D* are product flows. The absolute value of the profit index has no significance, but a change in profit index from a base operation point should represent the change in income for the plant.

A simulation of a depropanizer gives the steady state data plotted in Fig. 9 and 10. Fig. 9 shows the economic effects of propane concentration in the bottoms product. As the propane concentration increases in the bottom product, bottom product flow increases and internal reflux flow decreases as shown by equations above. Since, in this case, the bottoms product of the depropanizer is more valuable than the overhead product, total product value increases with the increase in propane concentration in the bottoms product. The decreasing internal reflux flow results in decreasing reboiler heat flow and decreasing utility cost. Requirements for condenser capacity, reboiler capacity and column liquid and vapor flow capacity also decrease with decreasing internal reflux flow; therefore, the optimal value for propane concentration in the bottoms product is the maximum impurity specification as shown in Fig. 9.

Fig. 10 shows the economic effects of isobutane concentration in the overhead product. As the isobutane concentration increasees in the overhead product, bottom product flow decreases and internal reflux flow decreases as shown by the equations above. The resulting decrease in total product value, combined with the opposing decrease in utility costs, causes a maximum in the profit index curve of Fig. 10.

NOMENCLATURE

- A_1, A_2 Arbitrary constant internal reflux loading curve
- A_3 Unit cost of reboiler heat
- A_4 Unit value of bottom product, \$/mol
- A_5 Unit value of overhead product, \$/mol
- B Bottoms flow rate, mols/hr.
- B_A Accumulation of bottoms product, mols/hr.
- B_I Internal bottoms product flow, mols/hr.
- B_H Mol fraction heavy key in bottoms product
- B_L Mol fraction light key in bottoms product
- C_P Heat capacity of reflux, Btu/(mol) (°F)
- D Overhead product flow rate, mols/hr.
- D_A Accumulation of overhead product, mols/hr.
- D_I Internal overhead product rate, mols/hr.
- D_H Mol fraction heavy key in overhead product
- D_L Mol fraction light key in overhead product
- F Feed flow rate, mols/hr.
- F_H Mol fraction heavy key in feed
- F_L Mol fraction light key in feed
- F_{LL} Mol fraction lighter than light key in feed
- G_T Column dynamic function, top section
- G_B Column dynamic function, bottom section
- H_V Heat of vaporization, Btu/mol
- H_S Heating value of steam, Btu/mol
- L Internal liquid flow rate below feed tray, mols/hr.
- O_B Bias signal from bottoms analyzer
- O_D Bias signal from overhead analyzer
- O_O Bias signal supplied by operator
- P_I Profit index, \$/hr.
- Q_B Heat added through reboiler, Btu/hr.
- Q_T Heat removed through condenser, Btu/hr.
- R External reflux rate, mols/hr.
- R_I Internal reflux rate, mols/hr.
- T_O Temperature of overhead vapor, °F
- T_R Temperature of external reflux, °F
- *TPV* Total product value, \$/hr.
- U_C Utilities cost, \$/hr.
- V Vapor rate above feed tray, mols/hr.
- V_F Fraction of feed which is vaporized
- V_B Vapor rate below feed tray, mols/hr.
- $b_1, b_2 \ldots$ Constants in linearized equation for estimating B_L
- $c_1, c_2 \ldots$ Constants in linearized equation for estimating D_H
- ΔC Change in product composition resulting from change in reflux by an amount ΔR around reflux operating point R.

LITERATURE CITED

1. Wherry, T. C. and Parsons, J. R., "Guide to Profitable Computer Control," *Hydrocarbon Processing*, Vol. 46, No. 4, April 1967, pp. 179-182.
2. Funk, G. L., and Smith, D. E., "Estimating Economic Incentives for Computer Control Systems: An Application Approach," IEEE 24th Annual Petroleum and Chemical Industry Conference, September 1977, Dallas, Texas.
3. Rutledge, R. W. Personal communications within Applied Automation, Inc., Work to be published.
4. Hobbs, J. W., "Unifying Concepts for Fractionator Product Composition Control," 1977 Joint Automatic Control Conference, San Francisco.
5. Perry, R. H., and Chilton, C. H., editors, *Chemical Engineers' Handbook*, 5th ed., McGraw-Hill Book Co., 1973, pp. 22.122-22.128.
6. Badger, W. L., and McCabe, W. L., *Elements of Chemical Engineering*, McGraw-Hill Book Co., 1936, p. 345.
7. Fenske, M. R., *Industrial and Engineering Chemistry*, Vol. 24, 1932, p. 482.
8. Clay, H. A., et al., "Effect of Load and Pressure on Performance of a Commercial Bubble Tray Fractionating Column," *Chemical Engineering Progress*, Vol. 50, No. 10, October 1954, p. 517.
9. Buss, R. A., and Griffin, D. E., "Modern Control Concepts for Gas Processing," 54th Annual GPA Convention, March 1975, Houston; Also *Oil & Gas Journal*, Vol. 73, No. 20, May 19, 1975, pp. 113-115.
10. Fair, J. R., and Bolles, W. L., "Modern Design of Distillation Columns," *Chemical Engineering*. Vol. 75, No. 9, April 22, 1968, pp. 156-178.
11. Smith, B. D., *Design of Equilibrium Stage Processes*, McGraw-Hill Co., 1963.

A Better Way to Control Composition

With a distillation column example, presented are the justifications for and economics of composition control using gas liquid chromatographs. Also included is an in-depth description of the control systems sample data controller, tuning considerations and special integral logic controller for minimizing oscillations.

B.D. Stanton, Tex-A-Mation Engineering Inc., La-Porte, Tex. and **M.A. Sterling,** Honeywell Inc., Fort Washington, Pa.

CLOSE COMPOSITION CONTROL in petrochemical and petroleum processes is more essential each day due to the large increases in utility and feed stock costs. Conventional methods such as controlling temperatures, pressures, levels and flow rates to maintain product purity are no longer adequate. On-line analytical instruments are essential if optimal use of utilities and feed stocks is desired. This article presents methods for automatic composition control of hydrocarbon products from a distillation tower using on-line analytical instruments such as gas-liquid chromatographs.

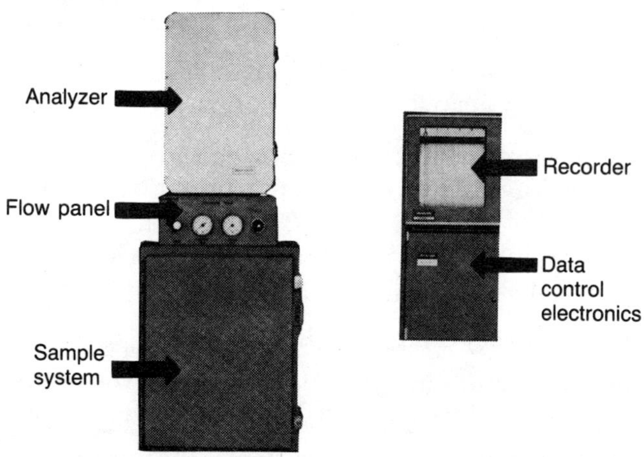

Fig. 1—Typical gas-liquid chromatograph with accessories.

Measurement device. Before control there is measurement. The gas-liquid chromatograph (GLC) provides a direct percentage measurement of components in a hydrocarbon stream. This instrument is widely accepted in automatic control of distillation towers. A modern GLC is shown in Fig. 1. This illustration shows the sample system, flow panel and analyzer which are normally mounted in the field, and the data control electronics and recorder which are usually located in a less harsh environment, such as the control room.

In this article a fluidized catalytic cracking unit gas plant depropanizer is used to illustrate the benefits of automatic composition control. The measuring device is a GC (gas-chromatograph), since the lighter hydrocarbons being analyzed are easily vaporized. Also, gas samples are more easily handled than liquid samples. The measurement techniques and control theory described here are general in nature and can be used for almost all distillation tower GLC control applications. Also, for the purposes of this discussion, the GLC is considered a black box which puts out sampled-data-type electrical signals which are proportional to the percent of the components in the stream. These measured components are called the key or control components. Signals from the GLC may also be printed out in the form of a spectrum, as illustrated in Fig. 2. This is a typical C_2 to C_5 spectrum showing the volume percent of each component analyzed. Selected key components of this spectrum are trend recorded and used as control signals.

Process to be controlled. The depropanizer tower in Fig. 3 separates propanes (C_3s) and lighter components from butanes (C_4s) and heavier components. The C_3s and lighter compose the top product; the C_4s and heavier, the bottom product. For many years these towers have been controlled by adjusting heat input (+q) and heat withdrawn (–q), using temperature and pressure measurements to maintain approximate composition of the products. Historically, pressure in the top of a tower is controlled by varying cooling water to the overhead condenser. Pressure control here is essential for safety, and technically it sets one of the variables which controls separation. Temperature is controlled

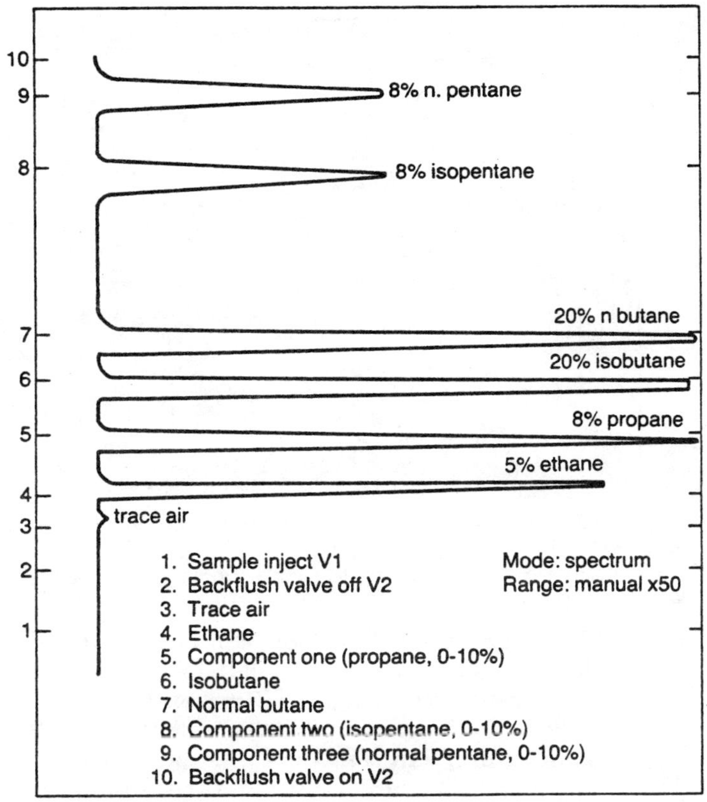

Fig. 2—C_2 to C_5 spectrum.

by heat input to the feed preheater and reboiler. Temperature measurement with fixed pressure is a rough indication of a stream's composition. Reflux to the tower is set by an operator at a high, safe value and the system operates with little problems. From time to time, the operator takes samples of the top and bottom products and sends them to a control laboratory for analysis. Within 8 to 16 hours, the operator will have the results of the analysis, and based on this analysis he will usually adjust the temperature controller's setpoint. This procedure has worked well for many years and is still in use in many plants today. This acceptance raises the question: why change to a measurement and control system which could cost up to $20,000 installed; approximately 10 times the installed cost of the temperature and pressure control loops? Also, the additional equipment is more complex and could cost up to 10 times as much to maintain.

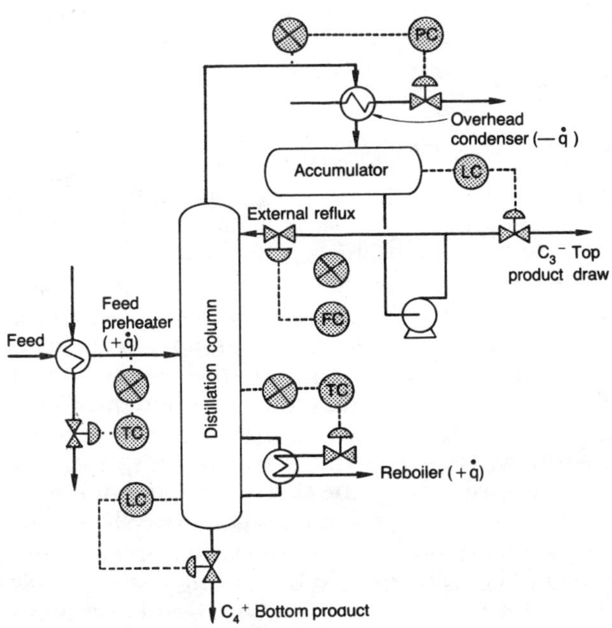

Fig. 3—Typical depropanizer tower.

Justification for composition control. Let us look at some of the answers to these questions. First, the cost of feed stocks has dramatically increased in the past five years. Therefore, it is extremely important to recover the maximum amount of on-specification products from the operated feed stock. Second, the cost of energy has skyrocketed in the past five years and will be much higher in a few more years. Third, process designs are more complex. Fourth, greater control system density is common. Operators who in the 1940s watched up to 20 to 30 controllers are now responsible for several hundred control loops. These operators now have little time for sample grabbing and running composition analyses. Fifth, modern plant operations are going under computer control, which means optimizations are run more frequently and operating objectives are continually changed. As a result, product specification and feed flow rates are constantly changing and composition analysis, which is performed once every 8 to 16 hours, is inadequate.

Great progress has been made in sample systems and on-line analyzers. In the 1950's systems were normally unreliable and cost more to buy and install—approximately $20,000 versus today's high reliability and lower cost of $15,000—and they were sold without sample systems and without such features as auto zero. Because of technical advances in electronics and physical chemistry, today's analyzer is lower in cost than the GLCs of the 1950s—$7,000 versus $12,000. They are also more easily installed, again reducing costs. Some analyzers can be mounted outdoors in the process area. Also, they are capable of being controlled by an optimizing computer. This means that the old statement, "Yes, I would like to have an on-line analyzer, but I can't see a payout or keep it running," no longer applies. In addition, the computer systems and algorithms which could best use these analyzers for automatic control did not exist. However, with the advent of microprocessor-based controllers and control computers, both of which are widely accepted today, control schemes which make maximum use of on-line analyzers are an engineering reality.

Economics of composition control. In reviewing the economics of using an on-line analyzer for composition control on a typical depropanizer tower (Fig. 3), the following assumptions are made:

- Reboiler heat input is 40 MM Btu/hr.
- The top key component is isobutane (iC_4).
- Temperature in the bottom of the tower is controlled by reboiler heat input.
- Temperature in the top section of the tower is controlled by reflux and that heat load is directly proportional to reflux flow rate.

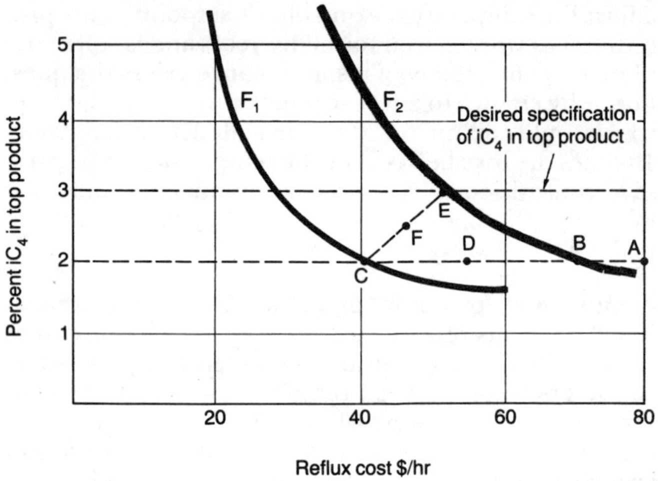

Fig. 4—iC_4 loss vs. reflux cost.

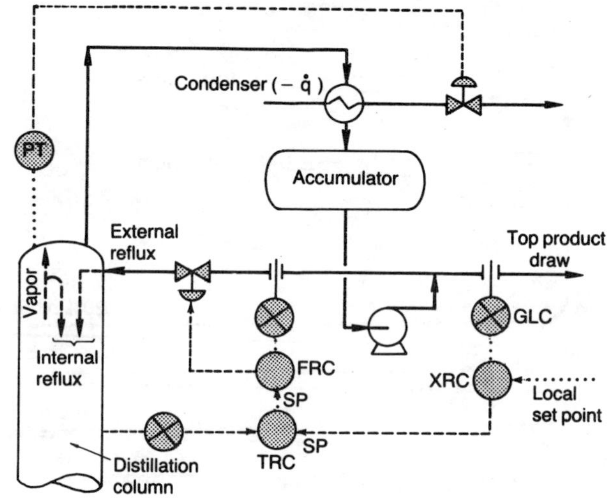

Fig. 5—Simplified single-loop multi-cascade control system for iC_4 losses.

►Heat costs $2.00/MM Btu.

Fig. 4 shows a curve of iC_4 loss in the top product as a function of reflux cost for two different feed rates ($F_1$and F_2) which span normal feed variations. Assuming top product specification is 2 percent iC_4, the operator in the "good old days" of $0.18 per MM Btu energy would set the reflux at the very safe operating point "A" and basically forget the column control. The column operated satisfactorily until feed increased to the point at which the column flooded, a safety relief valve would go off, or some other abnormal operation would occur. With today's energy costs, this would mean an approximate $1,920/day fuel bill. If an on-line analyzer controlled reflux flow rate, the flow rate would vary from points "C" to "B" in Fig. 4, normally running at an average at point "D" and saving approximately $500 per day in energy costs. In 40 days savings in energy alone would pay out a $20,000 analyzer installation. This does not include savings due to reduced operator attention, reduced control laboratory analysis, and reduction in off-specification product. Further savings also can be realized if an optimizing computer is running on line. Such a computer compares the cost of iC_4 losses from the top of the tower with the energy costs. It then shifts the analyzer controller setpoint to the best possible value, following a line similar to that at points "C," "F" and "E." This gives additional savings of approximately $1,500 per month using today's product costs. The key to these savings is accurate, reliable, closed-loop composition control.

Closed loop composition control system. A simplified diagram of a single-loop multi-cascade control system for the iC_4 losses from the top of the depropanizer is shown in Fig. 5. The flow controller (FRC) regulates an exact liquid volumetric flow for each setpoint. It eliminates disturbances from changes in the pump discharge pressure due to top product draw changes or column pressure changes. The flow control system is quite fast, taking only 5 to 10 seconds to line out. It uses a standard proportional/integral controller, which normally has a low gain (0.5 to 1.0) and a fast integration time (0.3 to 0.03 minutes per repeat). This control is well understood and easily configured.

The temperature controller (TRC) should be very sensitive (20 to 50°F span) and the tray in the column where the measurement is made should be selected to give a good indication of iC_4 loss out of the top of the tower. This is possible because column temperature is related in a non-exact and non-linear fashion to composition. The temperature control will line out in 2 to 8 minutes. Temperature control is also extremely useful; it fights feed disturbances in 5 to 30 minutes and moves reflux in the correct direction at an early point in time. If temperature in the top of the column rises, reflux rate increases. This action will reverse as the top temperature in the column goes down. Also, it will quickly sense changes in internal reflux, which is a function of external reflux and its temperature. This is especially true if the measuring element has a narrow span (20 to 50°F for 0 to 100 percent output). The temperature controller should be proportional only and tuned with a gain of 2 to 5. Proportional offset caused by the absence of reset is minimized by the high gain and is not critical because the analyzer controller will move the setpoint to compensate for this offset. However, if the analyzer controller is switched off, the temperature controller should be reconfigured to have a small amount of integral action (0.1 minutes). This is done manually by the operator, or automatically if a computer is used.

The analyzer controller is slow and will line out in 1 to 3 hours. It cannot fight fast disturbances. Therefore, the flow and temperature controllers relieve the analyzer controller of having to fight disturbances in the range of 1 to 30 minutes. The analyzer controller, however, is still essential to continuously bring product specification to the desired level. As mentioned before, the analyzer setpoint is available to an on-line optimization program which works at a much slower rate of 8 to 24 hours. The reason for the slow optimizing rate is that if product specification or feed rate to a distillation tower is changed too often by an optimal program, the optimization will become confused by its own moves. Good operators make a minimum number of changes and wait to see the results before making another.

The GC analyzer controller is a sample data device

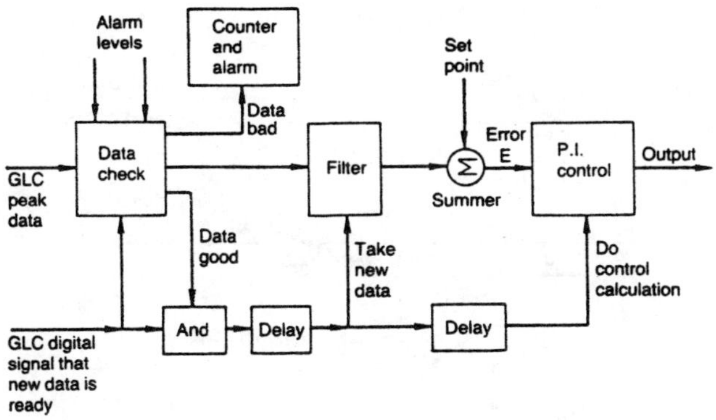

Fig. 6—Sample data controller.

which uses proportional and integral control functions. Its input information is not continuous like the flow and temperature controllers. The GC takes a sample, runs an analysis, and then outputs the results. Samples are normally taken once every 2 to 5 minutes. This means that the best information the controller has can be 2 to 5 minutes old. Because of this, a sample data controller, as shown in Fig. 6, is used with the GC analyzer. The sample data controller does not work continuously, but makes a computed control move after each new datum input.

First, the sample data controller checks the new datum from the analyzer to see if it is realistic when compared with past data. Alarm limits are set by the process engineer specifying the percent concentration that a component normally will increase or decrease. If the datum is outside limits, it is rejected and the controller works on the older data and indicates that the new datum is bad. This step is very sophisticated and requires a great deal of knowledge about the process and the analyzer. For example, from experience it may be known that iC_4 concentration in the top product normally goes up rapidly (1 percent/minute), but when the iC_4 concentration decreases, it does so slowly (0.1 percent/minute). A rapid increase would be accepted as normal data and a rapid decrease would be rejected. Also, if experience indicates that 90 percent of the analyzer failures gave consistently low readings of approximately 0.5 to 0.1 percent of the key iC_4 component, and if normal analysis of the key component is 2 percent ± 1 percent, concentrations below 0.5 percent would be rejected and a report of possible analyzer problem would be given to the operator so he could watch the system closely. If this problem persists, it may be necessary for a mechanic to be called, and the operator would place the column in a safe, less economic operating position.

Second, the datum, after it is checked for validity, is filtered and compared with the setpoint to develop an error signal for the controller. Filtering smooths out the new datum. Sampled data often steps back and forth with small amplitude changes caused by sample stratification or other unknown causes.

Third, the controller takes the error signal, multiplies it by proportional and integral factors and outputs the result from this calculation. Control action is based on the most up-to-date, error-checked, filtered input datum. The controller then waits for a new piece of datum. Because the controller is paced by the analyzer, it does not continue to integrate the old error signal over the entire sampling period as do the flow and temperature controllers. Because the composition process is so slow (1 to 3 hours), the controller does not have to hurry, but can wait for the relatively fast sample data from the analyzer. Fig. 7 shows a step-by-step diagram of how the special control system operates:

- A recorder outputs the bar graph of the key components
- The peak picker in the GLC selects and picks the peak of the key component desired
- This datum is compared to the last good sample and the alarm limits set in the controller
- If the datum is bad, it is ignored and an alarm may be sounded. The controller holds its last good output for control action
- If the datum is good, it goes through the filter
- Proportional and integral control are then executed using the latest valid datum.

Additional control sophistication can be added to the sample data controller by modifying the integral control function. Very long dead times and large time constants are characteristic of composition control systems. The controller with a continuous linear integral control function tends to make the loop oscillatory. This oscillation is not important if the system lines out within 5 to 10 minutes, but can be undesirable if the cycle takes 8 to 12 hours to damp out at a period of 1 hour/cycle. By reducing this line-out time to 1 to 2 hours, the integral logic system is sometimes a useful addition to the control.

Control functions of an ordinary PI (proportional and integral) controller and the special logic controller

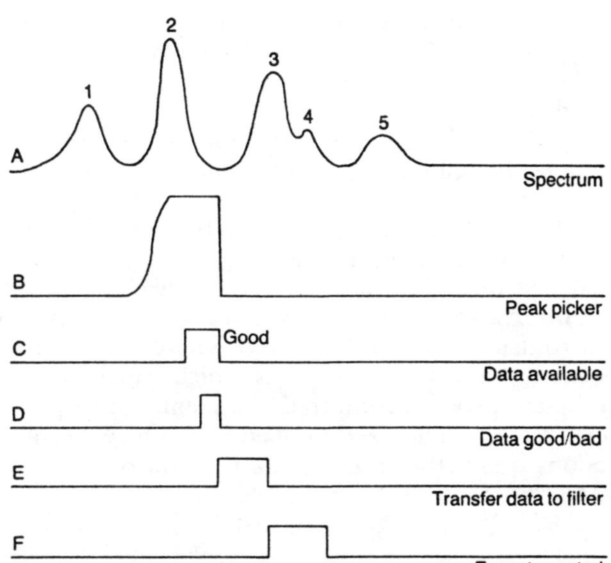

Fig. 7—Sample data controller operation.

are diagrammed in Fig. 8. Curve A shows a load change which causes concentration of the key component to rise. The output response of PI and special logic controllers, shown in curves B and D, moves from 0_1 to 0_2 to compensate for the load change. As shown in curve C, the error stops rising shortly after the controller output reaches point 1. As the error returns to zero, the proportional output starts to cut back. However, at that point where the error curve starts to turn, the error is still positive. With the conventional PI controller, integral action continues to push the output up and drives it too far above the desired value. The error will swing negative and an oscillatory response will result as shown in curve B. A computer monitoring the rate of error change would either switch the integration off, or switch to a much lower integral control function as the error slope changed sign. The result: overshoot is minimized and would be acceptable.

The special logic controller, as shown in curve D, runs the integrator as long as the slope times the error (*E*x*S*) is positive (conventional PI control); when it is negative, the controller will hold or reduce the integral time constant by a factor of 10 or greater. Therefore, as the error curve turns toward zero the slope is negative and the special logic controller stops integrating the error and holds during this time, thereby preventing a large overswing. This is most important when line out takes from 3 to 5 hours.

As stated previously, the composition analysis process is slow. Tests[1] were conducted to see how fast the GC should sample. Samples were run in an infrared (IR) analyzer which gave a continuous analysis, and a GC which gave a sample data analysis. Each used the same sample line dead time and the same distillation tower composition dynamics. Fig. 9 shows a typical curve of the GC sample time versus the integrated controllers error squared, which is the method selected for rating performance of the control system. This curve indicates when sample time control is too long. An error signal is zero when the measurement (or process variable) is equal to the setpoint. The error is squared to give the positive and negative errors the same weight, and to penalize larger errors more than small errors. Based on the data of Fig. 9, rates of 2 to 5 minutes are completely acceptable for distillation tower control. For 5-minute and larger sample rates, automatic control deteriorates rapidly.

Fig. 10 shows the depropanizer column using a GC analyzer for controlling composition of the top product. This is a minimum system commonly used when top composition is important and feed flow rate changes are slow and small in amplitudes.

Where feed flow rates are constantly changing, local optimal control, such as shown in Fig. 11, may be used. Two outside inputs are required. These are the top and bottom product specifications which can be set by a complete process optimization system. This optimization system will not be described here. Once the specifications are set, the local optimization controller can start its work.

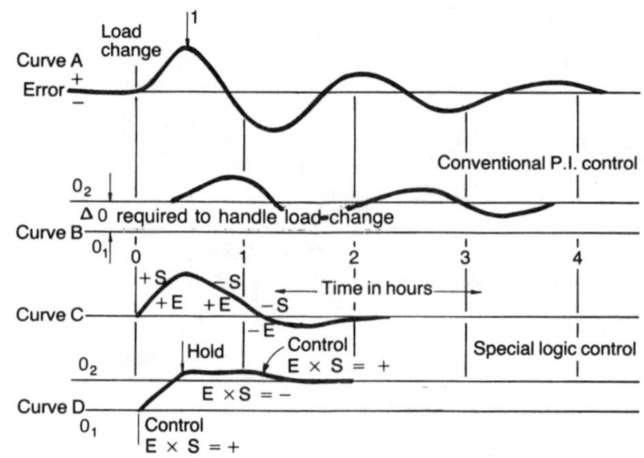

Fig. 8—Control functions of a conventional PI and the special logic controllers.

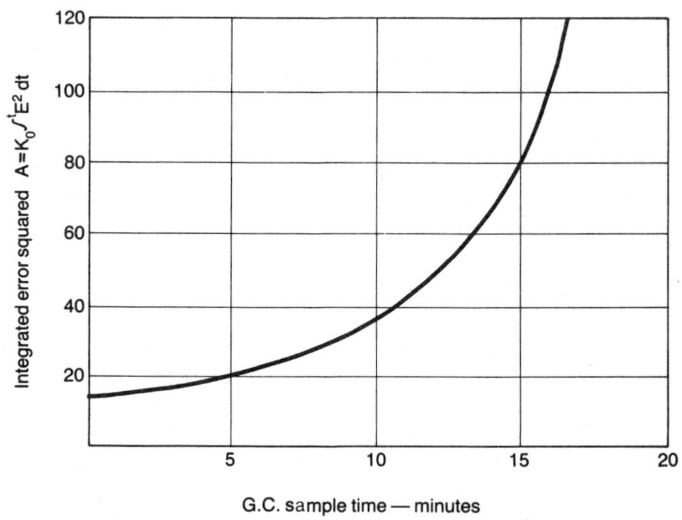

Fig. 9—Rating control performance based on G.C. sample time.

Discussion of variable pressure constraint control. There are three levels of control on this column: first, the front-line controls for pressure, level, flow and temperature; second, the feed preheat computer and composition controls; and third, the constraint or optimal controller (columns often operate against a constraint or limit).

This column uses a partially flooded condenser, which means heat removal is controlled by adjusting level in the condenser (or heat transfer area in contact with the fluid). Heat input to the reboiler is controlled by pressure. Reboiler and feed heat inputs must be enough to drive vapor over the top of the column for reflux and top product draw. If the top product flow rate is increased, level in the top condenser will drop, causing more heat transfer area to be exposed. More overhead vapor from the column will condense, lowering the temperature, causing column pressure to drop. The pressure controller would call for more heat to the reboiler, causing more vapor to go over the top of the column and a new stable condition would be realized.

If the top temperature is too low, iC_4 concentration in the top product will also be too low. The temperature controller would then increase the top product draw, causing more heat input to the reboiler, driving more iC_4 up the column. The bottom temperature controller

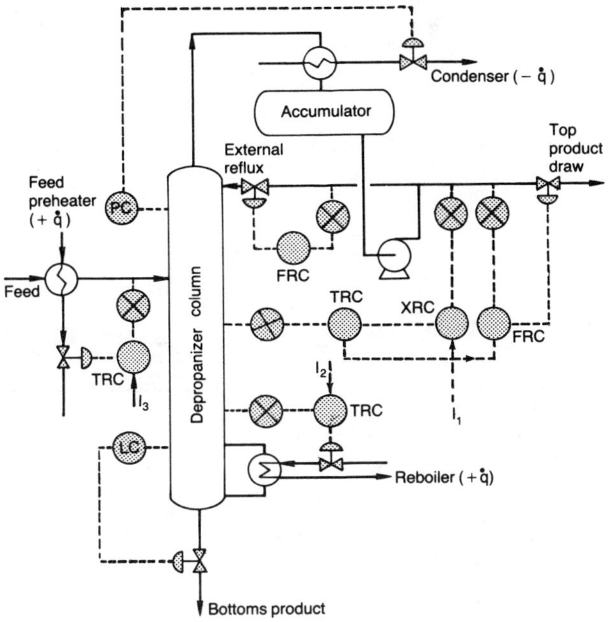

Fig. 10—Depropanizer column with G.C. for controlling top product composition.

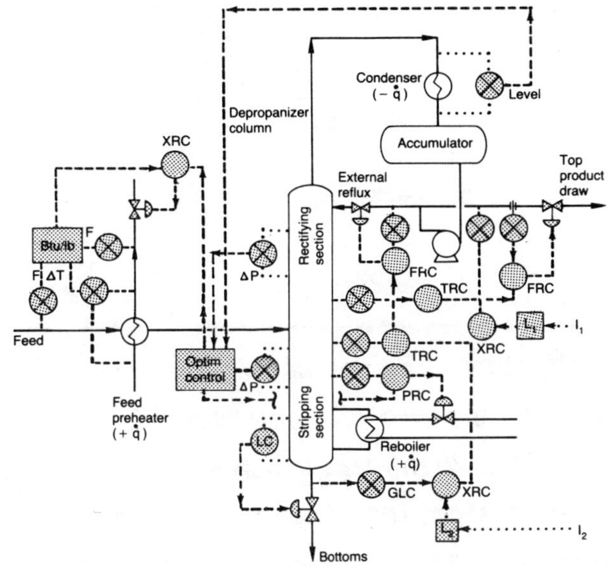

Fig. 11—Optimal control for changing feed flow rates.

works in a similar fashion. If bottom temperature is too low, it would indicate light ends (or too much C_3) in the column bottom. The controller would increase reflux, which would lower the top temperature and cause heat input to the reboiler to change as described above. The increased vapor and liquid flow rate would increase separation efficiency. It must be noted that bottom temperature control is not operating totally on the long reflux cascade time lag, but does feed to the reboiler via the pressure controller. There must be careful tuning of top and bottom temperature controllers (proportional only) because of serious interaction between the two. For these controls to work, the pressure control, which is fast compared to the temperature controllers, must be automatic. The bottom level control maintains the column material balance. As stated earlier, if pressure is maintained at a fixed level, temperature is a crude composition controller.

Composition controllers set the temperature controllers' setpoints to make process control more exact. I_1 and I_2 in Fig. 11 are the specifications set by an overall on-line optimal controller. The specifications must be realistic; therefore, limiters (L_1 and L_2) are used to allow the column to operate as designed. The optimal or constraint controller operates the column at its lowest pressure to conserve energy. It does this by lowering the pressure until a limit in the ability to condense vapor is detected by a level transmitter across the overhead condenser. When this level reaches 10 to 20 percent (determined by theoretical calculations), the controller stops lowering the pressure. The optimization system then looks at the stripping (lower portion of column) and rectifying (upper portion of column) sections' share of the load. To test this, the feed preheat (Btu/lb) is changed until condenser level begins to rise, permitting further pressure decreases. The correct feed preheat which gives the lowest pressure is found. A problem that may arise is the column may flood because of tray loading at the lower pressure, therefore, differential pressures across selected sections of the stripping and rectifying sections are measured, and when either of them reaches a predetermined high level, the lowering of pressure stops. These are constraints to the control scheme. With variable pressure operation, as just described, it has been found that heat savings of up to 10 percent ($5760 per month for a 40 MM Btu/hr column) can be realized. If specifications are to be held with variable pressure operation, composition must be controlled by on-line composition analyzers, because temperature is no longer a rough indication of composition.

ACKNOWLEDGMENT

Originally presented at the 25th Southeastern Conference and Exhibit of ISA, Mobile, Alabama, April 17-19, 1979.

LITERATURE CITED

[1]Stanton, B.D. and Bremer, A., "Composition Control of Distillation Columns," ISA presentation, 1962.

Analyzer Steam Tracing . . . Do It Right

Maintaining analyzer sample vapor at a high temperature all year round is much more difficult than just winterizing. Here's how to design, install and maintain a tracing system to insure your sample will be correct

E. W. McWhorter, Consultant, Longview, Texas

AN ON-LINE GAS ANALYZER for vapor with a dew point above ambient temperature requires heat tracing to keep the sample stream hot to the analyzer. Preventing condensation of a hot vapor year-round is much more difficult than more familiar steam-tracing applications, such as winterizing instruments. If you pay attention to some fine details, your system will work right the first time, saving you endless trouble later on.

Prefabricated steam-traced bundles are often the best way of handling this problem. Electric heating may be impossible due to a hazardous atmosphere and steam may be more convenient and economical. Furthermore, prefabricated bundles save you the expense and uncertainties associated with tracing and insulating a sample line in the field.

Typical requirements. A complete coverage of the subject of steam-tracing analyzer lines would require a textbook. For this discussion, however, let's limit ourselves to the fairly typical installation shown in Fig. 1. Although this is a process gas chromatograph, the steam tracing requirements are similar to those for many other instruments analyzing hot vapors near the dew point, such as flue-gas analyzers.

The sample stream is reduced in pressure to about 15 psig by a steam-heated regulator, which also serves to vaporize the sample if it is originally a liquid. The sample vapor flows continually at usually a few liters per minute through a couple hundred feet of stainless steel tubing to the analyzer. The analyzer is centrally located among several sample points that it analyzes in an automatic sequence.

Let's assume the dew point in the sample tubing is about 100° C and that all components of the vapor must be analyzed. That is, we must keep the vapor above this temperature all the way to the analyzer to prevent condensation of components that must be analyzed.

Why use a prefabricated bundle? A pre-insulated bundle should be used for this purpose. Otherwise, you'd have to make your own insulated and weatherproofed bundle. That might be okay for runs of 6 or 8 feet, if you happened to have the material on hand. But you've got to be sure the crew will stretch both tubes straight, tie them tightly together every few inches, and apply the right kind of insulation and weatherproofing in the right way. To make your own bundle along hundreds or thousands of feet of sample line would cost at least twice as much as the factory-made bundle. Furthermore, there might be any number of concealed defects where the sample vapor could condense, such as poor contact between the tubes, gaps in the insulation under the jacket, places where rainwater could leak in and accumulate and countless potentially leaky tubing unions. You could easily double your expense in just tracking down, ripping out and fixing these defective spots, after the analyzer is supposed to be in operation.

On the other hand, pre-insulated bundles provide uniform contact between tubes, insulation pre-engineered for the purpose, and a flexible weatherproof jacket. Such bundles vary in the number, size and composition of the

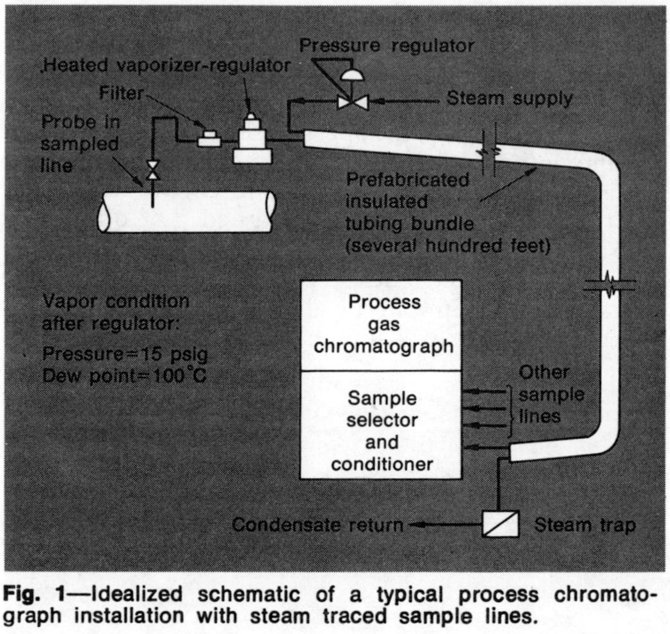

Fig. 1—Idealized schematic of a typical process chromatograph installation with steam traced sample lines.

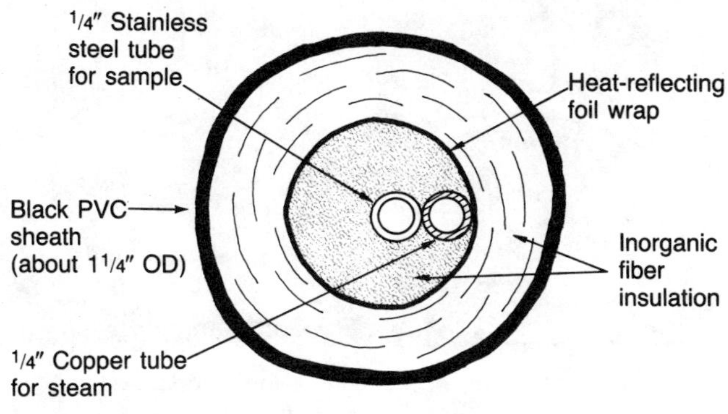

Fig. 2—Cross section of a typical pre-insulated tubing bundle.

tubes, in the insulating effectiveness and in the operational range of steam temperature. As a typical example, Fig. 2 shows a simplified cross-section of a bundle. This bundle has one ¼-inch stainless steel sample tube and one ¼-inch copper tracer tube. Pre-insulated bundles come in coils with lengths up to 300 feet.

Why is dew point tracing more difficult? Many vapor sample lines don't need to be traced at all—either no liquid will condense on even the coldest winter night, or else the condensed liquid doesn't need to be analyzed, so you can trap it out. Some other vapor samples have a rather low dew point—say, 5° C—so you need only a little more heat than you would for freeze protection, and only at the coldest times of the year. You can handle that kind of tracing pretty much as you would winterize any other instrument line.

Maintaining sample vapor at a high temperature in a long line, all-year-round, is much different. The 100° C dew point is much higher than ambient temperature, even on the hottest summer day, and it's much closer to the temperature of the tracing steam. That is, relative to the temperature of the vapor, you have a colder ambient temperature and colder steam than when you merely freeze-protect an instrument. There is a much greater chance of condensation at one point or another in the long line.

You cannot allow any condensation from the sample vapor—even momentarily at one or two points, followed by re-vaporization a little further down the line. For example, a percolating slug of liquid sample in a long line carrying water and ethanol vapor would turn your sample line into a sort of distillation column, concealing changes in composition that the analyzer is looking for.

Steam must be supplied continually, rather than occasionally as in winterizing. The steam-supply and condensate-return lines must be highly reliable. They are to be considered a process requirement rather than a dispensable auxiliary system. Therefore, the tracing system requires special attention to design, maintenance and personnel training.

It's tempting to ignore some of the details and expenses, because they go beyond the requirements for freeze-protection tracing. You may run into opposition on the cost and apparent complications of the system. But the extra cost and care may save you from the much greater cost and trouble (not to mention embarrassment) of endless tinkering and rework to get good samples to the analyzer.

Choice of steam pressure and tubing bundle. Steam pressure (temperature) and the type of tubing bundle must be chosen together. The supplier of the pre-insulated bundle can help in deciding on the steam pressure required to maintain the dew point in question, given your particular ambient temperatures. You'll be using saturated steam, because any superheat will be lost in the first few feet of tracing and may tend to overheat the bundle. At best, you're going to be pretty uncertain as to how low a steam pressure (and corresponding temperature) will suffice, so allow yourself a wide margin.

Use steam as hot as possible without exceeding the upper temperature limit of the traced bundle. The engineering and materials involved in typical bundles are such that they can handle 230 psig saturated steam (204° C) when the ambient temperature is 26° C. (We're assuming you don't have an upper temperature limit on the sample vapor. That would be another problem entirely—probably involving a special type of traced bundle with insulation between the tubes.) The expense of hotter steam and a higher-temperature bundle may be insignificant compared to the troubles of a system that doesn't work.

Finally, make sure the steam supply you choose is reliable. It can't be shut down or reduced in pressure for even a short time, or you may get a slug of liquid in your chromatograph.

Proper slope of traced bundle. The most important single thing you can do to assure that the sample tube is adequately heated is to carefully maintain enough slope at all points to drain condensate downward to the steam traps by gravity. Otherwise, condensate will tend to collect in low spots and upward runs in the tracer (except, of course, during the short intervals when the trap is dumping.) A slug of condensate in a low spot or upward run cools down rapidly—even though the bundle still feels hot—and it may allow the sample vapor to begin condensing there.

A slope of half an inch per foot may not be enough, and there's no slope too great. (You get the best heating in downward vertical runs.) In planning the slope, keep in mind that the tracer tube is spiraled around the sample tube to assure good contact. Also, the bundle may not lay perfectly flat. So allow a wide margin for the slope.

Unfortunately, structures in a typical plant run mostly horizontally and vertically. It's tempting to forget the slope requirement for your analyzer lines and lay them on top of existing horizontal structures, or hang them in drooping swags from the sides of horizontal members. Instead, you should install sloping trays to carry the bundles along horizontal distances.

Furthermore, if you have a "horizontal" run that's long enough to require intermediate trapping and resupply points, you'll have to use an up-and-down, zigzag path as shown in Fig. 3. Similarly, wherever the sample line needs to turn upward on its way to the analyzer, the steam and condensate must flow in the opposite direction in order to run downward to the trap. This sort of arrangement may look ridiculous, but it sure is cheaper to do it right the first time than tear it out and do it right later.

The rule is simple: Never make steam and condensate flow uphill—not even for a fraction of an inch, and not even if you've got enough pressure to push it.

Trap spacing in long sample lines. Long sample lines need to have more than one tracer run, each with its own steam supply and trap. The supplier of the pre-insulated bundle can help you decide how long you can make any one tracing run. About 100 feet from supply to trap is a typical maximum. The longer the run, the more condensate accumulates at the trap end between dumps of the trap, and the cooler that end becomes.

Fig. 4 will help you understand this fact. Although engineering data on the operation of tracers are very scarce, you can easily envision that a ¼-inch tracer is a very different kind of heat exchanger than a steam vessel or pipe with a vented float-trap. The so-called "steam trap" on a tracer doesn't really trap out condensate and retain steam. It's basically a periodic blowdown valve, and its dumping frequency is controlled by the rate at which it loses heat to the environment.

Therefore, no trap will keep condensate drained from the tracer at all times. It just blows out accumulated condensate periodically—along with some pockets of steam held by surface tension in between slugs of condensate in the tracer. That's why the trapped end of a tracer run gets colder than the steam-supply end.

If you wanted to keep the tracer as hot as possible throughout its length, you could provide an insulated reservoir for condensate right before the trap. But normally, you just keep the tracer run within a certain length limit, so that you don't have a very long section at the bottom end cooling down as it accumulates condensate.

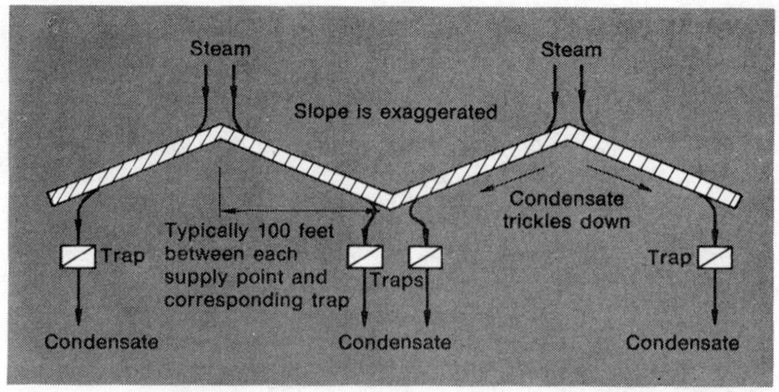

Fig. 3—Maintaining proper slope in a long horizontal run.

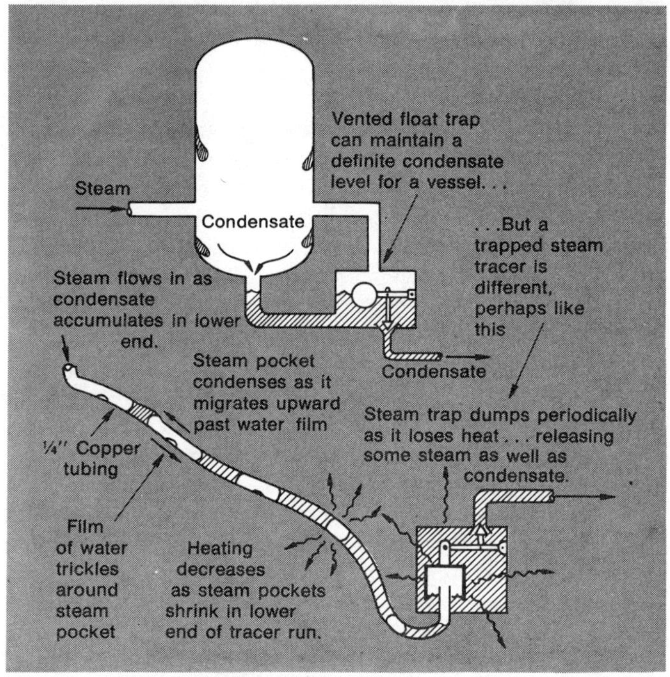

Fig. 4—A typical trapped steam tracer does not provide even heating at saturated temperature because condensate slugs accumulate in the tracer.

Planning the route and resupply points. Obviously, although pre-insulated bundles are easy to install, you can't just tell the tubing crew to run a bundle from there to here. You'll need to plan the route carefully, just as you would for proper installation of a hand-fabricated line. Furthermore, your drawings for the route and support tray should be rather detailed. Besides slope, you'll need to decide on convenient locations for trapping condensate and resupplying steam.

In planning the route, try to keep the trap and resupply points as accessible as possible—preferably at platforms if the bundle is elevated. Anything you can do to make maintenance and troubleshooting easier will keep your system operating better, with less down-time.

In locating steam traps, you need to be very familiar with the proper ways to install these devices. For example, it's all right to mount a steam trap above the end of the tracer that is serves, if necessary for accessibility. Just be sure you include a vertical "water seal leg" at least a few inches deep, between the tracer and the trap.

Use the right trap for the application. Carefully consider the capacity of the trap and other possible details such as the proper orifice size for the pressure differential involved. Use a type of trap known to give continuous, trouble-free operation for long periods.

Do not connect more than one tracer run to the same trap. Chances are one or more of the multiple tracers won't drain right. Similarly, don't put more than one trap on each tracer run. If you do, two or more traps will often open at the same time, interfering with normal drainage of condensate from the tracer between the traps.

Connections to traps and steam supply. In making connections to the tracer tubing, don't leave any length of sample tubing unheated, and seal the cut ends of the insulation jacket. Fig. 5 shows a possible arrangement at an intermediate trap-and-resupply point. Weatherproofing of exposed bundle ends is important to keep the insulation from becoming waterlogged and reducing heating efficiency.

Fig. 6 shows, in schematic form, a possible piping arrangement for supplying and trapping a tracer. To keep the incoming steam free of condensate (thereby preserving its heating value), you may want to include a drip leg and steam trap on the supply line. If the available steam is hotter than the maximum allowable temperature, you'll need a pressure regulator. In any system, a pressure gage is a good idea to help verify that the right pressure is being maintained. The strainer shown in Fig. 6 will help protect both the traps and the pressure regulator from becoming clogged.

As for the trap that drains condensate from each

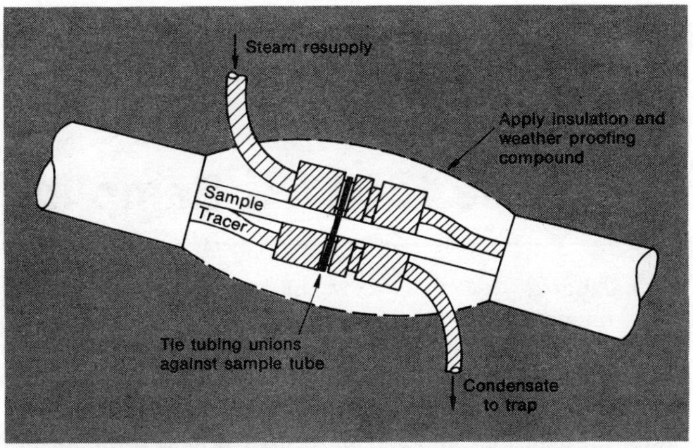

Fig. 5—Maintaining adequate tracing and insulation at trap and resupply points.

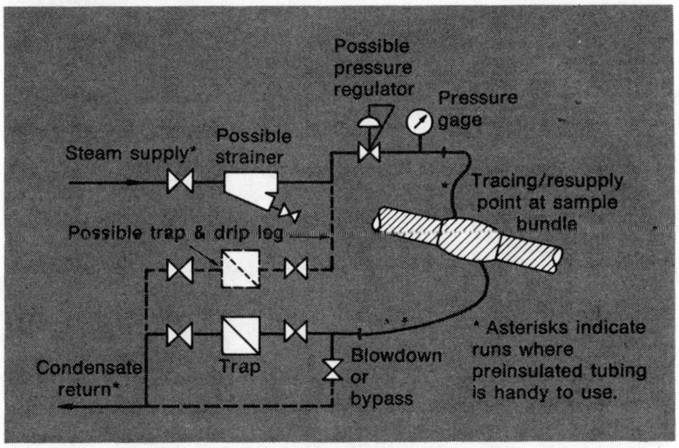

Fig. 6—Typical piping for trapping and resupplying steam for tracer (insulate all lines but not traps).

tracer leg, it's a good idea to provide a blowdown valve ahead of it or a bypass valve around it as shown in Fig. 6. This will allow the trap to be serviced without interrupting steam flow in the tracer. Shutting off tracing heat leads to complications. The sample line must be shut off to avoid getting liquid in the analyzer. After heat is restored, the sample line must be dried out before being reconnected to the analyzer—perhaps by flushing sample vapor through the hot line for a certain period.

As everywhere else in the system, the piping connections should be designed for reliability and ease of maintenance. An extra valve here or strainer there will pay for themselves the first time they save having to shut the system down.

Don't insulate steam traps; they're periodic blowdown valves, timed by the ambient temperature. If you insulate them, they won't blow down often enough to keep your lines hot.

Make sure the condensate return line is reliable. Be sure its pressure will always be properly low at maximum flow conditions. Verify that other equipment dumping into this system won't build up the back pressure so high that your tracing traps malfunction. Ordinarily, condensate return lines should be one to two sizes larger than the tracer to insure low back pressure.

In running your steam-supply and condensate-return lines, consider using pre-insulated tubing. It's available in several sizes from the same people who make the pre-insulated bundles. Using pre-insulated tubing saves the trouble and expense of applying insulation in the field.

When you've got more than one trap mounted at the same location (say, for several tubing bundles), take the trouble to design a neat, sensible, compact installation, in the fashion of a manifold. Otherwise, you may wind up with traps hung every which way, making maintenance more difficult and confusing.

Documentation, tagging and training. Prepare usable drawings for reference by maintenance people showing the location and relationship of all sample bundles, steam-supply valves and traps. Use a logical reference number for each supply valve and the associated trap. For example, AT-1-4-S3 might mean Analyzer #1, Stream #4, Steam-supply #3, and AT-1-4-T3 would mean the corresponding trap. Tie a metal tag stamped with the appropriate number on each supply valve and trap.

Furthermore, make sure all the operating personnel in the plant are indoctrinated as to the special nature of this particular tracing system. They must understand that the steam and condensate lines must never be turned off without special permission. If possible, paint the valve handles a special color and hang up big signs to remind people who haven't gotten the word. Even then, you'll still occasionally find valves turned off and your analyzer full of liquid.

Maintenance. Although steam is a beautiful and economical way to provide heat where you need it, steam traps aren't perfect devices. They sometimes clog up, jam open or closed, or freeze. Since you're trying to keep your analyzer system working 365 days a year, it's a good idea to set up a routine schedule for checking the operation of all steam traps—say, once a month. One handy way to verify that a trap is at least dumping periodically is to listen to it with a stethoscope.

If you don't check your system every so often, you might very well have a luke-warm or even stone-cold section of sample tubing in the middle of a long run and never know it. It would be full of percolating liquid that vaporizes again further down the line, so your analyzer results would be way off. If you treat the tracing system as a genuine engineering problem rather than an afterthought or nuisance, you will get the full benefits of this very practical method. It's all a matter of paying attention to details.

ACKNOWLEDGMENT

The author gratefully acknowledges the assistance of UniTherm Division, Samuel Moore and Co., in the preparation of this article.

System is Accurate, Precise for LNG Sampling

Heating value can be well defined in LNG transfer

W. R. Parrish, J. M. Arvidson and **J. F. LaBrecque,**
National Bureau of Standards, Boulder, Colo.

LNG SAMPLES may be obtained and analyzed from flowing streams with a precision of ±0.20% in the computed heating value. This value is derived from laboratory and field tests and is based on three standard deviations, including ±0.06% random error for gas analysis. The accuracy of the measurement depends only on the accuracy of the gas analysis; all tests show that a properly designed and operated sampling system produces no detectable systematic offsets in the computed heating values.

Since LNG is bought and sold on a heating value basis, accurate and precise LNG composition measurement is a necessity. At current prices, a composition uncertainty corresponding to a one percent error in the computed heating value means an inequity of nearly $65,000 for one 125,000 m^3 shipment.

Sampling LNG presents special problems because the component normal boiling points are so different—methane, the major component, has a boiling point of −161° C; while n-butane, the minor component, has a boiling point of −0.4° C.

Accurate LNG composition measurement requires a system containing

- A sampling probe which transfers a representative sample into the sampling system
- A sample conditioner which completely vaporizes the sample
- A gas analyzer which accurately and precisely analyzes the sample.

Finally, all operating variables must be carefully selected to insure that all system components perform properly.

Sampling systems fall into two broad categories—continuous and batch. Trego[1] describes a continuous LNG sampling system which is in operation at the LNG Terminal in Barcelona, Spain. He states that deviation between liquid samples taken upstream of a vaporizer and gas samples taken downstream of the vaporizer agree to within the error of the gas analysis (~ 0.5 to 0.9%). Cook[2] describes a batch sampling system; he reports[3] that the device gives good results for sampling LNG containing 95 mol % or more methane. However, the precision and accuracy have never been established quantitatively.

This paper describes a system for continuously sampling flowing LNG streams with a total uncertainty of ± 0.20% in the computed heating value. The system can be used in any LNG line where the pressure is at least 5 psi (32 kPa) above ambient conditions. The system design, dimensions and recommended ranges of operating variables are derived from a systematic study employing both laboratory and field tests. The laboratory work permitted rapid data accumulation under well controlled operating conditions using LNG-like mixtures of known composition. These tests considered 3 probes, 2 vaporizer designs and 10 operating variables. Sufficient data were taken to statistically establish the effect of each variable on the sampling precision and accuracy. Using the laboratory test results, a full scale sampling system was designed, built and tested aboard an LNG tanker and in the NBS flow facility.[4] Details and results of this work are presented elsewhere.[5]

SYSTEM DESCRIPTION

Fig. 1 shows a schematic of the full scale sampling system. All field tests were made with the sampling system having the dimensions given in Table 1. At this time it is not clear which of the dimensions are critical to the sampling system precision.

The sample is taken from the LNG line (in our tests a 3 in. (76 mm) pipe and flows through a side tap probe, past a stainless steel bellows block valve (T1) and through a 60 micron filter. A nitrogen purge line is attached to the system to inert the sampling system prior to startup and to back purge the filter if necessary. The LNG then flows through a stainless steel bellows needle valve (T2), which controls the sampling rate, and into the steam heated vaporizer. The vaporizer contains a coiled section of thin-walled stainless steel tubing and an impingement chamber. The impingement chamber causes a

TABLE 1—Dimensions of LNG sampling system (See Fig. 1 for system schematic)

Component	Length, inches (mm)	Outer Diameter, inches (mm)	Inner Diameter, inches (mm)
Line between test section & valve, T1	4.5 (114)	0.540 (13.7) (¼ in. Sch 80 pipe)	0.302 (7.67)
Line between valve, T1, and filter (contained a short radius 90-degree bend)	11 (279)	0.25 (6.4)	0.180 (4.57)
Line between filter and valve, T2	2 (51)	0.25 (6.4)	0.180 (4.57)
Line between valve, T2, and union at vaporizer	13.5 (343)	0.25 (6.4)	0.180 (4.57)
Vaporizer	36 (914)	0.13 (3.18)	0.101 (2.56)
Vaporizer steam casing (with flat welded ends)	10 (254)	6 (152)	5.5 (140)
Impingement chamber (with flat welded ends)	1.5 (38)	1.5 (38)	1.376 (34.95)
Accumulator	16 (406)	5.125 (130.2)	4.805 (122)

sudden change in the flow direction and speed to enhance vaporization of any entrained droplets. All of the cold lines, the steam lines and the vaporizer are insulated with fiberglass pipe insulation which is sealed with duct tape. After leaving the vaporizer, the sample goes to an accumulator. The accumulator acts as a time-averaging, or mixing, device to eliminate inhomogeneities caused by fractionation during vaporization. The homogeneous gas sample leaves the accumulator, passes through a manifold containing sample cylinders, a rotameter and a back pressure regulator before going to the vent system.

The sampling rate is set by adjusting the needle valve (T2) and the back pressure regulator. The back pressure regulator is extremely useful because it maintains a constant pressure in the sample cylinders.

TEST RESULTS

Fig. 2 shows the computed heating values for one series of tests made aboard the LNG tanker *El Paso Consolidated*. In these tests liquid samples were taken upstream of the shipboard steam vaporizer. Analyses of these samples were compared with those of gas samples taken from the completely vaporized stream. A laboratory gas chromatograph was used to determine the sample composition. The average heating value of the gas samples was 0.04% higher than that of the liquid samples. Also, the estimated standard deviation for both sets of samples was 0.04%. This precision is comparable to the estimated standard deviation of 0.05% in the heating value for gas analysis alone. A statistical analysis of computed liquid densities gave an estimated standard deviation of 0.02% for both the liquid and gas samples. This test shows that, within the precision of the gas analysis, samples taken from the liquid stream are indistinguishable from the samples taken from the totally vaporized gas stream.

Table 2 lists the variables evaluated in the laboratory and field sampling tests and groups them according to their effect on sampling precision and accuracy. Each variable's effect on sampling precision was measured by repeated sampling under controlled conditions. The accuracy was determined in the laboratory tests by sampling mixtures of known composition. The only verification of accuracy in the field tests was made by comparing the liquid sample compositions with samples taken from the totally vaporized LNG stream. The test results indicated that for uninsulated liquid lines the heat leak causes back-flashing of nitrogen (and probably methane) which enriches the samples in the heavy (C_2^+) components. However, insulating the line with a ½ to 1 inch (12 to 25 mm) thick layer of fiberglass sealed with duct tape is adequate to eliminate the heat leak problem. No other variable tested affected sampling accuracy. Therefore, by using insulated lines, the accuracy of the composition measurement depends only on the accuracy of the gas analysis.

Five variables were found to adversely affect sampling precision. Both laboratory and field tests showed that below a certain sampling rate, the sampling precision diminishes; this minimum sampling rate varied between various sampling systems. A mechanism for the sampling rate's importance is unknown. For the sampling system described here, the sampling rate should be greater than 20 standard liters per minute of gas.

Laboratory results clearly indicated that a small diameter tube vaporizer was better than a vaporizer with a large cross sectional area and that rapid sample vaporization gave the best results. However, there is no advantage in having the vaporizer outlet temperature above 100° F

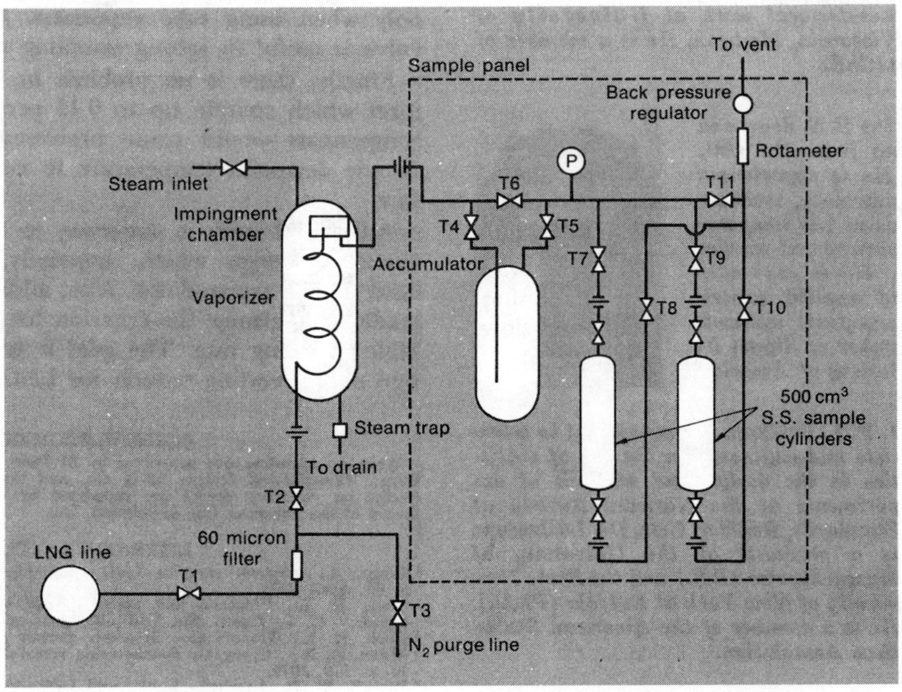

Fig. 1—LNG sampling system.

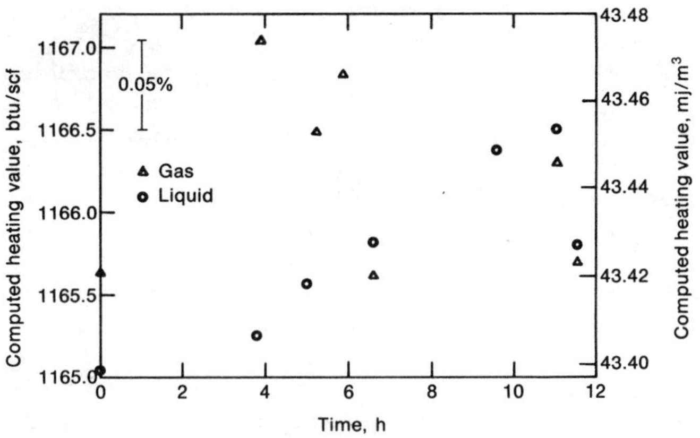

Fig. 2—Computed real gas heating values of shipboard data.

TABLE 2—Variables evaluated for their effect on LNG sampling precision and accuracy

Variable effecting both sampling precision and accuracy:
- Heat leak to liquid sample line

Variables affecting sampling precision but not accuracy:
- Sampling rate
- Vaporizer design
- Time averaging the vaporized sample
- Sampling rate transients
- Probe

Variables not affecting sampling precision and accuracy:
- Temperature and pressure at the sampling point
- Pressure differential between the sampling point pressure and the equilibrium LNG pressure
- Flow rate past the sampling point
- Pressure drop in the liquid sample line between the sampling point and vaporizer
- Composition of the liquid being sampled

(38° C). Both steam and electrically heated vaporizers produce comparable results. An impingement chamber is unnecessary when using an electrically heated vaporizer. This is because the electrically heated unit vaporizes the sample under constant heat flux conditions. The steam vaporizer transfers heat with a constant wall temperature. This is less effective due to the film boiling.

The need to time-average the vaporized sample in an accumulator arises from fractionation during vaporization. The minimum residence time necessary to obtain precise results depended upon the sampling system; residence times of 45 to 120 seconds gave good results.

Field tests indicated that taking samples within 30 minutes of a major change in sampling rate diminishes sampling precision. However, minor fluctuations and slow drifts in sampling rates did not affect the results.

The side tap probe was found to be more reliable than an upstream facing pitot tube. For unknown reasons, the pitot tube occasionally gave erratic results.

The conclusions that certain variables did not affect sampling precision must be considered valid only over the ranges tested; however, this range usually included the typical operating ranges encountered during custody transfer.

There is no reason to expect the temperature and pressure to affect sampling precision provided the conditions are far from the critical point of the LNG.

At some low value the difference between the pressure of the sampled liquid and the saturation pressure of the LNG (i.e., subcooling) in combination with heat leak will adversely affect sampling precision; however, this effect was not found at a pressure differential of 3.9 psi (25 kPa) (this corresponds to ~ 0.5 K subcooling).

LNG flow rates corresponding to Reynolds numbers ranging between 1,500 and 50,000 did not affect sampling precision; it is unlikely that there would be any adverse effect on precision at higher Reynolds numbers.

The fact that a large pressure drop in the liquid sample line is unnecessary eliminates the need for a liquid flow control valve. This conclusion is known to be valid only when using tube vaporizers. However, the needle valve is useful in setting sampling rates.

Finally, there is no problem in sampling LNG mixtures which contain up to 0.15 percent C_6^+. The heavy components would cause problems if they raised the sample dewpoint temperature to near ambient temperature.

Additional work is underway to develop an improved vaporizer design which, hopefully, will eliminate the need for an accumulator. Also, additional tests are being made to determine the criterion for the minimum acceptable sampling rate. The goal is to provide general design and operating criteria for LNG sampling systems.

ACKNOWLEDGMENT

Work on sampling was supported by El Paso Marine Co., Columbia LNG Corp., Consolidated System LNG Co. and Southern Energy Co. Current studies on vaporizer design are supported by the Pipeline Research Committee of the American Gas Association, Inc.

LITERATURE CITED

[1] Trego, J., "System samples LNG accurately," *Hydrocarbon Processing*, p. 83, April 1976.
[2] Cook, H. L., "Method and apparatus for sampling refrigerated volatile liquids," U. S. Patent No. 3,487,692, January 1970.
[3] Cook, H. L., Transco Co., Houston, private communication, May 1977.
[4] Mann, D. B., "Cryogenic flowmetering research at NBS," Cryogenics, Vol. 11, p. 179, 1971.
[5] Parrish, W. R. Arvidson, J. M., and LaBrecque, J. F., "Development and evaluation of an LNG Sampling Measurement System," NBSIR (to be published).

Test Olefin Feed by Micropyrolysis

Using a laboratory micropyrolyzer, this simple reproductive test measures ethylene yield and furnace coking rate for various feedstocks

A. J. Martens, J. Glas and **J. L. Gaultier,**
Naphtachimie, Lavera, France

AN ACCURATE CHARACTERIZATION of the ethylene feedstock has become very important as naphtha cost and demand continue to increase. At least two factors must be measured:

1. The olefin cracking yield, and
2. The furnace coking rate.

These two factors determine the economic interest in the fraction examined and can be related to its real cost in an overall model of the production unit.

We have devised a simple, reproducible test using a laboratory micropyrolyzer. These tests enable us to measure or calculate the two factors mentioned above from a normal gas chromatograph analysis of the fraction.

Using these tests in our laboratory since 1971-72 has helped us in optimizing profitability on two steam cracking units—100,000 and 500,000 tons/yr.

The literature is filled with references[1-4] on using hydrocarbon pyrolysis in characterizing feedstock. The

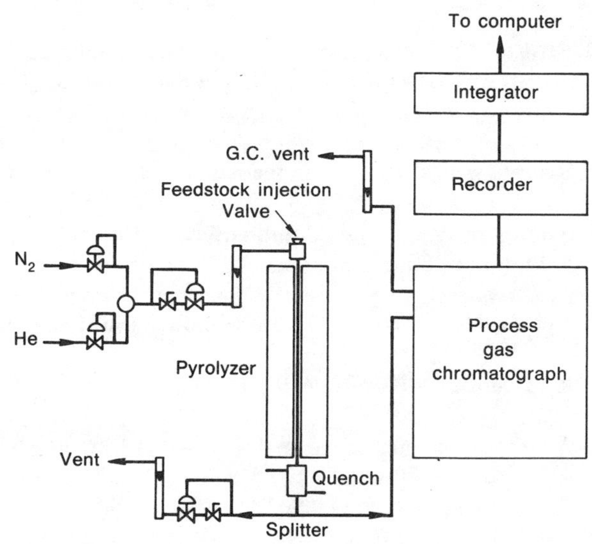

Fig. 1—Diagram for the testing apparatus for determining the olefin production capabilities and the coking possibilities of various feedstocks for ethylene plants.

TABLE 1—Micropyrolysis versus industrial steam cracking

Micropyrolysis	Feedstock	Cracking severity
800° C 0.18 sec resid. time	Current naphtha Sp. gr. 0.710/0.720 Mean I.P.N.A.: 33/36/23/8 wt %	Medium range for furnace K.S.F. = 1.6[6]
Cracking yield, wt % Hydrocarbons	**Micropyrolyzer**	**Industrial furnace**
Methane	13.4	13.4
C_2 (total)	26.5	26.5
C_3 (total)	14.9	15.5
Ethane/ethylene	0.26	0.21

Note: Fig. 1 shows a diagram and description of the apparatus.

quantative measurement of coke formation is generally not discussed although this is a very important economic factor when cracking heavy fractions.

The tests proposed here are simple, fast (1 hr. max.) and require only standard laboratory material (oven and gas chromatograph). It is possible to adapt a system (as described in Fig. 1) for on-line process analysis.

Test No. 1. The principle is to inject a few microliters of the sample into the oven at regular intervals. After cracking, the sample is analyzed by a process chromatograph. The entire operation is automatic permitting performing a series of analyses quickly. With normal gas chromatographic analysis,[7] hydrocarbons formed during pyrolysis can be measured: methane, ethane, ethylene, propane, propylene and the C_4 fraction (detailed analysis or overall evaluation).

By varying the residence time, the temperature and the temperature profile of the oven it is possible to very closely approximate the results of an industrial furnace (see Table 1). The data in Table 1 were obtained on an industrial furnace operating at medium severity feeding L.D.F. naphtha.

Procedure. The chromatographic analysis is carried out on two groups of columns at 130° C and takes about 15-20 minutes using these criterion:

- Sample size, 2 to 5 μl of feedstock
- Micropyrolyzer tube, 2 mm internal diameter, total length 35 cm (or 60 cm)
- Vector, N_2 (or He) with (or without) diluting steam
- Apparent residence time, 0.1 to 0.6 sec
- Average residence time in hot zone, 0.15-0.20 sec
- Temperature, 720° C/860° C
- Cracking severity range, from K.S.F. = 0.2 up to K.S.F. = 4 in conventional K.S.F. scale (pentane cracking[6])
- Average number of successive measurements: 4 to 6

The procedure is:

Group 1. Two columns in series; first column 0.50 m long, 4 mm ID by 6.0 D (about 1/8 inch ID) using Chromosorb W 60-80 and 10% Carbowax 400. The second column is 1 m long by 4 mm ID by 6.0 D filled with Porapak Q 50-80.

Group 2. A column 4 m long by 4 mm ID by 6.0 D using Porapak Q 50-80.

In Group 1, each chromatographic column has specific functions: the Carbowax 400 column forces the water beyond the C_4 and the short Porapak Q separates in groups the C_4 from the C_5 and C_3.

In Group 2, the single Porapak Q column fractionates the components from C_1 to C_3 in the following sequence: air, methane, ethylene, ethane, propylene and propane. The sample is injected on both groups of series columns. Group 1 is disconnected from Group 2 when all components from C_1 to C_3 have been allowed to flow through the latter. The C_4, water and matters heavier than C_4 are stored in Group 1. Then all the C_4 component is extracted from Group 1 before the water and heavy matters are purged by reversing the flow, component C_1 to C_3 being extracted from Group 2 and measured. This analysis is carried out by an isothermal process at 130° C (270° F).

Calibration of responses is made by means of a standard mixture including the evaluated products in proportions corresponding to the real contents recorded in the samples. This standard mixture is volumetrically prepared and weighed, then checked by a chromatographic analysis using a gas density balance as a detector.

It is possible to use another chromatography analysis system according to the range of products to be evaluated.

The repeatability of the entire first test for an average of four successive measurements (duration 1 to 1.2 hr.) is between ± 3% and ± 4%, given at the 95% level of probability.

Test No. 2. This same system permits the coking rate of the sample to be evaluated by a second test. To do so, the whole of the pyrolysis system described above is applied and several samples (6-12) from the same batch are injected and pyrolyzed for the same severity as in the cracking test. By increasing the size of the sample (25 to 30 μl) as well as the internal surface of the reactor tube, it is possible to obtain a measurable coke deposit after several successive pyrolyses. The size of the deposit is directly related to the coke formation of the sample examined.

By taking as a reference (base 100) a pure hydrocarbon (nonane) it is possible to draw up a Coke Index scale. We have confirmed empirically that this index is correlated to the coking rates obtained in steam cracker furnaces. It is obvious that this test is less accurate than the preceding one, for the coking phenomena are highly complex and the gas-solid exchange processes involved are still little known and poorly controlled. The repeatability obtained for an average of 4 successive tests (duration about 1 hr.) is roughly ± 10%, given at the 90%

TABLE 2—Tests for current naphtha L.D.F.

	Naphtha No. 1	Naphtha No. 2	Nonane
Sp. gr. (g/cc)	0.695	0.743	....
Normal paraffins	36.00 wt %	33.80 wt %	....
Isoparaffins	35.10	28.30	....
Naphthenic	20.80	21.20	....
Aromatic	8.10	16.70	....
Test No. 1*			
C_2 yield	25.50 wt %	21.40 wt %	40.90 wt %
C_3 yield	14.80	12.40	16.70
C_4 yield	10.40	10.00	8.20
Test No. 2*			
Coke Index	105	226	100 (ref.)

* Pyrolysis conditions: 800° C, residence time 0.18 sec.

TABLE 3—Tests for some hydrocarbons. Experimental conditions: 800°C, 0.18 sec (K.S.F. = 1.6)

	3-Methyl-nonane	Pentane	Cyclopentane
Test No. 1			
C_2 production	12.30 wt %	42.30 wt %	17.30 wt %
C_3 production	13.10	21.30	15.40
C_4 production	24.20	5.90	1.00
Test No. 2*			
Coke Index	160	25	40

* Ref: Nonane = 100 for K.S.F. = 1.6

TABLE 4—Tests for various temperatures. Feedstock, Naphtha L.D.F., sp. gr. 0.717, I.P.N.A. = 32/36/23/9, residence time, 0.18 sec. (Test No. 1)

Temperature	800° C	820° C	840° C
Severity (K.S.F.)	1.6	2.3	3.2
C_2 (total)	26.5 wt %	28.3 wt %	30.1 wt %
C_3 (total)	14.9	13.3	11.4
C_4 (total)	8.9	6.2	4.2

TABLE 5—Coke Index and run length. Furnace conditions: medium severity 1.5 < KSF < 2; temperature 800—820° C; wt steam/naphtha = 0.5. Feedstock range: Naphtha L.D.F. Sp Gr 0.680/0.740

Mean Coke Index during the run	Run length (range days)
100	180/200 days
135	90/100 days
175	50/60 days
200/210	30/40 days

level of probability. It is possible to improve this result by increasing the measuring time. However, this degree of accuracy is highly sufficient for present industrial requirements concerning the evaluation of the coking rate.

Examples for naphtha L.D.F. In case of naphtha, this technique enables the steam cracker feedstock to be classified in great detail for both the specific cracking yield and the coking. Table 2 gives some typical results for a medium cracking severity (K.S.F. = 1.6).

The No. 1 naphtha batch has a good ethylene content and very little coke. The No. 2 batch is heavy with

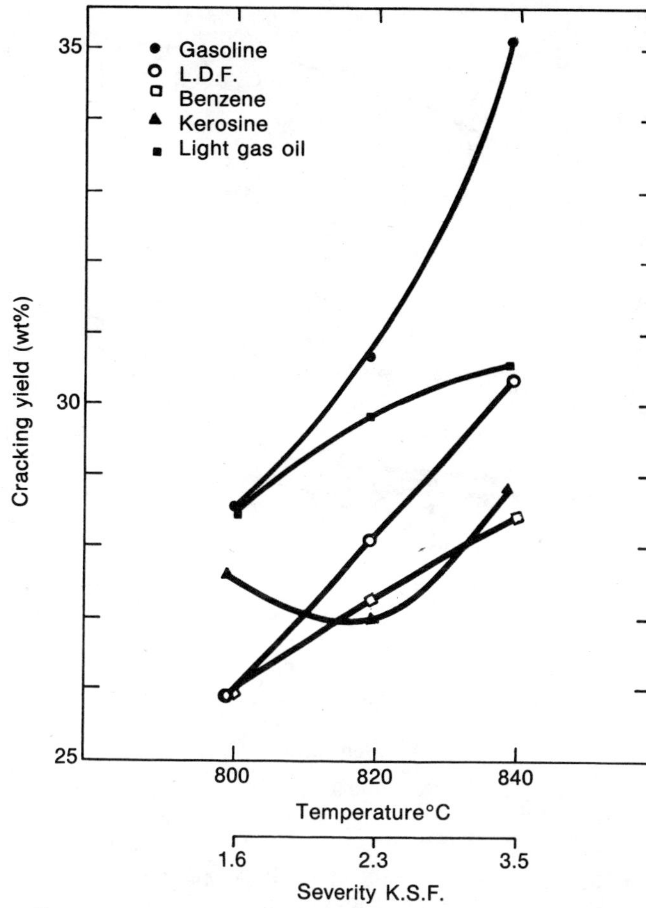

Fig. 2—Correlation of Test No. 1 showing the ethylene formation expected from various feedstocks.

a low yield and above all it has a very high Coke Index which in practice would require lowering the working temperature of the industrial furnace using such a feedstock.

It is also possible to obtain a good approximation of the experimental results, such as those in Table 2, by a rough calculation based on the detailed composition of the cut and on complete tables similar to Table 3.

The behavior of the feedstock at various cracking severity levels can also be easily obtained by this method, and the ethylene and propylene production yields for different temperatures also can be predicted (Table 4).

The Coke Index obtained by Test No. 2 is directly related to the coking rate of steam cracker furnaces (tube and transfer line exchangers). We were able to establish empirical correlations linking the run life between two decokings, expressed in days, to the average index of the feedstock used during this same period. Table 5 gives the results showing this relationship which was established after several years observing steam cracker furnaces.

Various petroleum distillates. The speed of this technique enables a great many batches of feedstock to be tested as well as the different simple hydrocarbons of which they are composed. Test No. 1 shows the potential yield of the fraction while Test No. 2 indicates

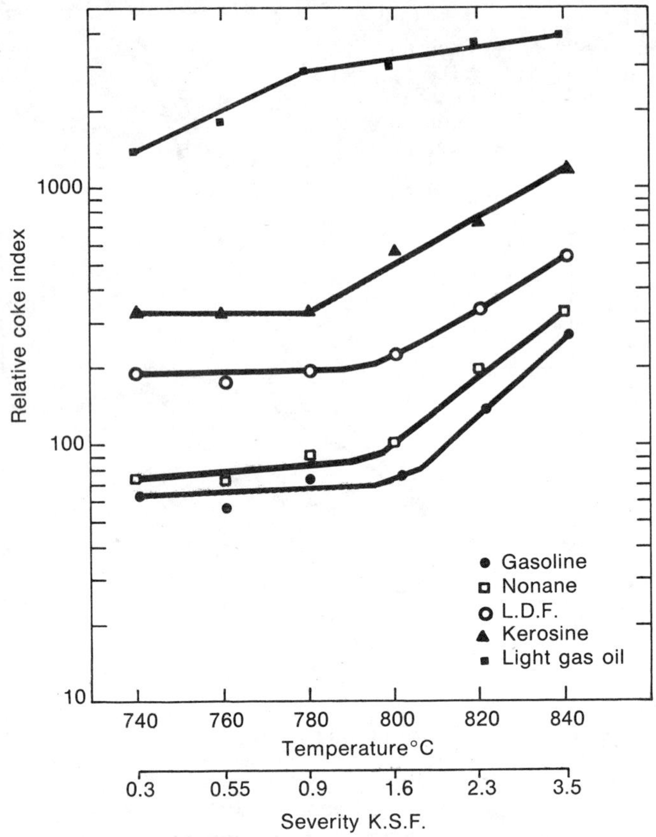

Fig. 3—Test No. 2 correlated to show the Coke Index for various steam cracker feedstocks.

TABLE 6—Distillates tested

	Gasoline	Naphtha L.D.F.	Naphtha topping	Kerosine	Light gas oil
Sp. gr. 15° C g/cc....	0.684	0.717	0.780	0.788	0.840
C/H (atoms)........	0.44	0.45	0.49	0.50	0.57
Distillation range, °C.	39—126	50—165	152—181	156—229	200—337
Overall Aromaticity, %*..............	1.5	4.6	9	6.4	7.5

* Proton N.M.R. measurement as (Aromatics H/Aliphatic H) %

whether or not the fraction can be used without being held up by excessive coking.

With these two criterion in mind, the distillates and their properties shown in Table 6 were studied.

Fig. 2 represents the evolution of the potential ethylene content for the different distillates examined under Test No. 1. The adverse effect of a high overall aromaticity on the ethylene production yield (naphtha fraction 0.780) can be seen. The good results of the gas oils which give a reasonable potential ethylene content (more than 30%, above 820° C K.S.F. = 2.3) should also be noted.

The combination of the results of Test No. 1 and No. 2 allows a good balance to be obtained between productivity and coking rate.

In Fig. 3 it is particularly worth noting the very high coking rate of the gas oil due (according to a more detailed study of each hydrocarbon) to the heavy linear paraffins (C_{12} and above) which have a very high specific coking rate (formation of tars).

The production/coking balance seems better for the kerosine (sp gr 0.788) than for the lighter gas oil (sp gr 0.840) which coke too much.

Conclusions. The technique which has been described is a fast and simple way of testing ethylene feedstocks.

Two indices are defined to measure the C_2, C_3 and C_4 cracking yield and the coking rate.

With these two indices the economic interest of the distillate, or even the crude oil, under examination can be thoroughly evaluated.

These indices may also be used for characterizing the naphtha in overall models of the plant, as we have done in Lavera, for a steam cracker producing 500,000 tons/yr. of ethylene.

This technique also permits the various naphthas on the market to be evaluated and purchasing policies guided accordingly. Lastly, the average coking rate of the feedstock being used can indicate the approximate run length between two decoking operations.

LITERATURE CITED

[1] Cramers, Keulemans, *Journal of Gas Chromatography*, p. 58 Feb. 1967.
[2] Fanter, Walker, Wold, *Analytical Chemistry*, p. 2168 Vol. 40, DRC 68.
[3] Illes, Pleszkats, Szepesy, Erdol un Kohle, p. 201 Vol. 4 (1969).
[4] Bibliography of Vapor Phase Thermal Fragmentation, S. F. Sarner[102] Journal of Chromatography Science, Vol. 10, Feb. 1972.
[5] eco, Journal of Chromatagraphy Science, p. 36, Vol. 16, Jan. 1978.
[6] Zdonick, Green, Hallee, *The Oil & Gas Journal*, p. 91 Feb. 1968.
[7] Martens, Glas, Mangin, Chromatographia, p. 467 Vol. 5 (1972).

Part IV.

Compressors and Turbine Control

Parallel Compressor Control . . . What Should Be Considered?

New process control computers can help but first the problems must be recognized

A. E. Nisenfeld and **C. H. Cho,**
Fisher Controls Co., Marshalltown, Iowa

CONTROL OF most compressor systems was not a problem until the energy crisis; control of systems with parallel compressors has always been a problem.

Control system engineers are recognizing that every process or unit operation is entitled to some design attention of its own, no matter how many times an apparently identical system has been implemented. This is true when we design control systems for centrifugal compressors and doubly so when these compressors are operating in parallel.

There are many parallel compressor applications in processing industries. Most common, perhaps, are refrigeration compressors in ethylene and propylene plants. Frequently three or more compressors are in parallel providing suction to a second pair of compressors in parallel. This configuration (Fig. 1) illustrates control problems which are uniquely associated with parallel compressors. In each set of compressors, one is a spare in standby having automatic startup controls.

Consequences. Bringing one compressor up to speed when it must pump against head established by a parallel compressor(s) is difficult and dangerous. If the "coming-up" compressor is tied into the discharge manifold too early, it can be thrown into very violent surge. If one compressor is failing or being shut down, bringing up the "new compressor" too late will cause the process to be starved for gas. Tying in too early can also cause reverse rotation of the compressor going down.

Perhaps the most serious problem, with due respect to the uniqueness of surge control requirements for parallel compressors, is balancing the load among compressors. The authors have seen perhaps 35 or 40 different parallel compressor systems and have yet to see two compressors in any of these trains with identical performances. At the same speeds, these compressors pump different quantities of gas and have different surge and stonewall points. (Stonewall is that phenomenon where a compressor can no longer pump the quantity of gas fed to it—sometimes called "choke.") Without proper balancing controls, the more efficient compressor often carries more than its share of the load and the less efficient compressor operates with its recycle valve (spillback) wide open, because the efficient compressor is taking all the feed.

From Fig. 1 we can readily see why surge control is difficult. Spilling back to the suction of one parallel compressor obviously affects the suction conditions and quantities of gas at other compressors.

Before control system design, process requirements and compressor operating characteristics must be understood. Process requirements determine the controlled variables; compressor characteristics dictate the strategy for capacity regulation.

Objectives of control. The primary objective of a control system is to meet and maintain system design specifications for thruput and discharge pressure in response to process demand variations. The system must provide compressor surge protection and orderly startup and shutdown. For parallel compressors the control system should be able to bring another compressor on line in parallel with one or more operating compressors.

Compressor control can be divided into two major functions:

Control related to process demand (capacity). Compressors are generally used for a process requiring a variable quantity of gas at constant pressure. As flow demand changes, control systems should respond to this change and maintain thruput by one or more of the following:

- Varying inlet pressure.
- Varying speed.
- Varying inlet vane angle.
- Varying spillback flow.

Varying inlet pressure is not the most economical because when we drop inlet pressure, the compressor uses more energy to reach discharge pressure. However, capacity of a centrifugal compressor is often controlled by placing a control valve in the suction line. With this control valve, compressor inlet pressure can be changed to obtain a new operating curve for compressor capacity regulation with a constant discharge pressure. Fig. 2 shows two sets of compressor curves for two different inlet pressure conditions. To obtain the change in flow rate

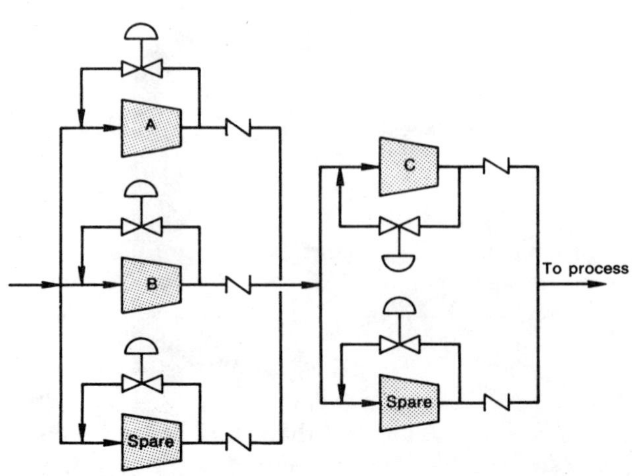

Fig. 1—Parallel compressor configuration.

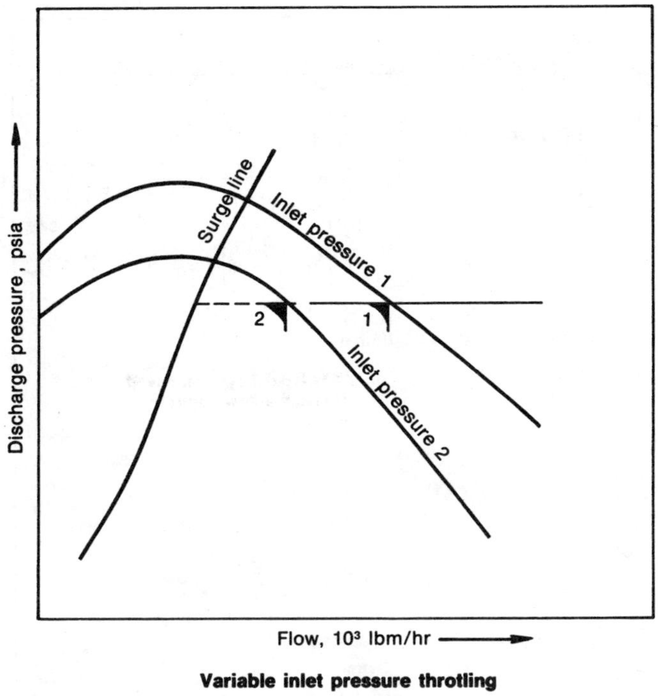

Fig. 2—Variable inlet pressure throttling.

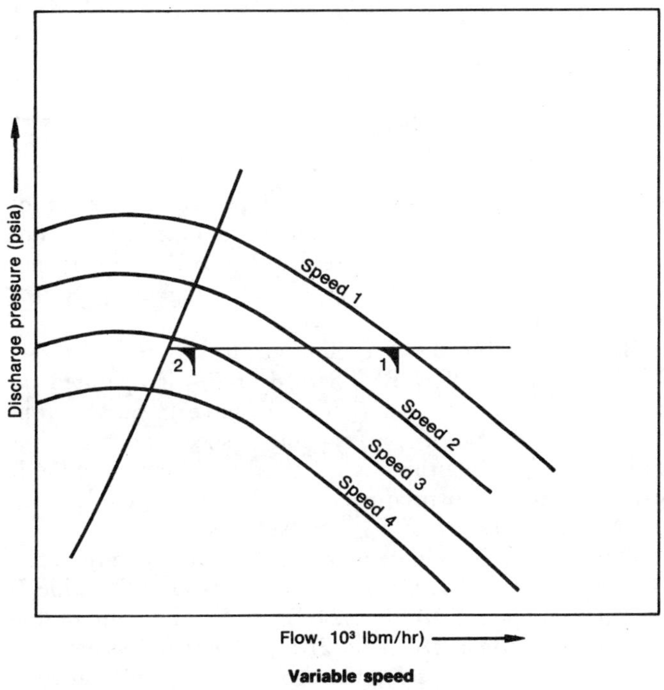

Fig. 3—Variable speed pressure throttling.

demand from point 1 to point 2, inlet pressure must be throttled by approximately 4.6 psi, assuming a compression ratio $\frac{(P_{Discharge})}{P_{(Suction)}}$ of 10.

Varying inlet vane angle to control capacity is not significantly more economical than varying inlet pressure, and it is less reliable. In this method, a set of adjustable guide vanes on the inlet to the compressor stage is used to pre-rotate or counter-rotate the gas stream with respect to impeller rotation. Therefore, this method unloads or loads the stage to lower or raise discharge head, at a constant speed and flow. Advantages of this method over varying inlet pressure are:

▶ The variable inlet vane can lower or raise (bi-directional) discharge pressure.

▶ It provides capacity regulation with less power consumption since it doesn't directly throttle flow.

Varying the speed of a compressor is the most economic means of capacity control, but it is not practiced as often as it could because of mechanical problems with speed transmitters and governors. Since the pressure ratio developed by a centrifugal compressor is related to impeller tip speed, variation of discharge pressure may be plotted as a percent of design speed. For example, Fig. 3 shows that a two-to-one flow turn down ratio can be achieved by varying maximum design speed by 20 percent. This method is also energy effective since power input is reduced by the square of the speed. One disadvantage is the sensitivity of capacity to speed change as shown in Fig. 3.

Varying spillback (recycle) flow is frequently used because it seems to be the safest. However, it is costly because gas is compressed and then dropped back to suction pressure. At a given speed, a compressor will pump a fixed quantity of gas against a fixed head. If a spillback valve is opened, net forward flow of gas will be reduced; but total flow through the compressor will remain constant and above the surge point.

Control related to surge protection. The purpose of an anti-surge control system is to protect the compressor and process from surge. It may be a separate system which overrides normal capacity controls or it may be part of the capacity control system.

The common anti-surge systems (Fig. 4) include the following:

Minimum flow, probably the simplest form of surge protection, uses a predetermined minimum flow as setpoint for the surge protection controller. For this system to work properly, the compressor must be operating at constant speed with constant gas properties; suction pressure and temperature must be virtually constant.

Minimum flow with automatic speed control provides surge protection by manipulating compressor speed to hold a minimum flow.

Flow-speed system uses compressor speed to determine the setpoint of the spillback flow controller. Therefore, the surge point is a function of compressor speed.

Guide vane position resets minimum flow controller. In a constant speed compressor, inlet guide vanes are often used for flow regulation. The surge point is a function of guide vane position as well as compressor speed. Guide vane position can be used as a remote setpoint to a spillback flow controller to compensate for change in surge point.

Flow-ΔP control is the best of the widely used protection systems because it is independent of compressor speed and suction conditions. The locus of the surge line and set point of the anti-surge controller, is determined by the following equation:

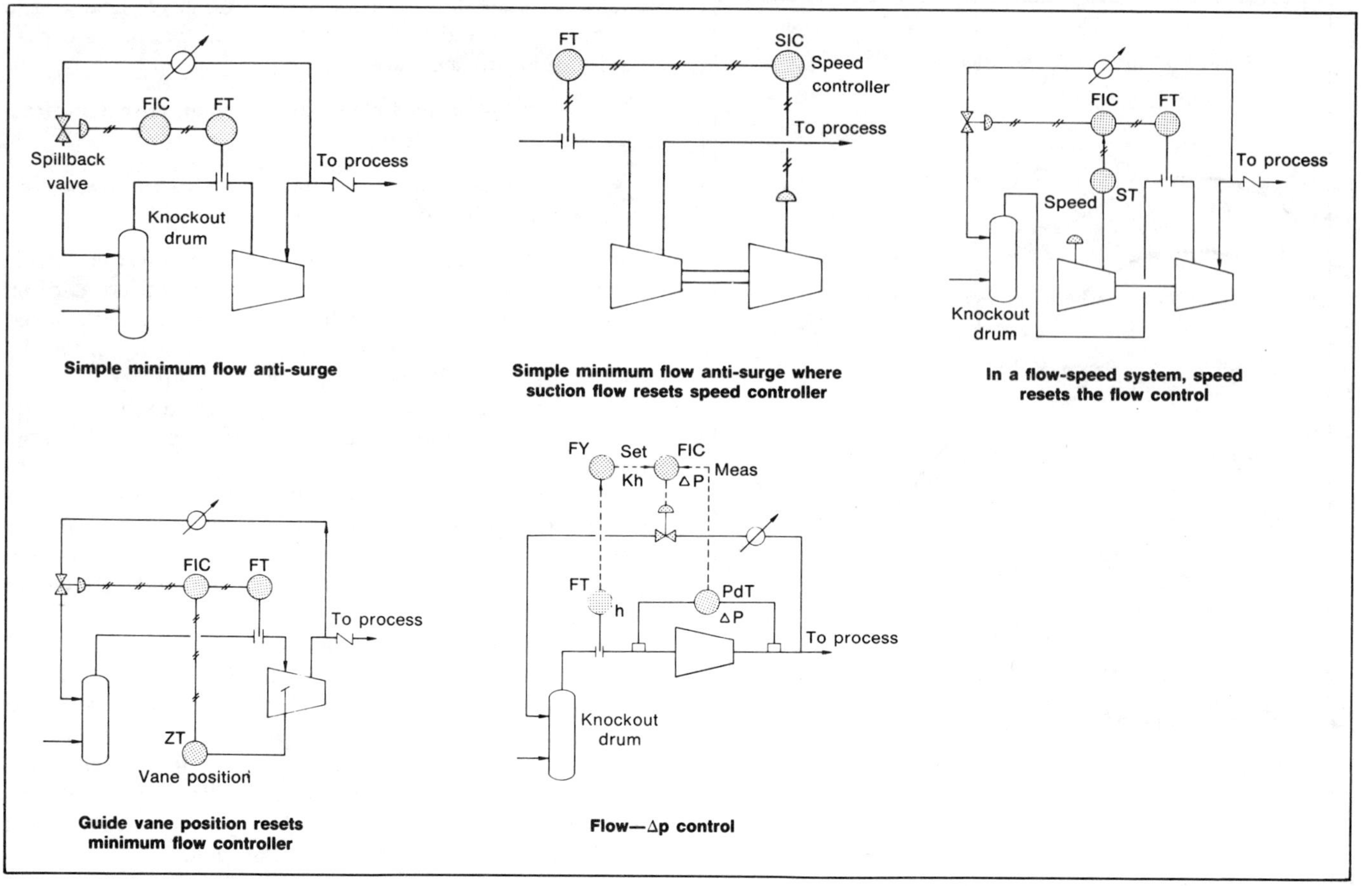

Fig. 4—Common anti-surge systems.

$$\Delta P = kh + b$$

where:

ΔP = Compressor discharge pressure-compressor suction pressure
h = Differential pressure across an orifice in the compressor suction
k = Proportionality constant
b = Bias to match control line origin with surge line origin.

Peculiarities of parallel compressors. The best way to design controls for parallel compressors is by simulation using a math model of the compressor and the process. The type of simulation required is a function of system complexity and nature of probable disturbances. For a simpler system, simulation can be by manual, graphic methods using compressor maps with simple process system curves. Large hybrid computer systems are required for very complex parallel systems.

A two compressor system, both machines operating continuously with 60 percent of total maximum flow as their design base, is the most straightforward example of parallel compressors. Let's examine specifics of load allocation control for this system.

There are two approaches: balance the load equally, or operate at maximum over-all efficiency. The difficulty of load balancing can be seen in Fig. 5 where the performance curve of one machine at a given speed is superimposed on the performance curve of the other at the same speed.

Surge points are different; while machine A operates safely at inlet flow B1, the surge protection system of machine B will be in operation. Both machines cannot pump flow B1 at discharge pressure P_{D2} at the same speed. Machine A will have to operate at a higher speed for equal load distribution.

But, if machine B is recycling gas to stay out of surge, it will not pump flow B1 forward at P_{D2} if operated on the indicated speed line. It too will have to be sped up. Since the surge point is a function of speed, among other things, it is possible there is no speed at which machine B can be operated to produce a net flow of B1 cubic feet/minute at a pressure of P_{D2} atmospheres.

One solution is to have machine A pump some quantity between A1 and B1, say A1 + 0.3 (B1-A1). Machine B would then pump B1 + 0.7 (B1-A1). The sum of the individual productions is 2B1. A control system was designed (Fig. 6) to solve the problem. Only the capacity controls on one machine are shown; no anti-surge controls are shown.

The demand controller is shown as a manually set FIC-1, but it can be a pressure controller, somewhere in the process, setting FIC-2 through characterizer f (N) or through FIC-1.

The characterizer compensates for the speed difference between the two machines. This control system was designed and calibrated by trial and error and has been in operation since 1970.

Efficiency became an important consideration when tests showed that the compressor system was using about 36 percent more steam than intended. The "safe" solution implemented maximized production through the less energy-efficient machine. Furthermore, depending on load,

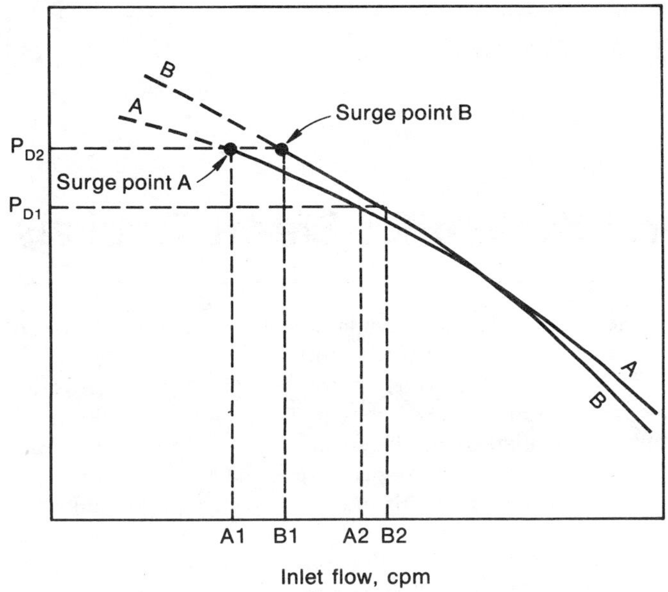

Superimposed performance curves

Fig. 5—Superimposed performance curves.

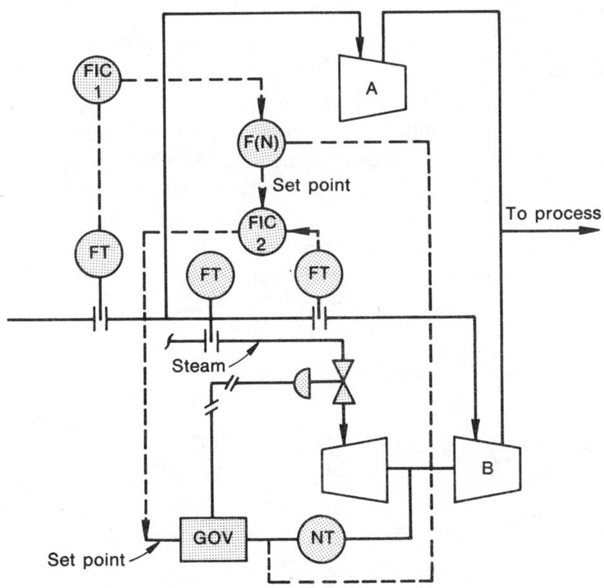

Load balancing control system

Fig. 6—Load balancing control system.

it was more economical to maximize thruput through A one time and B another. Occasionally one machine had to be run full out while the other had a slight amount of recycle.

In an effort to design an energy efficient system, manual graphic simulation was tried, without success. The system was finally simulated using a small, hybrid computer, using control strategy developed on a digital computer. Analog controller implementation was possible but impractical because it was too complex and unreliable.

Anti-surge considerations. Fortunately, effect of recycle flow on suction conditions in this system was not great. If spillback from one machine changed the common suction pressure significantly or rapidly, the simple control system of Fig. 14 would have trouble. When this occurs capacity control must be re-designed to counter this interaction. This can mean a complete, on-line change in control strategy.

Why on-line? Because economics require a strategy that cannot be executed when one machine is near its surge point. The near-to-surge strategy is uneconomic. Control strategies to cover different sets of operating conditions are desirable. The correct strategy can be selected on-line automatically. Such systems have been designed and implemented with a digital computer.

Automatic start-up. As previously explained, another important consideration in operating parallel compressors, is starting up a spare. Discharge pressure of machine A may be high enough to keep the check valve in machine B's discharge closed while B is trying to start up. If this check valve is located downstream of the anti-surge spillback, the anti-surge system may open the spillback valve to keep the compressor from surging. However, in order to force the check valve open and allow machine B to come on stream machine B's discharge pressure must be raised by closing the spillback valve. Will the machine surge? Probably.

This situation has to be handled by a carefully synchronized reduction of machine A P_D and an increase of machine B P_D. This has been done with analog controls, but it is difficult. Digital execution is recommended.

Suppose a check valve is placed inside the spillback take off? This is often done when there may be large volumes of gas in the spillback loop. A large volume of gas trapped inside the check valve could cause reverse rotation during compressor trip out. Compressor B would surge continuously unless the spillback valve opened and passed gas from the other machine.

At best, we are back to the problem of forcing the check valve to open. At worst the check valve will oscillate with the spillback loop. Once this cycle starts, nothing can be done except to take the spare down and try again.

There have been some successful attempts at automating on-line, spare start-up using relay logic or programmable logic controllers (PLC's) in combination with conventional analog controls. Although some of these systems have worked, today we see their implementation on digital computers.

These problems are not serious in every parallel compressor system. If they were, we would have a more complete answer by now.

There is no universal technique for designing compressor control systems, especially for parallel compressor systems. We cannot provide cookbook solutions but want to make you aware of potential problems.

The authors strongly recommend that control systems for parallel compressors be designed by dynamic simulation. We further recommend that these simulations be done on a hybrid computer.[1,2,3] Cost of simulation will be paid back many times by improved efficiency and safety of the resultant control system.

LITERATURE CITED

1 Nisenfeld, A. E. and Schultz, H. M., Analog Simulation Aids in Control System Design.

2 Nisenfeld, A. E. and Marks, A., Dynamic Simulation Prevents Start-Up Problems.

3 Nisenfeld, A. E., et al, "Better Compressor Controls," *Hydrocarbon Processing*, April 1975.

ACKNOWLEDGMENT

Based on material copyrighted by and reprinted with the permission of the Instrument Society of America.

New Extraction Control System for Steam Turbines

A new approach to automatic extraction pressure control has been developed for steam turbines driving process compressors. Here are the details

L. M. Anderson, Steam Turbine Division,
Turbodyne Corp., Wellsville, N.Y.

STEAM AT A CONSTANT pressure can be extracted from an intermediate stage of a multistage steam turbine by adding a pressure control system to the usual speed governor system. To control these two parameters simultaneously, two sets of steam control valves are required: (1) the high pressure valves which control the inlet flow to the turbine, and (2) the extraction valves which are located immediately after the point of extraction and conrol the flow of steam to the low pressure stages of the turbine. The function of the control system is to position both sets of valves, under varying conditions of horsepower and extraction system demand, in such a way as to maintain the set governor speed and extraction pressure. To accomplish this, it is necessary to relate the positioning of the inlet valves and extraction valves so that a change in power demand, or a change in demand for extraction steam, can be accommodated without disturbing the other controlled variable. The conventional method of relating the inlet and extraction valve positions is by a system of levers. This mechanism accepts position signals from the speed governor and pressure controller, proportions these inputs according to the design requirements of the particular turbine and provides a position output to each of the two sets of valves.

For steam turbines driving process compressors, it is becoming increasingly popular to use a pneumatic controller as a speed governor, in place of the conventional mechanical-hydraulic (fly-ball) governor. In the pneumatic systetm, a speed signal is obtained from a magnetic proximity pickup on the turbine shaft. An electro-pneumatic transducer then converts the electric pulses into an air signal which is proportional to the rotative speed of the turbine. The principal advantages of such a system are:

- The speed measurement is made without physical contact with the turbine rotor. This completely eliminates the reduction gearing required for mechanical governors.
- No moving parts are associated with the generation of the speed signal, and no lubrication is required.
- Any high quality, commercially available pneumatic controller can be used for the speed governor. Instrument air is generally available in process facilities which use this class of turbomachinery.

It is also common practice to use a pneumatic controller for extraction pressure control. The output of each of these controllers is in the form of an air pressure signal. In the past, these two pressure signals have been converted to position signals and, using a system of linkages, combined to meet the flow and power characteristics of the particular turbine. The output motions of the mechanical analog are then reconverted to proportional air pressures, which are used as positioning signals to each set of steam valves.

PNEUMATIC EXTRACTION CONTROLLER

A new approach to interconnecting the speed and extraction pressure control signals has been recently developed which completely eliminates the mechanical linkages. Pneumatic computing relays and other pneumatic devices are used to combine the speed and pressure signals and deliver directly, the proper output positioning signal for each set of valves. A block diagram of such a pneumatic control system is shown in Fig. 1, with the computer section enclosed by dashed lines. Operation of the control is shown in the following examples:

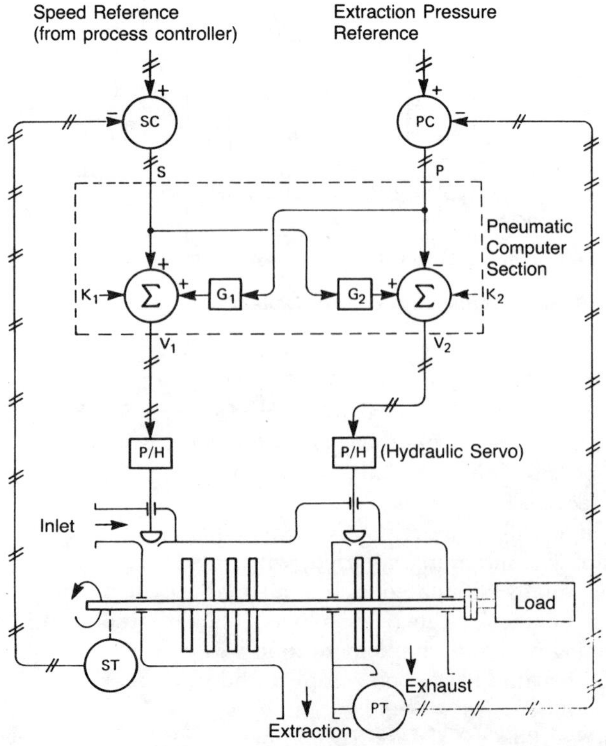

Fig. 1—Extraction control block diagram.

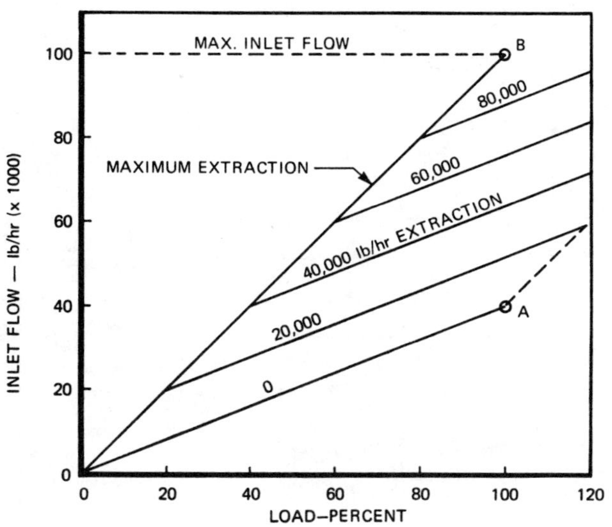

Fig. 2—Extraction turbine performance curve.

1. Assume an increase in power demand on the turbine, with the extraction flow remaining constant. As this happens, the speed controller reacts to increase its output signal. This increase in air pressure signal feeds directly through a computing relay in the extraction control system to increase the signal to the inlet valves, moving them in the opening direction. At the same time, in order not to upset the extraction pressure, this change in signal must also open the extraction valves to pass this increase in steam flow through to the turbine exhaust. The proportion of extraction valve opening is determined by the gain setting of the computing relay, and must be based upon the relative capacities of the inlet and extraction valves and the power capabilities of the turbine stages controlled by these valves.
2. Assume an increase in extraction flow demand while the power demand remains constant. As the steam pressure tends to drop in the extraction line, the pressure controller reacts to increase its output signal. This signal increase moves the inlet valves in the opening direction but, at the same time, it must move the extraction valves in the *closing* direction. The amount of movement of each set of valves must be in the correct proportion to insure the total horsepower output of the two turbine sections is not changed. Thus, there is no speed change to upset the speed controller. Here again, gain settings of the computing relays determine the amount of valve movement in response to the change in output signal from the pressure controller.

OPERATING EQUATIONS

The basic equations for the operation of the pneumatic computer are:

$$V_1 = S + G_1P + K_1$$

$$V_2 = G_2S - P + K_2$$

where:

S = signal from speed controller
P = signal from pressure controller
G_1 = pressure signal gain
G_2 = speed signal gain
V_1 = positioning signal to inlet valves
V_2 = positioning signal to extraction valves
K_1 and K_2 = fixed biasing pressures.

Gain values for the multiplying relays (G_1 and G_2) are easily calculated from a turbine performance curve such as the example shown in Fig. 2. Performance curves for extraction turbines are typically presented in this manner, with power plotted against inlet flow, and with extraction flow rates represented by diagonal lines. Point "A" on this curve indicates an inlet flow of 40,000 lb./hr., and, since it lies on the 0 extraction line, the extraction control valves must also be passing 40,000 lb./hr. This is the maximum flow that the extraction valves can pass, and is so indicated by the dashed line to the right of point "A". Thus, it can be seen that, with 0 extraction, as the load is increased from 0 to 100 percent, the inlet valves are required to open 40 percent, while the extraction valves must open 100 percent. Since this action is in response to demands from the speed controller, the speed signal gain is calculated as follows:

$$G_2 = \frac{\Delta V_2}{\Delta V_1} = \frac{100}{40} = 2.5$$

Similarly, the pressure signal gain can be calculated by considering the required valve actions as the extraction flow rate is increased from 0 to the maximum flow of 100,000 lb./hr., with no change in power demand. This action is in response to demands from the extraction pressure controller and is represented on the curve by moving from point "A" to point "B". Note that at point "B", the inlet valves are passing 100 percent flow, and since the full inlet flow of 100,000 lb./hr. is being extracted, the extraction valves are fully closed. Thus, in going from point "A" to point "B", the inlet valves have gone from 40 percent open to 100 percent while the extraction valves have moved from fully open to completely closed. The pressure signal gain then becomes:

$$G_1 = \frac{\Delta V_1}{\Delta V_2} = \frac{100\text{-}40}{100} = 0.6$$

The requirement that two sets of valves travel in opposite directions during changes in extraction flow rates is handled by feeding the summing computer in the negative direction as indicated on the block diagram.

Values of the biasing pressures K_1 and K_2 are selected to keep the output signals within standard pneumatic component ranges.

LIMIT CIRCUIT

Under some operating conditions, the design limits of the turbine may prevent it from reaching the desired criteria of both the speed and extraction pressure controllers. For example, the extraction flow demand may be high at a time when the horsepower requirement is low. Under these conditions, extraction flow could be met only if the flow through the high pressure section of the turbine was in excess of that required to produce the power demand. Referring again to the extraction curves in Fig. 2, this limit is represented by the line from the origin to point "B". At this line, the design limit of the turbine has been reached, and demands for extraction steam must yield to demands from the speed controller. In other words, if the design limit of the turbine does not allow sufficient extraction flow to satisfy the demand of

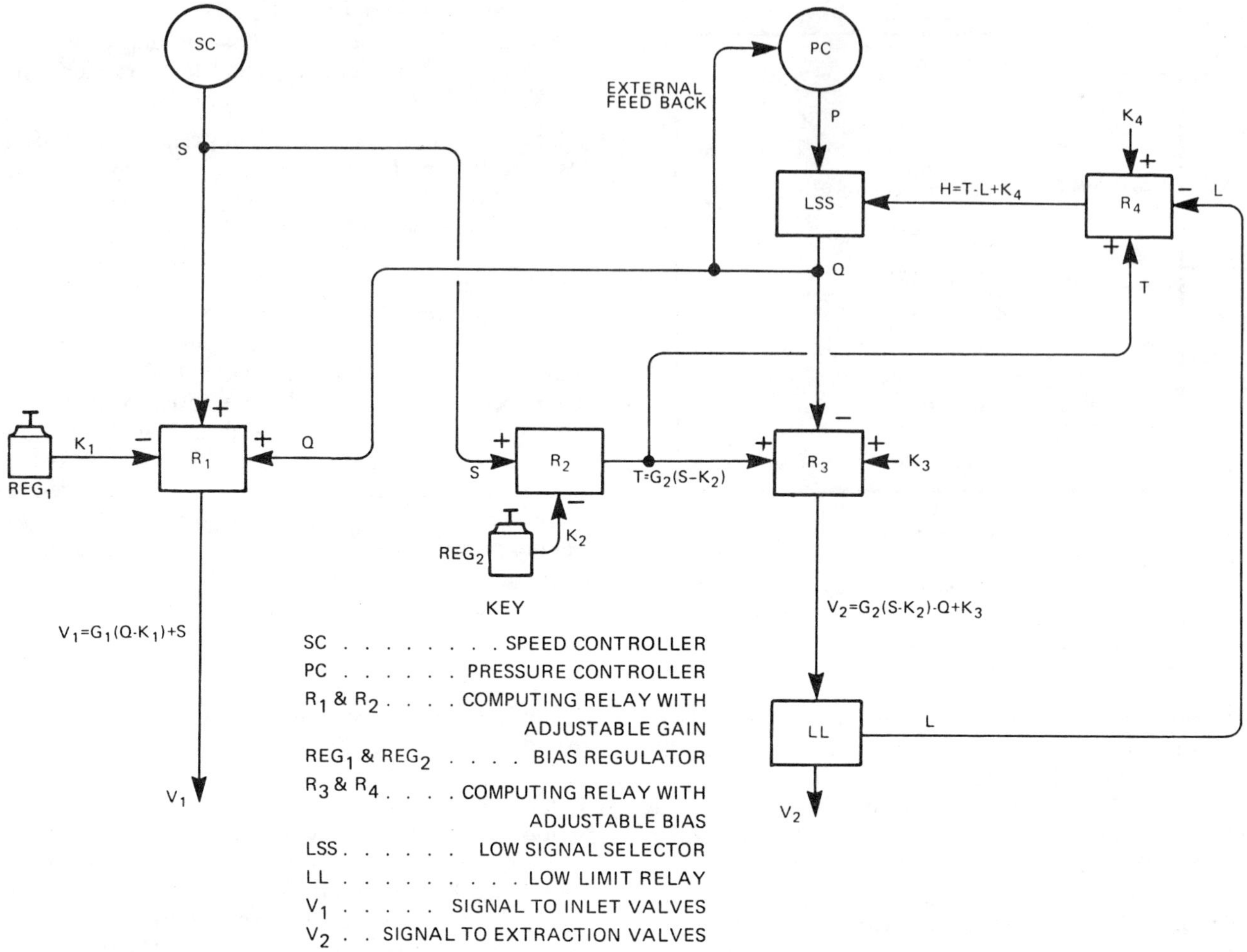

Fig. 3—Pneumatic control circuit, interconnected speed and extraction control system.

the pressure controller, controller action must be superseded in such a way that it cannot upset proper action of the speed controller.

Fig. 3 shows the complete pneumatic computer circuit, including the limiting feature. It can be seen how the limit circuit computes a superseding signal "H" as a function of low limit signal "L" and a modified speed controller signal "T". Value of signal "L" will usually be set to correspond with the fully closed position of the extraction valves, in order to allow full extraction operation. Signal "T" is substantially proportional to the horsepower being produced by the turbine. Thus a variable limiting signal "H" is computed to act as an override on pressure controller signal "P", as a function of the load demand. Signal "Q" is the lower of the two signals, "P" or "H". From the "Q" signal, an external feedback is provided for the pressure controller. This is to prevent reset windup of the controller when the limiting circuit is active.

CONCLUSION

The advantages seen for this system are:

- Signals are combined without the use of levers, linkages and pivots, which inherently have friction, lost motion and nonlinearities.
- With the exception of the final hydraulic control element, the complete control system is separate from the turbine. This simplifies inspection and maintenance procedures on the turbine. Isolation of the control components also lends itself to relatively easy protection in semi-enclosed or outdoor installations.
- The system uses proven pneumatic components of high reliability. Components are not subject to wear associated with mechanical members. Should maintenance be required, components are standard and should be familiar to instrument technicians. Over-all calibration of the system is accomplished in one-half hour by minor set screw adjustments.
- Valve positioning ratios are easily changed by adjusting pneumatic computer gains. This permits easy readjustment to accommodate changes in operating conditions. However, once the system has been calibrated for a given set of conditions, no adjustments of any kind are required, and the computer components can be completely contained in a tamper-proof enclosure.

Part V.

Utilities Control

Energy Management: More than Boiler House Control

There are three major areas for managing energy—energy production, energy distribution and the final energy user. Interaction of these areas must be considered for an over-all energy management scheme. Here's what to consider

A. Eli Nisenfeld, Fisher Controls Co.,
Marshalltown, Iowa

THERE ARE three major areas for "managing" energy. The first area in which we are concerned with energy is its production. Fig. 1 shows a boiler producing steam and a generator producing electricity. There is incentive to manage the production of energy or to conserve energy in its production.

Next is the distribution of energy (Fig. 2). Very often ignored in discussions of energy management, it is the link between the production and the final user, the third category. The distribution system in Fig. 2 has an 80, 20, 10 and a 2 atmosphere level and ultimately a vent line, which should normally be kept closed, but unfortunately is very frequently open. There are several ways for the steam at the highest pressure to work its way down to the steam at the lower pressure. It is desirable in the distribution of steam to keep the letdown valves closed, because the potential to do work is lost when steam is let down across a valve.

Fig. 3 shows examples of users and points up an interesting fact: A turbo/generator, which was previously identified as a producer of energy, is now a user of energy, steam. This one example by itself is a hint of the type interactions which exist between energy usage, production and distribution.

The effect of interaction can be seen by examining constraints. Boiler response is a constraint to effective control. For example, a boiler may be able to respond at the rate of a 10% change in production of steam per minute. If the user's demand for steam changes 20% per minute, there is a problem.

The problem is illustrated in Fig. 4. The line with the greater slope represents two curves. Circles represent user's needs and they are changing at the rate of about 1,000 or 1,500 pounds per half minute. The distribution system is a very well controlled system and is able to change a bit faster than the demands of the user. If the steam supply responded as rapidly, then there would be no problem. But the line with the lesser slope is the rate at which the boiler responds. The process cannot wait for the boiler to respond. It needs its steam right away.

The result can be demonstrated by the example in Fig. 5. A steam production system is producing steam at 90% of design. The distribution system is capable of getting steam from the producer to the user as fast as required. The user is using only 85% of design, but occasionally needs 90% of design steam very rapidly. In fact, too rapidly for the boiler to respond. Therefore, the distribution system must vent 5% of steam production, so when the user demands additional steam, the distribution system supplies it by closing the vent, giving the boiler time to repond.

Another example of waste is shown in Fig. 6, here we have a system with a waste heat boiler. Steam is made by sending reactor effluent through a three-way valve into the boiler. After the effluent leaves the boiler, it is water cooled to meet downstream temperature requirements.

The steam from the waste heat boiler goes into the distribution system and the distribution system also receives steam from the boiler house. Demand of steam is

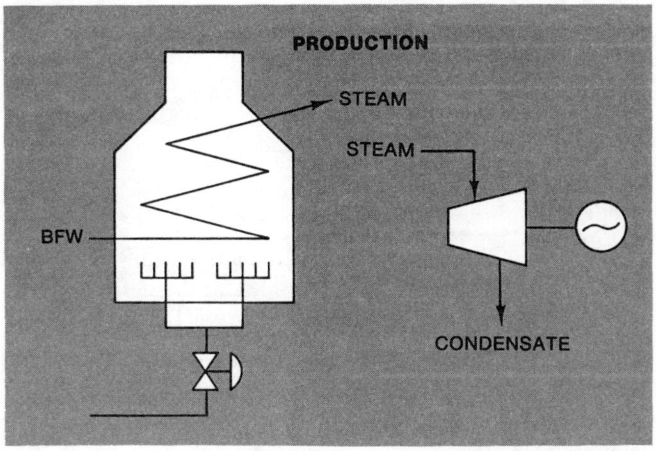

Fig. 1—Energy producing boiler and turbogenerator combination.

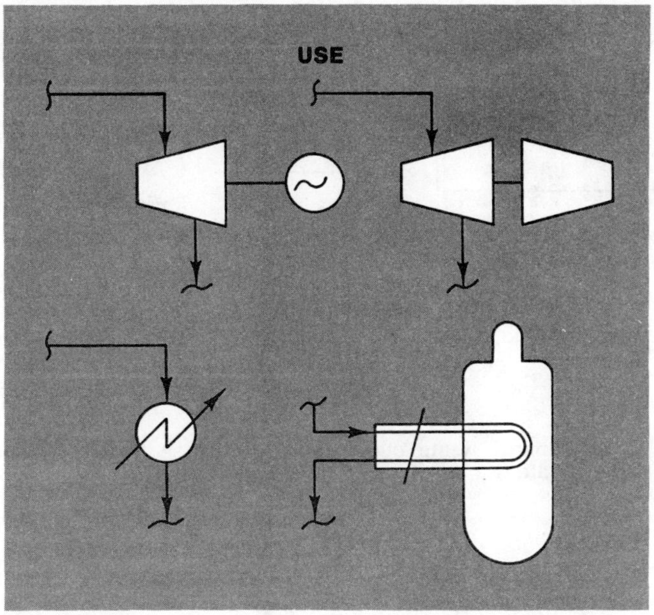

Fig. 3—Steam energy users.

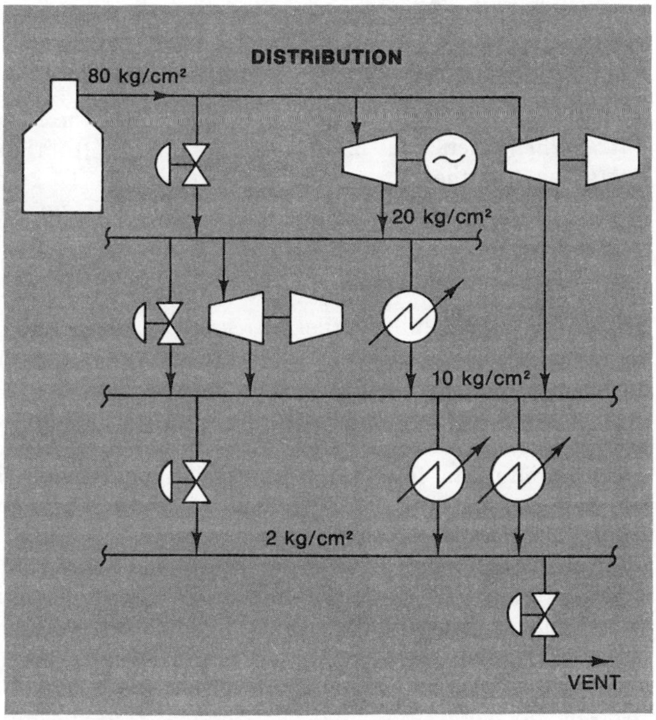

Fig. 2—Typical energy distribution scheme.

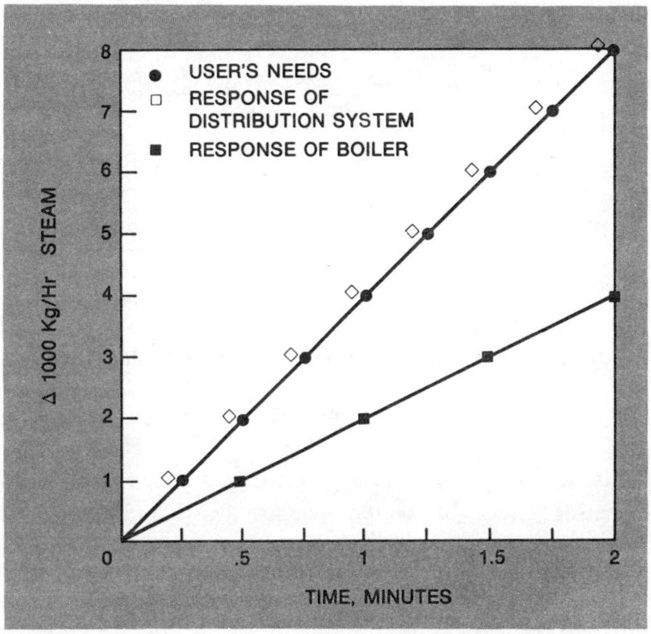

Fig. 4—Plot showing steam demand and system response time.

measured by a pressure controller which is usually part of the distribution system. This controller insures that the correct quantity of steam is in the distribution system. If a user increases demand for steam, pressure will tend to drop and, ideally, the pressure control system will adjust the three-way valve sending more of the reactor effluent through the boiler, producing more steam. (The percentages in this figure represent design values. It was intended that 5% of the heat in the waste stream would be sent to water with 95% of the waste heat being recovered as steam.)

Steam produced in this waste heat boiler is 10% of the total design steam. The user is using 90%, and the boiler house is producing 80%. Now, impose upon this situation the fact that the waste heat controls do not have an adequate response to the changes of steam required by the user. The result is that the system rejects more heat from the effluent to the water. Only 5% of the steam comes from waste heat; 85% of the steam must come from the boiler house. The waste heat boiler tends to be slower and the boiler house can pick up the loads in steam rapidly enough. Fig. 7 shows a less desirable

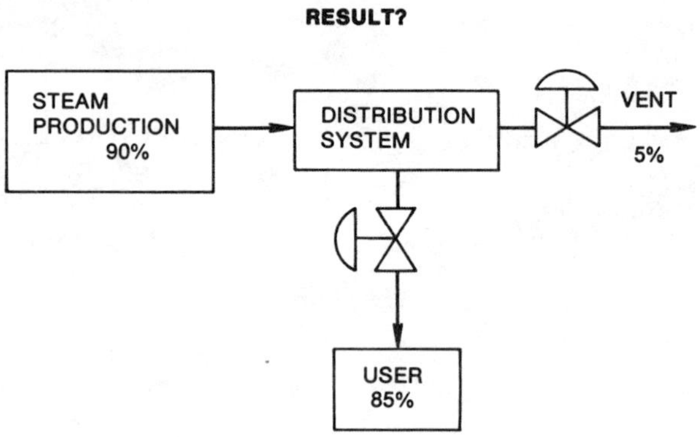

Fig. 5—Energy waste due to boiler response time and user demand rate.

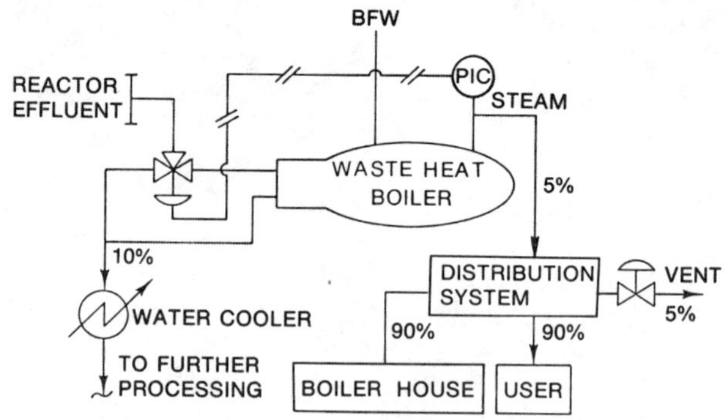

Fig. 7—Energy waste by venting steam and rejecting heat energy to water.

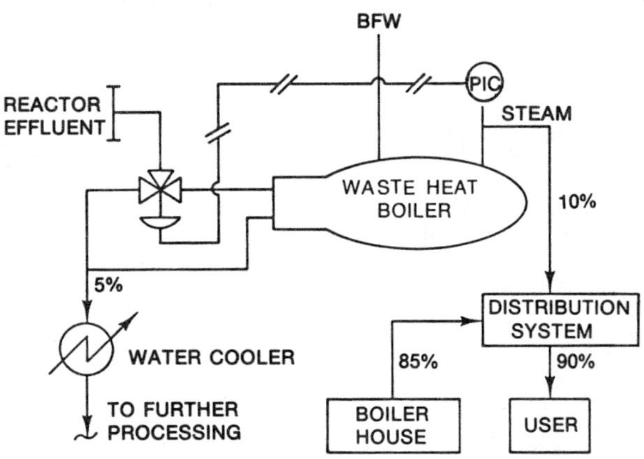

Fig. 6—Energy waste from inadequate control system response time.

solution. Energy is wasted by venting steam as well as by rejecting energy to water. We are wasting 10% of the potential heat by sending it to the water codler and making 90% from the boiler house. And even then venting steam. Capital investment for the waste heat boiler and the instrumentation and controls are lost because the system is venting the equivalent quantity of steam that comes from the waste heat boiler.

These are examples of how the response of systems can impose constraints upon attempts to conserve or manage energy. For the management of energy, we look at the production area. One energy management strategy is to calculate the efficiencies of a set of boilers operating in parallel and allocate the steam load among those boilers, so as to get greatest over-all efficiency. This was done in one system and despite the fact that calculations showed the boilers were operating more efficiently, they were still using more fuel.

We found that the least efficient boiler had the fastest dynamic response.

Because this boiler was the least efficient, the computer had actually backed it down to some minimum firing rate. When there was a change in demand from the users, this was the last boiler to be brought up. To compensate for the degraded dynamic response from the power house, the operators had adjusted controllers and the system was now venting more steam than the load allocation algorithm could save in the steady state. As a result, the client's bill for fuel increased.

The problem was solved by recognizing that, just as the processes areas interact, so must the controls.

The plant has steam production and controls, a distribution system and controls and users and controls. Steam production interacts with distribution, the distribution interacts with the user, but the controls do not interact with each other.

The solution was to provide information about the needs of the users to both the distribution and production areas and to have direct control interaction between distribution and production areas.

This system now operates with the vent closed, and over-all efficiency of the boiler system was significantly raised.

Final system design could have been done on the drawing board. Procedure for such a design technique is described below.

The first thing to do is determine disturbances that users must cope with. This must be done in every control system design. Determine the amplitude of the disturbance, rate of change and how often does this disturbance come. Then, estimate the response of process control loops. This includes the basic "user" control sys-

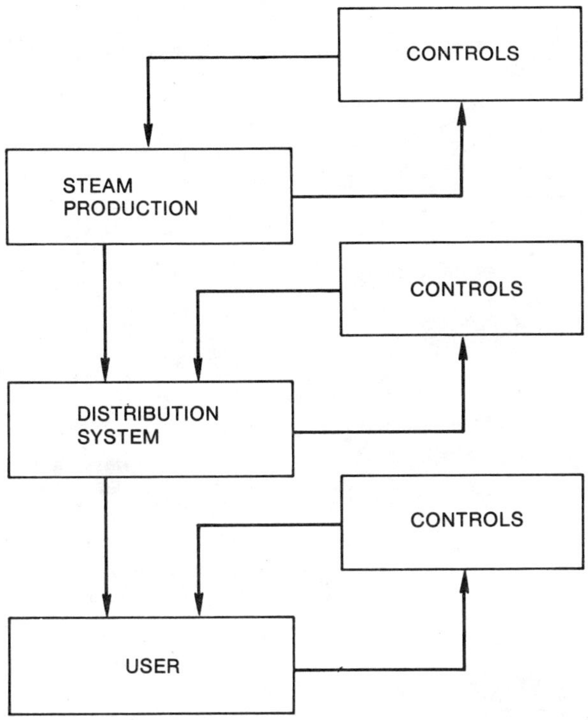

Fig. 8—Block diagram showing basic controls.

tems. It also involves the response of combustion controls, demand controls.

After estimating response of process control loops, respone of the distribution system and response of the energy producers, compare process response to production response plus distribution response. This evolution will identify the need for additional control intelligence between various levels of control.

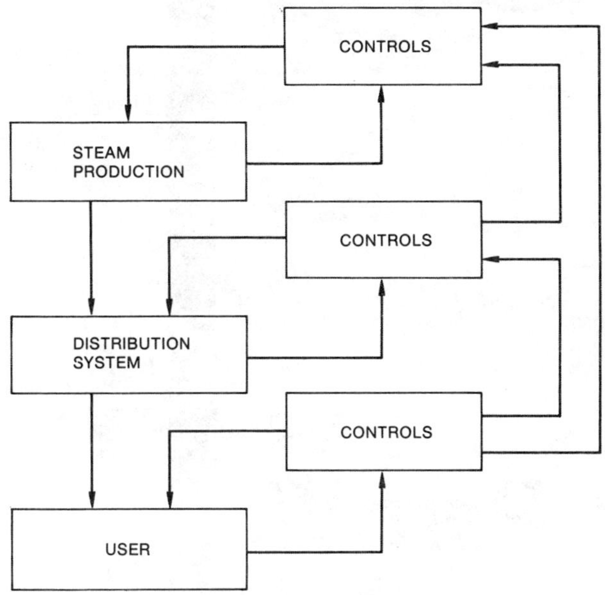

Fig. 9—Block diagram showing levels of communication among different levels of control.

The block diagram of Fig. 8 is for the basic controls; Fig. 9 shows levels of communication among the different levels of control which may be required.

Once the need has been identified, design of the control systems is almost routine.

Much of the content of this article seems obvious. However, it is the obvious which is most often omitted from our engineering efforts. As the complexity of the process increases, the danger from omitting this engineering work is great.

Part VI.

Pressure Relief and Storage Tanks

Relief Valves . . . What Needs Protection?

If relief valves are not located to be most effective, the results can be disastrous

P. B. Frankland, Davy Powergas Ltd., London

A "MUST" in all chemical processing plants and refineries is proper use and selection of relief valves. By means of relief valves, equipment is protected from rupture; fire hazards are reduced; safety to personnel is increased, and continuous plant operation is assured.

The importance of relief valves to the public welfare has grown to such an extent that all cities, counties, states and the federal government have specific relief valve codes that are strenuously enforced and regulated. Plants are often forbidden to start up because of not adhering satisfactorily to a relief valve code governing that particular locality.

Relief valves in general are discussed in the API Recommended Practice RP 520 and the ASME "Unfired Pressure Vessel Code." Mentioned as causes for over-pressure are fires, or other unexpected sources of heat. However, to properly diagnose a need for a relief valve, engineers should consider over-pressure due not only to the above causes, but also to:

- Process variations
- Abnormal process operation
- Equipment failure.

Fire considerations. Fire is number one in the minds of safety engineers in petroleum refineries and chemical plants because it is more prevalent and devastating.

If equipment containing liquid is exposed to fire, it is possible for the heat to vaporize the liquid. Pressure developed in the vessel will be a function of temperature and composition of the liquid-vapor state. To keep pressure down an escape for the vapors generated must be provided otherwise, it is possible to develop excessive pressure and rupture the equipment. Types of equipment to consider that may be exposed to fire are heat exchangers, fractionating columns, storage vessels and operating vessels.

In general, when applying relief valves, equipment that is connected by a system of adequate piping not containing valves which can isolate any item may be considered as one unit. If items can be separated from one another by valves in connecting piping, then each item must be considered as a unit with its own individual relief valve or valves.

As an example, in Fig. 1-A, the items are not separated by valves; therefore, PSV-1 can be used to protect V-1, V-2, and E-1. In Fig. 1-B, V-1 and E-1 are separated from V-2 by a valve; therefore, PSV-1 can be sized to relieve V-1 and E-1, and PSV-2 sized to relieve V-2.

Blocked outlets and inlets. Liquid filled lines that can be blocked in and are exposed to the sun should be protected by relief valves. The sun beating down on bare pipe can heat entrapped liquid, and create enough pressure to easily rupture piping. Relief valves should also be fitted to lines containing LPG and all traced lines.

If the cold side of a heat exchanger is liquid filled and can be blocked, it should have a relief valve. Referring to Fig. 2-A, it is quite possible to close the block valves on the cold side of the heat exchanger and at the same time have hot liquid go through on the "hot" side. In this situation, the "hot" side can transmit heat to the "cold" side, expand the block-in liquid and easily rupture the shell.

A fired-tube heater with liquid oil in the coils, if the coils can be blocked in, requires relief protection. In Fig. 2-B, the heater can very easily have the coils blocked in and heat applied at the same time. The coils can be blocked in by closing the two block valves "a" and "b"; or by closing the outlet block valve and the check valve

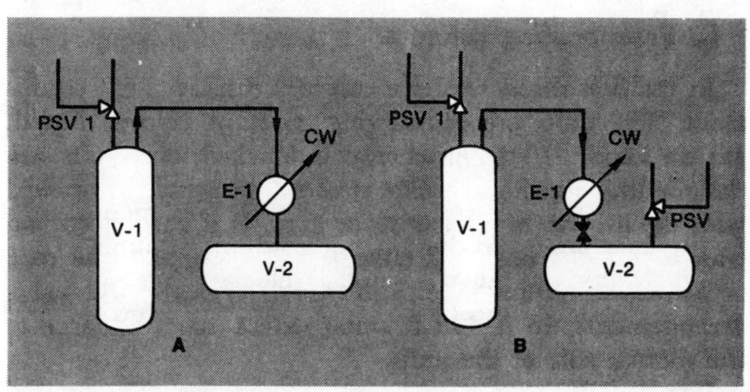

Fig. 1—(A) Interconnected equipment which cannot be isolated. (B) Interconnected equipment which can be isolated by a block valve.

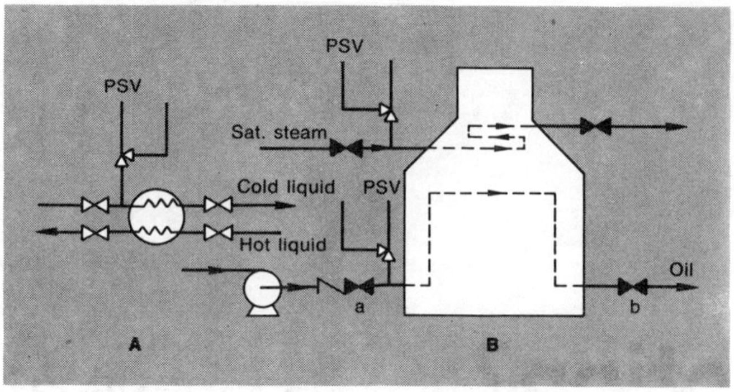

Fig. 2—(A) Heat exchanger protected on cold side. (B) Fired tube heater protected by relief valves on tube side.

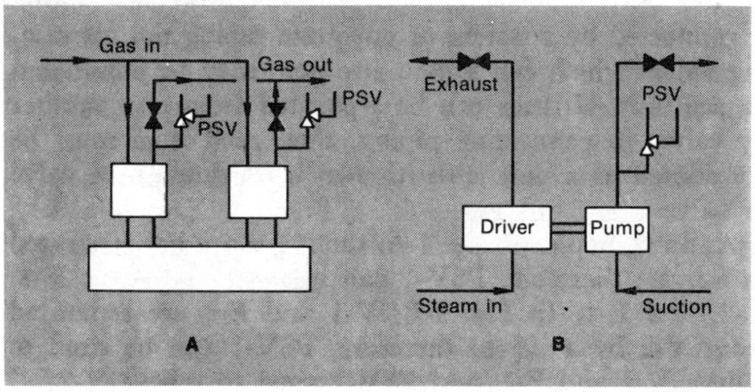

Fig. 3—(A) Reciprocating compressor protected on the discharge side. (B) Reciprocating pump or gear pump protected on the discharge side.

preventing any reverse flow; or by having the outlet block valve closed with the pump pumping liquid through the relief valve. Because of the latter condition, it is preferred that the relief valve be placed on the coil inlet side. This also protects the coils in the event of coking.

In Fig. 2-B, it is very possible to start up the heater with the valves on the steam coils closed with liquid condensed in the coils. Because of this possibility, it is necessary to insert a relief valve on the super-heater steam coil. A city inspector in a large city once refused to permit a plant from starting up until a relief valve was inserted for just such an application.

Block outlets can apply to:

A. Reciprocating compressors
B. Reciprocating pumps
C. Gear pumps
D. Steam exhaust side of turbines
E. Fractionating columns.

In the first three, we have constant displacement equipment. The units are discharging constant volume liquid, gas or vapor. If the discharge is blocked, the units are designed so that fluid cannot reverse but must go forward with the motion of the piston or gear. If it cannot go forward, then the unit will either stall or rupture the case or associated piping. Fig. 3 illustrates typical relief valve arrangements. In A and B, relief valves may discharge to the suction side of the units.

Fig. 4 refers to a turbine driven pump with a relief valve on the steam exhaust side of the turbine. The reason for this arrangement is that the exhaust side of a turbine is invariably designed for a much lower pressure than the inlet side. If the inlet side of the turbine were to be open and the outlet side blocked, then the exhaust pressure would be the same as the inlet.

Overpressure can also be developed in fractionating columns (Fig. 5) when the overhead vapors are blocked and the source of heat to the column continues. This condition can produce overpressure and necessitates the use of a relief valve on the column.

Loss of reflux can also cause overpressure to the column since it acts as a coolant. Reflux failure may be due to pump failure, instrument failure or loss of cooling water.

Instrument failure. Another possibility to guard against is instrument failure. Instrument failure can be attributed to the instrument itself failing, or loss of instrument air or power to one or all the instruments.

Instrument failure usually causes control valve "freezing" in an open or closed position. Suppose, in Fig. 6, that the stripping to the column is at 10 kg/cm², the maximum design of the column is 3.5 kg/cm², and the control valve, FRCV, freezes in an open position. If the overhead condenser is not designed to condense the full capacity of steam through the control valve, or if the block valve between the condenser and accumulator is closed, then it is possible to overpressure the column with 10 kg/cm² steam.

Entrance of a fluid or vapor from a high pressure system to a low pressure system may sometimes be a source of equipment overpressure. In Fig. 6-B, absorption of wet gas with lean oil takes place at 30 kg/cm². Rich oil leaves the absorber through a level-control valve to the deethanizing column, which operates at 13.5 kg/cm² and has a maximum design pressure of 17 kg/cm². The only thing that separates the absorber from the deethanizer is the control valve which reduces the pressure (and controls the level) down to the operating pressure in the deethanizer. If the control valve were to freeze in an open position, it is possible to feed an excess of high pressure liquid and gas to the deethanizer. If the deethanizer is not designed to handle this high rate, then relief protection must be provided.

Other possibilities to consider as the basis of instrument failure are:

- Steam or hot oil control to a reboiler
- Cooling water control on a condenser or cooler
- Steam control to a turbine
- Pressure control on a fractionating column
- Steam boiler controls.

Abnormal process operation. Abnormal process operation also has to be considered in the application of relief valves. The human element is just as fallible as equipment elements.

As an example, in Fig. 2-A, manually shutting off the block valves on the cold side of the exchanger and opening the block valves on the hot side. In Fig. 2-B, manually shutting off the block valves on the steam coil containing water, and keeping the heater fired. Manually shutting off discharge valves on the compressor and pump in Fig. 3-A and 5. These examples are just a few in which it is possible that manual operation can cause overpressure of equipment, and in all cases must be considered for sizing relief valves.

Equipment failure. Plant cooling water pump failure, fractionating column reflux pump failure and tube rupture of heat exchangers are the most important of process equipment failures causing overpressure.

The pumps may fail due to loss of power or mechanical breakdown. Exchanger tubes are subject to splitting and consideration should be given to the source of high pressure and its effect on the low pressure side. If, for example, a tube splits and the high pressure is on the tube side, then it is possible to overpressure the shell side, shell piping or associated equipment. The same consideration should be given if the high pressure were on the shell side.

Water hammer. Rapid shut-off of a valve in a liquid line or shut-off of a pump may induce water hammer in piping. Water hammer fundamentally is pressure waves transmitting back and forth in a liquid filled pipe.

Sometimes this pressure may rupture piping or equipment. To protect piping and equipment from water hammer, a relief valve should be used. Refer to API 521 paragraph 2.3 for solution to fast response systems.

Vacuum relief. Equipment (if not designed for vacuum service) has to be protected against underpressurizing as well as overpressurizing. If atmospheric pressure on the outside of the vessel is greater than the pressure on the inside, it is possible to collapse the walls.

Full or partial vacuum in equipment may be caused by any one of the following:

- Pumping more liquid out of a vessel than is entering
- Draining liquid from a vessel (water, such as after a hydrostatic test, which is most common) without venting
- Injecting cold liquid into a vessel after steaming it out or continuing reflux after reboiler steam failure.

The first two are quite obvious. Withdrawing liquid without replacing the displaced volume will produce a vacuum. In the third, if the vessel is full of steam and cold liquid is injected into the unit, then the liquid will condense the steam, considerably reducing the volume occupied by the steam and produce a vacuum.

Pilot operated relief valves. A pilot operated relief valve is not operated directly from line pressure, as with the conventional type, but via a pilot valve which introduces a snap action to the main relieving valve. With conventional valves an overpressure of 10 percent is required, but a P.L.O.R.V. can be set up to 2 percent, this has obvious advantages when high pressures are involved.

Other advantages include adjustable blowdown (typically 2 percent) which means a saving of product, large valve capacity and repetitive operation. The disadvantages are long delivery and price (50-100 percent higher) and they must be used with clean fluids or the pilot valve may become blocked.

Isolation valves. Isolation valves should be provided when a rupture disc is used in combination with a safety/relief valve if local regulations permit.

Stop valves in conjunction with safety/relief valves are

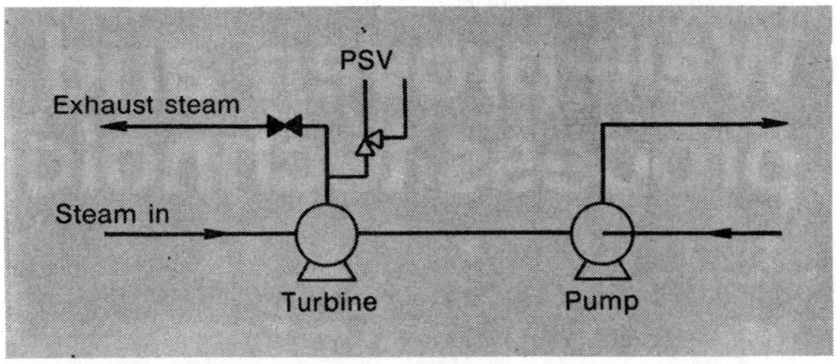

Fig. 4—Turbine driver protected on the exhaust side.

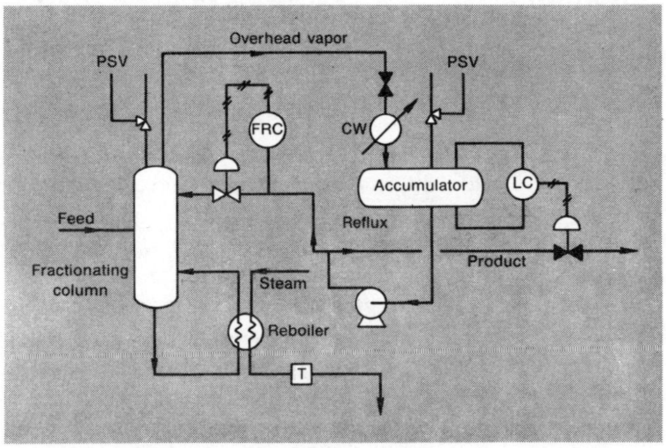

Fig. 5—Typical fractioning unit.

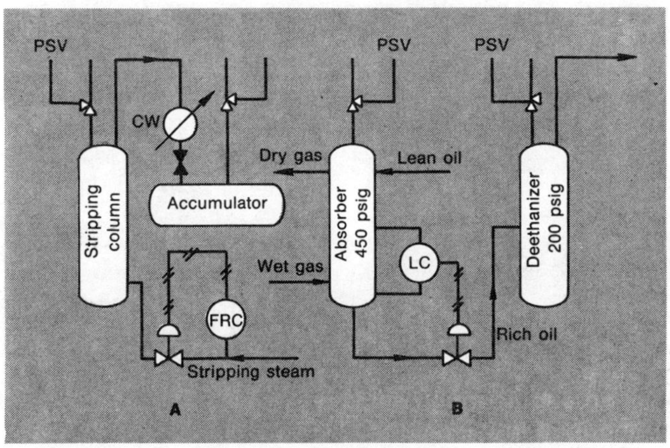

Fig. 6—(A) Stripping steam to column. (B) Absorption unit.

only applied if replacing a safety/relief is deemed necessary and is expected to cause unacceptable production loss, operational upset or air pollution in which case a second (spare) relief valve should be installed and a suitable interlock of block valves provided to ensure that the plant is fully protected at all times.

References: API RP 520 Part I, API RP 520 Part II, API RP 521, ASME "Unfired Pressure Vessel Code", DPG Standard 5540 STD/S1.

The typical examples cited will take care of approximately 95 percent of all safety/relief requirements in refineries, chemical plants and power plants.

In all probability the safety equipment may never be used—which is good; but, should it be used only once, then it more than pays for itself in rate of return on the investment.

Design for Better Safety Relief

Here's how to provide a valve, header and flare network that will protect against reasonable emergencies at minimum cost. The first step is a resolve to apply high-level engineering judgment

J. H. Jenkins, P. E. Kelly and **C. B. Cobb,**
The Pace Co. Consultants & Engineers, Inc., Houston

DESIGN OF THE SAFETY RELIEF system for a new processing plant is frequently one of the least rigorous aspects of the total design package. Analysis of the safety system is normally one of the final phases of the project when schedules and costs are tight. Consequently, there is a tendency to use short cut methods. These usually result in overdesigns. For example, some designers simply add all of the predicted relief loads together and size the header system and flare for one massive, catastrophic failure. Other designers divide the total by some conservative preconceived number to account for the fact that all units will probably not relieve at the same time.

The objective of a safety relief system design is to provide a valve, header and flare network that will protect against reasonable emergencies at a minimum cost. The designer is not so much concerned with avoiding emergencies through interlocking shutdown devices or automatic deluge systems, as with avoiding overpressuring of units during an emergency.

Rigorous analysis. Proper design of the safety valves, headers and flare entails rigorous analysis and may demand as much engineering judgment and expertise as the most sophisticated reactor design. Such an analysis will not only protect the processing units and the health and safety of operating personnel, but may substantially reduce the cost of required equipment. The "safe" system does not necessarily demand huge headers and flares, but rather adequate equipment in the right places.

A hypothetical FCCU is analyzed to illustrate how header and flare size requirements can be affected by a rigorous analysis. As shown in Fig. 1, the unit consists of a typical fluid cat cracker system and a gas plant including an absorber/stripper, sponge absorber, debutanizer, depropanizer and C_3 splitter. (Auxiliary units and minor equipment have been omitted in this example.) Design data are for illustration only.

SETTING THE BASIS

The most critical aspect of the design is to develop a definitive, reasonable set of contingencies or probable

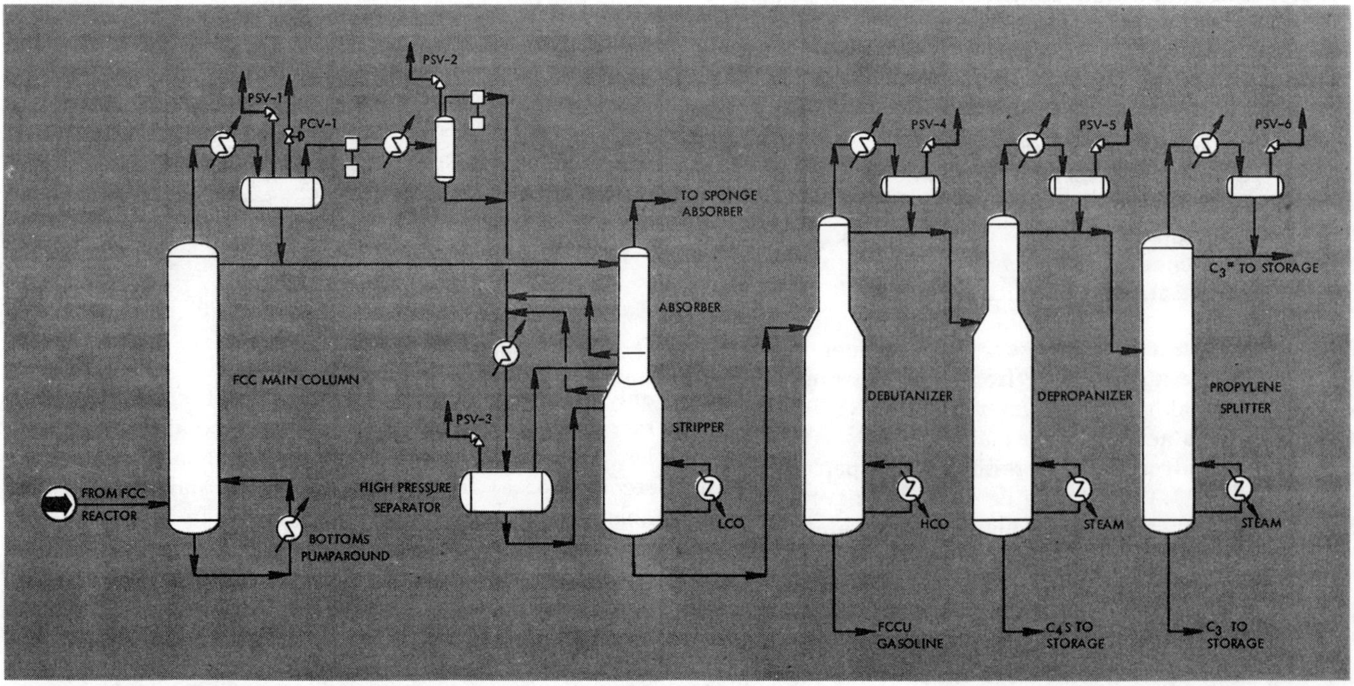

Fig. 1—This typical FCCU (gas plant) is analyzed to show how header and flare size requirements can be affected by rigorous analysis. Auxiliary units and minor equipment have been omitted. Design data are for illustration only.

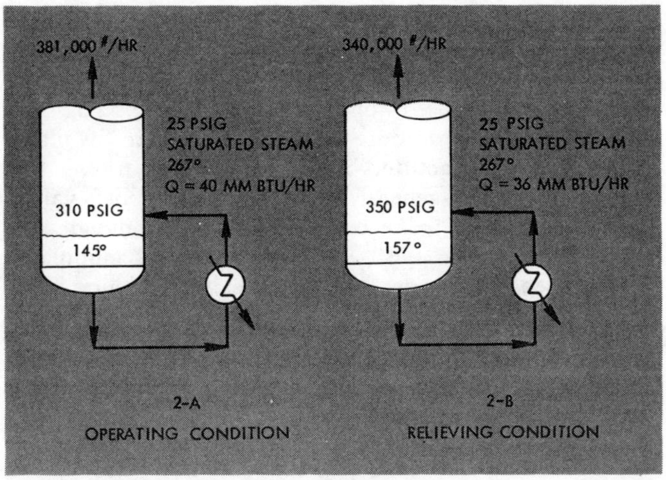

Fig.2—Relieving rates are compared here in terms of operating versus relieving conditions.

emergencies based on plant history and over-all experience. In the same way that an architect designs a building for specific wind loadings, earthquake factors and soil conditions, the safety designer must determine which emergencies have a reasonable chance of occurring and which emegencies are virtually impossible. A designer who simply assumes that all relief valves will open simultaneously may well end up with a "tail that wags the dog."

Gather data. First, all of the pertinent design data on the unit must be accumulated. These include:

- Process flow sheets with rates, pressures, temperatures, stream compositions and physical properties
- Mechanical flow sheets
- Maximum allowable working pressure (MAWP) and design conditions for all items of equipment and outline drawings showing skirt heights, dimensions, insulation specifications, etc.
- Exchanger and furnace data sheets
- Plot plans and plant maps
- Pump and compressor characteristics.

In addition, specifications for critical control valves, electrical layout drawings and utility battery limit conditions may be required.

Make a list. Next, a contingency list is developed showing all potential causes of overpressuring in the unit. In addition to normal plant emergencies such as cooling water or power failure, one must be aware of special contingencies that may be unique to a particular unit, such as a reactor runaway. For the hypothetical FCCU, the following contingencies apply:

- Cooling water failure
- Power failure
- Instrument air failure
- Individual control valve failure
- Fire
- Blocked outlet and operator error.

TABLE 1—Calculated maximum loads

Unit	Maximum Predicted Rate
FCC Main Column	450,000 Lb/Hr 60 M.W. @190°F
Interstage Receiver	30,000 Lb/Hr 42 M.W. @240°F
HPS	100,000 Lb/Hr 70 M.W. @300°F
Debutanizer	150,000 Lb/Hr 54 M.W. @190°F
Depropanizer	90,000 Lb/Hr 43 M.W. @160°F
Propylene Splitter	347,000 Lb/Hr 42 M.W. @140°F

DESIGN CONSIDERATIONS

Individual unit analysis. The next step of the safety relief design procedure is to analyze individual units to determine how they will react under emergency conditions. Every piece of equipment that can generate a vapor or liquid load under any of the contingencies must be analyzed. Although this analysis is not detailed, no load is too small to be considered.

A fatal accident was caused at one Gulf Coast refinery when an exchanger was blocked in on the tube side while heating media circulated on the shell side. Thermal expansion of the liquid caused the head to blow off, killing one man and injuring others. The disaster could have been prevented by a single ¾ in. x ¾ in. relief valve.

Calculating relief loads is not a simple exercise. In general, the calculations fall into three classes.

- **Unbalanced heat loads.** A classic example of an unbalanced heat load would be a cooling water failure in the propylene splitter. Since the reboiler will continue to generate vapor while the condenser is no longer working, the column will quickly pressure up and relieve. The rate of relief is not necessarily equal to the operating reboiler vapor rate, however; rather, the relief load should be calculated at relieving conditions. As shown in Fig. 2, the effect of increased pressure limits the thermal driving force in the reboiler and lowers the relieving rate. In some columns, this effect can reduce the calculated relief rate to a fraction of the normal vapor rate.

- **Block-in loads.** Inadvertently closing the outlet of a unit that is being fed or heated will frequently generate a relief load. Again, it is important to calculate the relief load at relieving conditions. For example, blocking in the vapor side of the high pressure separator in our hypothetical gas plant will result in a vapor load from both the stripper reboiler and compressor. Instead of combining the operating rates, the designer should consult the wet gas compressor curve to determine the actual compressor rate at relieving pressure and then add the unbalanced heat load from the stripper.

- **Fire.** Fire load is defined as the amount of vapor that can be generated by a fire directly under the vessel.

Calculations are based on unit geometry and insulation. Insulation cannot be assumed to reduce fire loads unless it can withstand direct flame impingement and cannot be torn off by high pressure fire hoses. No reduction may be allowed for polyurethane unless it has been fireproofed by an exterior shield.

The designer should regard calculated loads as a good approximation but not as an absolute value. Although it is possible to more accurately define relief loads by analyzing the operation at relieving conditions, it is impossible to predict exactly how a unit will react under emergency conditions. Table 1 shows maximum calculated loads for the FCCU units.

Safety devices. When calculating relief loads, the contribution of any safety device such as shutdown switches or solenoid valves should be neglected. As a general rule, the designer should assume that all safety devices that must make a positive action to prevent overpressure (that is, a valve closing or a pump being shut down) will fail. If, however, the positive action of a safety device will add to the relief load, the designer must assume that it will operate properly.

An example of this would be an automatic turbine spare on a reboiler furnace charge pump. Since a power failure would shut down an electrical pump, the column power failure load would be low. If, however, the spare started properly, there would be recirculation in the reboiler and the vapor relief load would be increased. Although it may seem overly conservative to assume that several shutdown devices will fail simultaneously, it should be noted that most shutdown valves and solenoids go virtually unattended for years and are subject to corrosion and sticking.

Valve sizing. After vapor loads are calculated from each unit for all of the various design contingencies, individual safety relief valves can be selected. Obviously, the highest predicted rate is used for each unit. Valve sizing is a straightforward procedure and methods are available from valve manufacturers and in trade literature.

There are two general types of safety relief valves, conventional and balanced. Because of its design, the conventional valve "sees" any back pressure accumulated in the discharge header so the relieving pressure is affected. For example, a conventional valve set at 350 psig that is relieving into a closed header system with 10 psig back pressure caused by loads from other units will not open until the vessel is at 360 psig. Since normal practice is to allow only 10 percent of set pressure "build-up" in a unit for most emergencies, choosing a conventional valve may force the designer to increase header pipe size to limit back pressure.

The second type of safety valve, the balanced valve, does not "see" the accumulated pressure and can relieve at design rates even when back pressure is 30-50 percent of set pressure. Since the balanced valve is only marginally more expensive, it is generally good practice to specify this type in any service with potential high back pressure in the header.

Analyzing simultaneous failures. Having sized the valves and calculated relief loads for all contingencies, the flare and header system is now sized. It is obviously unreasonable to add up the maximum load from each unit, then size the header for the entire load. In the same way that the architect does not plan for a hurricane at the same time as an earthquake, the safety designer assumes that only one thing will go wrong at a time. If, for example, the control valve on the FCC column

TABLE 2—Safety relief valve list

Valve Number	Location	Set Pressure PSIG	Max. Back Pressure, PSIG	GENERAL EMERGENCIES LBS/HR: Power Failure	Cooling Water Failure [1]	Inst. Air Failure	Fire
PSV-1	FCC Main Column	33	15	No Load	170,000—60 M.W. 190°F	No Load	No Load
PSV-1	FCC Main Column Atmospheric Vent	36	—	No Load	No Load	No Load	No Load
PCV-1	FCC Main Column	Opens at [2] 25	—	10,000—50 M.W. 140°F	150,000—60 M.W. 190°F	200,000—45 M.W. 130°F	25,000 —70 M.W. 200°F
PSV-2	Interstage	80	35	No Load	15,000—42 M.W. 240°F	No Load	5,000 —60 M.W. 160°F
PSV-3	High Pressure Separator	225	100	No Load	50,000—40 M.W. 240°F	No Load	100,000 —70 M.W. 300°F
PSV-4	Debutanizer	200	90	No Load [3]	75,000—54 M.W. 190°F	No Load	5,000 —70 M.W. 200°F
PSV-5	Depropanizer	275	130	20,000—43 M.W.	130°F	No Load	10,000 —43 M.W. 180°F
PSV-5a	Depropanizer Atmospheric Vent	285	—	70,000—43 M.W. 130°F	15,000—43 M.W. 130°F	No Load	No Load
PSV-6	Propylene Splitter	300	140	40,000—42 M.W. 140°F	40,000—42 M.W. 140°F	No Load	10,000 —44 M.W. 160°F
PSV-6a	Propylene Splitter Atmospheric Vent	310	—	133,000—42 M.W. 140°F	133,000—42 M.W. 140°F	No Load	No Load
PSV-7	Storage Sphere	100	30	No Load	No Load	No Load	210,000 [3] —48 M.W. 100°F
	TOTALS TO CLOSED HEADER	—	—	70,000—44 M.W. 140°F	475,000—57 M.W. 180°F	200,000—45 M.W. 130°F	210,000 —48 M.W. 100°F

NOTES TO TABLE 2—(1) Assumes 50 percent cooling water failure; (2) PCV-2 is a pressure control valve on the reflux drum that opens to the header if drum pressure exceeds 25 psig. As a safety device that must make a positive response to protect the unit, it must be backed up by safety valves. That is, in sizing the valves, one must assume PCV-2 will fail. In sizing the header, it should be assumed that it will open and increase total header loadings; (3) Each sphere is a separate fire zone, so this load is not cumulative to other fire loads.

failed open, there is no reason to assume that the debutanizer cooler would fail at the same time.

It is important to notice, however, that one failure may precipitate another, so an analysis should be made to determine how the units and their utility systems interrelate and how the units relate to each other. For example, a total power failure can cause a total cooling water failure if the circulation pumps are electric. If, however, cooling water is supplied by two pumps in parallel, one electric and one turbine, an instantaneous total cooling water failure is not normally considered a valid contingency. For any contingency that would cause the loss of one pump, 40-50 percent of the total cooling water failure load is assumed.

Fire zones. Instead of adding all of the fire loads together to obtain the total vapor load caused by a fire, the plant is divided into separate areas or "fire zones." It is not necessary to size the header for a fire that spreads into several zones since a fire of such magnitude would destroy sufficient process piping to keep vessels from overpressuring and exploding. Fire zones are usually 2,500 ft.^{2} but may be defined by dikes or deluge systems that would prevent the spread of a fire. For example, each propane, butane and propylene sphere would be a separate fire area if adequate dikes separated each vessel.

Atmospheric release. One of the oldest and simplest means of reducing header and flare requirements is by venting light hydrocarbons to the atmosphere under emergency conditions. Before designing a system with atmospheric vent, however, several criteria should be met:

- **Emergency venting of hydrocarbons is allowed by local codes.** The designer should first contact local air pollution agencies to find out if emergency release of nontoxic hydrocarbons is allowed. Although some states demand that any hydrocarbon release be contained in a closed header, other states will allow atmospheric venting under emergency conditions.

- **The hydrocarbons pose no threat of fire.** Vapor should be released at the highest possible elevation and maximum velocity. Air dispersion models similar to those used in pollution modeling can be used to ascertain that hydrocarbon levels are well below combustion limits before reaching ground level or nearing fired heaters. Because such computer models rarely account for updrafts, downdrafts and plant geometry, a substantial margin should be allowed.

- **Noise level at grade will not be excessive during venting.** The high velocities necessary for proper dispersion of hydrocarbons into the atmosphere also result in considerable high frequency noise. Insulating the outlet header can reduce this substantially.

- **Valves will not leak.** Some operating companies install rupture discs under safety valves to guarantee that continuous venting does not occur. Normal prebulged rupture discs should not be used since they tend to fragment and could plug the safety valve. A reverse buckling type disc that is cut by stationary blades is recommended.

In many states, the operating company must report any emergency atmospheric venting to pollution control boards, so the entire strategy is jeopardized if "emergencies" are too frequent. A high percentage of these releases can be eliminated by installing a second, smaller valve sized for 10-20 percent of the total load. The set pressure of this valve is several pounds lower and the discharged is tied into the closed system. Most minor upsets can thus be released to the closed system and the atmospheric valve will open only under truly catastrophic conditions.

Header and flare sizing. As shown in Table 2, results of the analyses should be tabulated to determine the design contingency rate, molecular weight and temperature. Pressure drop calculations for sizing the main header and network should be made from the flare tip backwards. The header is sized such that accumulated back pressure in the header is lower than the lowest allowable back pressure of an individual valve.

Significant improvements have been made in flare design in the last several years. Previously, the maximum vapor rate up the flare was limited to approximately 200 ft. per second for most gases. At higher velocities, the combustible vapor was not sufficiently mixed with oxygen near the tip and the flame could lift or even blow out. Given this limitation, very large flares were required for large loads of low molecular weight gas. Improvements in tip design now afford good mixing and stable combustion at far higher velocities. It is important to note, however, that pressure drop in the tip may be limited by back pressure considerations. Flare vendors can supply additional information on pressure drop and maximum velocities.

SINGLE EMERGENCIES LBS/HR			
Blocked Outlet	Reflux Failure	Single Unit Cooling Failure	Other
No Load	170,000—60 M.W. 190°F	200,000—60 M.W. 190°F	N.A.
No Load	No Load	50,000—60 M.W. 190°F	N.A.
50,000—50 M.W. 140°F	150,000—60 M.W. 190°F	200,000—60 M.W. 190°F	25,000—45 M.W. 130°F
10,000—42 M.W. 240°F	N.A.	30,000—42 M.W. 240°F	N.A.
150,000—48 M.W. 140°F	N.A.	10,000—40 M.W. 240°F	415 gpm—.63 S.G. 200°F
150,000—54 M.W. 190°F	150,000—54 M.W. 190°F	150,000—54 M.W. 190°F	N.A.
20,000—43 M.W. 130°F	20,000—43 M.W. 130°F	20,000—43 M.W. 130°F	N.A.
70,000—43 M.W. 130°F	70,000—43 M.W. 130°F	70,000—43 M.W. 130°F	N.A.
40,000—42 M.W. 140°F	40,000—42 M.W. 140°F	40,000—42 M.W. 140°F	N.A.
307,000—42 M.W. 140°F	307,000—42 M.W. 140°F	307,000—42 M.W. 140°F	N.A.
10,000—48 M.W. 80°F	N.A.	100,000—48 M.W. 20°F	20,000—48 M.W. 80°F
150,000—48 M.W. 140°F	150,000—54 M.W. 190°F	400,000—60 M.W. 190°F	25,000—45 M.W. 130°F

Knock-out and seal drums. A knock-out drum is required at the base of the flare to collect any liquid released into the header. For higher molecular weight mixtures, a condensing drum, in which a portion of the vapor is condensed by circulating water or an internal cooler, may be installed within battery limits. Vapor to the flare is reduced, thereby lowering header and flare requirements. In large systems that serve many units, it is often economical to install excess surge capacity and recover vented gases and liquids. Complete discussions on different types of drums and sizing are readily available in the literature.

It is also necessary to provide a means of sealing the flare from the header since cooling in the header after a release can pull air into the closed system. There are several sealing systems in use.

- **Water seal drums.** The oldest method of sealing employs a drum half full of water with the vapor inlet line submerged. This type of unit is becoming less popular since water from the drum poses a disposal problem.

- **Molecular seals.** Several vendors offer a molecular or "dry" seal as a means of sealing the flare. A small natural gas purge is required. The molecular seal functions by trapping a volume of the purge gas in an invertical compartment, preventing air from entering the flare stack by buoyancy effects.

- **Purge gas seals.** In the past, many units were sealed with a continuous flow of natural gas up the flare. A purge velocity of 1-3 ft. per second is required. In lieu of higher energy costs, a simple purge gas system without a supplementary seal is not economic and is rarely used today.

- **Internal seals.** Several vendors now offer a stationary seal requiring a small gas purge that keeps air out of the flare by making the purge gas turbulent. Because the low purge gas rate is not sufficient to equalize header pressure after a release, such units should be used in conjunction with a control system to increase the purge gas rate as the header cools.

Moving parts are not allowed inside a flare header, so no type of check valve can be used to seal the flare.

FINAL DESIGN

As shown in Table 2, the design case for the hypothetical gas plant is 50 percent cooling water failure. The maximum allowable back pressure in the header is 35 psig as limited by PSV-2 on the interstage receiver. Assuming a pipe run of 2,000 ft. to the flare area, the major equipment requirements are:

- 24-in. main header
- 20-in. I.D. flare stack with a riser height of 175 ft.
- 14-ft. I.D. x 30-ft. T-T double entry knockout drum
- Stationary seal requiring 9 Mscfd natural gas purge.

In comparison, a safety relief system designed to handle the maximum unit loads simultaneously would be much larger:

- 36-in. main header
- 30-in. I.D. flare stack with a 250-ft. riser
- 20-ft. I.D. x 40-ft. T-T double entry knockout drum
- Stationary seal requiring 20 Mscfd natural gas purge.

By defining a reasonable basis, employing atmospheric venting in proper places, and by calculating relief loads at relieving conditions, the designer can provide a safety system that offers protection against vessel explosion yet costs far less than a system that is not rigorously designed.

ROLE OF THE OPERATOR

The one factor deliberately not considered in all of the calculations, approximations and assumptions of safety system design is the operator. A well trained operator can reduce a very large potential load that could overpressure and destroy a unit by simply turning off reboiler steam or shutting down a pump. It is therefore important to educate operators in proper responses to overpressuring in the same way that fire prevention is taught. Even with the finest in safety equipment and design, no plant is safer than its personnel.

LITERATURE

This article covers in a general manner the engineering principles that should be considered in designing a safety relief system. Before beginning a safety system design, one should become familiar with the American Petroleum Institute's "Recommended Practices," 520 and 521. Other sources of interest include:

1. Heitner, Irving, "A Critical Look at API RP 521," *Hydrocarbon Processing,* November 1970.
2. Klooster, H. J., Vogt, G. A., and Braun, G. F., "Optimizing the Design of Relief and Flare Systems," *Chemical Engineering Progress,* January 1975.
3. Lurden, D. E. "Requirements for Safe Discharge to Atmosphere," paper presented at American Petroleum Institute 28th midyear Refining Div. Meeting, May 1963.
4. Reed, Robert D., "Design and Operation of Flare Systems," *Chemical Engineering Progress,* June 1968.
5. Isaacs, Marx, "Pressure-Relief Systems," *Chemical Engineering,* February 1971.

Control Blowdown to the Flare

Happily this design concept can benefit the environment and pocketbook all at the same time!

B. Paruit and **W. Kimmel,** Brown and Root, Inc., Houston

Flare systems designed for controlled blowdown can improve environmental protection and save money. Peak noise levels can be reduced, smoke production abated, space saved, and investments cut. But design for controlled blowdown requires extensive calculations while its application is limited to special situations.

Emergency relief systems usually vent gases through an unrestricted system. This conventional design minimizes venting time but maximizes system size and attendant costs. Controlled blowdown design, on the other hand, limits and extends peak venting rate (Fig. 1). Consequently noise levels are less, smoke abatement is more efficient, radiation levels are reduced and equipment spacing is minimized.

As flaring rate increases, nominal equipment size (due to hydraulic considerations) and spacing requirements (due to radiation and dispersion considerations) must increase. Added costs for nominal equipment size, while large, are usually overshadowed by the added costs due to spacing requirements to avoid high radiation flux on equipment and personnel. This is especially true if the cost of "Real Estate" is high. The best example of this is on offshore platforms where the use of a controlled blowdown design can often eliminate an intermediate flare bridge platform costing several million dollars.

Controlled blowdown is not realized at the expense of any safety requirements or reduction in system reliability. The system is designed to blowdown the plant in the required time without interferring with the relief pressure system. In fact, the system is made up of simple, well tested and proven equipment to ensure highest reliability and an attractive cost.

A disadvantage of the concept is that it requires extensive calculations to make sure that the hardware has been properly sized. Also it does not usually present any significant incentive for use with units having a small gas inventory.

This concept has been applied at three offshore platforms. In these particular cases, because of the very large gas inventories, the economics overwhelmingly favor controlled blowdown.

THE SYSTEM

The main design requirements for a controlled blowdown system are to:

- Be fail-safe
- Add to the relief system without creating interference
- Use commercially available components of the highest quality
- Be independent of normal instrument supply
- Eliminate dead time at the beginning of blowdown
- Avoid overshoot at the beginning of blowdown
- Control flow at a constant maximum rate over a wide range of source pressure for as long as possible
- Return the system to standby condition after use.

Such a system is shown in Figs. 2 and 3. The control blowdown system is located off the relief system (Fig. 2) to avoid any interference with the relief system, pressure safety valves, ruptured disk, etc. The reliefs discharge directly to the flare header such that a closed blowdown

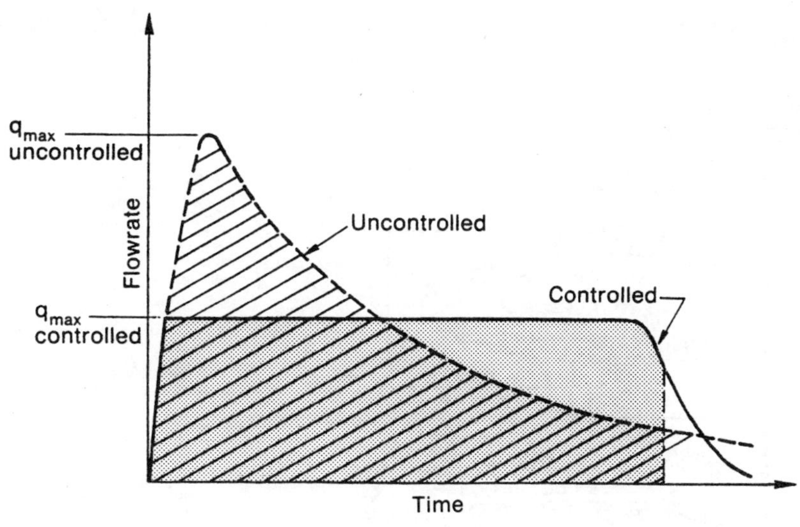

Note: areas under curves are equal.

Fig. 1—Blowdown can be controlled to avoid initial peak flow.

control valve cannot block in the plant relief system.

The system is made up of a main control valve (PCV-1) with its positioner, a restriction orifice (RO), a pressure indicator controller (PIC), two pressure sensors (PS-1 and PS-2), a test relay and a surge bottle. Each component is designed to provide a specific feature.

The main control valve (PCV-1) regulates the blowdown flowrate to keep the pressure in the flare header close to its desired value. The main control valve is closed when not energized.

The restriction orifice (RO) is sized for the design flare flowrate and delivers a pressure equal to the design pressure in the flare header at the beginning of the blowdown. This restriction orifice minimizes dead time at the beginning of blowdown. It also ensures system depressurization in case of a mechanical failure of the main control valve.

Two pressure sensors (PS-1 and PS-2) detect the blowdown header pressure and flare header pressure, respectively.

The Pressure Indicator Controller (PIC) senses the pressure in the flare header and sends a signal to the main control valve (PCV-1). The PIC sets system pressure such that the back pressure in the flare header corresponds to the desired blowdown flowrate to the flare. The set pressure is low enough to prevent derating the relief valves which discharge to the flare header.

The test relay is used to minimize the probability of mechanical failure of the main control valve. This relay exercises the main control valve and verifies the operability of the main control valve without interfering with the controller logic.

The surge bottle provides an adequate supply of air for operation independent of the normal air supply.

OPERATION

Before blowdown starts there is no flow and pressure in the blowdown header (PS-1) and flare header (PS-2) is low. The PS-1 senses very low pressure and does not deliver instrument air. The main controller has no power supply so the main control valve is unloaded (closed). This will avoid overshoot at the beginning of the Emergency Shutdown (ESD).

When the blowdown starts the ESD valve on line from main plant opens and flow rushes through the restriction orifice. This orifice is sized to set the desired back pressure in the flare header at the beginning of blowdown. The PS-1 senses high pressure in the blowdown header. The PS-2 senses high pressure in the flare header causing the main controller to receive power.

Flow continues to go through the restriction orifice and is supplemented by flow through the main control valve which is adjusted to maintain the desired flare header back-pressure. As the pressure in the blowdown header decays with time, the PIC increases the opening of the main control valve to keep the pressure in the flare header as close as possible to its desired value.

As long as the main control valve, PCV-1 is operating, the flow rate going to the flare is constant and the system behaves as a zero order system with respect to time.

BLOWDOWN TIME

Actual calculations of blowdown time require dynamic analysis. However, a simplified approach to sizing the

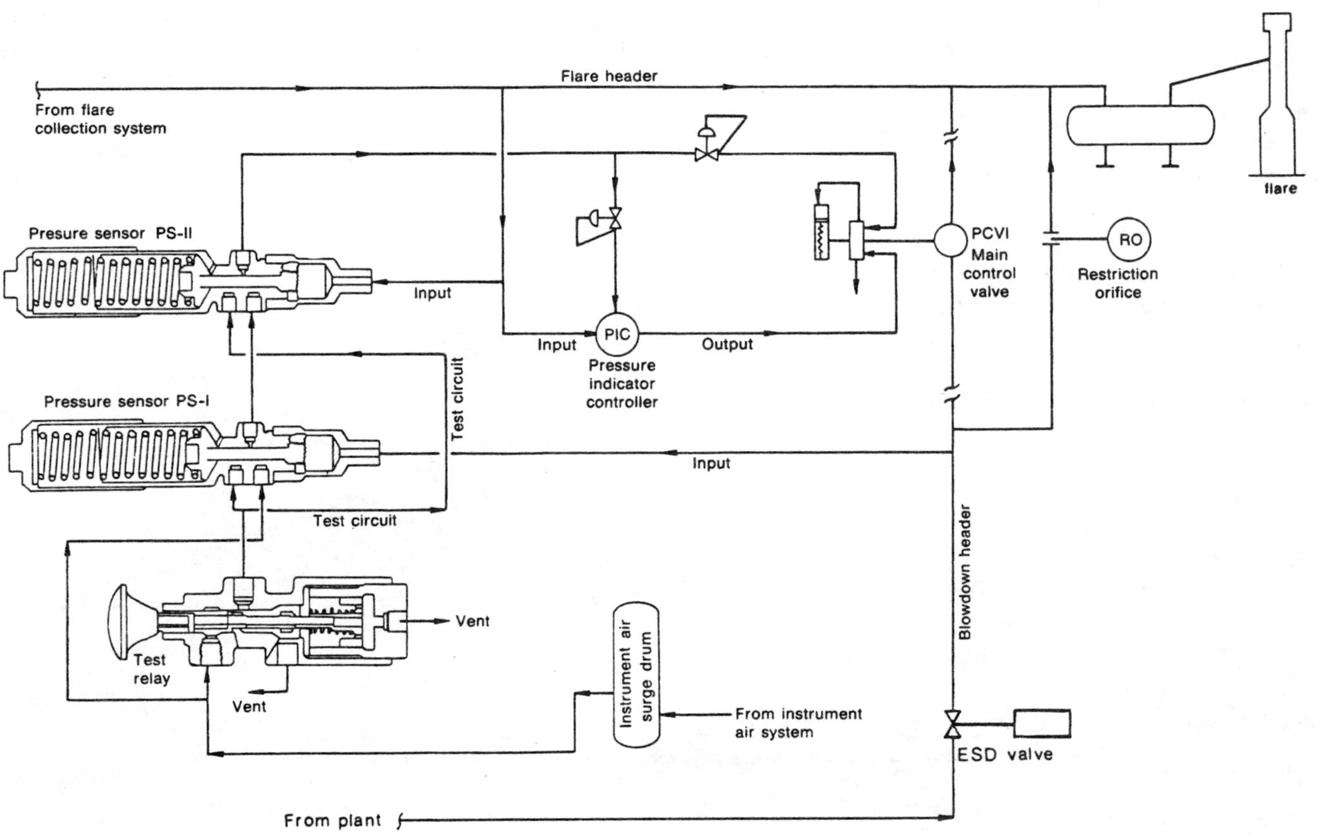

Fig. 2—Arrangement of the control system for controlled blowdown.

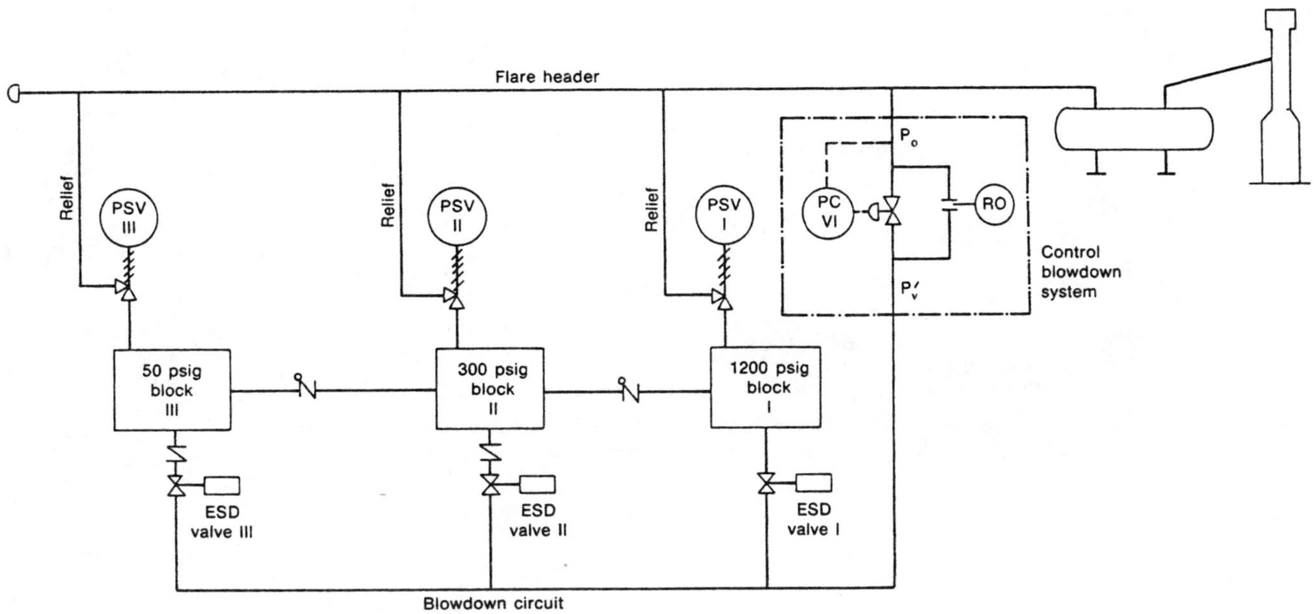

Fig. 3—Arrangement of a controlled blowdown system in a unit.

major components from its operating pressure down to a given pressure (P_f) is possible. (Refer to Fig. 3)

►Evaluate each volume V_1, V_2, V_3 to get actual gas volume. Convert each volume to standard conditions.

►Pick a maximum number for a first approximation of gas flowrate during blowdown and size flare header and flare tip. This gives the back pressure to be maintained downstream of the main control valve. This pressure must not exceed allowable flare header pressure.

►Size the main control valve and orifice in order to have the desired back pressure (P_o) for the chosen blowdown flowrate. The valve should be regulating on the widest possible pressure range.

►Find the upstream pressure of the valve (P^1) when the control valve is wide open. P^1 is the pressure at which the system stops to react as zero order system.
If $P^1 \leq 2P_o$ = regular flow through an orifice.
If $P^1 > 2P_o$ = exponential decay until upstream pressure is $2P_0$.

►Calculate the time θ_1 it takes for Block I to have its pressure equalizing Block II.

►Calculate the time θ_2 it takes for Block I and II to have their pressure reaching P^1.

►Calculate the time θ_3 it takes for Block I, II, III to have their pressure reaching the final desired pressure P_f. As long as $P > 2P_0$, the flow is an exponential decay.
When $P \leq 2P_0$, blowdown flowrate is given by a flow through an orifice—step wise and iterative calculations are needed to calculate θ_3.

►Add θ_1, θ_2, and θ_3 and check if the total blowdown time is in accordance with the blowdown required time. If not pick up another blowdown flowrate and start calculation from the second step.

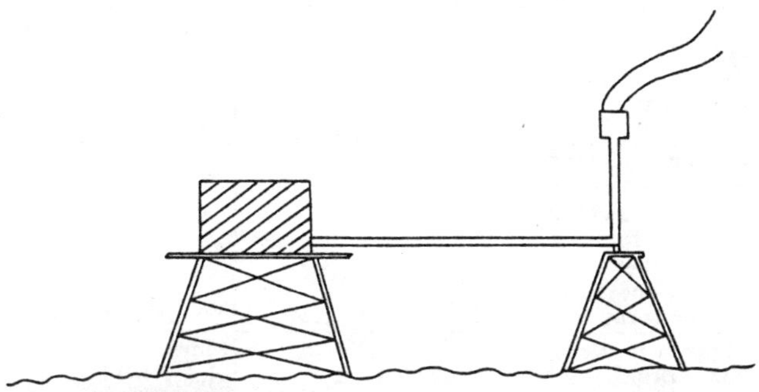

Assumptions

— Initial pressure, PSIG	1,200
— Final pressure, PSIG	150
— Total volume of the system, A.C.F.	10,000
— Total gas inventory @ 1,200 psig, A.C.F.	10,000
— Allocated blowdown time, min.	15
— Maximum radiation rate on highest point of platform, btu/hr, sq. ft.	2,000
— Wind velocity, mph	20
— Velocity @ tip of flare, FPS	400
— Gas flowing	Hydrocarbon vapor
— Compressibility factor @ 1,200 psig	0.72
— Compressibility factor @ 150 psig	0.92
— Initial gas temperature, °F	60
— Final gas temperature, °F	10
— Heat of combustion, BTU/C.F.	1,700
— Molecular weight of gas	30
— Maximum distance between two support platforms, ft.	300

Fig. 4—Example assumptions.

COMPARISON BETWEEN CONTROLLED AND UNCONTROLLED BLOWDOWN

A simple system consisting of a single pressurized vessel in an offshore location is investigated to outline the economic impact that controlled blowdown can have.

The system chosen is represented by Fig. 4 and the assumptions shown there. All calculations are based on API RP 521.

In this example, it is possible to derive a simple rela-

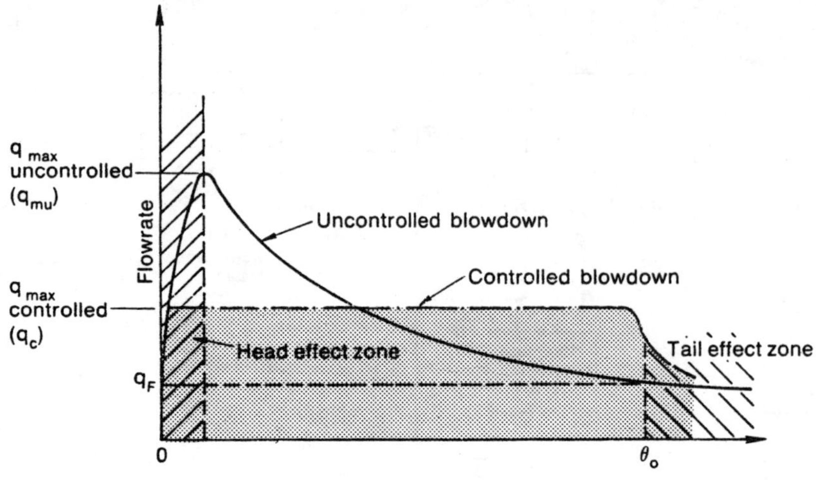

Fig. 5—Typical blowdown flow-time relationships.

TABLE 1

	Controlled blowdown	Uncontrolled blowdown
Peak flare flowrate MMscfd	99.6 MMscfd	230 MMscfd
Distance from midpoint to flare to reach 2,000 Btu/hr. sq. ft. radiation	240 feet	360 feet
Required distance at which the flare stack should be located from platform if 10 ft higher than the highest point on the platform	About 340 feet	About 500 feet
Support needed for 2,000 Btu/hr. ft.^{2}	Flare platform plus 1 flare boom	Flare platform plus 1 bridge tripod
Savings	1 tripod platform 160 feet of pipe bridge support (Several millions of dollars)	

tionship (See Box) between the peak flare flow rate for uncontrolled and controlled blowdown. This relationship is:

$$\frac{Q_{max} \text{ uncontrolled}}{Q_{max} \text{ controlled}} = \left[\frac{\int}{\int - 1}\right] \text{In} \int$$

where $\int$ = absolute starting pressure over absolute final pressure.

To derive this equation the effects occurring at the very beginning and at the end of the blowdown have been neglected. This corresponds to neglecting the head effects (flow rate going from 0 to maximum value) and tail effects on the blowdown flow rate curve. This is shown on Figs. 5 and 6.

In this example the initial pressure is 1,200 psig and the final pressure is 150 psig and the system has to be depressurized in 15 minutes. The above equation shows that the peak flare flow rate will be 2.3 times larger for the uncontrolled blowdown compared to the controlled blowdown.

In Table 1, a comparision between the controlled and the uncontrolled blowdown is made. This comparison

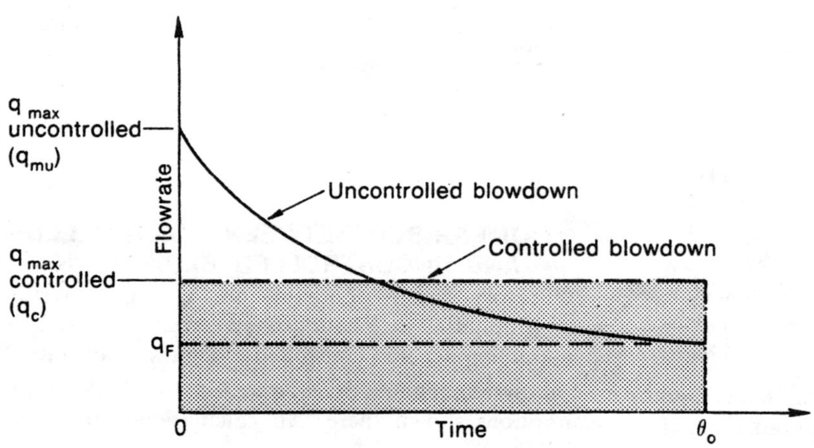

Fig. 6—Approximate blowdown flow-time relationships used for derivation.

RATIO OF CONTROLLED AND UNCONTROLLED PEAK FLARE FLOWRATES

FLARE FLOWRATES

Refer to Fig. 4, 6 and 7.

The uncontrolled blowdown flowrate is an exponential decay described by

$q = ae^{-b\theta}$ a, b unknown

with stated assumptions, the boundary conditions are:

@ $\theta = 0$ $q = q_{mu}$ (q max uncontrolled)
@ $\theta = \theta_0$ $q = q_f$ (q final)

this yields $$a = q_{mu};\ b = \frac{1}{\theta_0} \ln\left(\frac{q_{mu}}{q_f}\right)$$

replacing $$q = q_{mu}\left[\exp - \frac{\theta}{\theta_0} \ln\left(\frac{q_{mu}}{q_f}\right)\right]$$

by definition,

$$Q = \int_0^{\theta_0} q(\theta)\, d\theta$$

this gives

$$Q = \frac{q_{mu}}{\frac{1}{\theta_0} \ln\left(\frac{q_{mu}}{q_f}\right)}\left[1 - \exp\left\{-\ln\left(\frac{q_{mu}}{q_f}\right)\right\}\right]$$

using $G = \dfrac{q_{mu}}{q_f}$

and $Q = q_c\,\theta_0$

where

$q_c =$ maximum control blowdown flowrate

$$q_{mu} = q_c\left[\frac{G}{G-1}\right] \ln G$$

Since operation is in the choke flow region for uncontrolled blowdown the relationship between flowrate and pressure is given by

$q = \alpha\, C_v\, P$ (α proportionality constant)

therefore

$$q_{mu} = \alpha\, C_v\, P_{in}$$
$$q_f = \alpha\, C_v\, P_f$$

$$G = \frac{q_{mu}}{q_f} = \frac{P_{in}}{P_f} = \hat{\int}$$

finally,

$$\frac{q_{mu}}{q_c} = \left[\frac{\hat{\int}}{\hat{\int} - 1}\right] \ln \hat{\int}$$

outlines the differences which will lead to a very significant cost difference between the two systems.

Location of the flare stack will be 500 feet from the main platform for the uncontrolled blowdown compared to 340 feet for the controlled blowdown. This is based on the same height of the flare stack (10) feet above the highest point on main platform. This 160 feet difference in offshore flare piping will have an impact on the economics because of the cost of the required structural support facilities, bridge platforms or other types of platforms especially in deep water.

In systems containing several pressure levels, the above equation is somewhat more complicated because of interacting systems. The impact on the economics due to a controlled blowdown versus an uncontrolled blowdown in case of interconnecting systems will be even more significant than for the simple example developed here.

NOMENCLATURE

- a proportionality constant
- b proportionality constant
- c subscript for controlled
- C prefix for curve
- C_v flow coefficient for valve
- f subscript for final
- G flowrate ratio $\frac{q_{mu}}{q_f}$
- n number of moles
- P instant pressure
- P_{in} pressure of the system before blowdown
- P_o pressure downstream of the control blowdown valve—desired set pressure in flare header
- P^1 upstream pressure of the control blowdown valve when the control valve is wide open
- P_f pressure of the system at the end of the blowdown
- Q total amount of gas released during blowdown
- q instant gas flowrate during blowdown
- q_c maximum gas flowrate during controlled blowdown
- q_f final gas flowrate for uncontrolled blowdown at the end of blowdown
- q_{mu} maximum gas flowrate during uncontrolled blowdown
- R gas constant
- T temperature
- V volume
- Z compressibility factor
- X proportionality constant
- θ time
- θ_0 allocated time for blowdown
- $\hat{\int}$ pressure ratio, $\frac{P_{in}}{P_T}$

ACKNOWLEDGMENT

The authors would like to acknowledge for their comments and assistance S. Burk, M. Ligon, J. May, and a special thanks to G. Millar for his guidance and support throughout this work. Also A. Wassermann for his suggestions and help for the instrumentation system.

Size Relief Systems for Two-Phase Flow

Take the thinking out of the calculation. Use this easy-to-follow procedure

S. H. Richter, Arthur G. McKee & Co., Cleveland, Ohio

PRESSURE PROFILES for two-phase relief and blowdown flows can be calculated using a theoretically derived equation. The calculation procedure is applicable for any fluid of reasonably low viscosity for which intermediate vapor and liquid properties can be generated as a function of pressure by means of an adiabatic flash program. The basic working equation is incremental such that the profile can be generated by repeated application of the equation with a computer making small pressure steps.

The two-phase flow is treated as a homogeneous fluid with no-slip between phases because a more sophisticated approach is not required for most relief and blowdown flow problems. This assumption is felt to be safe because:

- Relief and blowdown headers are designed to slope towards a relief knockout drum without pockets
- Risers on relief systems are normally upstream of relief valves and therefore handle only a single phase flow.
- Risers on blowdown systems are found only at the very beginning where flow is mostly all liquid; flashing only starts to occur homogeneously throughout the flow cross-section.
- The negative effect of phase separation on the pressure drop in long horizontal lines and across fittings is assumed to be cancelled out by the other assumption that flashing is instantaneous and in equilibrium at the fluid pressure.
- A more sophisticated calculation approach, based on more complex flow patterns, is mostly not required because of the transient occurrence of relief and blowdown flow problems.

CALCULATION PROCEDURE

The procedure[3] is designed to provide a handbook type, easy-to-follow approach for design engineers to use when faced with two-phase relief or blowdown problems.

A typical simplified relief and blowdown system is given in Fig. 1. The relief system consists of:

- A liquid filled process vessel (No liquid/vapor separation occurs at the inlet of the relief valve)
- A relief valve, where capacity and setting are established by process considerations
- A relief knockout drum for liquid/vapor separation, and
- A relief header system, connecting the relief valve outlet with the relief knockout drum.

The task is to determine the relief valve size and to find the pressure profile in the relief header for the required relief flow rate in order to check if all back pressure limitations at various junction points can be met.

The blowdown system consists of:

- A process vessel with liquid inventory
- A blowdown drum for liquid/vapor separation, and
- A blowdown header system with or without optional flow restriction (in the case presented with a short two-inch lateral line).

The task is to determine the maximum possible blowdown flow rate for given inlet and outlet pressures and for a given pipe header configuration.

Two flow regions may be involved: Two-phase flow below the bubble point pressure and single phase liquid flow above the bubble point pressure.

A simplifying assumption has been made for both systems: the pressure in the relief/blowdown drum is assumed to be atmospheric and the actual variable built-up back-pressure from the vent stack is neglected. This may not always be permissible.

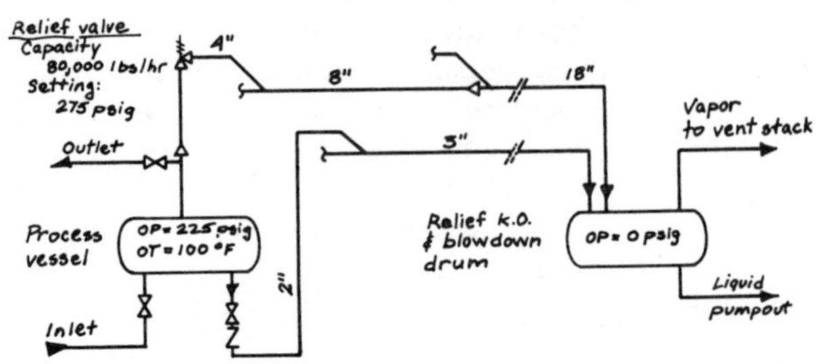

Fig. 1—Typical relief and blowdown system.

TABLE 1

System	Line size, in.	Measured length, ft.	Equiv. length for fittings, ft.	Total equiv. length, ft.
Vapor relief.............	18	500	300	800
	8	100	30	130
	4	50	25	75
Liq. blowdown...........	3	500	50	550
	2	100	50	150

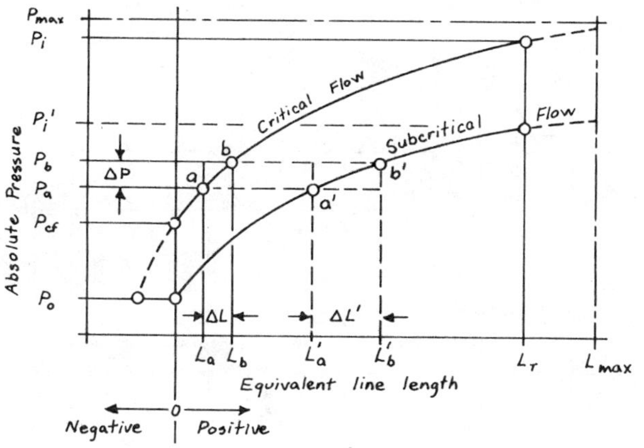

Fig. 2—Generalized pressure drop profile for two phase flow assuming homogeneous and no-slip flow.

Line sections are shown (Table 1) for illustration purposes.

WORKING EQUATION

Elevation differences are neglected and assumptions are made that flow is homogeneous with no phase separation (no-slip) and that flashing is instantaneous and fluid is in vapor-liquid equilibrium at every pressure level (Fig. 2).

Paige's equation[1] is suggested for use in determining the line pressure drop and the critical flow outlet pressure:

Flow distance = Pressure drop − Acceleration loss

$$\Delta L \doteq \frac{2\,D}{f}\left[\frac{144\,g}{v}\ \Delta P\left(\frac{A}{w}\right)^2 - ln\left(\frac{V_a}{V_b}\right)\right] \qquad (1)$$

Where:

A	ft²	= Flow area
D	ft	= Line inside diameter
f	——	= Moody friction factor (Avg. between "a" and "b")
g	ft/sec²	= Acceleration factor (= 32.2)
ΔL	ft	= Equiv. line length increment (between "a" and "b")
L_T	ft	= Total actual equiv. line length
L_{max}	ft	= Equiv. line length limit stop ($\geqslant L_T$)
ΔP	psi	= Pressure increment (between "a" and "b")
P_a	psia	= Outlet pressure for line increment
P_b	psia	= Inlet pressure for line increment
P_o	psia	= Line outlet pressure
P_{cF}	psia	= Critical flow outlet pressure
P_i	psia	= Line inlet pressure
P_{max}	psia	= Inlet pressure limit stop ($\leqslant$ Bubble point)
v	ft³/lb.	= Spec. volume of fluid (Avg. between "a" and "b")
V_a	ft³/lb.	= Spec. volume of fluid at point "a"
V_b	ft³/lb.	= Spec. volume of fluid at point "b"
w	lbs/sec.	= Flow rate of mixture in line

This equation gives the incremental equivalent length of a pipe of inside diameter D, which is required to cause a pressure reduction from P_b to P_a in a homogeneous two-phase fluid, flowing with a mass velocity of w/A, under simultaneous expansion from specific volume V_b to V_a.

By repeated stepwise application of this equation for a constant mass flow rate, the summation of all incremental results will yield a total equivalent line length required for a pressure reduction between any two pressure levels.

The equation is valid over reasonably small pressure steps. For larger pressure steps the results become less accurate because the mean specific volume and the mean friction factor for each pressure step is assumed to be the arithmetic mean between each incremental inlet and outlet point. This is not exact mathematically since the pressure profile for two phase flow is not following a straight line as for incompressible flows. But for small pressure steps the deviation is negligible because of other much more influential simplifying assumptions as instantaneous flashing and homogeneous no-slip flow.

The following considerations are now necessary:

- For pure liquid (incompressible flow), the density is constant and the logarithmic term reduces to zero. The equivalent length interval ΔL will in all cases be positive. There is no critical flow condition.

- For pure vapor or a mixture of liquid and vapor (compressible flow), the density is a function of pressure. The equivalent length interval ΔL will be positive for small density changes, but will become negative at some point where the density changes become larger for constant pressure increments and constant mass velocity.

The point where the calculated equivalent length interval changes from positive to negative is the critical flow condition where $\Delta L = 0$. At this point the total available incremental pressure difference is used up completely by acceleration of the fluid due to the incremental density change, and no pressure difference is left to overcome the additional line friction.

This occurs at the pressure P_{cf}, which is the critical flow outlet pressure.

A time-sharing computer program can be used for the step-wise repetition of this calculation in small pressure increments from the pipe section outlet pressure up to the pipe section inlet pressure and for the addition of all positive equivalent pipe length increments ΔL. The resultant ΔL of the incremental equation will be zero at critical flow conditions and positive at subcritical flow conditions. Any negative ΔLs mean super critical flow, which is a practical impossibility. The minimum outlet pressure of a particular equivalent pipe section can never be lower than the critical flow pressure. Therefore, only positive ΔLs are additive in this program.

The mean specific volume and the mean friction factor in the incremental working equation are obtained as arithmetic means between the individual values for the incremental inlet and outlet pressure points P_b and P_a. The specific volume and the viscosity of the two-phase mixture for each particular pressure point is calculated according to De Gance and Atherton[2] for no-slip flow.

From the viscosity of the mixture at P_a the computer calculates the Reynolds number at P_a and from the Reynolds Number, the Moody friction factor at P_a according to Colebrook's equation.

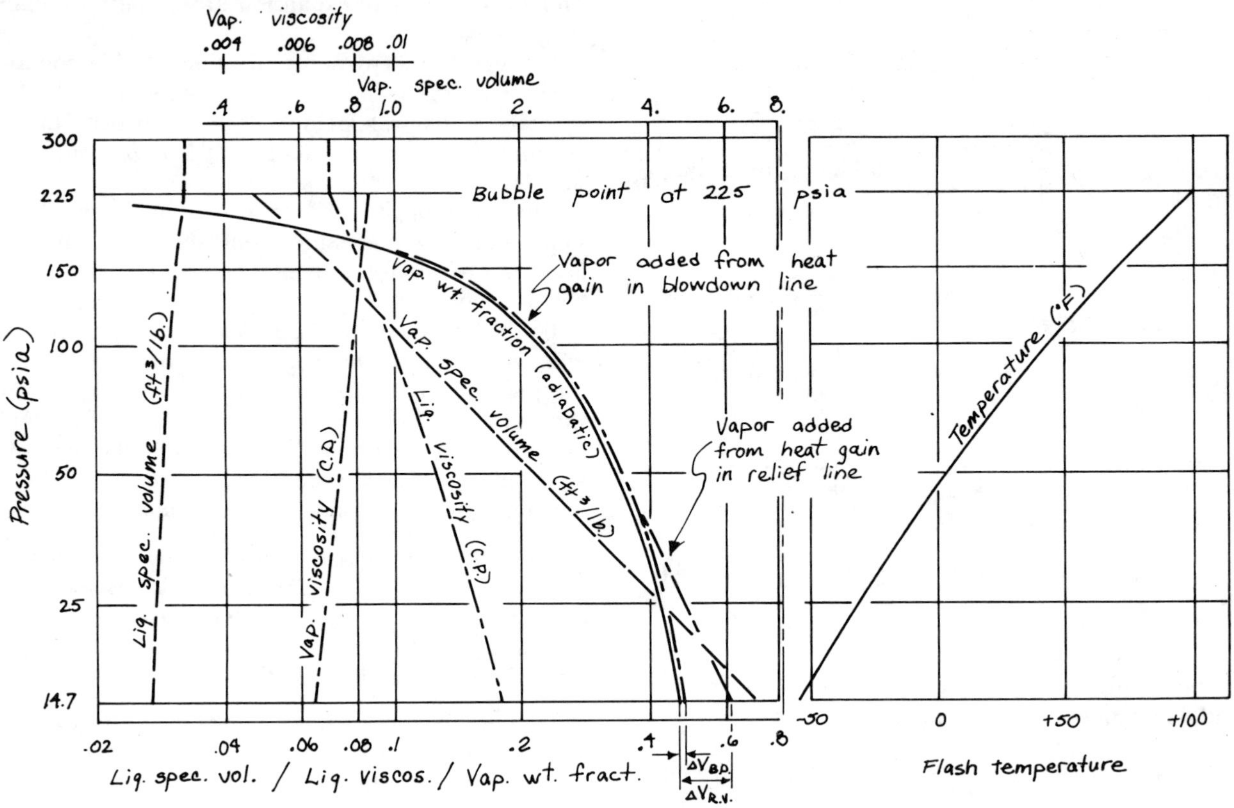

Fig. 3—Typical fluid data developed from series of adiabatic flashes (for propylene at 100° F).

Inputs. In order to obtain the variable specific volume and friction factor data for the vapor-liquid mixture at various pressure steps, the input of a number of points is required, consisting of:

- Specific volumes of liquid and vapor
- Absolute viscosities of liquid and vapor
- Vapor weight fraction in total stream.

All of these should be handled as functions of absolute pressure. From these data points the computer calculates any intermediate point by means of a least-squares curve fitting subroutine.

The input properties should be obtained in advance by running a series of flash calculations for the fluid under consideration.

Fluid properties. Fig. 3 shows typical fluid data, developed from a series of adiabatic flashes by means of a separate flash program. The presented example is based on liquid propylene at 100°F.

Before these data are fed into the computer, it is important to check them for discontinuities since they would upset the computer's curve fittings ability and bring distorted calculation results. The calculation should not be continued beyond the bubble point, although the equation is perfectly valid for pure liquids and pure vapors (within a reasonable viscosity limit for liquids and within small pressure steps for vapors).

Discontinuities are easily detected by plotting the flash data on logarithmic paper as shown in Fig. 3.

The plotted curve for the adiabatically flashed vapor weight fraction will become particularly helpful later when heat loss or heat gain of the flowing fluid are estimated. The effect of this heat transfer is then superimposed on the adiabatic flash curve without introducing discontinuities.

In Fig. 3 there are two heat gain curves superimposed on the adiabatic weight fraction line: one for the relief header flow and the other for the blowdown flow.

Heat gain generates a lot more vapor in the relief flow than in the blowdown flow, because of the larger temperature differential between surrounding air and the relief flow. The later gains most of its heat at a lower pressure and evaporation temperature than the blowdown flow.

The larger pipe surface area for the relief flow contributes further to the heat gain.

For curve continuity and simplicity the vaporization line due to heat gain for the relief flow is drawn as a straight line from the calculated atmospheric end point tangentially to the adiabatic vapor fraction line. In most relief problems this simplification has very little affect on the outcome and is therefore justified.

Typical worksheet. Fig. 4 shows a typical worksheet consisting of a computer flash flow calculation printout sheet with superimposed hand notations.

Selection of the pressure increment for calculating the equivalent length of pipe requires some careful consideration because it sets the number of calculation steps between the given outlet and inlet pressures. Generally it is advisable to use larger pressure increments of 10, 15, or 20 psi for a first trial and then to repeat the calculation in narrower pressure ranges with smaller pressure increments in order to obtain a better fix on the critical flow outlet pressure, P_{cf}, and on the line inlet pressure P_i, for a given equivalent line length.

Comparing the change of magnitude for the upstream specific volume of mixture between pressure steps with the change of magnitude in the upstream friction factor shows clearly that the choice of the arithmetic mean value over any other type of mean value has considerable influence only in the case of the specific volume, but not in the case of the friction factor. Thus, the mathematics could be improved by developing a different formula for the mean specific volume, but the practical improvement in accuracy is insignificant.

The program is designed to stop when the preset maximum equivalent pipe length (550 feet) is exceeded. The exact point where the desired equivalent length is reached is determined by interpolation. A horizontal line drawn across the table indicates where the sum of all positive equivalent length increments reaches 550 feet. The inlet pressure P_i for the three-inch pipe is interpolated to be 190 psia as noted to the left of this line.

Values of the equivalent length increments turn from negative to positive between 45 and 60 psia upstream pressure. This is the critical flow outlet pressure region for the three-inch pipe. A recalculation of this region with smaller pressure differentials (2.0 psi) shows that the critical flow outlet pressure P_{cf} is at 46 psia. This is indicated by another horizontal line. The equivalent length of the total pipe system is set at zero at this point.

Suppose that it is necessary to go up to or beyond the bubble point pressure using a different line size but the same flow rate and fluid properties. This illustration is shown as a following calculation starting with 190 psia back pressure and ending with 225 psia bubble point

```
PROBLEM IDENTIFICATION:
BLOWDOWN FROM 225 psig TO 0 psig, TRY FLOW: 80,000 LBS./HR.

NO. OF DATA POINTS FOR PROPERTY CURVES AS A FUNCTION OF PRESSURE
6

ENTER 6 PRESSURE VALUES
14.7, 25., 50., 100., 150., 225.

ENTER 6 VALUES OF LIQUID SP. VOL. (CU. FT./LB.)
.0264, .0271, .0282, .0299, .0314, .0333

ENTER 6 VALUES OF VAPOR SP. VOL. (CU. FT./LB.)
7.042, 4.213, 2.156, 1.105, .7383, .4810

ENTER 6 VALUES OF LIQUID VISCOSITY (CP)
.180, .155, .127, .100, .088, .072

ENTER 6 VALUES OF VAPOR VISCOSITY (CP)
.00645, .0068, .00735, .0080, .00845, .0090

ENTER 6 VALUES OF VAPOR WEIGHT FRACTION
.4684, .4208, .3391, .2250, .1333, .000

OUTLET P, MAX INLET P, (PSIA), PRESS INC. (PSI), MAX EQ. LENGTH (FT.)
15.,        225.,           15.,              550.

TOTAL FLOW (LB./HR.), PIPE ID (INCHES), PIPE ROUGHNESS, ETA (FT.)
80000.,             3.068,            .00015
```

	UPSTREAM PRESSURE (PSIA)	UPSTREAM SP. VOL. OF MIXT. (CU. FT./LB.)	UPSTREAM VISCOS. OF MIXT. (CP)	UPSTREAM REYNOLDS NUMBER	UPSTREAM FRICTION FACTOR	EQ. LENGTH INCREMENT (FEET)	SUM OF EQ. LENGTH INCREMENTS (FEET)	SYSTEM EQ. LENGTH (FEET)
	15.00	3.232	0.007	0.2283E+08	0.0173	......	......	
	30.00	1.435	0.009	0.1928E+08	0.0173	−19.25	0.0	
	45.00	0.871	0.010	0.1678E+08	0.0173	− 5.25	0.0	
P_{CF} = 46.00								0.0
	60.00	0.595	0.011	0.1472E+08	0.0173	3.67	3.67	
	75.00	0.432	0.013	0.1293E+08	0.0174	11.85	15.52	
	90.00	0.325	0.015	0.1133E+08	0.0174	20.51	36.03	
	105.00	0.250	0.017	0.9885E+07	0.0174	30.31	66.33	
	120.00	0.195	0.019	0.8578E+07	0.0174	41.78	108.11	
	135.00	0.154	0.022	0.7395E+07	0.0174	55.53	163.64	
	150.00	0.122	0.026	0.6322E+07	0.0174	72.32	235.96	
	165.00	0.096	0.031	0.5353E+07	0.0174	93.17	329.13	
	180.00	0.076	0.037	0.4477E+07	0.0174	119.59	448.72	
P_i (3″ line) = 190.00								550.0
	195.00	0.059	0.045	0.3689E+07	0.0174	153.80	602.51	

```
END OF PROBLEM.
TO RUN NEW PIPE ID, OR TOTAL FLOW, ENTER '1'.
TO RUN NEW OUTLET OR MAX INLET PRESSURE, PRESSURE INCREMENT, OR MAX EQUIV. LENGTH, ENTER '2'.
TO RUN ENTIRELY NEW PROBLEM, ENTER '3'.
TO STOP CALCULATIONS, ENTER '4'.
2

OUTLET P, MAX INLET P, (PSIA), PRESS INC. (PSI), MAX EQ. LENGTH (FT.)
190.0,     225.,           5.,               150.

TOTAL FLOW (LB./HR.), PIPE ID (INCHES), PIPE ROUGHNESS, ETA (FT.)
80000.,             2.067,            .00015
```

	UPSTREAM PRESSURE (PSIA)	UPSTREAM SP. VOL. OF MIXT. (CU. FT./LB.)	UPSTREAM VISCOS. OF MIXT. (CP)	UPSTREAM REYNOLDS NUMBER	UPSTREAM FRICTION FACTOR	EQ. LENGTH INCREMENT (FEET)	SUM OF EQ. LENGTH INCREMENTS (FEET)	SYSTEM EQ. LENGTH (FEET)
P_o (2″ line)	190.00	0.064	0.042	0.5852E+07	0.0191			550.0
	195.00	0.059	0.045	0.5476E+07	0.0191	5.94	5.94	
	200.00	0.054	0.048	0.5113E+07	0.0191	6.56	12.50	
	205.00	0.050	0.051	0.4762E+07	0.0191	7.25	19.75	
	210.00	0.046	0.055	0.4424E+07	0.0191	8.02	27.78	
	215.00	0.042	0.060	0.4098E+07	0.0191	8.88	36.66	
	220.00	0.038	0.065	0.3784E+07	0.0191	9.85	46.50	
Bubble Point	225.00	0.034	0.070	0.3481E+07	0.0191	10.94	57.45	607.5

Above Bubble Point, for Liquid only:

$$\Delta P_{avail.} = 239.7 - 225 = 14.7 \text{ (PSI)}:$$

$$\Delta L_{liq.} = \frac{\Delta P_{avail.} \times 100}{\Delta P_{friction} \text{ (PSI/100')}} = \frac{14.7 \times 100}{36} = 40.83 \text{ (ft.)}$$

System Inlet — 239.70 (PSIA) — 98.28 — 648.3

Fig. 4—Flash flow calculation.

pressure, using smaller pressure increments of five psi and a new pipe inside diameter of 2.067 inches.

Inspecting this calculation printout shows that the flow is subcritical at 190 psia. Therefore, the outlet pressure for the two-inch pipe is equal to the back pressure superimposed by the three-inch pipe downstream. At the 190 psia upstream pressure level, write into the column under "System Equivalent Length" "550 feet" in continuation with the three-inch diameter downstream pipe.

The calculation stops at the bubble point pressure of 225 psia in accordance with the input instructions. This pressure is on the same level with 57.45 feet of accumulated equivalent length of two-inch pipe, equal to about 607.5 feet of equivalent system length.

Suppose that the system inlet pressure is 239.7 psia. The pressure differential between 239.7 psia and 225 psia is 14.7 psi and is used up by single phase liquid flow friction loss. The equivalent length differential of two-inch pipe required to do this is 40.83 feet.

Adding this equivalent length differential for the liquid flow to the 57.45 feet equivalent length differential for the two-phase flow shows a result of 98.28 feet of equivalent length of two-inch pipe. This is required to bring the fluid pressure up from 190 psia outlet pressure to 239.7 psia inlet pressure. The total system equivalent length is approximately 648.3 feet.

The blowdown system, Fig. 1 and Table 1, consists of 550 equivalent feet of three-inch pipe and 150 equivalent feet of two-inch pipe. This is 51.72 feet of two-inch pipe more than required to use up the given pressure difference between the process vessel and the blowdown drum with an 80,000 lbs./hr. mass-flow rate.

Thus the initial flow assumption of 80,000 lbs./hr. is too high and the calculation must be repeated with a lower flow assumption until, by interpolation, the exact flow rate is found. In the example, the answer is 70,500 lbs./hr. which requires exactly 150 equivalent feet of two-inch pipe in addition to the 550 equivalent feet of three-inch pipe (Figs. 9 and 11).

EXAMPLE PROBLEM

Assume that it is necessary to design a relief system to serve the process vessel as shown in Fig. 1.

Relief system. The first task is to size the relief valve using the recommended practice given in API-RP 521,[4] Section 3.18 A, and in API-RP 520.[5] These recommend separate calculation of orifice area requirements for vapor and liquid portions of the flow and the addition of the two in order to arrive at a total orifice area requirement.

The quantity of vapor should be determined by adiabatically flashing the total flow down to the critical flow outlet pressure or to the back pressure, whichever is higher. The first step is to determine the critical flow outlet pressure at the orifice. This is done as discussed before. Assume:

- The relief valve is fully open. This is true as long as the inlet pressure is at the accumulation level and no choking due to excessive back pressure occurs at the outlet.
- The nominal effective orifice area of the relief valve, A_{sv}, is converted to an equivalent orifice diameter as follows:

$$DE = \sqrt{\frac{4}{\pi} A_{sv}}$$

Assume: Liquid propylene at 100° F
Capacity req.: $W_T = 80{,}000$ lbs./hr.
Set pressure: $P_S - 275$ psig
Accum. pressure: $P_1 = 1.1 \times 275 + 14.7 = 317.2$ psia

Try valve size:	4 M 6	3 L 4	3 K 4
Orifice area, Asv, sq. in.	3.60	2.853	1.838
Equiv. orif. diam., $DE = \left(\frac{4}{\pi} Asv\right)^{.5}$, in.	2.14	1.906	1.53
Orif. crit. outl. press., Pcf, psia	91.5	113.	165
Wt. fract. flashed at Pcf	0.240	0.20	0.104
Vapor relief			
Vapor flow at Pcf: W_V, lbs./hr.	19,200	16,000	8,350
Vapor temp. at Pcf: T, °R	498	512	537
Vapor comp. at Pcf: Z	0.88	0.86	0.81
MW	42.1	42.1	42.1
C_p/C_v	1.16	1.156	1.151
C	333.	332.	331.
K	0.975	0.975	0.975
For back press. on balanced valve: K_b	1.0	0.93	0.62
$A_v = \frac{W_v}{C\ K\ K_b\ P_1}\sqrt{\frac{Z\ T}{MW}}$ sq. in.	.605	.545	.426
Liquid relief			
Liquid flow at Pcf: W_L, lbs./hr.	60,800	64,000	71,650
Liquid Sp. Gr. at Pcf: G	0.54	0.53	0.505
Liquid vol. flow at Pcf: gpm	225.2	241.5	283.8
$P_b = Pcf - 14.7$, psig	76.8	98.3	150.3
K	0.62	0.62	0.62
K_p (for 10% overpress.)	0.61	0.61	0.61
K_w (for back press.)	0.88	0.81	0.62
K_v (for viscosity)	1.0	1.0	1.0
$A_l = \frac{gpm}{38\ K\ K_p\ K_w\ K_v}\sqrt{\frac{G}{1.25\ P_s - P_b}}$ sq. in.	.801	.964	1.628
Total orifice required			
$A_{req.} = A_v + A_l$, sq. in.	1.406	1.509	2.054
Valve tried is	too big	OK	too small
because of	$A_{req.} << A_{sv}$	$A_{req.} \leq A_{sv}$	$A_{req.} > A_{sv}$

Note: Relief valve manufacturer should be consulted before final selection of relief valve size and type for the described service.

Fig. 5—Relief valve sizing for flashing liquid.

The equivalent orifice diameter is then entered into the computer program instead of a pipe inside diameter.

The computer output yields the critical flow outlet pressure for the relief valve in that pressure region where the equivalent length increments turn from negative to positive values.

For trial and error assume three relief valve sizes (Fig. 5). The normal orifice areas, equivalent orifice diameters, and the critical flow outlet pressures for the orifices are also shown.

Find the fluid temperature and the weight fractions flashed at the respective critical flow outlet pressures using Fig. 3. From this the respective vapor and liquid flow rates are calculated.

In this case where the built-up back pressures exceed 10 percent of the set pressure, balanced relief valves must be used.

The calculation of the required orifice area for vapor flow is made in accordance with formula (C-1) of API-RP 520.[5] Note that the back pressure sizing factor, K_b, becomes very influential for the smaller relief valves with the higher critical flow outlet pressures.

The calculation of the required orifice area for liquid flow is made in accordance with formula (C-7) of API-RP 520.[5] Here again the capacity correction factor due to the back pressure, k_w, is very influential for the smaller valves.

At the bottom of Fig. 5 the total required orifice area, $A_{req.}$, is shown. When this required area is compared with the available orifice area, A_{sv}, at the top of each column, it is clear that the "K" orifice is too small, the "M" orifice is too big, and the "L" orifice between the two others is the right choice. This pre-selection is, of course,

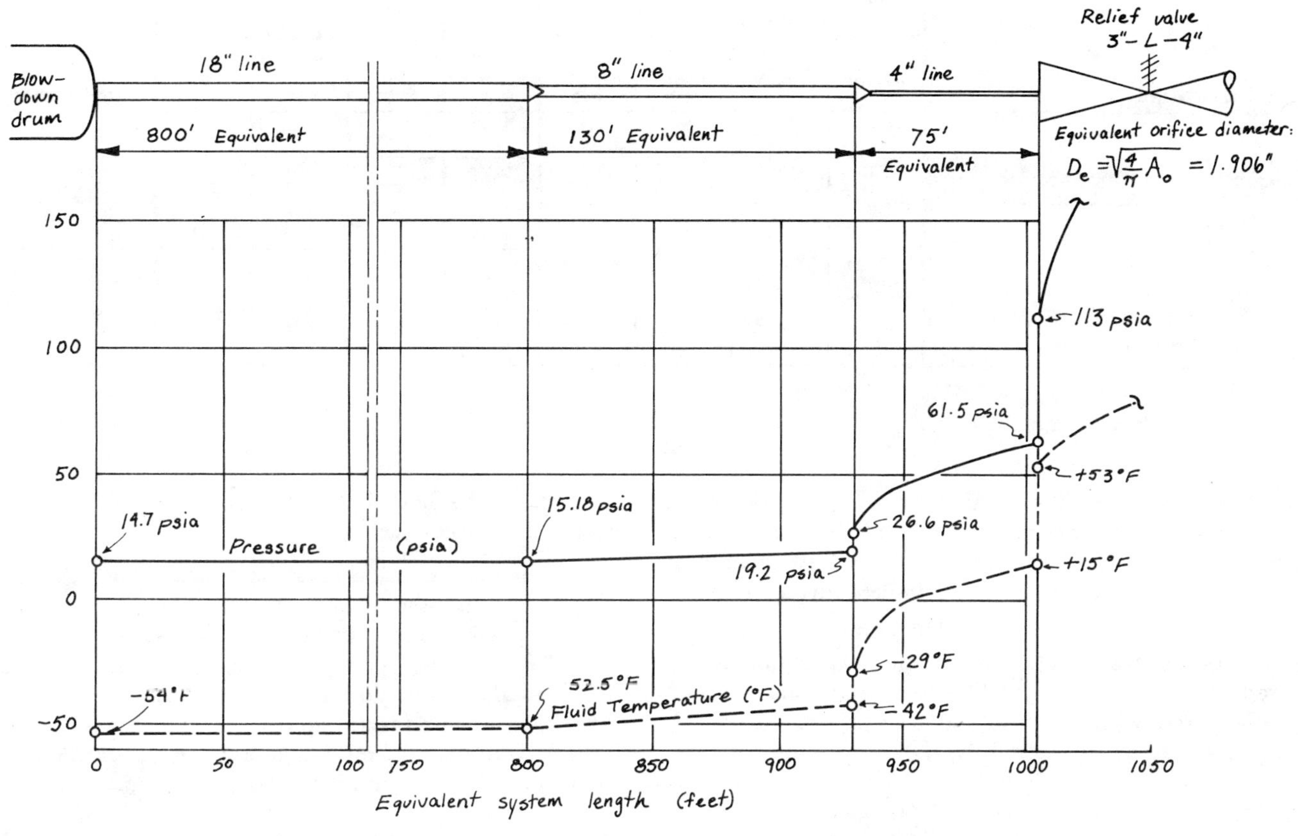

Fig. 6—Pressure and temperature profile in a relief system.

subject to review by the relief valve manufacturer.

P-T profile. Fig. 6 shows the Pressure and Temperature (P-T) Profile for the pipe configuration (Fig. 1) when subjected to 80,000 lbs./hr. total flow of fluid with properties given in Fig. 3 for adiabatic flash. The pressure profile is calculated using the computer as discussed above. The profile reveals critical flow outlet conditions at the outlet of the relief valve and at the outlet of the four-inch pipe section. The fluid temperatures have been plotted in accordance with Fig. 3 for the respective pressure levels. Using these temperatures and assuming a logarithmic temperature profile, it is possible to evaluate the influence on the flowing fluid of heat gain through the pipe wall.

Assumptions:
- Pipe not insulated and no ice build-up
- Pipe wall temperature ≈ Fluid temperature
- Temperature of surrounding air = 100° F
- Wind velocity = 20 mph

Heat transfer coefficient:

$U_o = h_o \approx 6$ Btu/hr. ft.² °F

Consider pipe sections separately:

Pipe sections	18" Pipe	8" Pipe	4" Pipe
Measured pipe length: ft.	500	100	50
Pipe outside diam.: in.	18.	8.625	4.5
Heat transfer area, A_o: ft²	2,356	226	59
Mean temp. diff.: °F			
Fluid temp. in, t_{in} °F	−52.5	−42.	+15.
Fluid temp. out, t_{out} °F	−54.	−52.5	−29.
$MTD = \dfrac{(100 - t_{out}) - (100 - t_{in})}{\ln\left(\dfrac{100 - t_{out}}{100 - t_{in}}\right)}$	153.2	147.2	105.5
Heat gain: Btu/hr $\Delta Q = U_o \times A_o \times (MTD)$	2,165,650	199,600	37,350
Total, $\Sigma \Delta Q$: Btu/hr		2,402,600	

Vapor generated due to heat gain:

$$W_{heat} = \frac{\Sigma \Delta Q}{\text{Latent Ht. @ 14.7 psia}} = \frac{2{,}402{,}600}{190} = 12{,}645 \text{ Lbs/hr.}$$

$$\text{or } \frac{12{,}645}{80{,}000} = 0.158 \text{ wt. fract. of total flow}$$

Total vapor arriving in blowdown drum:

	wt. fraction	lbs/hr.
Due to adiabatic flash	0.4684	37,472
Due to heat gain (max)	0.1580	12,645
Total vapor (max)	0.6264	50,117

Fig. 7—Maximum heat transfer to relief flow.

Maximum heat transfer. The maximum heat transfer from the atmosphere to the relief flow is estimated as shown in Fig. 7. The three pipe sections (18, 8 and 4-inch) are treated separately. The heat transfer areas are calculated from measured pipe lengths (Table 1) and from the pipe outside diameters. As driving forces, the logarithmic mean temperatures between air and fluid are used.

The maximum heat gains of each pipe section are added up and divided by the latent heat of the fluid at outlet pressure conditions. The result gives the quantity of vapor generated due to maximum heat gain, which may arrive at the outlet point in addition to the vapor flashed adiabatically.

In the example, the heat gain vaporizes another 16 percent of the mass flow in addition to the 47 percent vapor flashed adiabatically. This additional vapor is added tangentially as a straight line to the vapor weight frac-

Line identification	Pipe dimensions: Inside diameter inch	Measured length feet	Equiv. length fittings feet	Total equiv. length feet	Total flow (estim.) lbs/hr	Pressure (psia): Crit. flow outlet	Actual at outlet	Actual at inlet	Case description
18" Line	17.124	500	300	800	80,000	....	14.7	15.18	Adiabatic flash
18" Line	17.124	500	300	800	80,000	....	14.7	15.34	Max. heat gain
8" Line	7.981	100	30	130	80,000	....	15.18	19.22	Adiabatic flash
8" Line	7.981	100	30	130	80,000	....	15.34	20.6	Max. heat gain
4" Line	4.026	50	25	75	80,000	26.6	26.6	61.5	Adiabatic flash
4" Line	4.026	50	25	75	80,000	30.3	30.3	64.0	Max. heat gain
Relief valve orifice									
Size / Sq. Inch	Equiv. diam. in.							Accum. press.	
M 3.60	2.14	....	....	....	80,000	91.5	91.5	317.2	Adiabatic flash
L 2.853	1.906	....	....	....	80,000	113.	113.	317.2	Adiabatic flash
K 1.838	1.53	....	....	....	80,000	165.	165.	317.2	Adiabatic flash

Fig. 8—Relief system calculation results.

tion curve (Fig. 3). This is another simplification. Most of the heat is gained in the 18-inch pipe at a pressure level of about 15 psia and the actual effect of heat gain on the pressure drop would be less severe than the one calculated from a straight tangent line.

Calculation results. In Fig. 8 the results of the repeated pressure drop calculation for maximum heat gain are compared with the results of the pressure drop calculation for adiabatic flash only. The pressure differences for the outlet or inlet points of each pipe section are insignificant. Therefore, there is no need to show the pressure profile for maximum heat gain superimposed on the pressure profile for adiabatic flash in Fig. 6.

The investigation is for a cold fluid gaining heat under the assumption that the additional vapor generaton might significantly increase the back pressures in the relief header.

A hot fluid, losing heat through the pipe wall, will flow easier and the actual back pressures will be less than calculated adiabatically. Therefore, it is conservative to neglect the heat loss problem for hot fluids in relief systems, in most cases.

One note of caution must be added: The vapor quantity to be disposed from the relief knockout drum may not be equal to the vapor quantity arriving at the knockout drum. If, for example, a flashing cold liquid is relieved while a hot liquid is already present in the relief knockout drum, instant additional flashing of the cold liquid could significantly increase the vapor quantity to be disposed from the relief knockout drum.

Blowdown system. The actual two-phase blowdown flow rate is found by trial and error as demonstrated (Fig. 4). The results are diagrammed for the temperature and pressure profile (Fig. 9). The same propylene fluid properties are used as for the relief system Fig. 3. Fig. 9 shows clearly how the assumed larger and smaller flow rates undershoot and overshoot the actual equivalent length of the example blowdown piping system.

Remembering the curve discontinuity at the bubble

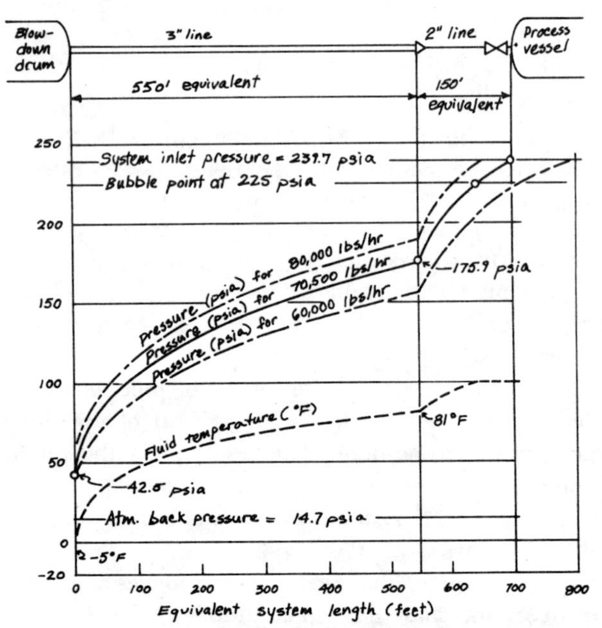

Fig. 9—Pressure and temperature profile in blowdown system.

Assumptions:
- Pipe not insulated and no ice build-up
- Pipe wall temperature ≈ Fluid temperature
- Temperature of surrounding air = 100° F
- Wind velocity = 20 mph.

Heat transfer coefficient:

$U_o = h_o \approx 6$ Btu/hr. ft.^{2} °F

Consider pipe sections separately:

Pipe section:	3" Pipe	2" Pipe
Measured pipe length: ft.	500	100
Pipe outside diam.: in.	3.5	2.375
Heat transfer area, A_o: ft.^{2}	458	62
Mean temp. diff.: °F		
Fluid temp. in, t_{in} °F	+81	+100
Fluid temp. out, t_{out} °F	− 5	+ 81
$MTD = \dfrac{(100 - t\ out) - (100 - t\ in)}{\ln\left(\dfrac{100 - t\ out}{100 - t\ in}\right)}$	50.3	~ 9.5
Heat Gain: Btu/hr $\Delta Q = U_o \times A_o \times (MTD)$	138,225	3,535
Total, $\Sigma \Delta Q$: Btu/hr	141,760	

Vapor generated due to heat gain:

$$W_{heat} = \frac{\Sigma \Delta Q}{\text{Latent Ht. @ 14.7 psia}} = \frac{141{,}760}{190} = 747 \text{ Lbs./hr.}$$

$$\text{or } \frac{747}{70{,}500} = 0.0106 \text{ wt. fract. of total flow}$$

Total vapor arriving in blowdown drum:

	Wt. fraction	Lbs./hr.
Due to adiabatic flash	0.4684	33,023
Due to heat gain (max.)	0.0106	747
Total vapor (max.)	0.4790	33,770

Fig. 10—Maximum heat transfer to blowdown flow.

Line identification	Pipe dimensions: Inside diameter inch	Pipe dimensions: Measured length feet	Pipe dimensions: Equiv. length fittings feet	Pipe dimensions: Total equiv. length feet	Total flow (estim.) lbs/hr	Pressure (psia): Crit. flow outlet	Pressure (psia): Actual at outlet	Pressure (psia): Actual at inlet	Case description
3″ Line	3.068	500	50	550	80,000	46.0	46.0	190.	Adiabatic flash
	3.068	500	50	500	70,500	42.5	42.5	175.9	Adiabatic flash
	3.068	500	50	550	60,000	36.0	36.0	157.0	Adiabatic flash
	3.068	500	50	550	70,500	44.0	44.0	176.8	Max. heat gain
	3.068	500	50	550	70,000	42.5	42.5	176.0	Max. heat gain
2″ Line, Part I	2.067			57.45	80,000	....	190.	225.	Adiabatic flash
Part II	2.067			40.83	80,000	....	225	239.7	Liq. above bubble pt
Total	2.067	<100	50	98.28	80,000	....	190.	239.7	Total
2″ Line, Part I	2.067			97.0	70,000	....	175.9	225	Adiabatic flash
Part II	2.067			53.0	70,500	....	225	239.7	Liq. above bubble pt
Total	2.067	=100	50	150.0	70,500	....	175.9	239.7	Total
2″ Line, Part I	2.067			166.03	60,000	....	157.0	225	Adiabatic flash
Part II	2.067			74.6	60,000	....	225.	239.7	Liq. above bubble pt
Total	2.067	>100	50	240.63	60,000	....	157.0	239.7	Total
2″ Line, Part I	2.067			93.37	70,500	....	176.8	225	Max. heat gain
Part II	2.067			53.0	70,500	....	225	239.7	Liq. above bubble pt
Total	2.067	<100	50	146.37	70,500	....	176.8	239.7	Total
2″ Line, Part I	2.067			95.86	70,000	....	176.0	225	Max. heat gain
Part II	2.067			54.4	70,000	....	225	239.7	Liq. above bubble pt
Total	2.067	≈100	50	150.26	70,000	....	176.0	239.7	Total

Fig. 11—Blowdown system calculation results.

point of the fluid, the pressure drop in the two-inch pipe is calculated in two parts (Fig. 11).

The flow rates indicated (Fig. 9) are based on adiabatic flashing. They are expected to be different for a fluid gaining or losing heat along its flow path. The question is how much the difference amounts to.

Maximum heat transfer. Fig. 10 shows how the maximum heat transfer from the atmosphere to the blowdown flow is estimated. The assumptions and calculations are made in the same way as for the relief flow (Fig. 7). Because of the smaller pipe surface areas and the higher flow pressures and temperatures, the heat gain of the blowdown flow is considerably less than that of the relief flow. The vapor generation rate in the blowdown flow due to heat gain brings only a one percent increase. The heat gain of the blowdown flow occurs in a much higher pressure region than that of the relief flow, and therefore, the vapor generated due to heat gain must be superimposed (Fig. 3) on the weight fraction line for adiabatically flashed vapor in a different manner than for the relief flow.

Blowdown system calculation results. Fig. 11 summarizes results of the repeated pressure drop calculation for maximum heat gain and compares the results with the pressure drop calculation for adiabatic flash only. The blowdown flow rate for the maximum heat gain flow is found by trial and error in the same manner as for the adiabatic flash flow. All trials are recorded. There are three on the adiabatic flash flow and two on the maximum heat gain flow. The resultant blowdown flow rates are 70,500 lbs./hr. for the adiabatic flash flow and 70,000 lbs./hr. for the maximum heat gain flow. This is an insignificant flow decrease due to heat gain.

A hot fluid would lose heat in order of magnitude of the temperature difference between fluid and air. In this case the winter minimum air temperature should be used in order to be conservative and to arrive at the largest possible blowdown flow rate.

The heat loss will cause the fluid flow rate to increase because of the condensation of vapor, and a neglect of the heat loss for blowdown systems may therefore not be the conservative approach.

The same note of caution applies with regard to the vapor quantity leaving the blowdown drum as already mentioned in the preceding discussion on relief flow with regard to the vapor quantity leaving the relief knockout drum.

ACKNOWLEDGMENT

Presented at the 43rd Midyear Refining Meeting of the American Petroleum Institute, Toronto, Canada, May 8-11, 1978.

LITERATURE CITED

1 Paige, Peter M., "How to Estimate the Pressure Drop of Flashing Mixtures," *Chemical Engineering*, August 14, 1967, pp 159-164.
2 De Gance, A. E. and Atherton, R. W., "Chemical Engineering Aspects of Two-Phase Flow" (8 parts) *Chemical Engineering*, March 23, 1970 to February 22, 1971.
3 Richter, S. H., "For Flashing Fluids, Try Control Valve Sizing by Computer" *Petro/Chem Engineering*, December, 1968.
4 API RP521; "Guide for Pressure Relief and Depressuring Systems" September, 1969, American Petroleum Institute, New York, N.Y.
5 API RP520; "Design and Installation of Pressure Relieving Systems in Refineries, Part 1—Design," Fourth Edition—December, 1976, American Petroleum Institute, New York, N.Y.

Blanket Tanks for Gas Control

Improved approach saves energy while avoiding pollution

D. M. Gammell, Engineering Consultant, Arcadia, Calif.

GAS BLANKETING TANKS and loading racks for emission control is effective when using improved techniques. Safety is greatly enhanced, operator attention is minimized and problems previously associated with blanketing are minimized with instrumentation methods now available. Vapors previously lost to the atmosphere are recovered for use and energy is conserved over all.

Improved practices now in operation are being evaluated by field tests to establish empirical design criteria for future applications.

SYSTEM CONCEPT

Natural gas is used for blanketing in a system (Fig. 1) installed experimentally in a southern California refinery and in a similar system installed for recovering vapor from tankage only. Net vapor recovery amounts to about 3 percent of the refinery's fuel demand, although 25 tanks and 2 gasoline truck top-loading racks are involved. This level of usage is established under some of the most stringent conditions in the United States where the average ambient temperature change between day and night is maximum.

The system is less expensive to install and operate than commercial vapor recovery units and eliminates surveillance of vapor collection and emission rates. Net vapor recovered may either be burned as fuel or traded to a utility or fuel consumer.

Tank inhalations actuate self-contained regulators, usually at 0.25 ounces per square inch (osi) vacuum, admitting blanket gas into connected tank ullages before tank vacuum relief valves actuate, usually at 0.5 osi vacuum. One regulator can be used for several adjacent tanks if interconnecting pipe sizes are adequate. Sour crude oil or light straight run distillate vapors are prevented from contaminating finished products or sweet intermediary storages by a normally closed butterfly valve

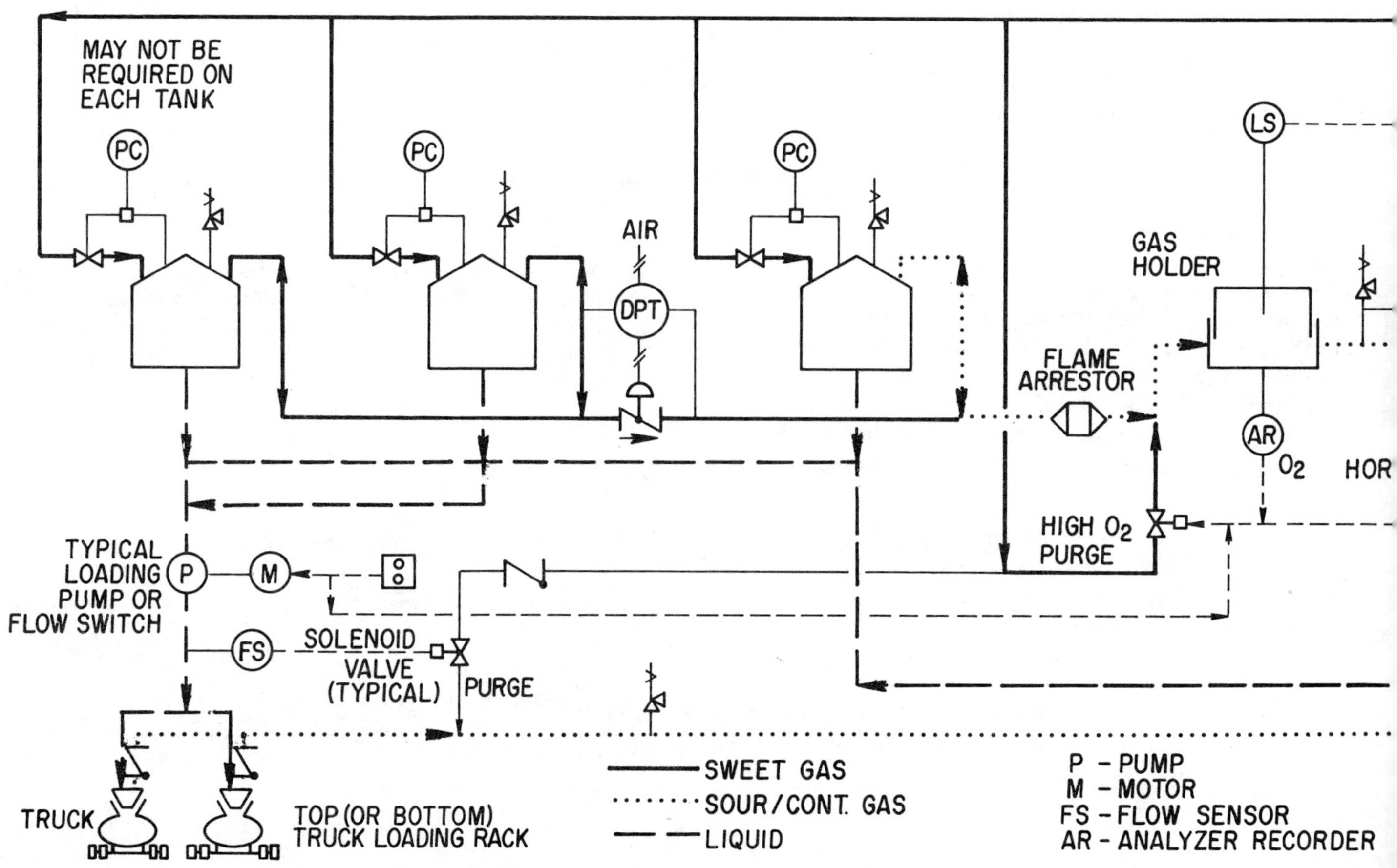

Fig. 1—Gas blanketing system for recovery of fuel.

which opens fully when the upstream pressure exceeds that downstream by 0.1 inch of water.

Tank exhalations breathe into gas holders by the very small difference between the weight of the tank cone roofs (0.5 osi) and that of the gas holder diaphragms (0.2 osi). This pressure difference motivates net exhalations into the gas holder, even through the differential check valve which prevents intertank contaminations and the flame arrester, without lifting the pressure relief valves on the storage tanks, which are usually set at 0.5 osi. A fan, and/or a heavier roof than one of 3/16 inch steel plate, would provide a greater pressure differential, but such has not been found to be necessary in most cases.

A variable volume gas holder is provided so that the compressor and evaporative condenser need not operate continuously, but only to empty the holder as far as tankage is concerned. When the gas holder diaphragm reaches almost full capacity, its level switch starts the compressor and condenser. Surplus holder volume is provided for the special case where net tank exhalations into the holder plus loading rack vapors exceed the compressor capacity. When the holder is evacuated and the diaphragm drops to a low capacity, the level switch stops the compressor and condenser unless the loading rack is in operation. Compression is necessary only to enter the refinery fuel gas system, usually about 40 psig. Any condensation is returned to storage. Flowmeters are provided to record blanket gas usage and the net gain realized in vapor recovery.

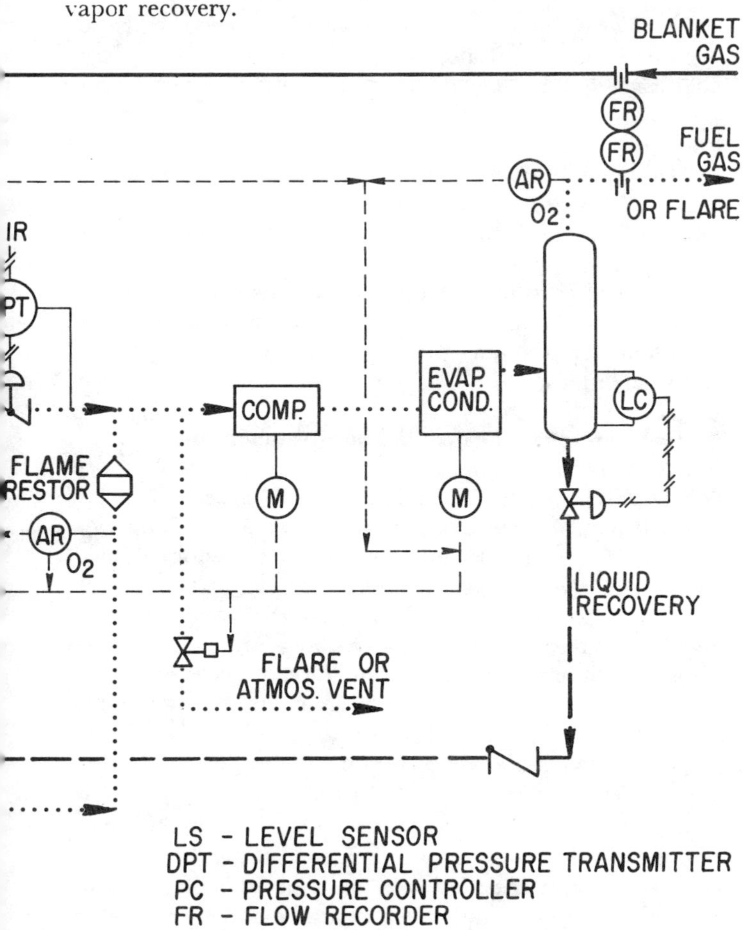

TABLE 1

Circumstance	Floaters	Blanket (Fig. 1)	Blanket with a commercial package
No blanket gas economically available	Yes	No	Rack only
Inert blanket gas economically available	No	Yes	No
Combustible gas but no dispositions economically available	Maybe	No	Maybe
Combustible gas and dispositions economically available	No	Yes	No

All tanks in the vapor recovery system must be airtight at operating pressures and tank PSVs must operate properly without leaking. An oxygen analyzer continuously records oxygen concentrations in the gas holder, where no oxygen should ever be present.

Loading racks. Fig. 1 also illustrates how truck, rail or ship loading systems can be safely gas blanketed in conjunction with either associated or unassociated tankage to essentially eliminate atmospheric vapor emissions during loading operations. The same compressor and condenser can be utilized if its size is adequate for both tank and loading rack loads.

Because empty vehicle compartments may contain some air, blanket gas is purged through the vapor collection header to the compressor whenever the loading rack is in operation. If oxygen concentrations above some predetermined level occur, an oxygen analyzer actuates a warning to the loading rack attendant who may then elect to do nothing for 10 or 20 seconds in hopes that the warning will cease; stop the compressor and condenser and vent loading rack vapors to flare or atmosphere until satisfactory oxygen levels are reached, (the compressor and condenser will automatically start again and the vent automatically close) or switch off the loading pumps that feed the rack.

A compressor shutdown signal from the loading rack overrides that from the level switch on the gas holder, which is another reason for surplus volume in the holder. Safety relief to atmosphere is provided to prevent pressure build-ups which could structurally damage vehicle loading compartments, loading connections, or etc.

A normally closed butterfly check valve prevents loading rack vapors from entering the gas holder, so that oxygen recorded there is known to come from tankage only.

SYSTEM IMPROVEMENTS

Oil refiners have been averse to gas blanketing because sensitive operating controls and diligent attention were necessary to avoid structural tank damage or gas leakage, explosive conditions could develop with devastating results, tank turnarounds were difficult and vapors from one tank could contaminate the liquid in another tank. Some refiners have phased out blanketing systems for floating roofed tankage and commercial loading rack vapor recovery packages. However, since both of these installations have vapor emission factors above zero, a second look has been taken at the totally enclosed system outlined above.

First of all, gas blanketed tankage is basically much safer than fixed roof tankage without blanket gas. This is true for crude oil, intermediate refinery products and finished products. A hydrocarbon gas mixture without hydrogen which is too rich to burn is about 40 times easier to maintain than one that is too lean to burn. Very reliable, sensitive, self-calibrating oxygen analyzer-controllers have been strategically located in the system illustrated by Fig. 1 to continuously record and control (in two cases) oxygen concentrations.

When oxygen concentration in the gas holder reaches only 10% of the upper explosive limit (UEL), for instance, blanket gas automatically purges the holder until less than that concentration is obtained. If the holder fills before the oxygen concentration is diluted sufficiently, it will be relieved to atmosphere. If oxygen concentration in the loading rack vapor collection header reaches 30% of the UEL, an alarm is actuated for the attendant and, simultaneously, a greater than normal blanket gas purge rate is automatically admitted to the header. A third oxygen analyzer is located after condensations are removed because such removals increase oxygen concentrations in the gas phase. This analyzer-recorder backs up the other two and warns operators of concentrations over 50% of the UEL.

Task pressure-vacuum relief valves do need to be inspected periodically for tight, but not sticky, seat rings-to-knife edges, as should those on any fixed roof tank, and the rim vents and bleeder vents on any floating roof tank. It is perhaps more imperative that tankage in gas blanketed service be generally air tight. Hand-operated butterfly dampers may be installed at each tank vapor collection roof flange to prevent blanket gas losses during tank gaging and sampling operations, but such is not required by code. Modern instrumentation (Fig. 1) has been found to be reliably responsive, including that which prevents contaminations between tanks.

Turnaround procedures have been formalized to prevent a tank from ever containing a combustible gas mixture while it is selectively taken out of and placed into service in the vapor recovery system.

APPLICATIONS

New tankage and tankage-loading installations should consider the gas blanket alternative in view of current ambient air quality regulations, air pollution rules and the availability and disposition of blanket gas.

Gases for blanketing may come from a number of sources; but they should be sweet and not otherwise detrimental to the liquid in the tanks:

- **Combustible blanket gases** (net exhalations to fuel gas)
  Commercial gas from gas utility companies
  Natural gas from private fields
  Treated refinery gas having no hydrogen
  Vaporized liquid petroleum gas (LPG) when absorption not critical.

- **Inert blanket gases** (net exhalations to flare)
  Compressed flue gases
  Nitrogen from inert gas generators
  Bottled nitrogen or CO_2
  Compressed CO_2 exhausts having no hydrogen.

Combustible gases offer the advantage of total heat recovery from net exhalations, which include hydrocarbon vaporizations from the stored and loaded liquids. Inert gases offer the advantage of easier and safer shutdowns for entering into the tanks. The use of combustible gases is usually more economical if such gases are available.

From the oil refiner's point of view, the more blanket gas at profitable fuel oil equivalent prices that is burned necessarily for processing, the more oil there is available to process. From the bulk oil terminal operator's point of view, where fuel gas may not be needed or may not be easy to trade off to a local user, the most economical system that satisfies the environment is best. From the public utility commission's point of view, commercial gas should not be used if any other gas is available. From the air pollution agencies' points of view, the best system is that which reliably emits nothing to the atmosphere.

On balance, the ideal solution is dependent upon the owner's need for fuel, or his ability to trade-off net exhalations, and the long term availability and cost of the blanket gas (Table 1).

Net exhalations from inert gas blankets would normally be flared, but flaring would not normally be economical for net exhalations from combustible gases because of their initial fuel value. Perhaps the largest obstacle towards the use of commercial (natural) gas for blanketing is the likely need for approval by the PUC.

If sweet combustible gas is available at costs which are less than fuel oil on a heating value and operating basis, then new gas blanketed tankage should take precedence over new open floating roof tankage where the heating value of net exhalations can be used or traded-off. Where heavy precipitation occurs, new gas blanketed tankage may be mandatory, even with the use of inert gas, such as at Valdez. Where floating roof seals need extensive repair, an evaluation should be made to compare the cost of seal repairs to that of covering the existing tanks with a sealed roof on a scheduled turnaround basis, and then gas blanketing them to satisfy environmental and, perhaps, public relations.

Design criteria. Formal methods for determining net gas blanket inhalation and exhalation design capacity requirements are currently very arbitrary and suspect. The best method for such determinations, in this author's opinion, are empirical formulations derived from historical field-monitored data which are gathered at an actual gas blanketed tankage system. In the absence of such information, computations on tank breathing are largely based upon the derivations and formulations developed in the American Petroleum Institute bulletins for average annual evaporation losses. Modifications thereto have been made for enclosed gas blanket atmospheres and for momentary operating capacities in order to establish design criteria. Reasonableness checks have been made by independent approaches that substantiate the conservativeness of those derivatives, and the sensitivity of some variables that inherently are assigned arbitrary values have been evaluated.

Tankage. Net exhalations from a vapor recovery system containing several tanks is the concurrent result of individual breathing, V_b, and working, V_w, exhalations. Since tank breathing results from ambient temperature changes, all tanks tend to breathe in and out together. Working exhalations, however, result from independent tank filling rates less emptying rates, expressed as V_w' and V_w'' in cubic feet per minute (cfm), respectively. The net effect, Q, in cfm is:

$$Q = \Sigma\,[(V_b + V_w' - V_w'') \text{ for each tank}]$$

The amount of hydrocarbon vapor in Q would be something less than that from cumulative concurrent true vapor pressures in each tank by the amount of non-equilibrium stratification in the tanks.

Breathing exhalations can be approximated by reducing the API formula for average annual breathing losses from cone roof tanks to atmosphere to an instantaneous rate of vapors plus blanket gas, using perfect gas laws and considering an average daily quantity to occur evenly for 8 of 24 hours exclusively. Maximum monthly average daily temperature changes should be determined from data published by the nearest weather bureau. Exhalations are assumed to be 15° F warmer than bulk liquid temperatures due to their residence time in steel tank ullages, vapor recovery piping and the gas holder.

$$V_b = 3.5 \times 10^{-5} \frac{\rho_L(475 + T_3)P(T_2)^{0.5}(D)^{1.73}(H)^{0.51}}{M_v(p)^{0.32}(P\text{-}p)^{0.68}}(F)\ (C)$$

Working losses are simple volume displacements in and out of each tank, unless tank outages cause V_w'' to exceed exhalations $V_b + V_w'$, which results in net inhalations for the system. In this special case, simple outage volume displacement is offset by vaporization to accommodate Raoult's Law, and:

$$V_w'' = \frac{(G)\,(P\text{-}p)}{7.48(P)}$$

in cfm into tanks.

Loading racks. Total vapor loads from loading rack operations include direct volume displacements plus whatever entrainment and vaporization occur during loading operations. The rate of vaporization in each compartment being filled depends upon:

- Vapor pressure of the liquid
- Total absolute pressure of the compartment
- Amount of vapors already in the empty compartment
- Loading method.

Petroleum products are loaded into vehicle compartments by top (splash) methods, bottom (submerged) methods or combinations thereof. Truck and trailer loading racks may utilize any one of these three methods. Rail cars are usually splash loaded and marine vessels are submerged loaded.

If the compartments to be loaded are already saturated with hydrocarbon vapors, the rate of vapors expelled from the loading rack will only equal the rate of liquid displacement, no matter what loading method is used, disallowing for liquid entrainments. If the compartments are not saturated, however, the rate of expelled vapors will equal the rate of liquid displacement plus a rate depending upon the factors listed above. Submerged loading will cause relatively slight increases, while splash loading causes significant increases, depending upon the degree of unsaturation at the start of loading.[2] A 1973 publication claims that trucks average about 30% saturation at the start of loading.[3] With vapor recovery systems being required at unloading points since then, a 50% saturation was assumed for designing the system illustrated by Fig. 1.

The capacity of a combined tankage and loading rack system is limited by the compressor capacity. The minimum time for a truck and trailer, rail car or ship to be at the loading rack can be expressed as loading time, t_L, plus set-up time, t_s. The minimum set-up time allowed by the capacity of the vapor recovery system, without overloading it, is:

$$t_s\ (minimum) = \frac{V_L t_L}{R\text{-}Q} - t_L$$

wher V_L = cfm of emissions from the loading rack, including purge gas and t_L = time in minutes to load the vehicle. The higher the rack loading pump capacity, the more set-up time there is available. The number of vehicle loadings per hour, N, becomes:

$$N = \frac{60\ (R\text{-}Q)}{V_L t_L}$$

NOMENCLATURE

C = Small tank correction factor[1]
D = Diameter of tank, feet
F = Paint factor[1]
G = Liquid rate into or out of tank, gpm
H = Average annual outage of liquid in tank, feet
M_v = Molecular weight of vaporized tankage liquid
N = Number of vehicle loading set-ups per hour
P = Average annual atmospheric pressure, psia
p = True vapor pressure in psia of bulk liquid at temp. T_3
Q = Momentary net exhalation rate from tankage, cfm
R = Actual compressor capacity, cfm
T_2 = Largest daily temperature change °F in 30-yr. monthly averages
T_3 = Temperature of bulk liquid in storage, °F
t_L = Time to fill vehicle compartments, minutes
t_s = Time to set up a vehicle rig, minutes
V_b = Tank breathing exhalations, cfm
V_L = Loading rack vapors in cfm, including purge gas rates
V_w = Tank working losses, cfm
V_w' = Tank working exhalations, cfm
V_w'' = Tank working inhalations, cfm
ρ_L = Specific gravity of liquid as stored

LITERATURE CITED

1 API Bulletin 2518 on Evaporation Loss from Fixed-Roof Tanks dated 6/62.
2 Air Pollution Engineering Manual, U.S. Department of H.E.W., Public Health Service Publication No. 999-AP-40.
3 Hydrocarbon Emissions from Refineries, API Publication No. 928 dated 7/73.
4 API Bulletin 2513 on Evaporation Loss in the Petroleum Industry—Cause and Control, dated 2/59.

Part VII.

Maintenance

Improve Automatic Control System Reliability

Here is information on improving control system reliability by proper design, vendor selection, installation and startup. Control system maintenance requirements, instrument technician training and maintenance department organization are also extensively covered

R. F. Mahood and **J. E. Martin,**
E. I. du Pont de Nemours & Co., Inc., Orange, Texas

MODERN CONTROL SYSTEM technology is extremely complex and often completely integrated with high thruput, continuous processes. Most processes cannot be operated without their control systems and analytical equipment functioning correctly. Therefore, improving performance of the controls impacts directly and heavily on profits. Well designed and executed maintenance programs are described which have proven value in achieving higher control system performance levels and at the same time optimizing maintenance costs.

Are you a typical process industry plant? If so:

- Instruments and control systems represent 8 to 10 percent of your plant investment

- Maintenance of this portion of your plant requires 15 to 20 percent of your total mechanical manpower

- You will average 5 to 15 instruments for each item of process hardware

- The installed cost of a single instrument loop is currently over $5,000

- You can expect an average of about five years between major maintenance work on a well designed device supplied by a reputable vendor.

Because modern processes are highly dependent on automatic control systems, failures are often very expensive in terms of lost or off-quality production and sometimes present severe safety hazards. On the other hand, the cost of overmaintenance can also be very high. Much of the effort of those directly concerned with maintaining automatic control systems has been devoted to finding the best balance between these two extremes.

As this specialty has matured over the past 20 to 30 years, a number of techniques have emerged which can be used to successfully arrive at reasonably optimum maintenance systems. These techniques will be examined. A great deal of valid literature is available in each of the following broad areas so only those items of special significance, or those which have proven to be particularly advantageous will be highlighted.

DESIGN, CHECK-OUT AND STARTUP

Low maintenance cost and high performance of automatic control systems begins with good design. Unfortunately, many problems are identified and corrected after the fact which would never have occurred had adequate attention been given to initial design.

To help minimize these problems, include instrument maintenance people in the plant's design, construction and startup team. In selecting them, consider the following factors:

- 10 to 20 years' experience in first or second-line control equipment maintenance supervision

- One person full time for each $1 to $2 million of automatic control system investment

- Continuity throughout the life of the project

- Stature to get the respect of those with whom they work

- Management support of their design decisions.

Adequate numbers of qualified personnel will pay for themselves many times over.

Having assigned the necessary people, what will they look at? Some of the more crucial areas include power sources, plant layout and hardware selection.

Power sources. Two sources of power are used in all automatic control systems—air and electricity. Both must be as close to 100 percent in reliability and quality as the best available technology and hardware will permit. This means complete redundancy with at least two independent primary power sources, automatic throw-over systems and the necessary treatment to insure high quality.

Motor generator sets, battery banks with inverters, and alternate feeders with solid-state throw-over switches are proven techniques for supplying reliable electrical power to computers, analog electronics and interlock circuits. Even with this, because of the extremely low inertia of most electronic and electromechanical devices, expect at least one outage per year resulting from loss of power for more than four or five cycles.

Supplying reliable, high quality instrument air for pneumatic controls and automatic valves is usually expensive, but the technology is more mature. Provide two complete compression, drying and filtration systems operated in

Fig. 1—Inaccessible control valve.

Fig. 2—Properly located control valve.

parallel with each capable of carrying the full load. Locate them in separate buildings and provide independent motive power, such as electricity and steam. Use only galvanized piping from the discharge of the final filtration

Fig. 3—Operator console.

stage and be sure to blow it clear before putting it in service.

Dirt, moisture and oil, individually or in combination, are deadly to satisfactory operation of pneumatic control systems and valves. This equipment has orifices as small as 0.003 inch and the passages are seldom larger than 0.027 inch. The air must meet the quality standards shown in Table 1.

Finally, be sure rigid procedures are established and enforced to limit users of these power sources strictly to control systems. Otherwise outages from voltage or pressure dips, cross contamination and distorted wave forms are sure to occur.

Plant layout. Process equipment and piping arrangements are extremely important in minimizing future maintenance costs of automatic control systems. These are much easier to change in the design phase than after the facilities are installed. For example, Fig. 1 shows a large butterfly valve installed in a completely inaccessible location. When this valve fails—notice the word is when, not if—it will shut down a unit producing 400 MM lbs./yr. of a high value product. Furthermore, it will take at least 24 hours to repair or replace it. It is difficult to routinely inspect for pending failure. On the other hand, the valve shown in Fig. 2 is the same size and in as critical a service. However, it can be routinely serviced to prolong its life and predict its future use with a reasonable degree of accuracy so it can be removed and overhauled on a scheduled shutdown.

Properly located, sized and laid out shops, work areas, and central control rooms are vital to the long term efficiency and productivity of maintenance people. An adequately sized and equipped field shop for each major process should be located as close as possible to the Central Control Room. For this shop, 75 to 125 sq. ft./man is needed, exclusive of offices, change and rest rooms, etc. Since manpower forecasts tend to underestimate the number of people required to maintain a process, lean toward the higher number.

In addition, a satellite work area should be provided for each significant concentration of control equipment located more than about 500 feet from the field shop. One hundred and fifty to 200 sq. feet located in a standard electrical classification area usually will be adequate

Fig. 4—Rear view of the same operator console.

for a work bench, test equipment, tools and equipment reference files.

Central control room location, layout and arrangement from the viewpoint of one who maintains the hardware is always from the back of the panels, not the front. An example is a particular situation where the size and location of the control room building was finalized before the control systems were designed. From the front (Fig. 3), it is a prime example of excellent human engineering. From the back (Figs. 4 and 5), it presents a somewhat different picture. Conservatively estimated, the instrument maintenance work force in this area could be reduced by two to three people if closer attention had been paid to design of the central control room earlier in the project.

On the other hand, when continuing maintenance costs are given heavier weight, the results lend themselves to a more optimum situation (Fig. 6).

Hardware selection. Two guidelines should be applied to each design decision:

- K.I.S.S. (Keep It Simple, Stupid)
- Standardize

The drive to automate processes, coupled with the technology explosion in this field has tended to unnecessarily complicate control systems and add greater cost and difficulty to maintenance. To cite an extreme example, is it really necessary to have a control system so complicated that it requires 121 separate instruments and valves in a single loop? Be sure the plant's liaison team is constantly applying pressure to simplify—simplify—and then simplify some more!

Whenever possible, stay with well established vendors and proven hardware already in use on your site. Require that the engineer justify on solid technical grounds every excursion from this position he proposes to make. Completely investigate the financial soundness and maintenance support capability of each new vendor.

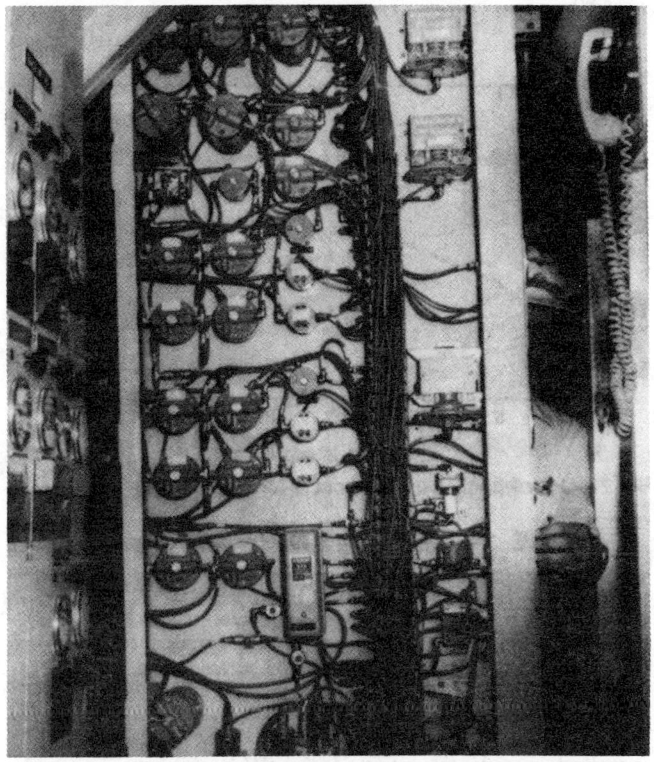

Fig. 5—Control system racks.

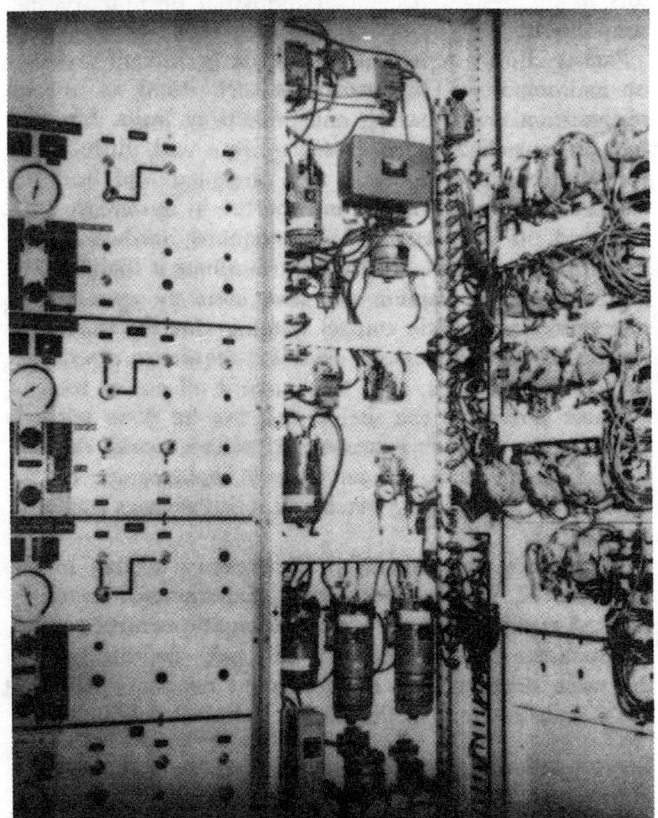

Fig. 6—Improved control system racks.

To cite another example, the vendor of a particular cathode ray tube was out of business and his written and verbal expertise lost before the equipment was even

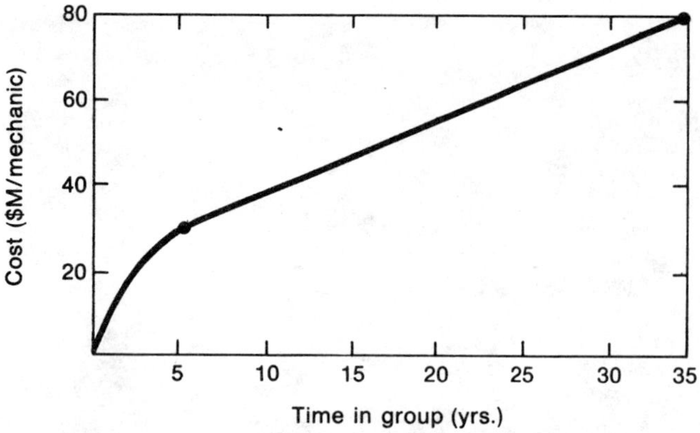

Fig. 7—Out-of-pocket training costs (anticipated by 1988).

TABLE 1—Instrument air quality standard

- Dew point at line pressure must be at least 10° C below minimum local recorded ambient temperature.
- Maximum oil or hydrocarbon content is 1 ppm.
- Maximum particle size permitted is 3 micrometers.
- Completely free of all corrosive, flammable and toxic gases.

installed, much less started up. Worse yet, this vendor used obsolete, commercial grade chips, bugs and transistors in his design, instead of industrial or military spec components.

Packaged units represent a particularly difficult problem for maintenance. This includes such items as process refrigeration units, boxing and palletizing units. Usually, they are specified by process engineers with little input from the control engineer; then designed and built by vendors who have little or no expertise in automatic controls and no knowledge of the standard hardware used in the process design. Further, the business is highly competitive and the vendor can save costs by specifying inexpensive automatic control systems. Get the vendor to use the standard hardware, even if it requires a surcharge. If that doesn't work, plan to change it all out as soon as the unit arrives at the site, if this can be done without violating the vendor's guarantee. If neither works, develop a contingency plan for an orderly replacement of the control hardware as the vendor supplied systems malfunction or fail.

To remain competitive, it is necessary to stay in the forefront of proven technology. Experienced, carefully selected and strongly supported automatic control systems maintenance liaison personnel will help do this and, at the same time provide the judgment necessary to avoid the pitfalls inherent in this highly technical field.

Check-out and startup. In following the installation of major new facilities, there are many things to be done. Equipment reference files must be set up, spare parts and equipment established, protected storage of hardware awaiting installation provided, training completed, procedures written, and on and on. This discussion will be limited to better ways to perform pre-startup function, calibration and safety checks.

Any preassembled units of significance should be completely physically inspected and function checked by control equipment maintenance liaison personnel at the vendor's site prior to shipment. Control panels, relay racks, some analyzers and most packaged units fall in this category. Since such checks increase the vendor's cost, requests for quotes should include this requirement. Even though a thorough check may consume the time of one or two people for a week or more, the payoff is high. Physical inspection by knowledgeable maintenance personnel will disclose any number of instances of control equipment tubed, wired, piped, or mounted in such a way as to be completely inaccessible for maintenance. Inadequate numbers or inaccessible locations of test points is another common error. Loose fittings, connections, mounting brackets, unrestrained wiring and tubing, metal shavings in air supply lines, wrong gage scales—the list of problems is very long. All these problems can be corrected quickly, easily, and at much lower cost at the vendor's rather than at the plant site.

Following correction of items found on the physical inspection, a complete function and calibration check of every item of control hardware should be run. This requires detailed written procedures, which the customer usually has to provide, and test equipment which the vendor usually can supply. These checks will disclose deficiencies such as inoperable instruments, leaks, discontinuities, calibration errors, etc., which can be corrected on the spot at vendor expense. Such checks also provide invaluable "hands on" training for plant maintenance personnel.

In addition to the cost savings realized by making such inspections and checks at the vendors, startup delays are often avoided. Control equipment is usually among the last to be completed and turned over by construction forces. Any delays avoided while placing the equipment in service pay handsome dividends in smooth, on-time startups.

But what about damage in shipment of this delicate control equipment? Experience has shown that, if proper packaging is used, carriers are carefully selected and protected on-site storage is provided prior to installation, with today's modern, rugged industrial equipment, there are few instances of actual damage prior to installation.

Virtually all on-site preinstallation checks should be omitted. Concentrate instead on a complete, thorough and accurate post installation function, calibration and safety check to pick up any errors, leaks or discontinuities. There are, of course, some special nonstandard items which can and should be shop checked before installation. But make these the exception rather than the rule by insisting that they be itemized, and such checks justified on an individual basis. Even where large numbers of items from a nonstandard vendor are used, selective diminishing sampling techniques can reduce the cost of shop checks and still build the required confidence.

With proven vendors standard hardware, which should comprise 60 to 80 percent of the control equipment, only 3 to 5 percent will need significant attention after installation and these corrections are usually minor. A further advantage of doing business this way is the training the maintenance personnel receive on the complete loop and its process interrelationships as such checks are performed.

TRAINING

Like any other area of automatic control systems maintenance management, the training program needs to be frequently monitored and updated to insure the highest possible return for each training dollar. Typically, from the time a person is hired to train in automatic control systems maintenance until he retires, $40M out-of-pocket will have to be spent on his training at today's prices. Ten years from now, assuming current inflation trends continue, this figure will approximate $80M (Fig. 7). Of this, about one-third will be spent in the trainee's first three or four years and the remainder will have to be spent while he is at top rate to keep him up-to-date on new technology and process changes.

Original training program. In 1975, a modern training facility was completed at one Du Pont site. It includes a TV studio, modern classrooms, and appropriate office facilities. All maintenance training programs on this site are administered by a training supervisor who is a member of the Employee Relations Department. Instructors are provided by the field groups needing the training and are line foremen or wage roll loaned for this purpose.

Prior to 1974, the curriculum for automatic control systems trainees had changed little in the past 25 years. It was the conventional, traditional, education approach. It started with math through trigonometry and progressed on through physics and a touch of unit operations. Then the electronics course was started, beginning with basic circuit theory. Concurrently, the instrument series was started which began with pressure and ended several years later with chromatography. Each instrument course explored in depth the theoretical basis for the various measurements, such as Bernouilli's Theorem for pressure, thermodynamics for temperature, etc. Then the hardware used to make the measurements and control them was studied. Lectures were supplemented with programmed instruction courses, videotapes and lab work to the extent this was possible, in an effort to provided a more stimulating training environment. Each trainee attended classes two or three half days per week for the first two years of progression and by the end of his four year minimum progression to the top rate had been in class between 700 and 800 hours.

Training of personnel already at the top rate was done as the field foreman had the time and inclination. Generally, the motivation was a new type of hardware or process revision which forced a training response. Those working in the more stable areas received little, if any, such training. Typically, at the beginning of each quarter, the foreman would select three to ten topics he felt his crew should be trained on and assign a man or ask an engineer to develop and present a one hour session on it. As would be expected, the results of this training effort were varied and the yawns prodigious.

Revised program. A preliminary review of this situation in late 1973 rapidly reached the conclusion that if there was any more efficient and effective way to train instrument maintenance personnel, it should be tried. An in-depth study of alternates produced a program simple in concept: get out of the academic education business and train people to do work in the field.

The two most radical changes required are:

- Minimize theory and increase "hands on" work under the instructor in a structured training environment

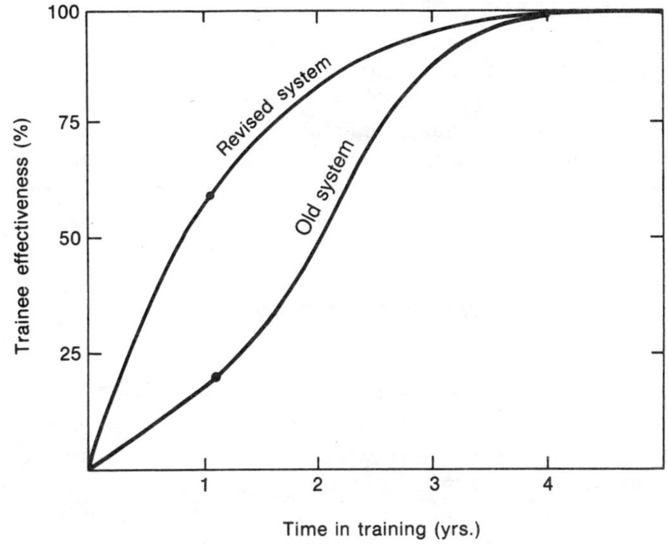

Fig. 8—Trainee productivity comparison.

- Instead of spreading training over two to four years, concentrate it into two or three 5-to-6-week blocks of full-time training.

Once these concepts are accepted and the decisions made to implement them, the way is open to a much more effective training program.

Results. Use of the new training system has resulted in major cost savings. Each trainee now receives 400 hours of classroom work, compared with about 800 under the old system. At present wage rates, this is a $4,500 savings per trainee. In addition, the trainee now starts field work able to accomplish assigned tasks on his own. He is doing productive work instead of working with more experienced mechanics most of the time. His productivity at the end of the first year is two to three times greater under the new program (Fig. 8).

Employe relations benefits have been an unexpected bonus. The hourly paid and first and second-line supervision are now fully sold on this program. The trainee is proud of what he can do and anxious to show both supervisors and coworkers his new skills. Supervision feels far less burdened by the training job and is freer to attend to his primary job of managing his area's maintenance.

MAINTENANCE ORGANIZATION

Maintenance philosophy. To arrive at an optimum organization for automatic control systems maintenance—in terms of configuration, size, skill levels, etc.—operating and maintenance personnel first must agree on their maintenance philosophy. "Breakdown" maintenance is usually unacceptable because of the production penalties involved. Maintaining all equipment in a "like new" condition is usually best. Further, any such philosophy must remain flexible so it can be easily reivsed to accommodate external influences, such as changing business conditions, varying employe relations situations, etc.

The key element in defining such a philosophy is that it must be written and communicated through the Maintenance and Operating organizations. Writing it requires going beyond platitudes and generalities and confronting head on and resolving some very basic issues. How much is appearance worth? How many outages of what duration are acceptable? What priority does installation of new or revised facilities carry?

Once a maintenance philosophy has been written, it then must be communicated and discussed throughout the Operating and Maintenance organizations—and others if they are involved. This way no misunderstandings arise and continuity is sustained, even with personnel changes.

Centralized instrument maintenance. In most situations, the benefits of the centralized, functional instrument maintenance organization, covering all operating areas on a single plant site, significantly outweigh the disadvantages. Since many of their functions overlap, it is also usually more economical, where labor relations permit, to combine the instrument and electrical maintenance functions into a single organizational entity at all supervisory levels. Some of the advantages of this organization include:

- More flexibility in resource allocation
- A larger, more economical base for training programs to establish and maintain specialized skills
- Minimum spare parts, spare equipment and test equipment and facilities
- Better and faster implementation of new concepts and exchange of new ideas and experience
- More effective standardization of test procedures, PM programs, planning and scheduling, etc.
- Better control of shift coverage
- The long term continuity of important programs is better assured.

With a centralized organization, most of the advantages of a decentralized or area instrument maintenance organization can be retained if all levels in the group are sufficiently sensitive to the needs of the operating people. This means knowing plant-wide priorities and schedules and understanding the needs of each product.

By organizing and managing this way, the best of both types of organizations can be utilized which will produce minimum cost and highly effective automatic control systems maintenance.

Multi-skills. Most instrument and electrical work requires only one person, provided he is trained, and required to perform, in many skills. It makes good economic sense and provides a much greater sense of personal satisfaction to the mechanic to operate this way.

The instrument man, for instance, is trained for the top level in his field so he can properly maintain everything from pressure gages to complex electronic and analyzer control systems. In addition, he is trained to perform other tasks to do the "whole job." Such work should include electrical, to the extent of working with confidence in circuits up to and including 110 volts, process pipe and instrument tubing, conduit and wire, fabrication, painting, insulation, light rigging and valves. Training to third or second-class levels in these allied skills is adequate to cover 75 to 90 percent of the organization's needs. Trying to move beyond these levels reduces the incremental economic benefits and increases employe relations' problems.

It is interesting to note that, at national levels at least, both building trades and industrial unions are accepting the approach of multi-skills. They, too, see the economics and personal satisfactions available by working this way and are becoming increasingly receptive to it.

Outside services. Generally, contracting all or a portion of the automatic control systems maintenance in a sizable plant with any significant degree of complexity is unprofitable at best and impossible at worst. Accordingly, it is recommended that 80 to 90 percent of this type work be done "in house" with only occasional supplemental contract maintenance.

There are good shops available for selected equipment, such as gages and automatic valves. However, this requires the equipment to be sent off plant, often at a significant time penalty. Some, but by no means all, vendors provide excellent on and off-site services for their own equipment, but again at the risk of unacceptable delays or at an unrealistic level of spare parts and equipment.

Instrument maintenance management consultants can provide valuable assistance by surveying the organization's performance, identifying areas where improvements can be made and recommending proven programs which will enhance the effort. Such programs usually include training, planning and scheduling, improving manpower effectiveness, cost and materials control, reliability improvements, PM programs, etc. These management tools are well defined and proven, and utilization of consultant services to help adapt them to your specific situation is an economical course of action.

PREVENTIVE MAINTENANCE—HOW MUCH?

Sooner or later all automatic control systems will fail. Assuming the maintenance philosophy calls for something better than breakdown maintenance, some type of ongoing maintenance program must be designed, implemented and monitored. Such programs usually include Preventive and Corrective Maintenance, and should include Predictive Maintenance. The care and feeding of analyzers, computers and other special hardware require particular attention. Literature is abundant in each of these areas but some additional comments are in order.

Preventive Maintenance systems are designed from historical data to service the hardware just prior to the point in time it becomes unusable. Modern control system hardware has few, if any, moving parts to wear out. Those which do are usually lightly loaded and overdesigned to the point that "wearout" is a matter of many, many years. Calibration drift beyond specification, deterioration of elastomeric components from dry rot, fatigue, and external factors such as mechanical and process damage are the most common causes of failure.

Most Preventive Maintenance systems for automatic controls have two major shortcomings:

- They are too conservative, thus cost too much
- They are not kept up-to-date with new technology.

To root out the conservatism and get more realistic PM schedules, first examine the PM frequencies. Minor maintenance scheduled more often than every two years or major overhauls more frequently than every four or five years should be suspect. Analyze the equipment history and see if the intervals can be extended. Next, classify all the equipment into critical or noncritical. Equipment is critical if it "adversely affects safety, continuity of operation, quality, yield, cost or environment, when it either fails abruptly, or is out of service more than two shifts." Then take the noncritical equipment out of the PM system altogether or, at most, schedule minor maintenance every few years to keep it up to minimum appearance and calibration standards. When this was done at one Du Pont site about 10 years ago, it was found that approximately 40 percent of the plant's instruments and valves were noncritical. Eliminating PM overhauls yielded maintenance savings of about $145M per year. An interesting side benefit disclosed that 2.5 percent of the 16,000 items then in service weren't needed at all and they were subsequently removed and the investment reduced.

Third, use a service severity index to extend PM intervals for individual critical locations. Table 2 is one such index. Experience has shown that external factors, such as those considered in this index, are the most significant to the life of any single installation. By applying this index at one site, time intervals between overhauls were doubled on 65 percent of the critical instruments and valves—a substantial cost savings with no adverse effect on safety or production.

Every PM system for automatic controls needs to be reviewed in depth once a year by persons who have intimate knowledge of the field situation. If this is not done, new types of hardware will not be included and newer PM techniques will not be utilized.

True Predictive Maintenance involves a measurement of suitable parameters on individual equipment and performance of maintenance only when these measurements show the instrument or valve is out of predetermined specifications.

Electronic systems and automatic valves are the two areas most amenable to this approach and work is currently underway at some locations to develop viable programs. Manual monitoring of selected test points on supercritical installations is a good first step. Computer monitoring on electronic hardware at predetermined frequencies has been tried with uncertain results at this time.

However, in the near future, at least on some control systems, techniques will be developed, monitoring frequencies specified, and limits established. Such an approach will be better and more cost effective than the present statistically based Preventive Maintenance systems and well managed plants should be moving in this direction.

Maintaining computers. Over the past 8 to 10 years, the number of process computer installations have increased at a geometric rate. Initially, this equipment was used primarily for data collection and operator displays with perhaps a few supervisory control systems thrown in. With time, training and new technologies, more and more processes are computer rather than operator controlled as loops are closed through the computer. As a result of this, chances are excellent that operators cannot run the process without the computer if it is more than five years old.

Concurrent with increasing process dependency on computers, there has been a decline in vendor maintenance support. This has occurred for a variety of reasons—some good, some bad—but it is a fact of life today. If you have computers, you should consider establishing an in-house maintenance capability for troubleshooting, parts replacement, spare parts storage, and shops and training facilities. You should also have an "on call" contract with your vendor when problems get beyond the capacity of your own people to solve.

Today, the economics of this approach will vary among locations, depending on the number of computers, who the vendors are, business conditions and the cost of process outages. There is no question, however, that technology is moving in this direction. It takes 2-5 years of lead-time to establish and train the necessary organization and support facilities—so now is the time to start moving.

TABLE 2—Service severity index

	Instruments Condition		Valves Condition	
	Good	Poor	Good	Poor
Process materials:				
Corrosion	0	5	0	10
Erosion	0	10	0	20
Fouling or plugging	0	5	0	10
Pulsating	0	10	0	5
Process temperature (extreme)	0	5	0	5
Environment:				
Vibration	0	30	0	15
Corrosive atmosphere	0	15	0	15
Dirty atmosphere (dirt, chem., etc.)	0	5	0	5
Temperature and/or humidity	0	5	0	5
Location:				
Exposure to mechanical damage	0	5	0	5
Inaccessibility to routing maintenance and patrol	0	5	0	5
Judgment of foreman:				
Experience by overhauls or maintenance	−25	+25	−25	+25

Severity index interval scale

Severity index	Interval (maximum 15 years)	
0- 25	Increase by	100%
30- 50	Increase by	50%
55- 75	Increase by	33%
80- 95	No Change	
96-125	Decrease by	33%

ACKNOWLEDGMENT

Based on a paper presented at the NPRA Refinery and Petrochemical Maintenance Conference, San Francisco, Feb. 13-16, 1979.

How to Pack High Temperature Valves

Tests show that sliding-stem valves in high temperature service can be cycled almost 50,000 times before packing leakage becomes significant. All that's required is a simple change in packing technique

Robert T. Wilson, Fisher Controls Co., Marshalltown, Iowa

STEAM LOSS through a leaking control valve packing box not only adds to the day-to-day operating energy costs of a process, it also increases the maintenance burden. While most sliding-stem type valves used in high-temperature situations incorporate some form of graphite-based packing, recent tests show that conventional methods of using these graphite packing rings are far from optimum.

For example, a standard globe valve with a one-half inch diameter stem and conventional, laminated graphite ring packing exhibited an unacceptably high leakage rate after 800 cycles against 600° F, 400 psig steam. With a new packing system we developed, this same valve operating under the same conditions, was well within leakage specifications after 45,000 cycles. The increase in packing life was over 5,000 percent.

The key to this success lies in the physical structure of the rings and the order in which they are placed on the stem. To understand the wide difference in performance between the conventional and the new systems, first consider the physical structures of commercially available graphite packing rings.

LAMINATES, RIBBONS AND FILAMENT BRAIDS

Laminated graphite rings are made from a large number of thin sheets of laminated graphite, stacked to the desired thickness and compressed to force the laminates tightly together.

Packing made from graphite ribbon is formed by winding the ribbon in a spiral around a mandrel that is the same size as the valve stem until it is large enough to loosely fill the packing box. Pressure is then applied to the two ends of the roll to compress it. In theory, the accordion-like pleats give a good seal and create lower friction than laminated rings. Either crinkled or smooth graphite ribbon can be used, but crinkled ribbon seems to be preferred.

Fig. 1—The three rings lying flat in the upper row are all standard graphite types. The leftmost ring in the foreground is a laminated graphite type with the laminates spread open. Ring standing on edge is a ribbon graphite type that has suffered extrusion; note raised rims around both OD and ID. The split ring in the right foreground is a braided graphite filament type.

Braided rings are just that; graphite filaments are braided into a rope-like structure that looks very much like black sash cord. The braid is cut to the proper length to make one turn around the valve stem. When inserted into the packing box, pressure then forces the two ends together and establishes the necessary seal at the OD and ID.

Graphite ring type has its advantages and disadvantages. Laminated rings give a good seal because the only possible leak paths are along the OD and ID surfaces. On the other hand, they sometimes "extrude" into the small space between the packing box and the packing follower, or between the valve stem and the follower. When this happens, the packing pressure is lessened at the OD and ID with greater tendency toward leakage. This can be prevented, however, by using "anti-extrusion" rings above and below the laminated graphite ring. These are designed to fit snugly in the box and around the stem to eliminate the "cracks" which allow packing ring extrusion.

Graphite ribbon wound rings, either crinkled or smooth, have the same disadvantages as laminated rings plus the additional disadvantage of leakage possibilities if the turns of the ribbon winding are not kept tightly com-

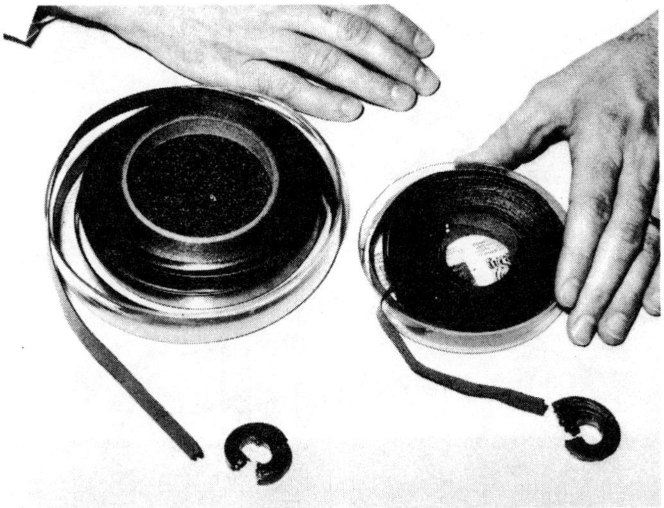

Fig. 2—Graphite ribbon rings. On the right is the crinkled ribbon starting material and a ring made from it. On the left is smooth ribbon and ring. Both have been cut open to show the direction of the laminations.

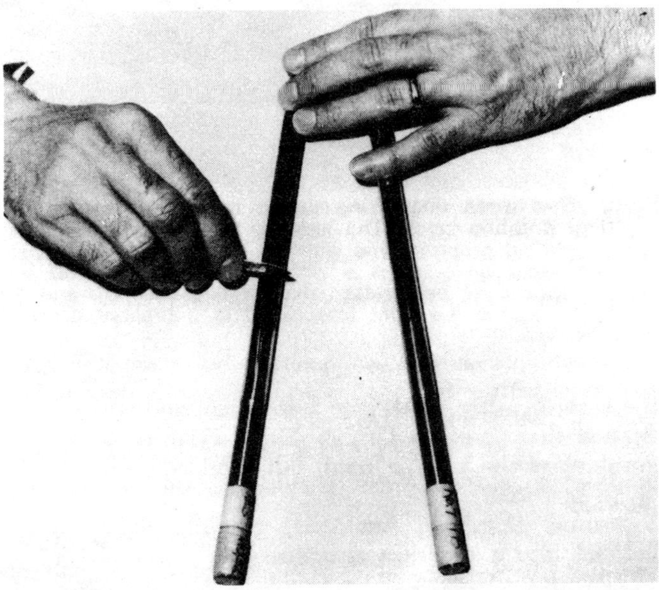

Fig. 3—Valve stem on the left is from a valve that had been lab tested using laminated rings. The transferred graphite particles can be scraped into a pile with a knife. Stem shown on the right had been in extended lab use with the braided graphite filament/laminated graphite combination. This stem is free of transferred particles and appears to be even more smooth than when it was installed in the valve.

pressed together. This can happen when wear at the ID allows the windings to come apart. Once a leakage path is well established between ribbon turns, it's difficult to close even with increased tightening of the packing follower bolts.

Braided graphite filament rings, under pressure, expand to fill the space between box and stem quite well, but the tiny openings inherent in the braided structure cannot provide as good a seal against leakage as either the laminated or ribbon rings. For this reason, braided rings find limited use.

Study started from scratch. There were some unanswered questions about ribbon graphite packing material when it was introduced to the market. One was why it didn't give the expected reduction in friction. One of the first steps in a program designed to answer this question was to observe what happened when a newly packed valve was stroked, cycle by cycle.

The packing used for this experiment was a set of standard ribbon-wound rings (crinkle ribbon) pre-compressed to increase their density. The rings were installed in a packing box using a standard packing ring sequence. In the first few strokes of the stem, friction became very high. However, on continued stroking, friction dropped. At this point leakage occurred, and while further tightening of the packing follower stopped the leakage, friction was then undesirably high in the many uneventful strokes that followed.

Trouble source: graphite particle transfer. On the premise that some kind of wear was occurring quickly in the break-in period, valves were taken apart at various stages of break-in and the packing rings and stem examined. Particles of graphite had deposited on the polished stem surface, darkening and dulling the surface area where ring and stem had been in sliding contact. Surprisingly, this graphite transfer was observed even when the stem had not been stroked. Each ring left its own imprint where it had been compressed against the stem.

Examination of valve stems that had been in service in the field for extended periods showed a heavy graphite build-up. The surface of the deposit was much rougher than the underlying stem surface—a condition which had

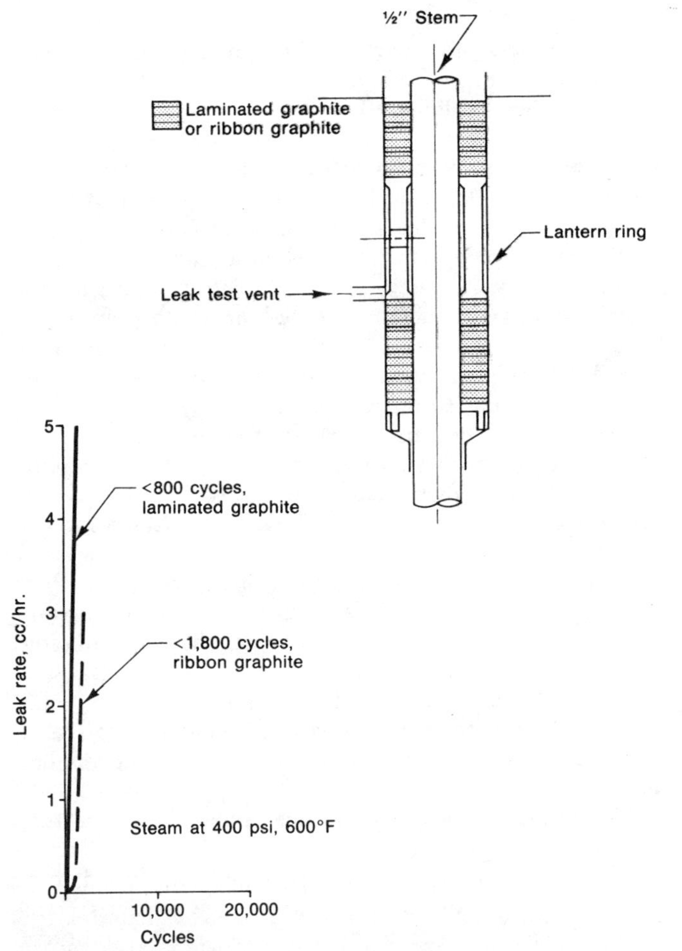

Fig. 4—Plot of valve stem cycles vs. leak rate through packing for laminated graphite or ribbon graphite only.

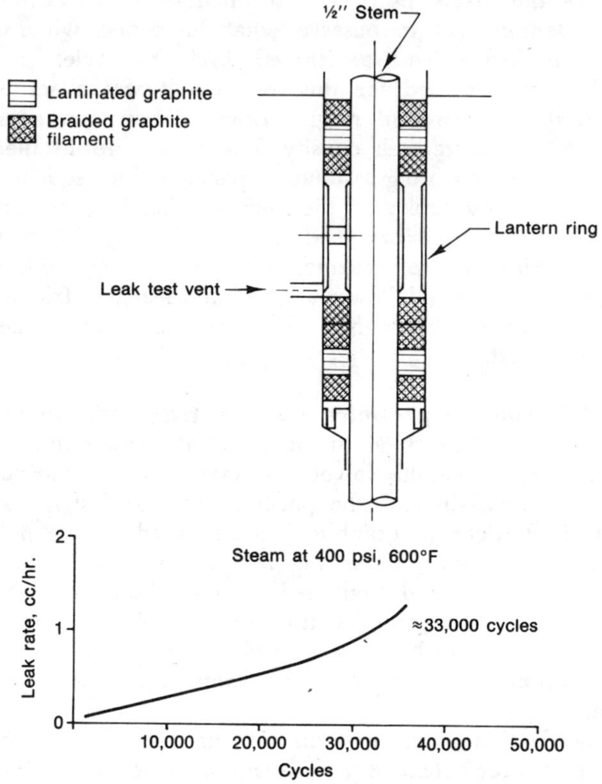

Fig. 5—Same as Fig. 4, except packing is laminated graphite sandwiched between braided graphite filament rings.

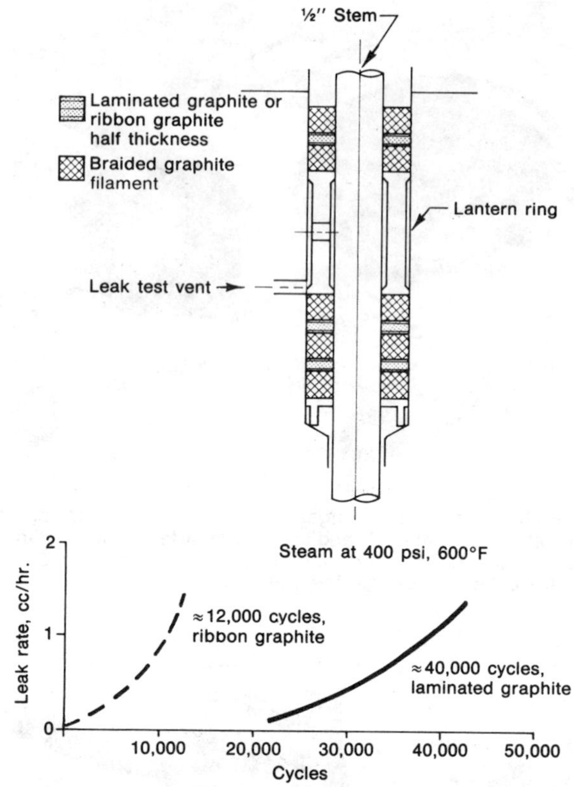

Fig. 6—Same as Fig. 4, except packing is a combination of half-thickness laminated or ribbon graphite rings and braided graphite filament rings.

to be conducive to wear. Further investigation revealed that both graphite ribbon-wound rings and laminated graphite rings exhibited the effect.

Eliminating particle transfer. The obvious answers to the graphite buildup problem were to either prevent the particle transfer in the first place or to find some means of cleaning the deposit off the stem as fast as it formed.

One attempt involved using tight-fitting solid graphite rings above and below laminated or ribbon rings. Although they behaved as very good anti-extrusion rings, they didn't help the particle transfer problem and showed no tendency to clean particles off the stem.

Surprisingly, the answer came from a separate study of anti-extrusion rings. One such ring was the graphite filament type—chosen because it could be fitted tightly to the stem and box surfaces. This set the stage for discovery.

Rings made of braided graphite filament did not show the graphite particle transfer effect. A systematic investigation was then set to test all reasonable combinations of graphite ring structures to see if the non-particle-transfer properties of braided graphite filament could be combined with the good sealing properties of laminated or ribbon graphite rings, thereby achieving the best of both worlds.

The result of this study was even more rewarding than expected. Not only did the braided graphite filament exhibit its non-transfer property, it also was found to be an excellent "clean-up" material for the particles transferred from the laminated or ribbon rings. Additional studies were then designed to verify this observation and to optimize the number of rings needed and their position on the valve stem.

Investigating packing ring sequence. To allow direct comparison of the great many packing systems tested, the dimensions of stem and packing box were standardized and, with a few justifiable exceptions, the number of rings used was kept constant. The packing system schematics shown in Figs. 4 through 7 indicate the packing ring sequences and the resulting cycles vs. leak rate plots.

The lantern ring shown in each schematic separates the four or five-ring major seal located at the bottom of the packing box from the minor, or secondary, seal located above the lantern ring. The lantern ring forms a chamber which allows testing for leakage through the side vent shown.

The major seal was always tested against 600° F, 400 psig steam. The minor seal prevents steam that penetrates the major seal from leaving the lantern ring chamber. The maximum allowable leakage rate typically specified for high temperature steam service is 0.5 cc/hr. for a 0.5 in. diameter stem.

Testing reveals dramatic effects. Results with either conventional laminated graphite or ribbon graphite packing rings are shown in Fig. 4. The packing sequence, starting from the high pressure end of the stem, is four graphite rings, a lantern ring and three graphite rings. As expected, leakage occurred during the packing break-in period. Even after normal retightening of the packing box follower, the leak rate limit was quickly exceeded. Tightening the packing follower bolts again may have brought the valve stem leakage within specification, but friction would have increased above the starting level.

Fig. 5 shows the cycles vs. leak rate obtained when braided graphite filament rings were used to replace

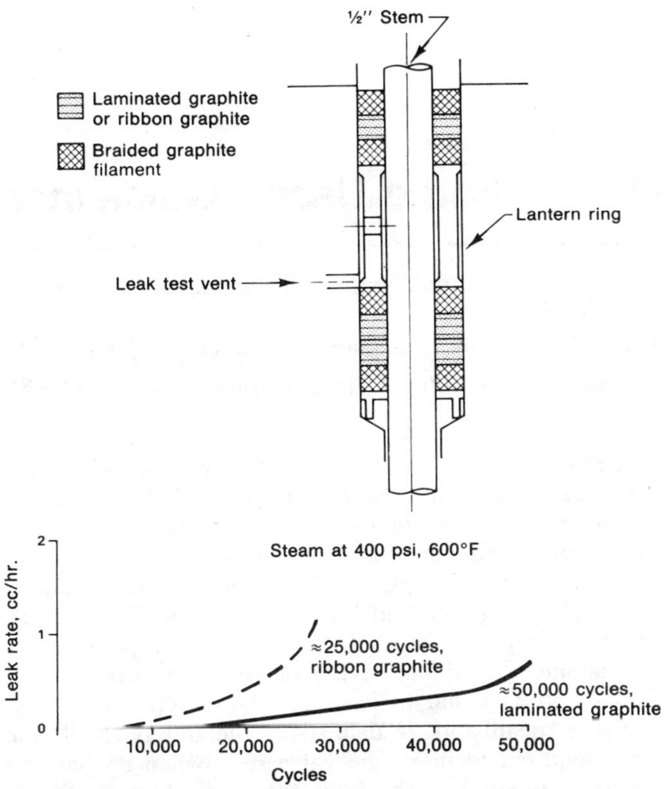

Fig. 7—Same as Fig. 4, except packing is a combination of four braided graphite filament rings with three laminated or ribbon graphite rings. This packing configuration, with laminated graphite, gave the greatest number of cycles with the lowest leaking rate—more than 50,000 cycles before the leak rate exceeded 1 cc/hr.

some of the laminated graphite rings. The 0.5 cc/hr. leak rate specification was not exceeded until after 19,000 cycles—an improvement of about 2,000 percent in packing life. Even at 30,000 cycles, leak rate was only 1.0 cc/hr. For this test, the laminated graphite rings (one above the lantern ring, one below) were positioned between braided graphite filament rings. Packing ring stress on these rings was the same as for the tests shown in Figs. 4, 6 and 7.

Another variation was to use laminated rings and ribbon rings reduced in thickness by half, in conjunction with graphite filament as shown in Fig. 6. Two, 1/8 in. thickness rings were used below the lantern ring chamber, and one was used above. At the 0.5 cc/hr. leak rate, reached after 30,000 cycles with laminated rings, packing life was increased about 3,600 percent. The theory here was to keep the graphite particle transfer to a minimum, consistent with adequate sealing, by reducing surface contact between the seal ring and valve stem.

The best of the packing sequences to date is shown in Fig. 7. Over 47,000 cycles were logged using laminated rings with braided graphite filament rings before the 0.5 cc/hr. leak rate specification was exceeded. After 50,000 cycles the leak rate was still only 0.75 cc/hr. (At 50,000 cycles, some valves may be considered near the end of their mechanical life due to general wear and tear.) The lower curve plotted in Fig. 7 shows the effect when using ribbon rings instead of laminated rings. The leakage specification is exceeded at about 23,000 cycles, and the 1.0 cc/hr. rate is exceeded at about 28,000 cycles. All packing cycle life listed, for all tests, is with no adjustment made to the packing follower bolts after start of test.

Stem roughening vs. stem polishing. Valve stems that had been in extended use in the field were brought into the laboratory and compared in surface appearance and roughness to the stems used in the tests. Field valves had used standard graphite ribbon or laminated graphite packing rings. In all cases, their sliding contact area was blackened, and the deposit was thick enough that the graphite coating could be scraped into a pile with a knife blade (Fig. 3).

On the other hand, the stems used in laboratory tests that had gone over 20,000 cycles were still extremely smooth with only slight discoloration. Interference colors were visible, which suggested that any coating present was very thin and very smooth. The slight coating seemed to be polished to a smoothness better than that of the original stem.

The contact surfaces of the field examples seemed blacker and rougher the longer they had been used, indicating that graphite particle transfer continues as long as the stem is stroked. In addition, putting new packing rings on a stem with this built-up deposit could lead to a shorter operating life than that provided by the originals.

Another factor related to stem surface roughness is the force necessary to start relative motion between stem and packing rings when the packing ring stress has been adjusted to allow for long cycle life and low leakage. In laboratory tests, the required force to start sliding motion for the laminated ring system shown in Fig. 4 was 25 percent greater than for the systems containing braided graphite filament rings.

During these hysteresis tests, another fact emerged: Ribbon graphite packing rings made from crinkled ribbon didn't show the lower friction expected of them when compared to laminated rings.

Final proof in the field. The results of the study are consistently repeatable in the laboratory. This, together with past experience in applying practical research to field use, provides a high degree of certainty that the packing system will be superior to those currently used.

By using the new graphite filament packing technique to advantage and by using TFE packing for conditions below 450° F, only two packing styles would be required for the majority of valve applications.

Control Valve Packing—Some Useful Guidelines

Many industrial process valves provide leak-free service for years while others show stem leakage in less than a month. Here are some of the reasons for the difference and a few hints on how to extend packing life

Allen Hays and **Greg Berggren,** Fisher Controls Co., Marshalltown, Iowa

PACKING SMALL manually operated valves that are not exposed to extreme environments is usually simple and effective, but complications may result when valves are large or operated at higher pressures and temperatures and are continuously being positioned by a valve actuator. Most process control valves are also called upon to handle a wide range of fluids, some of which can disintegrate packing materials that function well in ordinary applications. Improvement is needed in both packing design and packing materials to up-date the state of the art and solve packing problems in systems where previous experience is limited.

KEEP FRICTION LOW FOR PRECISE CONTROL

Control valve actuators are designed to cope with a reasonable amount of packing friction, but to achieve the most sensitive response of valve plug to the actuator, the lowest possible packing friction without leakage is desirable. As any homeowner do-it-yourself plumber knows, the way to stop leakage around the stem of a kitchen faucet is to tighten the packing by giving the packing nut an extra turn. The valve may operate a little more stiffly, but this is a minor trade-off; if needed, a larger handle can provide extra leverage. Most industrial control valves, on the other hand, are designed for use with an actuator that is sometimes limited in both the amount of force or torque it can provide and in the amount of "response lag" the system can tolerate without failure to provide good control. Extra force can be provided by going to actuators designed for heavy duty, but cost goes up accordingly. Far better to use proper packing and keep the friction low. Packing suppliers have responded by offering a number of packing compositions designed for low friction performance. Some examples:

- Braided asbestos impregnated with TFE, molded into split rings
- Asbestos fiber, lead wool, flake graphite, metal particles and neoprene binder
- Braided pre-shrunk yarn of Teflon fiber impregnated with TFE
- Molded rings of TFE, glass fibers and compounder
- Square braided blue African asbestos impregnated with TFE.

A new valve, or an old valve with new packing, will have higher friction until it's "broken-in." Valve design engineers use a plot of the force required to move the valve stem versus valve stem travel to show friction effects. With no friction, the plot for both opening and closing is the single straight dotted line, as shown in Fig. 1. With friction present, however, the position of lines for opening and closing are different; the degree of separation is directly proportional to friction. This difference, or hysteresis + deadband, is defined as the difference in the force required to move the valve stem when no packing is used compared to the force required when packing is in place.

New valves show larger hysteresis than nominal hysteresis exhibited during most of packing life. At some lower hysteresis value, leakage occurs, and packing must be tightened or replaced. The goal is to achieve a constant hysteresis as low as possible without leakage; at this point both the response of the valve trim to the actuator and packing wear-rate are optimum.

CONTROL VALVES REQUIRE SPECIAL ATTENTION

Some industrial control valves involve a sliding motion between the valve stem and the packing, rather than the

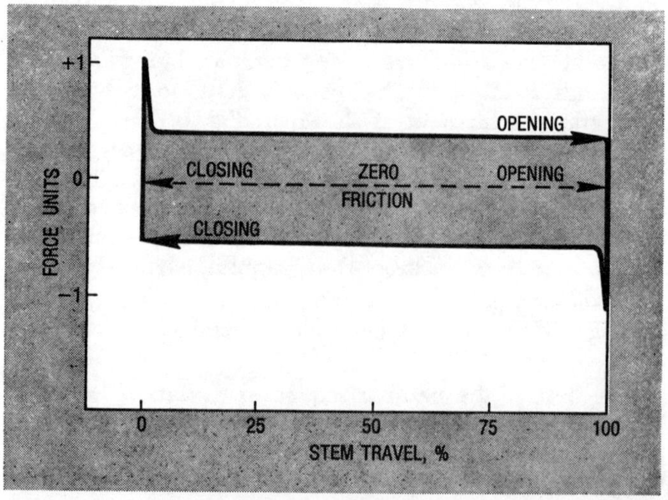

Fig. 1—Simplified drawing showing hysteresis + deadband effect caused by the packing in valve opening and closing. With no packing friction, force to open or close is practically zero. Horizontal solid lines represent dynamic friction, spikes at each end represent static friction which must be overcome before travel can begin. Colored area represents hysteresis + deadband, which can be very large if packing is tight and friction high.

ordinary rotary motion of stems in manually controlled valves and other control valve types. This sliding, or reciprocating, motion tends to shorten packing life and causes variations in friction as the packing wears. In a rotary stem valve, mating surfaces remain in tight contact and are mutually "polished in" for a tight seal, and the rotary motion does not tend to force packing out of its seat.

In reciprocating motion, every stem movement tends to disrupt the packing material due to the reversal in sliding direction that occurs with every stroke. Dirt or contaminating material picked up on the stem while it is exposed can be drawn back into the packing when the stroke is reversed. In cryogenic systems, a common problem with reciprocating valves is frost formation on the exposed stem; when this is subsequently forced into the packing, the wear-rate skyrockets.

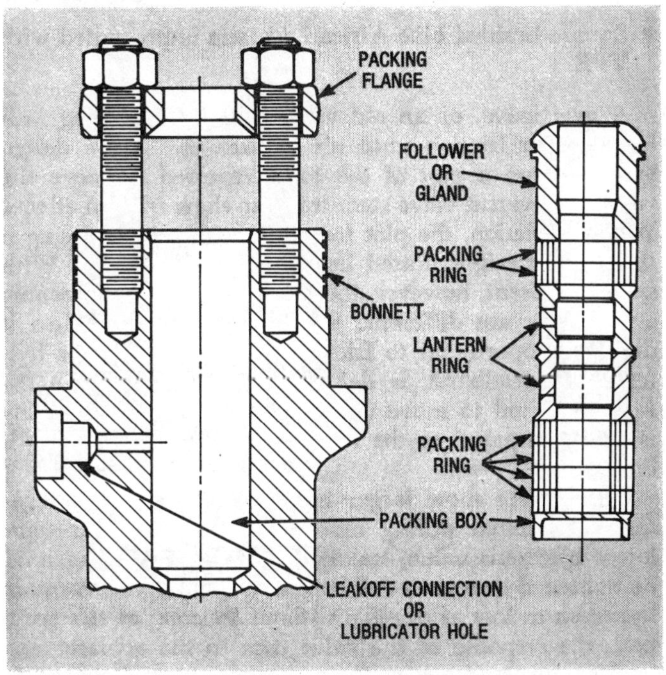

Fig. 2—Typical arrangement for jam type packing usually used with graphite or asbestos and graphite rings. On the left is the bonnet showing the packing box and packing flange. On the right is the stack of rings and follower that are assembled on the valve stem before inserting into packing box.

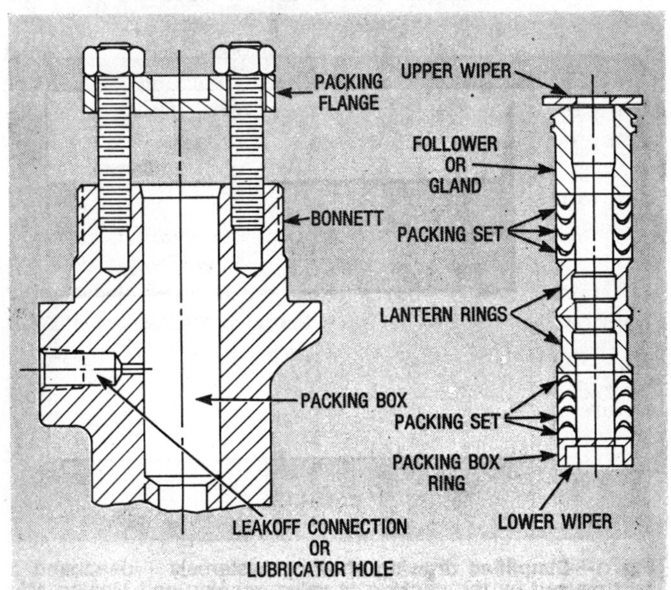

Fig. 3—Special TFE packing ring arrangement for a valve used in both pressure and vacuum service. Because TFE can be damaged by contaminants drawn into the packing by reciprocating stem motion, an upper wiper (usually made of felt) and a lower wiper (TFE) are added.

All else being equal, packing wear-rate is proportional to the friction and the total relative sliding motion between valve stem and packing. Because automatic control valves are being positioned frequently by the actuator, the amount of relative sliding motion is usually much higher than for manually operated valves. A control valve in constant motion ("hunting") because its controller is improperly adjusted can suffer greatly reduced packing life.

"Hunting" can also result when packing friction becomes too great. The overshoot that occurs when static friction exceeds dynamic friction may keep the valve stem in almost continuous motion. In some cases, repacking may be required once a month. Valves on low duty cycles may go years with nothing more than an occasional tightening.

PRE-FORMED PACKING RINGS ALLOW MANY OPTIONS

For general convenience in handling and to speed new valve assembly, packing material usually is available in molded or pre-formed rings (with either square or circular cross-sections), or V-rings. The inside diameter of the ring will be matched to the diameter of the valve stem and the outside diameter will allow a snug fit inside the packing box. There's no rule as to how many rings should be used; in general, the more rings, the better the seal but the greater the friction.

Another option is ring spacing. Lantern rings can be used to provide a space between packing rings without adding significantly to the friction. Using rings of different compositions on the same stem is still another option. Figs. 2 and 3 show some of the most commonly used packing ring, lantern ring and follower combinations used in many control valves.

Other variations are found in special applications. If low friction and chemical inertness are desired, Teflon or Kel-F packing materials can be used—provided temperatures are kept within reasonable bounds. These materials are not very compressible and are usually supplied as V-rings under spring-loading to achieve adequate sealing. Packing materials don't have to be used as pre-formed rings; we'll discuss some other options in a following section.

CHOOSING PACKING MATERIAL: THE SCIENCE AND THE ART

When a valve manufacturer receives an order he usually supplies his standard packing for that application. Sometimes the customer specifies a packing material he feels is suitable. If the valve manufacturer thinks there's a better choice, he'll offer recommendations based on past experience and knowledge of the strong and weak points of available packing materials.

Basic packing materials in widespread use today are asbestos, graphite and TFE. Asbestos is low cost and stable to about 800° F, but tends to give a high friction seal. Graphite may also have higher friction. Graphite is stable to 3,000° F in nonoxidizing service, but practical limits are 750° F for oxidizing service and 1,200° F for nonoxidizing service. It has high thermal conductivity and long service life, but requires higher gland pressures for leak-free operation. It is impervious to a wide range of liquids. TFE type materials are even more chemically inert and have very low friction properties, but they function best with a smooth stem finish (four to six microinches RMS). Temperatures must be not less than −40° F or more than 450° F.

To produce optimum properties, various materials can be mixed with each other. Two good examples of materials that are not suitable packing when used alone, but which result in better packing when used as impregnants, are soft metals and fiber glass. Soft metals provide better retainment and can provide cathodic protection in systems where stem or packing box corrosion is a problem. Metals or moly disulfide strengthens packing materials; they're often used in TFE.

The science of packing material selection lies in the knowledge of individual material properties—the art lies in how to mix and apply these materials for optimum performance.

PACKING FABRICATION: RINGS, STRINGS AND RIBBONS

As mentioned earlier, most control valve packing is furnished as pre-formed rings made to specific dimensions for the packing box and stem. The method by which the ring is formed as well as the physical form of the starting material(s) can have a significant effect on its performance. Starting materials can be bulk solids, strings, ribbons, braids, laminations or powders. Ring formation can be accomplished by winding, weaving, curing in a mold, compression, extrusion, punching or any other suitable technique.

Elastomer rings are usually molded to produce ring structures. Almost any elastomer can be used, as well as many plastics, such as TFE. If the material has insufficient resilience to maintain tight contact with packing box and stem, spring loading will provide this necessary feature.

Spring loading has advantages. The packing follower—once tightened down—doesn't require readjustment, more consistent friction properties are obtained and assembly is easier. A disadvantage is inherent; rings can't be replaced without disconnecting the stem. In some applications there is enough resilience in TFE V-rings to effect adequate seal without a spring.

V-rings should be installed so that the "V" is open toward the high pressure side of the seal. Also, a stem sliding away from the point of the "V" will show somewhat less friction than one sliding in the same direction as the point of the "V."

As in the early days, braided packing is still used when a ring of the right size is not available or repacking is desirable in the field without disconnecting the stem. Braids can be made of any useful composition, either plain or in impregnated combinations. Asbestos, TFE and graphite are all available in braided form, with many different combinations of the three or with other impregnated materials. These materials are less costly than preformed ring structures and are more versatile; as much braid as needed can be used to fill the space between packing box and stem regardless of dimensions.

Pure graphite packing materials were not available in the past, but in recent years technology has produced both graphite sheets and ribbon forms. These are now used as field packing or as pre-wound and pre-compressed ring structures. Circles of graphite sheets, stacked many layers high (with a hole punched through the center to allow stem penetration) are compressed to form a laminated graphite ring structure.

A more recent development in pre-formed graphite rings is ribbon graphite wound like a roll of tape to the proper diameter (leaving a hole in the middle to fit the stem). Compression causes the ribbon to cohere into a stable ring shape and also wrinkles the ribbon into accordion-like pleats along the surfaces that contact stem and packing box.

Split rings of different materials are commercially available for repacking valves in the field. Ring splits should be staggered to avoid a continuous leak path. Also, the material must be resilient enough to achieve a good seal across the split when the packing nuts are tightened.

HINTS AND KINKS FOR BETTER PACKING

Proper selection of valve design can minimize the packing problem. Cryogenic valves, or valves used at very high temperatures, are often fitted with extra-long bonnets to keep the packing box as far as possible from the heat or cold. To combat stem wear, hardened or chrome-plated stems can be provided.

Valve stem corrosion or pitting can occur in a situation in which two dissimilar metals and an electrically conductive liquid are in mutual contact. Water in the packing material may, in some cases, be all that is needed to start corrosive attack of the valve stem or packing box. When this is recognized as a potential problem the valve manufacturer should take precautions to assure dry packing. Packing materials combined with sacrificial metals are used to retard stem pitting.

Some packing materials require many more cycles of "break-in" than others. Be sure to check new valve installations frequently to guard against rapid decrease in hysteresis to the point of leakage. Progressively adjust packing pressure until constant hysteresis is obtained. Spring-loaded TFE type packing materials should not require break in.

Corrosion Monitoring . . . Select the Right System

Advances in electronics have lowered the cost and improved the application of corrosion monitoring. Many techniques are available. Here's how to select the right system for your corrosion problem

P. J. Moreland and **J. G. Hines,** ICI Ltd., Mond Division, Runcorn, Cheshire, England

A WIDE RANGE of corrosion monitoring techniques is now available (Table 1). These allow determination of total corrosion, corrosion rate, corrosion state, analytical determination of corrosion product and detection of defects or changes in physical parameters. Costs can be small where simple instrumentation and a few measurements are appropriate but in some instances may be extremely costly and require expert skills.

Much of the progress made in the past few years has been due to advances in electronics which have allowed multiprobe measurement and recording at a tolerable cost. Instantaneous feedback of corrosion information can be obtained from various parts of the plant, and can be fed to the plant control room and/or plant computer to permit control of the necessary process variable and provide corrosion control.

WHY MONITOR CORROSION?

Estimates have been made of the potential benefits of more extensive use of corrosion monitoring techniques, including possible developments, and of more effective use of the information. Very large potential savings have been suggested, such as 25 percent of the costs associated with maintenance (including production losses) in major industrial complexes. Such savings would be several millions of dollars annually in an individual complex.

Estimates of this type are obviously subject to error but the evidence that very substantial savings are possible, in principle, is solid. There is no doubt that the need to inspect major items at shutdown, and delays arising through rectification work not anticipated, cause a substantial part of the maintenance costs on a large plant. Major returns are possible from developments both in the large plant situation and in repetitive situations where the individual return is less spectacular.

Justifying corrosion monitoring. Corrosion monitoring may be used for a variety of reasons and it is often difficult to obtain the information needed to make a meaningful assessment of the benefits on a strictly economic basis. One difficulty in the development of a general economic case is that confidentiality limits the extent to which information about many of the best examples can be disclosed. Occasionally this is because the facts could be embarrassing, but typically the reason is simply that a successful application contributes valuable know-how. There are examples where proprietary processes (i.e., available for sale) depend in part on monitoring for their success, and others where knowledge gained by monitoring provides important competitive advantage.

A more common problem, however, is that the necessary economic information is simply not available. Existing accounting systems are rarely suitable for isolating the appropriate costs, and it can be expensive to devise a special system for this purpose. In any event, corrosion monitoring is typically only one of a number of actions taken in a given situation. Even if the total benefit were known with certainty, the proportion of this benefit which should be balanced against the costs of corrosion monitoring would be at best a matter of opinion. Therefore, the ability to compute the costs and savings due to corrosion monitoring with any degree of certainty is unusual. In

TABLE 1—Instrumentation for corrosion monitoring

Method	Measures or detects	Notes	Use
Linear polarization (Polarization resistance)	Corrosion rate is measured by the electrochemical polarization resistance method with two or three electrode probes.	Suitable for most engineering metals and alloys providing process fluid is of suitable conductivity. Portable instruments at modest cost to more expensive automatic units available.	Frequent
Electrical resistance	Integrated metal loss is measured by the resistance change of a corroding metal element. Corrosion rates can be calculated.	Suitable for measurements in liquid or vapor phase on most engineering metals and alloys. Probes as well as portable and more expensive multi-channel units available.	Frequent
Potential monitoring	Potential change of monitored metal or alloy (preferably plant) with respect to a reference electrode.	Measures corrosion state of plant directly, e.g., active, passive, pitting, stress corrosion cracking via use of a voltmeter and reference electrode.	Moderate
Corrosion coupon testing	Average corrosion rate over a known exposure period by weight loss or weight gain.	Most suitable when corrosion is a steady rate. Indicates corrosion type. Moderately cheap method with corrosion coupons and spools readily made.	Very frequent
Analytical	Concentration of the corroded metal ions or concentration of inhibitor.	Can identify specific corroding equipment. Wide range of analytical tools available. Specification electrodes readily used.	Moderate
Analytical	pH of process stream.	Commonly used in effluents. Standard equipment available though robust pH responsive electrodes such as antimony, platinum, tungsten can be preferable to glass electrodes. Solid Ag/AgCl is useful reference electrode.	Very frequent
Analytical	Oxygen concentration in process stream.	Useful where oxygen control against corrosion using oxygen scavengers such as bisulfite or dithionite is necessary. Electrochemical measurement.	Moderate
Radiography	Flaws and cracks by penetration of radiation and detection on film.	Very useful for detecting flaws in welds. Requires specialized knowledge and careful handling.	Frequent
Ultrasonics	Thickness of metal and presence of cracks, pits, etc., by changes in response to ultrasonic waves.	Widely used for metal thickness and crack detection. Instrumentation is moderately expensive but simple jobs contracted at fairly low cost.	Frequent
Eddy current testing	Uses a magnetic probe to scan surface.	Detects surface defects such as pits and cracks with basic instrumentation of only moderate cost.	Frequent
Infrared imaging (Thermography)	Spot surface temperatures or surface temperature pattern as indicator of physical state of object.	Used most effectively on refractory and insulation furnace tube inspection. Requires specialized skills and instrumentation is costly.	Infrequent
Acoustic emission	(a) Leaks, collapse of cavitation bubbles, vibration level in equipment. (b) Cracks by detection of the sound emitted during their propagation.	A new technique capable of detecting leaks, cavitation, corrosion fatigue and stress corrosive cracking in vessels and lines.	Infrequent
Zero resistance	Galvanic current between dissimilar metal electrodes in suitable electrolyte.	Indicate polarity and direction of bimetallic corrosion. Useful as dewpoint detector of atmospheric corrosion or leak detection behind linings.	Infrequent
Hydrogen sensing	Hydrogen probe used to measure hydrogen gas liberated by corrosion.	Used in mild steel corrosion involving sulfide, cyanide and other poisons likely to cause hydrogen embrittlement.	Frequent in petrochemical industry
Sentinel holes	Indicates when corrosion allowance has been consumed.	Useful in preventing catastrophic failure due to erosion at pipe bends, etc. Leaking hole indicates corrosion allowance has been consumed.	Infrequent

consequence, while those who know the background of a particular situation are able to appreciate the significance of a semi-quantitative argument, to construct a strictly economic assessment which is completely convincing to others is difficult. A number of examples have been described in the literature, and in particular in two recent symposia,[2,3] and the Handbook of Industrial Corrosion Monitoring.[1]

It is helpful to analyze some reasons for adopting monitoring techniques:

► **As a diagnostic tool.** The most common use of on-line monitoring is providing information for the solution of a corrosion problem. Knowledge of the pattern of behavior (e.g., the uniformity or time-dependence of corrosion rate) can be very valuable. Frequently several variables can be significant, and the ability to correlate corrosion with the significant variable under specific circumstances may be vital. This may provide the basis of a solution, but a protective system is best designed to counter the real conditions.

Corrosion monitoring is one of several tools available to the investigator, but is particularly valuable because it may provide information not easily obtained in other ways and thus lead to a better or speedier solution. Since the work will normally be under the direct control of a specialist, whose interpretative skill may be vital, relatively simple equipment is often adequate.

► **To monitor the effectiveness of a solution.** A logical extension of the diagnostic application is using corrosion monitoring techniques to establish whether a solution has been effective. This can be done simply by continuing the original investigation. More permanent installations are being used to increasing extent to provide longer-term assurance. Such equipment is likely to be more sophisticated, since the information is recorded with other operational data and interpreted, in the first instance at least, by staff with a more limited corrosion knowledge.

► **To provide operational or management information.** Corrosion can often be controlled by maintaining a single operational variable (e.g., temperature, pH, humidity) within limits determined by prior monitoring or other investigations. If the significant variable is measured for other reasons, this measurement can be used directly for corrosion control. If the variable is not otherwise measured, or in more complex areas where several variables interact, corrosion monitoring information can be used to control plant operation so as to minimize corrosion.

A similar application involves process optimization or examines the effect of variation of raw material. Any process change may have significant effects on corrosion, and corrosion monitoring techniques allow full-scale trials to proceed with minimum plant risk.

In all these examples the information is used by non-specialists, who may be non-technical staff. A high standard of equipment reliability is essential, and the display must be appropriate to the particular circumstances. The equipment is therefore likely to be more sophisticated, and costly, than in an installation controlled by a specialist. A detailed understanding of the corrosion behavior of the system is essential before such systems are designed. Further, clear and unambiguous instructions must be provided for action by the plant operators.

► **As part of a control system.** An extension of monitoring techniques to provide operating information is to use the monitoring information to control the plant, or parts of the plant. Potential measurements controlling anodic or cathodic protection systems are an example, as is the more recent development of using monitoring signals to control inhibitor or other additions to cooling water systems. Practical applications have been confined to systems whose behavior is relatively simple. Some progress has been made in handling more complex systems however, and the increasing use of computers for process control is likely to accelerate such developments. Clearly equipment used in this way requires a high standard of reliability, and some form of "self-testing" feature. A detailed understanding of the corrosion behavior of the plant is essential to successful control system design.

► **As part of a management system.** The data from corrosion monitoring can help management in several ways. It can supplement other inspection techniques, and improve maintenance and its co-ordination with production schedules. This approach is used to varying degrees in many industries. The same information can also be used in optimization and other investigations, though this is less widely appreciated. Computer and data handling developments have been an important stimulus in these areas, since a substantial mass of data has to be processed to a manageable form.

SELECTING A TECHNIQUE

Many techniques have been used for corrosion monitoring and other developments are clearly possible. Consequently, when a new application is considered, a problem arises in choosing the most appropriate technique. Each has strong points and limitations, and none is best for all situations.

Any monitoring technique can provide only limited information and the techniques should be regarded complementary rather than competitive. Where more than one technique will give the information required, the information is obtained in different ways. A cross-check can be very valuable and differences in detail can add meaning.

A corrosion monitoring technique rarely gives wrong information, unless of course the equipment is faulty. "Nonsense" results arise because the information is irrelevant in the corrosion sense. The polarization resistance method, for example, measures the combined rate of any electrochemical reactions at the surface of the test sample. If the main reactions are corrosion, the rate measured is the corrosion rate. If, however, other reactions are possible at comparable or greater rates, the measured rate includes the other reactions. Useful deductions can still be made provided it is recognized that the corrosion rate has not been measured.

Choosing a monitoring technique can be complex, requiring expert knowledge. This is true to a point, but in practice the choice for a given situation is usually clear once the requirements are defined. First, establish what type of information is needed. This necessarily involves an input from the management responsible for the plant in question. This decision will, in effect, decide which monitoring techniques should be used. It is helpful to recognize two monitoring categories. These are:

• **Where the primary objective is diagnosis in a new situation.** Typically the corrosion processes involved and the controlling parameters are uncertain. It may be difficult to decide on the most appropriate technique, but it is often advantageous to use more than one. Factors that actually prove to be significant are not always those expected.

One approach is to undertake a laboratory study to determine which parameters are likely to be important. This information can then be used to decide which techniques should be used and to aid in interpreting the results obtained for the plant. Alternatively, monitoring can be undertaken directly. The choice between these approaches depends on the availability of suitable laboratory facilities and staff with the necessary experience, and to what extent the problem is understood. In either situation check the information obtained by monitoring, by inspection before and after, or other means. Expert help is often needed in interpreting results and may be desirable in planning the work and selection of techniques. However, successful interpretation requires knowledge of the plant and process in question as well as expertise in monitoring techniques and corrosion.

• **Where the primary objective is to monitor the behavior of a known system.** Applications of this type often follow one of the diagnostic types. Alternatively the problem resembles other applications where monitoring has been used successfully. In either area the choice of technique is based on past experience. Expert assistance may well be unnecessary even in interpreting results, unless unusual features appear.

In addition to choosing the technique, decide the appropriate degree of complexity. Basic monitoring equipment for most techniques is relatively simple, comprising a probe (the sensing element) and a measuring instrument.

Equipment and labor costs are relatively modest, if only a few readings are required. The amount of information that can be obtained by this approach is limited, but may be sufficient. If not, additional probes can be installed and/or more complex instrumentation introduced to allow automatic scanning, automatic recording or regular readings from one or more probes. These more complex installations are more expensive, but much more information is obtained more conveniently and it is more likely to be practical to use the information for control.

Eight criteria to determine a technique are summarized in Table 2 and described as follows:

(1) **Time for individual measurement**—Some techniques provide information that is effectively instantaneous, while others are necessarily slower.

(2) **Type of information obtained**—Some techniques measure corrosion rate, others measure total corrosion, or the remaining thickness which is not exactly equivalent. Yet others provide information on the distribution of corrosion on the corrosion regime.

(3) **Speed of response to change**—Techniques which do not provide individual measurement quickly are obviously unsuitable where fast response is required. Not all techniques that provide effectively instantaneous information are, however, capable of a fast response. Where the measurement is rate or corrosion regime, a fast response can be obtained, but if the measurement is of total corrosion, remaining thickness or corrosion distribution, speed of response is limited by the technique's ability to discriminate between successive readings.

(4) **Relation to plant behavior**—Many of the more effective techniques provide information on the behavior of a probe inserted into the plant rather than the behavior of the plant itself. The information measures the corrosivity of the environment, from which plant behavior can be inferred. Other techniques indicate total corrosion in the system, with little or no indication of its distribution. Others give an accurate picture of a local corrosion pattern of the plant itself, but no information on what is happening elsewhere.

(5) **Applicability to environments**—A fast response is most readily obtained from electrochemical measurements which require an electrolyte environment; a high electrolytic conductivity is not always necessary however. Non-electrochemical measurements can be used in gaseous environments, or non-conducting fluids, as well as in electrolytes.

(6) **Type of corrosion**—Most corrosion monitoring techniques are best suited to situations where corrosion is general, but some provide information on localized corrosion.

(7) **Difficulty of interpretation**—Interpreting results is often relatively straightforward if the technique is used within its limitations. Interpreting results obtained by some techniques is, however, more difficult. This is true of all techniques if they are used near the limits of their applicability.

(8) **Technological culture**—Some techniques are inherently technically sophisticated. This limits their use to organizations with a strong technological culture. Most others are much less demanding in this respect.

In principle, available techniques could be ranked in an order of merit for each of the eight criteria. In practice the relative merits change with circumstances, so a formal treatment of this type is potentially misleading. The most useful general approach is to consider the strengths and weaknesses of the techniques individually. Table 2 provides a reasonable starting point.

Possible future developments. The most obvious line of development continues the past trend, whereby additional techniques are devised for particular purposes. A second trend likely to continue is refinement of the measuring element to achieve greater reliability or to more precisely simulate plant conditions. Many corrosion monitoring techniques, for example, utilize probes which are geometrically very different to plant surfaces and may behave differently for this reason. For some applications using a heated probe to simulate heat transfer surfaces is desirable. Another may be multi-purpose probes exploiting both polarization resistance and electrical resistance.[3]

Another obvious line of development improves measuring instruments by developments in electronic engineering. More sophisticated and reliable instruments can now be devised, often at lower cost. Some techniques will become available for plant use that have previously been confined to the laboratory because the available instruments were too delicate or could only be used by skilled operators.

Developments in electronics may be even more significant, by removing some practical limitations common to many existing techniques. There are two related possibilities, the inclusion of small computer facilities and modified forms of display.

Existing techniques often give only limited information and are practical only under restricted circumstances. Many also have the weakness that "nonsense" results are difficult to distinguish from valid ones and can cause the wrong conclusions to be reached. In other words, a measurement is obtained that is valid in itself but does not correlate with the quantity of interest.

These difficulties could be overcome by using computer technology, which should permit relatively inexpensive equipment incorporating logic and calculation facilities. The instrument might make several measurements, by the same or quite different techniques, and perform calculations based on algorithms that could be complex, displaying a derived quantity. Some, if not all, non-valid measurements could be rejected and the fact indicated.

The UMIST polarization impedance technique is an example in which measurements at a range of frequencies are analyzed to separate the component representing corrosion rate from other that may interfere with conventional polarization resistance measurements.[3] Analysis is possible only if certain conditions are fulfilled so that a test for validity can be incorporated. Developments of

TABLE 2—Characteristics of corrosion monitoring techniques

Technique	Time for individual measurement	Type of information	Speed of response to change	Relation to plant	Possible environments	Type of corrosion	Ease of interpretation	Technological culture needed
Electrical resistance	Instantaneous	Integrated corrosion	Moderate	Probe	Any	General	Normally easy	Relatively simple
Polarization resistance	Instantaneous	Rate	Fast	Probe	Electrolyte	General	Normally easy	Relatively simple
Potential measurement	Instantaneous	Corrosion state and indirect indication of rate	Fast	Probe or plant in general	Electrolyte	General or localized	Normally relatively easy but needs knowledge of corrosion. May need expert	Relatively simple
Galvanic measurements (zero resistance ammeter)	Instantaneous	Corrosion state and indication of galvanic effects	Fast	Probe or occasionally plant in general	Electrolyte	General or unfavorable conditions localized	Normally relatively easy but needs knowledge of corrosion.	Relatively simple
Analytical methods	Normally fairly fast	Corrosion state, total corrosion in system item corroding	Normally fairly fast	Plant in general	Any	General	Relatively easy but needs knowledge of plant	Moderate to demanding
Acoustic emission	Instantaneous	Crack propagation and leak detection	Fast	Plant in general	Any cavitation	Cracking, cavitation and leak detection	Normally easy	Crack propagation specialized, otherwise relatively simple
Thermography	Relatively fast	Distribution of attack	Poor	Localized on plant	Any. Must be warm or sub-ambient	Localized	Easy	Specialized and difficult
Optical aids (closed circuit TV, light tubes, etc.)	Fast when access available, otherwise slow	Distribution of attack	Poor	Localized on plant	Any	Localized	Easy	Relatively simple
Visual, with aid of gages	Slow. Requires entry on shutdown	Distribution of attack indication of rate	Poor	Accessible surfaces	Any	General or localized	Easy	Relatively simple but experience needed
Corrosion coupons	Long duration of exposure	Average corrosion rate and form	Poor	Probe	Any	General or localized	Easy	Simple
Ultrasonics	Fairly fast	Remaining thickness or presence of cracks and pits	Fairly poor	Localized on plant	Any	General or localized	Easy	Simple
Hydrogen probe	Fast or instantaneous	Total corrosion	Fairly poor	Localized on plant or plant or probe	Non-oxidizing electrolyte or hot gases	General	Easy	Simple
Sentinel holes	Slow	Go/no go on remaining thickness	Poor	Localized on plant	Any, gas or vapor preferred	General	Easy	Relatively simple
Radiography	Relatively slow	Distribution of corrosion	Poor	Localized on plant	Any	Pitting, possibly, cracking	Easy	Simple but specialized radiation hazard

this type can be envisaged with many other techniques.

Obviously corrosion monitoring is not a panacea, but since its adoption in the late 1950s certain industries have recognized the significant returns. The state of the art is such that a wide variety of corrosion monitoring tools are available and although the limitations of existing techniques need to be recognized, they are within the scope of future developments which should make the applications more practical for a wider range of users.

ACKNOWLEDGMENT

This article is based on material copyrighted by Her Britannic Majesty's Stationery Office, London, and presented at the NACE Corrosion '78 Conference, March 6-10, Houston.

LITERATURE CITED

1 Handbook of Industrial Corrosion Monitoring, HMSO, London, (1978) (U.S.A. agent: Pendragon House, Old Mystic, Conn., and Palo Alto, Calif.).
2 "Corrosion Testing and Monitoring," Inst. Corr. Sc. Tech Conference, Westfield College, London, March 30-April 1, 1977, Chameleon Press Ltd. (1977).
3 "On Line Surveillance and Monitoring of Process Plant" Soc. Chem. Ind. Symposium, City University, London, September 1977.

To Calibrate Transmitters Faster

This two-step method to calibrate transmitters is not only fast, but gives the improved accuracy needed for digital readouts

G. F. Riedmuller, Leslie Co., Parsippany, N.J.

INSTRUMENT transmitters, calibrated to communicate with analog display meters, may not send accurate enough signals to assure efficient use of more precise digital readouts. Yet the typical multi-step method for calibrating transmitters can be laborious. Therefore, a simple two-step procedure is suggested to shorten time and increase calibration accuracy.

Current method inadequate. The subjective readout of analog meter in practice is limited to ±2 percent accuracy. Usually an analog meter is not expected to provide pinpoint readings. Therefore, calibration, although desirable, is not critical.

When a digital readout meter is used, the numerical display implies additional accuracy and the transmitter needs to be calibrated more carefully. The most commonly used method to calibrate a transmitter for use with digital meters is time-consuming, costly and inherently inexact.

It is common practice to calibrate a transmitter to give its low output at zero pressure by setting the zero adjustment. Then at full pressure, the high output is set by the span adjustment. A problem arises when the span adjustment interacts with the zero adjustment so that when the pressure is reduced back to zero, the transmitter's low output is no longer at its proper output. Instructions often suggest to repeat these steps as often as necessary to get the desired outputs. The repetition is time-consuming, difficult and under some field conditions may be impractical.

An easy two-step procedure. Precision calibration of process control instruments can be readily achieved without repeated adjustments. A simple equation is given which makes it possible to set most transmitters and other instruments in two steps. As digital test and calibration equipment becomes more commonplace, this two-step procedure offers an easy way to shorten the time required for calibration and to give the accuracy needed by digital readouts.

The following equation determines an intermediate full pressure setting S_2 which, when zero is reset, brings the unit into complete calibration:

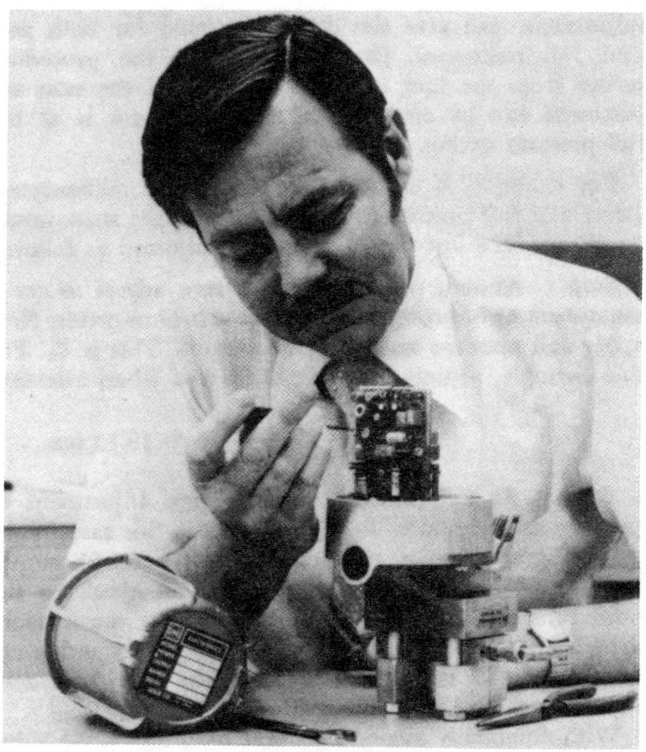

Fig. 1—Proper transmitter calibration is needed to take advantage of modern digital readouts.

TABLE 1—Full pressure output to be set by span adjustment
(Basis: Zero pressure output = 4 ma, final full pressure output = 20 ma)

S_1	0.0	0.1	0.2	0.3	0.4	0.5	0.6	0.7	0.8	0.9
15	21.82	21.77	21.71	21.66	21.61	21.57	21.52	21.47	21.42	21.38
16	21.33	21.29	21.25	21.20	21.16	21.12	21.08	21.04	21.00	20.96
17	20.92	20.89	20.85	20.81	20.78	20.74	20.71	20.67	20.64	20.60
18	20.57	20.54	20.51	20.48	20.44	20.41	20.38	20.35	20.32	20.30
19	20.27	20.24	20.21	20.18	20.16	20.13	20.10	20.08	20.05	20.03
20	20.00	19.98	19.95	19.93	19.90	19.88	19.86	19.83	19.81	19.79
21	19.76	19.74	19.72	19.70	19.68	19.66	19.64	19.62	19.60	19.58
22	19.56	19.54	19.52	19.50	19.48	19.46	19.44	19.42	19.40	19.39
23	19.37	19.35	19.33	19.32	19.30	19.28	19.27	19.25	19.23	19.22
24	19.20	19.18	19.17	19.15	19.14	19.12	19.11	19.09	19.08	19.06
25	19.05	19.03	19.02	19.00	18.99	18.98	18.96	18.95	18.94	18.92

TABLE 2—Full pressure output to be set by span adjustment
(Basis: Zero pressure output = 10 ma, final full pressure output = 50 ma)

S_1	0.0	0.1	0.2	0.3	0.4	0.5	0.6	0.7	0.8	0.9
45	51.43	51.40	51.36	51.33	51.30	51.27	51.24	51.20	51.17	51.14
46	51.11	51.08	51.05	51.02	50.99	50.96	50.93	50.90	50.87	50.84
47	50.81	50.78	50.75	50.72	50.70	50.67	50.64	50.61	50.58	50.55
48	50.53	50.50	50.47	50.44	50.42	50.39	50.36	50.34	50.31	50.28
49	50.26	50.23	50.20	50.18	50.15	50.13	50.10	50.08	50.05	50.03
50	50.00	49.98	49.95	49.93	49.90	49.88	49.85	49.83	49.80	49.78
51	49.76	49.73	49.71	40.69	49.66	49.64	49.62	49.59	49.57	49.55
52	49.52	49.50	49.48	49.46	49.43	49.41	49.39	49.37	49.35	49.32
53	49.30	49.28	49.26	49.24	49.22	49.20	49.17	49.15	49.13	49.11
54	49.09	49.07	49.05	49.03	49.01	48.99	48.97	48.95	48.93	48.91
55	48.89	48.87	48.85	48.83	48.81	48.79	48.77	48.75	48.73	48.71

$$S_2 = \frac{(\text{Live zero})\ (\text{Span})}{S_1 - (\text{Live zero})} + (\text{Span})$$

where S_1 is the initial output at full pressure and S_2 is the full pressure output setting which should be made by the span adjustment. Then a resetting of the zero adjustment will give the desired outputs for both zero and full pressures. The speediness of the procedure comes from the fact that the resetting of the zero adjustment can be made while the instrument is at the full pressure setting.

For example, if the live zero is to be 4 milliamperes (ma) and full pressure is to give 20 ma, the span would be 16 ma. The instrument would be adjusted as follows:

Step 1. At zero pressure, set the zero adjust to get 4 ma output as measured by a digital volt-ohm-meter. Now apply full pressure and read the output. This is S_1. For this example, assume S_1 reads 21.06 ma. Then calculate

$$S_2 = \frac{(4)\ (16)}{21.06 - 4} + 16 = 19.75 \text{ ma}$$

Step 2. At full pressure, set the span adjustment to get an output equal to S_2, or 19.75 ma in this case. Without changing pressure, reset the zero adjust for an output of 20 ma. The calibration is complete. A check on the procedure can be made simply by reducing the pressure to zero and noting the output is 4 ma. The equation accounts for the interaction between the span and zero adjustments.

With commonly available pocket calculators, the value for S_2 can be solved easily. For instruments with an output range of 4-20 ma, the value of S_2 to compensate for various readings of S_1 can be taken directly from Table 1. The first column of Table 1 gives values for S_1. The next column gives appropriate values for S_2. Subsequent columns of Table 1 give intermediate values for S_2 at added increments of S_1.

Returning to the example: with an S_1 of 21.06, the value for S_2 given in Table 1 would be between 19.76 (for $S_1 = 21.0$) and 19.74 (for $S_1 = 21.1$). Interpolation gives S_2 *of* 19.75.

Another common range for instruments is 10 to 50 ma. Table 2 gives appropriate values for S_2 for these instruments. Also the equation can be used in which the live zero is 10 ma and the span is 40 ma.

This method has been applied in the production of electronic transmitters at Leslie Co. and has provided a great savings in time, as assured precise calibration of all instruments and has proved useful in startups and in field service work.

Part VIII.

Other Ideas in Control and Optimization

Guide to Distillation Pressure Control Methods

Column pressure is one of the most difficult distillation variables to control. Presented is a basic discussion of column pressure control and a description of 21 different control methods showing advantages, disadvantages and applications

T. G. Chin, Chevron Research Co., Richmond, Calif.

THERE IS NEED for better understanding of distillation column pressure control. Our own experience supports the contention of two authorities in the field: Shinskey wrote, ". . . column pressure can be one of the most difficult variables to control."[1] And Buckley said, "The column overhead system is generally more complicated than either the feed system or the bottoms system."[2]

This article consists of two parts: The first includes an overview, the terminology, diagram notes and a guide to selecting a pressure control method. The second part begins with an index to the methods (Table I) followed by discussions of each method.

This discussion of pressure control methods hopefully will:

► Facilitate future study

► Help designers select methods and evaluate alternatives

► Facilitate computerizing the design procedure.

An overview. To begin, let us look at the broad picture. It is possible to control the pressure of any system containing vapor by varying temperature, volume, or mass. This possibility is implied in the following expression, an altered form of the general equation-of-state given in Reference 3:

$$P = f\,(V, T, M)$$

In a distillation column, we cannot control pressure, *P*, by varying vapor volume, *V*, or vapor temperature, *T*. We are normally stuck with varying vapor mass, *M*. When we do this, we are regarding pressure as a measure of vapor inventory and can think of pressure control as inventory ("*M*") control. (This may bring to mind that column pressure is analogous to tank level.)

There are two ways to affect vapor inventory and, hence, control pressure in a continuous process: Vary the rate of vapor inflow (or rate of generation), and vary the rate of vapor outflow (or rate of condensation).

This discussion is limited as follows:

- We have included mainly methods based on varying vapor outflow and condensation rate but have also included one method using gas addition. We have not discussed varying vapor generation rate because this has been rarely used in our experience.

- We have included systems that might be used in petroleum and petrochemical plants. We have not tried to include additional systems that are unique to other industries.

- We have not discussed atmospheric pressure operation because pressure control is not involved.

Methods of varying the rate of condensation are described briefly by Rademaker, et al.[8] They present a table of literature references covering 14 pressure control methods and 21 authors. Among them is Boyd,[9] who in 1948 presented a classification of pressure control methods that is still instructive. (To improve our understanding of what column pressure control includes, we should remember that pressure control and manipulating condenser duty are not identical, even though related. Pressure may be controlled by other means, and condenser duty may be manipulated to control such variables as reflux temperature and reflux drum level.)

Terminology. The control methods are divided into five groups:

A. Vary net vapor rate from the column or reflux drum

B. Vary overhead condenser area by changing condensate or coolant level

C. Vary temperature difference between coolant and process

D. Vary heat flux of overhead condenser (Heat transfer coefficient multiplied by temperature difference)

E. Other methods.

Features such as control valve location, condenser type, and the method of varying flux distinguish one method in a group from another.

The 21 methods we elected to discuss are listed in Table I.

We discuss behavior and suitability of the pressure control methods as they apply to the following three "cases" of net vapor rate (Table I, Column 3).

Case I: Always positive; the overhead is partially condensed.

Case II: Always zero at steady state but could be negative during transients; the overhead is totally condensed.

Case III: Sometimes positive and sometimes zero at steady state but could be negative during transients.

(Although net vapor rate must be either positive or zero at steady state, we can get a negative vapor rate but only as a transient. This happens when the condensing rate is higher than the combined vapor generation and vapor feed rates. Unless a negative net vapor rate is offset by adding gas, column pressure will decrease.)

Figure notes. In some instances, behavior of the pressure control system will depend upon whether condensate enters the reflux drum above or below the liquid level. For condensate entering above the liquid level, the figures always show a line to the top of the drum—no dip-tube is shown and none is implied; for entering below the liquid level, the figures always show the line entering the drum at the bottom for clarity. A dip-tube could be used.

The equalizing line shown on some diagrams keeps the reflux drum and the column at the same pressure.

The condenser symbol consisting of a circle and a rectangle superimposed means the condenser may be either liquid-cooled or air-cooled.

Selecting a control method. We have not devised a complete guide to selecting a pressure control method but offer the following:

- For vacuum operation, use Method A-1c. For pressure operation, decide which of the three net vapor rate cases applies.

- For Case I (net vapor rate always positive), use one of the Group A methods, i.e., vary net vapor rate.

- For Case II (net vapor zero at steady state), choose a method from among Groups B, C, and D, considering factors such as the ones discussed below. We frequently use Method B-2.

- For Case III (net vapor rate positive or zero at steady state), use Method E-1 (Method A-1a or A-1b plus a method from Groups B, C, or D), or use Method D-3A or E-2 alone. (With a Group B method, reflux temperature may be more variable than with most other methods.)

Design notes. When the steady state vapor rate is zero, rate of condensation in the reflux drum can sometimes change so fast it upsets column pressure.

Potential for such upsets is high when condensate is highly subcooled, and the vapor has a narrow condensing temperature range. Under these conditions, condensation will change almost instantly if the liquid surface is agitated. All the variable area methods except B-6 produce subcooled condensate.

To avoid upsets in potentially sensitive situations, velocity of streams entering the reflux drum must be dissipated so the liquid surface will remain undisturbed. Vapor bypass piping and equalizing lines should be designed to avoid buildup of condensate that could discharge into the drum and agitate the liquid surface.

If the condensing temperature is high compared to ambient and the condensing range narrow, condensation on the walls of the reflux drum may be high enough to justify insulating the upper part of the drum.

The pressure equalizing line across the condenser should be large enough so that the pressure drop across it will be negligible. We have found that an equalizing line with at least 20 percent of the cross-sectional area of the column overhead vapor line is adequate.

ACKNOWLEDGMENTS

The author is grateful for the direct and indirect help of his colleagues, especially Bruce E. Powell and Arthur Alston of Chevron Research Co.

TABLE I—Index to methods

Method	Description	Case where used
A	Vary net vapor rate	
A-1a	PC valve in net vapor line	I
A-1b	PC valve in compressor spillback line	I
A-1c	PC valve in jet ejector spillback line	I
B	Vary condensing area	
B-1a	PC valve at condenser inlet: liquid-cooled	II
B-1b	PC valve at condenser inlet: air-cooled	II
B-2	PC valve at condenser outlet	II
B-3	PC valve in bypass	II
B-4	PC valve in liquid distillate	II
B-5	Two control loops	II
B-6	Control valve on coolant	II
C	Vary temperature difference	
C-1	Liquid-cooled condenser	II
C-2	Air-cooled condenser	II
C-3	Evaporative condenser	II
D	Vary heat flux	
D-1a	Vary coolant rate, control valve	II
D-1b	Vary coolant rate, fan pitch	II
D-1c	Vary coolant rate, louvers	II
D-2	Vapor bypass	None
D-3a	Throttle total overhead, drum vent open	II, III
D-3b	Throttle total overhead, drum vent closed	II
E	Other	
E-1 = A-1 + (B, C, or D)	Combination method	III
E-2	Make-up bleedoff	III

METHODS FOR CASE I (NET VAPOR RATE POSITIVE)

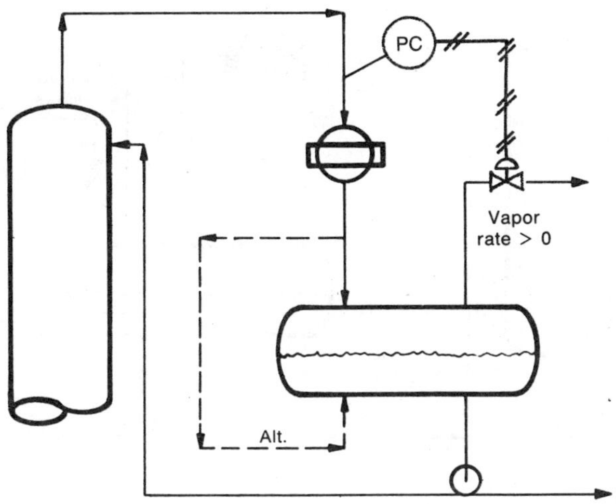

FIG. 1

Method A-1a (Fig. 1)
A: Vary net vapor rate.
1: No make-up gas.
a. PC valve in net vapor line.

Operation: The controller directly affects vapor inventory and hence, column pressure.

Advantages: This method is simple, and direct.

Disadvantages: None, generally, although response is slow when net vapor rate is low.

Where used: Method A-1a is usually the best choice when net vapor flow is always positive, even if small.

Configurations: Although Method A-1a can be configured in many ways, we prefer mounting the condenser above the reflux drum and arranging the piping so condensate and vapor flow freely into the vapor space of the reflux drum. This reduces the chance of liquid hammer and unsteady liquid and vapor flows. A sour water stripper feeding its vapor to a sulfur recovery plant is one example where steady vapor flow to and from the reflux drum is desirable.

Alternatives: Method D-3 (throttling the total overhead stream) is generally unattractive.

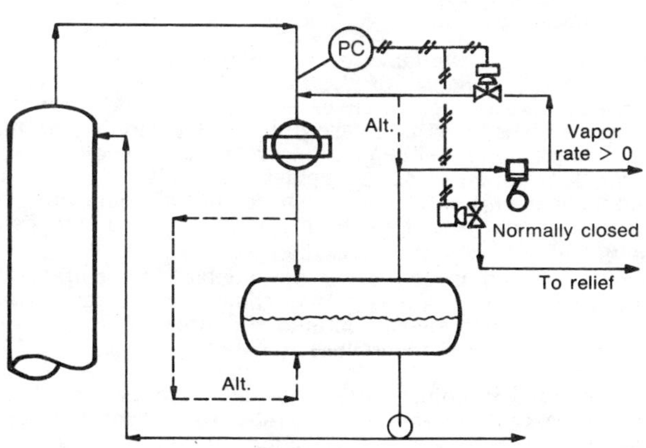

FIG. 2

Method A-1b (Fig. 2)
A: Vary net vapor rate.
1: No make-up gas.
b: PC valve in compressor spillback line.

Operation: The controller directly varies net vapor rate and, hence, column pressure. The line to relief allows the column to operate when the compressor is inoperative.

Where used: Method A-1b has been used on petroleum refinery crude unit atmospheric columns and on catalytic cracker main fractionators. These columns operate at low pressures and send their vapors to higher pressure equipment.

Configurations: See Method A-1a.

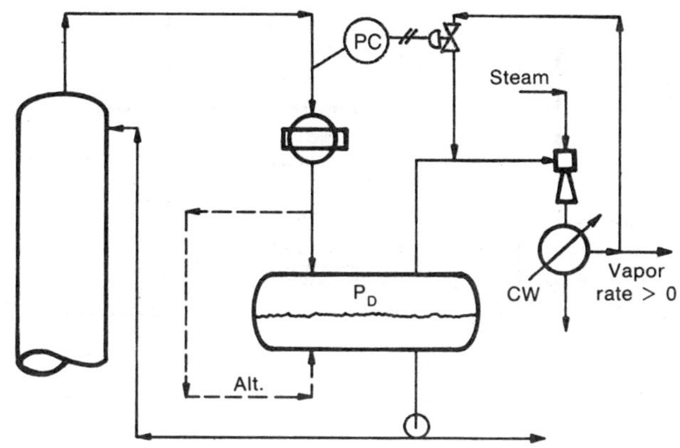

FIG. 3

Method A-1c (Fig. 3)
A: Vary net vapor rate.
1: No make-up gas.
c: PC valve in ejector spillback line.

Operation: The PC varies spillback rate to affect the process vapor rate to the ejector suction and, hence, maintain column pressure. The ejector removes the air which leaks into the distillation system along with the process vapor which saturates that air.

Where used: This is the most common method for vacuum distillation. The analogous system for pressure distillation is Method A-1b.

Configurations: The ejector is simple and reliable under proper conditions. Motive steam must be of constant pressure. Ejector suction must not be throttled. Excessive throttling causes the ejector to become unstable (i.e., "breakback").

If allowable variation in absolute pressure of the column is less than changes in atmospheric pressure, an absolute pressure measuring element should be used.

There are three ways to preclude measurement errors due to accumulation of condensate in the pressure lead: (1) mount the instrument at the same level as the tap, and make the lead horizontal; (2) mount the instrument above the tap, and provide a small leak to air-purge the lead; (3) mount the instrument above tap. Make the lead sufficiently large and steep so condensate will not accumulate. When warranted, insulate or heat the lead.

METHODS FOR CASE II (NET VAPOR RATE ZERO)

Method B-1a (Fig. 4)
B: Vary condensing area.
1: PC valve at condenser inlet.
a: Liquid-cooled condenser.

Operation: Controller varies area available for condensing by varying vapor rate. As control valve closes, pressure in the condenser decreases and liquid level in condenser rises.

Advantages: Simple. Pressure in the drum and column is the same. Useful when distillate cannot be contaminated with a blanketing gas.

Disadvantages: Operation harder to understand than some alternative methods. Condensate piping must be designed for gravity flow. This requires attention to condenser elevation, control valve location and size and piping. Control valve must be larger than for Method B-2.

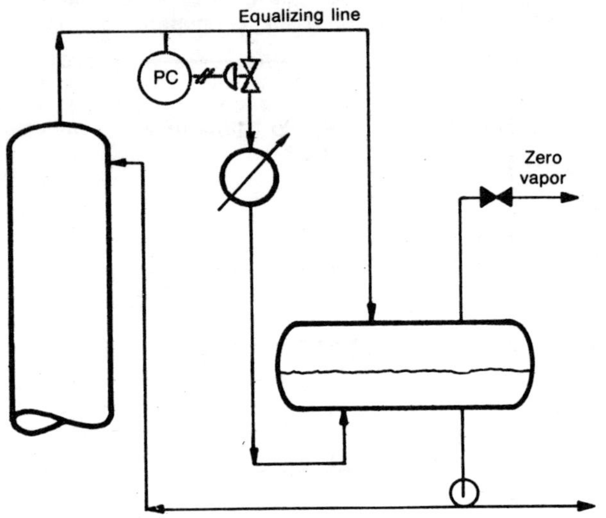

FIG. 4

Comments: We prefer Method B-2 to B-1.

Note: Condensate must enter reflux drum below the liquid level so that changes in condenser pressure will affect condenser liquid level. Although the piping configuration of B-1a and B-1b is similar to D-3b, the principle of operation is different.

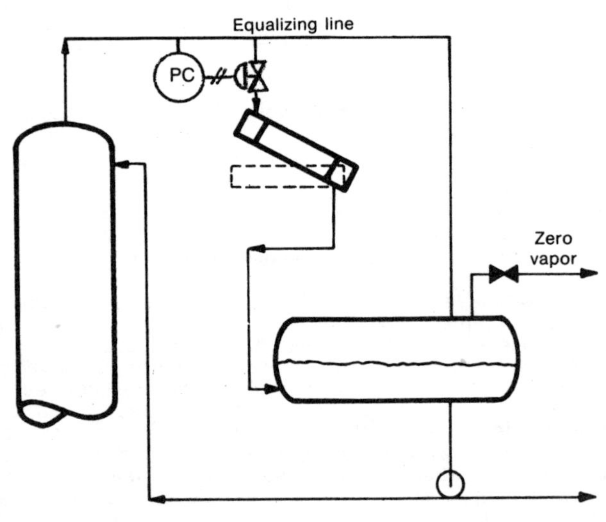

FIG. 5

Method B-1b (Fig. 5)
B: Vary condensing area.
1: PC valve at condenser inlet.
b: Air-cooled condenser.

The Method B-1a discussion applies. Two comments specific to this method (B-1b) should be added:

1. Since condensation occurs in the tubes, some tube-pass arrangements are more suitable than others. A single tube pass works well.
2. Some designers prefer that the air-cooled condenser be single tube-pass and sloped, rather than horizontal, for all variable-area (Group B method) designs.

Method B-2 (Fig. 6)
B: Vary condensing area.
2: PC valve at condenser outlet.

Operation: Controller varies area available for condensing by varying condensate flow from the condenser.

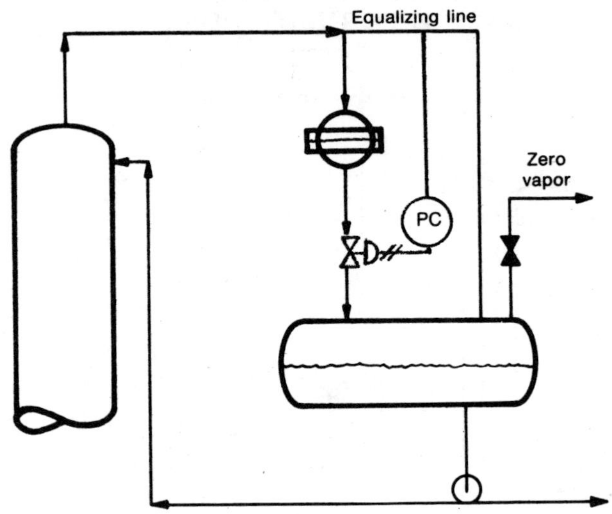

FIG. 6

Advantages: Same as for Method B-1a.

Disadvantage: Condensate piping must be designed for gravity flow. This requires attention to condenser elevation, control valve location and size, and line size and configuration.

Comments: We prefer Method B-2 to B-1. It is easier to understand, and it requires a smaller control valve. Note that (1) the control valve is located at the condenser outlet rather than the inlet and (2) the condenser outlet line may enter the reflux drum below the liquid level. We prefer to enter above the liquid level so drum level will not affect level in the condenser.

The designer should locate the control valve as near as possible to the reflux drum to assure maximum static head to promote flow of condensate and to minimize the flow effects of liquid in the vertical piping downstream of the control valve.

Alternative: Hollander[4] says his Method 2, which is Method B-2 without an equalizing line, should not be considered. We agree. Without an equalizing line, reflux drum pressure would be indeterminant and might be unsteady. Pressure variations could upset flow of condensate both into and out of the reflux drum.

Method B-3 (Fig. 7)
B: Vary condensing area.
3: PC valve in bypass.

Operation: In this variable area method (commonly called the "hot vapor bypass" or "submerged condenser" system) the condenser is located near the ground and carries a level of condensate. The reflux drum is located higher than the condenser, high enough to provide pump suction head. The condensate must be subcooled. Some vapor bypasses the condenser and enters the drum above the liquid level. As this vapor condenses, it raises the temperature of the liquid surface.

Liquid surfaces in the drum are cooler than surfaces in the condenser. This causes enough difference in vapor pressures to move condensate up into the drum. This pressure difference is usually 3 psi or less.

Control action: If column pressure falls, the pressure controller increases the vapor bypass rate. This raises the liquid surface temperature and, thus, vapor pressure in the drum. This raises the level in the condenser, reducing the area for condensing. This reduces the condensation rate and tends to restore column pressure.

Advantages: Condenser is easily accessible for inspection and repair; condenser structural support is cheaper than for an elevated condenser, control valve is smaller than for other methods.

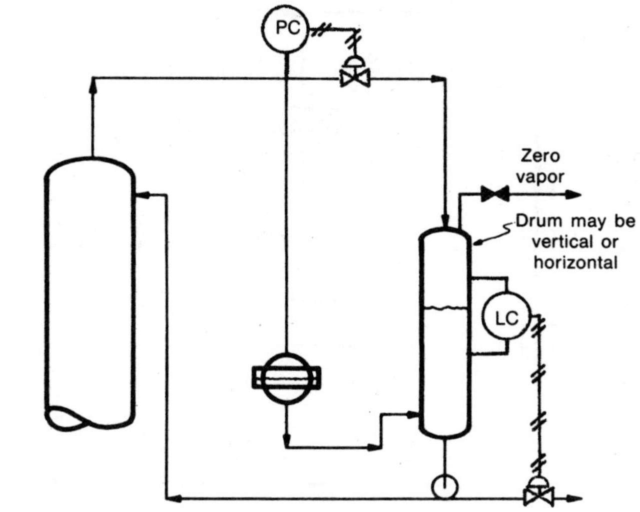

FIG. 7

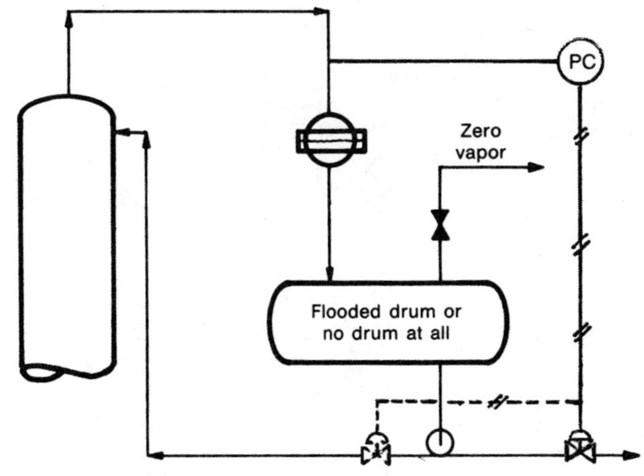

FIG. 8

Disadvantages: The subcooling area required in the submerged condenser makes it more expensive than an elevated condenser of equal condensing capacity. Whistler[6] wrote that in one application with a wide boiling range stock, the heaviest portions condensed first and separated from the vapor. This fractionation within the condenser generated a noncondensible vapor at the condenser outlet. Good performance requires a steady condensation rate in the reflux drum. This means the liquid surface must be unagitated and the upper part of the reflux drum must be insulated. In one application where the distillate had a narrow boiling range, very severe pressure and reflux drum level upsets occurred whenever the reflux drum surface was inadvertantly agitated. The design basis (vapor bypass rate and required subcooling) are empirical and not amenable to calculation. Operations and others often are hazy on how the system operates.

Variations: Sometimes the pressure in the reflux drum rather than in the column is controlled. This precludes poor control that could arise due to any "reverse reaction" of column pressure to control valve action. Since difference between reflux drum and column pressure is small, controlling reflux drum pressure is satisfactory.

Comments: This method has been controversial for several reasons. First, the merits of locating the condenser at ground level versus higher are hard to quantify. Second, it is not easy to understand how the method works. Third, the design is empirical. And finally, some installations have failed, yet others are operating successfully.

Method B-4 (Fig. 8)
B: Vary condensing area.
4: PC valve in liquid distillate.

Operation: This operates like other variable area methods, except the reflux drum is always flooded.

Advantages: A level control system for the reflux drum is not required. In some instances the reflux drum itself is omitted resulting in a faster quality-control response.[7]

Disadvantages: Changes in distillate flow rate caused by the pressure controller may be objectionable unless the distillate flows directly to storage.

Where used: This system has been used in natural gasoline plants.

Alternative: Shinsky[7] states that the pressure controller should vary reflux or total condensate rather than distillate, except when distillate flow is relatively high and the distillate goes directly to storage. Why? If the manipulated variable is distillate flow (1) if it is smaller than reflux flow it will have less effect on column pressure; (2) heat-load variations upset the column external material balance and, hence, product quality; and (3) heat-load variations upset the column fed by the distillate.

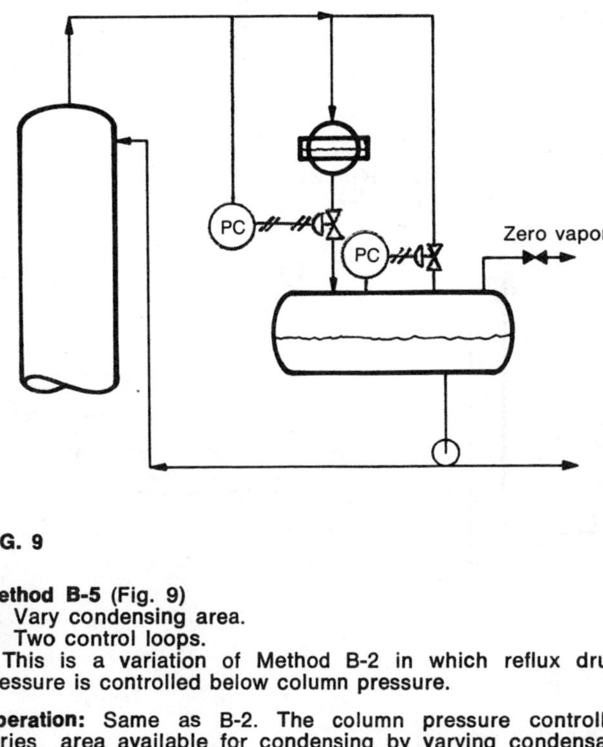

FIG. 9

Method B-5 (Fig. 9)
B: Vary condensing area.
5: Two control loops.

This is a variation of Method B-2 in which reflux drum pressure is controlled below column pressure.

Operation: Same as B-2. The column pressure controller varies area available for condensing by varying condensate flow from the condenser.

Where used: Sometimes Method B-2 (control valve in the condenser outlet) cannot be used because desired condenser elevation is too low to assure flow of condensate into the reflux drum by gravity. (Typically refinery air-cooled condensers are placed in the overhead pipeways, about 15 ft. above grade. Frequently, the top of the reflux drum approaches that elevation leaving little elevation difference to promote gravity flow.) This is when Method B-5 may be appropriate. Our one experience with this method has been good. The condenser is single-pass and sloped.

Variations: One might consider avoiding possible dynamic interaction between the two controllers of B-5 by using only one controller operating the two control valves on split-range. We have no experience with such a system and it could be troublesome.

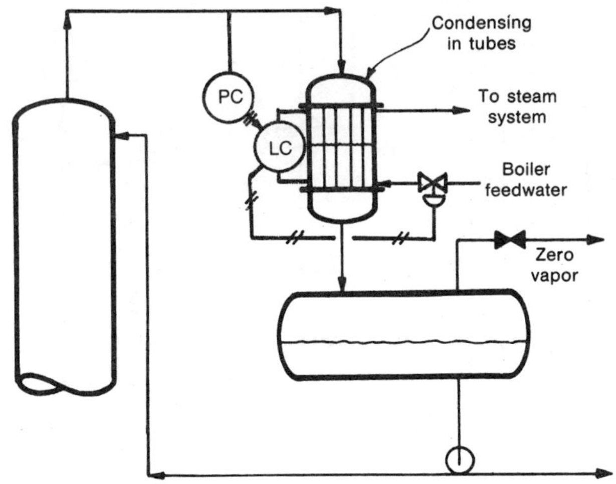

FIG. 10

Method B-6 (Fig. 10)
B: Vary condensing area.
6: Control valve on coolant.

Comments: Not common in petroleum refining. We would favor an alternative variable-area method such as B-2 unless excessive subcooling would freeze the stock or cause other problems.

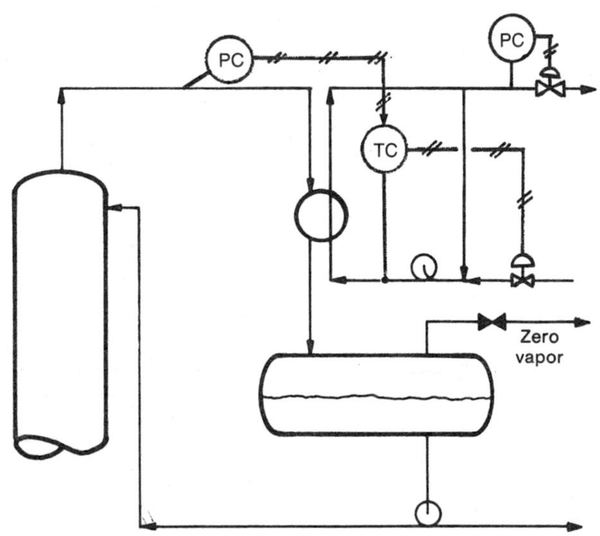

FIG. 11

Method C-1 (Fig. 11)
C: Vary temperature difference.
1: Liquid-cooled condenser.

Operation: The temperature controller varies the rate of fresh coolant so as to hold the temperature of the blend of fresh and recirculated coolant entering the condenser. The column PC resets the TC.

Advantages: Rangeability and linearity are superior to Method B-3 ("hot vapor bypass").[7] Heat is recovered at the highest temperature level because the heat transfer coefficient and area are kept at their maximum.[7] (Advantageous only if hot coolant is used for heating.)

Disadvantage: This system will probably respond slower than variable area methods.

Comments: This method is analogous to Method C-2.

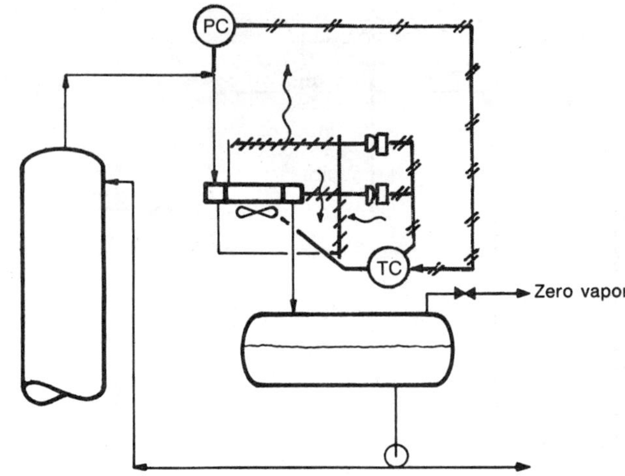

FIG. 12

Method C-2 (Fig. 12)
C: Vary temperature difference.
2: Air-cooled condenser.

Operation: The temperature controller varies the ratio of recirculated air to fresh air to control temperature of the blended air entering the condenser. The PC resets the TC.

Disadvantages: Fresh air and recirculated air must be mixed well so air to the condenser will have a uniform temperature. Otherwise, pressure control may be erratic even if solids buildup is successfully avoided. Operator attention will be needed if changes in required condenser duty, ambient air temperature, or winds cause the automatically positioned recirculation louvers to reach one extreme position or the other.

Where used: Air recirculation can prevent solids buildup on the process side of condensers. We commonly use it in freezing climates when the process vapor contains steam. Also, we use it to avoid hydrate formation.

Comment: This method is analogous to Method C-1.

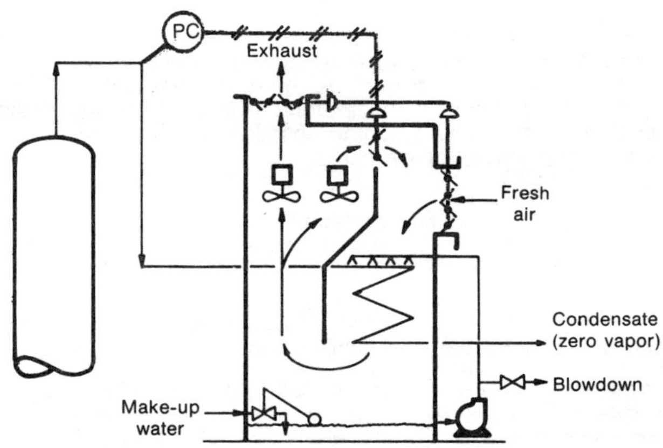

FIG. 13

Method C-3 (Fig. 13)
C: Vary temperature difference.
3: Evaporative condenser.

Operation: Vapors are condensed by a water spray. The water is cooled by evaporation. Evaporation rate and, hence, cooling water temperature is varied by adjusting recirculation of moist air.

Comments: This is a cooling tower where the condenser is internal rather than external. A similar type of cooling tower with multiple internal coolers and condensers has been used in oil refineries but is no longer common because it has become more attractive to pipe cooling tower water to process equipment rather than pipe process fluids to the cooling tower.

Shinsky[7] discusses the thermodynamics and heat transfer of the evaporative system. He also lists the following advantages over dry-air cooling: It is possible to

- Lower air flow and, thus, lower fan horsepower
- Use multiple passes on the process side, and reduce total area
- Achieve much lower condensate temperatures
- Control the unit with dampers because air flow and size are smaller.

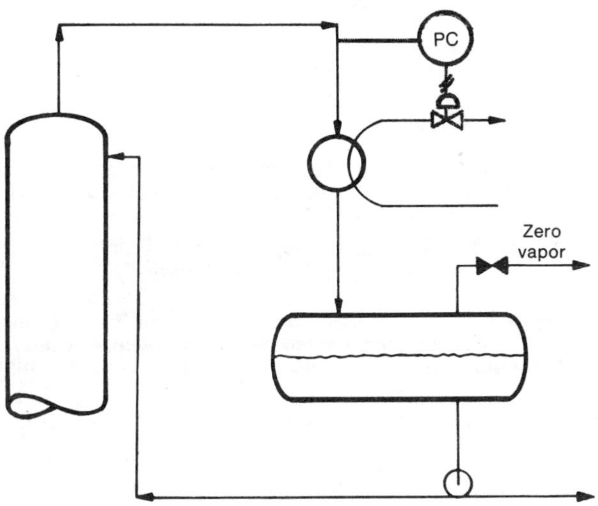

FIG. 14

Method D-1a (Fig. 14)
D: Vary heat flux.
1: Vary coolant rate.
a: Control valve.

Operation: The controller increases flow of coolant to decrease pressure and vice versa.

Disadvantages: Response is often slow because condenser lag is in the control loop.

Can promote fouling, especially when water is coolant. Water fouls faster as its velocity is lowered and it gets hotter in the condenser.

Where used: Limited to applications where exiting coolant won't get too hot. We have used 120° F as the limit for cooling tower water.

(Hollander[4] says that throttling cooling water should be considered only when overhead vapor line temperature is less than about 120° F. A guide line of this type is useful because the exit cooling water may heat up to the vapor line temperature when condenser area is ample (clean tubes, cold cooling water, low overhead rates). But one needs to examine the condensation curve. If condensing temperature decreases sharply with initial condensation and decreases slower thereafter, maximum cooling water temperature may be much less than the vapor line temperature. In such instances higher vapor line temperatures are safe).

Comment: Throttling appears to be economical because it minimizes coolant rate. But throttling does not reduce the cost of coolant facilities (e.g., cooling tower) because they must have the capacity to meet maximum condenser duty.

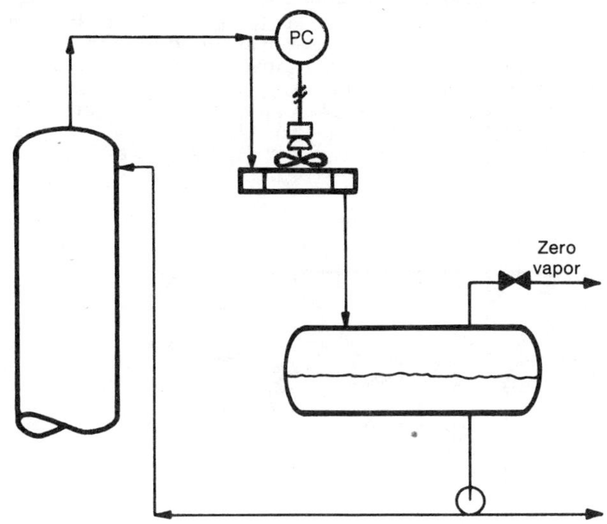

FIG. 15

Method D-1b (Fig. 15)
D: Vary heat flux.
1: Vary coolant rate.
b: Fan pitch.

Operation: The controller varies fan pitch to control pressure.

Disadvantage: Fan must be shut down to service fan pitch adjusting mechanism.

Comments: Duty will be less affected by precipitation in an induced-draft fan condenser than in a forced-draft unit. This should be considered along with cost, maintenance, capacity on power loss, short circuiting air, etc., in selecting the type of condenser.[5] The fan blade pitch-adjusting mechanism should have low hysteresis and should include a positioner. Pitch travel can be adjusted to give enough negative pitch to offset natural convection so low condenser duties can be reached. If the condenser has more than one fan, all fans need not be variable pitch. This method is similar to using louvers (Method D-1c) and analogous to throttling liquid coolant (Method D-1a).

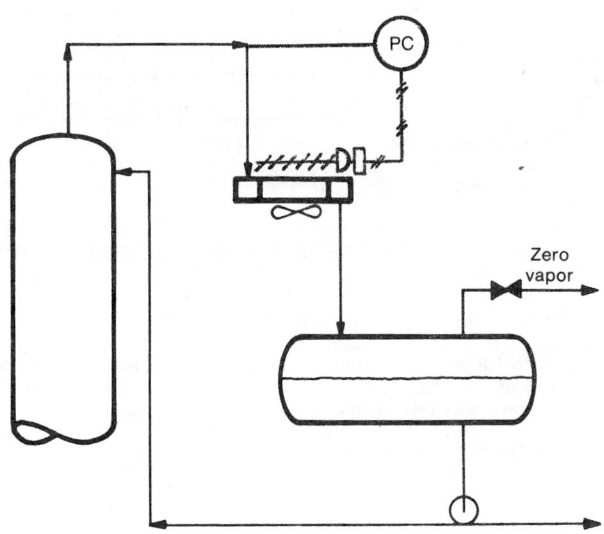

FIG. 16

Method D-1c (Fig. 16)
D: Vary heat flux.
1: Vary coolant rate:
c: Louvers.

Operation: The controller varies louver angles to control pressure.

Comments: If louvers are placed above the condenser, they afford rain protection. Although a set of louvers functions as a control valve, louvers seem to be more variable in quality than control valves. For good, reliable performance, a set of motor-operated louvers cannot be made from louvers designed merely for manual operation. Attention must be paid to wear resistance, friction, hysteresis, and motor-operator power. This method is similar to using pitch (Method D-1b) and is analogous to throttling liquid coolant (Method D-1a).

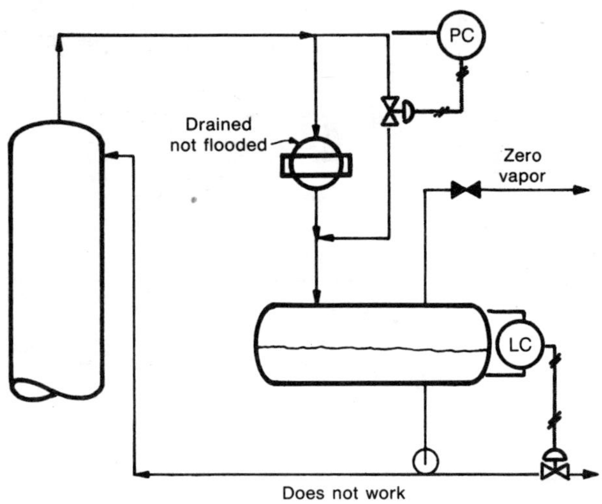

FIG. 17

Method D-2 (Fig. 17)
D: Vary heat flux.
2: Vapor bypass.

Behavior: This method does not work for zero net vapor. We have witnessed a failure and offer the following explanation based on the steady state. Remember, we have the following: the condenser drains freely into the vapor space of the reflux drum where pressure is less than in the column. Subcooling is slight. The bypass may be mixed with condensate or enter the drum separately.

Vapor condenses in the condenser and at a lower rate in the reflux drum. Condensing rate in each of these two locations is equal to the product of heat transfer coefficient, condensing area, and mean temperature difference. Control valve changes would affect the condensing rate negligibly in the condenser and just slightly in the reflux drum. Suppose the system were at equilibrium and the control valve opened wider. Reflux drum pressure and, hence, mean temperature difference would increase. Thus, opening the bypass would increase the total condensing rate, not decrease it as required by design.

Comments: This method looks like Method B-3, but there are two differences: (1) In Method B-3 the condenser is partly flooded. In this method (D-2) the condenser runs drained; (2) Method B-3 works, this method will not.

Regarding the condenser bypass, Rademaker, et al[8] (page 480) state: "It is rather obscure how this method works." One obfuscating factor is the method as described above does not work. Another is the label "condenser bypass" is too general. It has been applied to methods that could work, e.g., the flooded condenser where net vapor rate is zero and to the drained condenser where net vapor rate is positive.

Method D-3a (Fig. 18)
D: Vary heat flux.
4: Throttle total overhead.
a: Drum vent open.

Operation: The controller simply throttles the total overhead stream to hold column pressure.

Note that the condenser, operating at constant pressure, automatically condenses all vapor which enters it. As inlet vapor rate decreases, the heat transfer coefficient decreases,

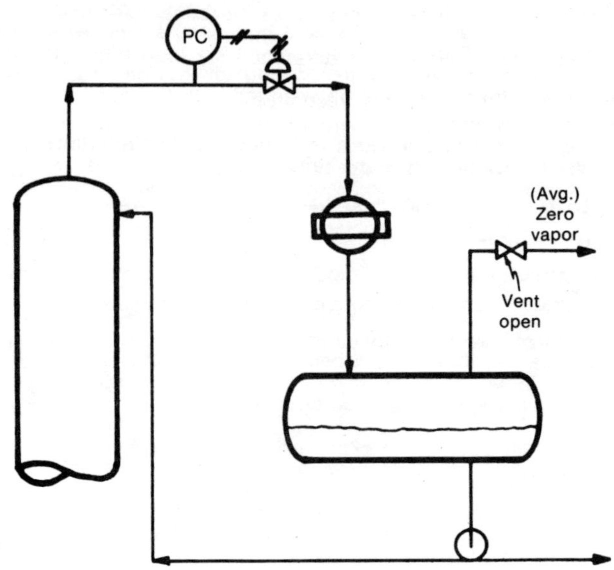

FIG. 18

because concentration of uncondensibles in the condenser increases. Unconderisibles flow into and out of the condenser via the reflux drum vent.

Advantages: Simplicity; fast response, reflux drum and condenser need to be designed for atmospheric pressure operation only.

Disadvantages: Control valve is larger than for alternative control systems; condenser must be larger because it operates at atmospheric pressure, reflux pump must generate higher head.

Where used: More attractive when overhead line is small. It is hard to imagine this method being justified on an oil refinery column where vapor line diameter may be 3 ft. or more. Method D-3a is suitable for both case II and Case III.

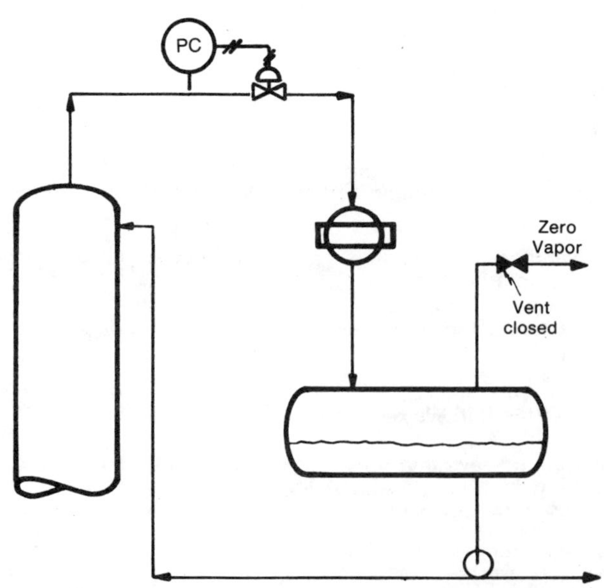

FIG. 19

Method D-3b (Fig. 19)
D: Vary heat flux.
3: Throttle total overhead.
b: Drum vent closed.

Operation: As in Method D-3a, the controller holds column pressure simply by throttling the total overhead stream. The condenser automatically condenses all vapor which enters it. As inlet rate decreases, pressure at the condenser and reflux drum decreases. This decreases condensing temperature and, thus, the mean temperature difference for heat transfer.

Advantages: Simplicity, fast response.

Disadvantages: Control valve is larger than for alternate control systems; condenser and reflux drum pressure is variable. This affects required pump head and may dictate designing the condenser and drum for vacuum service, system requires separate pressure safety valves for column and reflux drum.

Where used: Same as Method D-3a except for case II only.

Comment: In this method, condenser pressure will usually be below column pressure; but this does not mean more condenser area is required. When maximum condenser duty is required, the control valve will be wide open so that condenser pressure will approach column pressure.

Note: Method D-3b for controlling column pressure is the same as the method of controlling heat input to a steam-heated heater or reboiler. The reflux drum and level controller in Method D-3b perform the same function as the steam trap in the steam system. It is condensation pressure that is the manipulated variable in both systems.

METHODS FOR CASE III (NET VAPOR RATE SOMETIMES POSITIVE AND SOMETIMES ZERO)

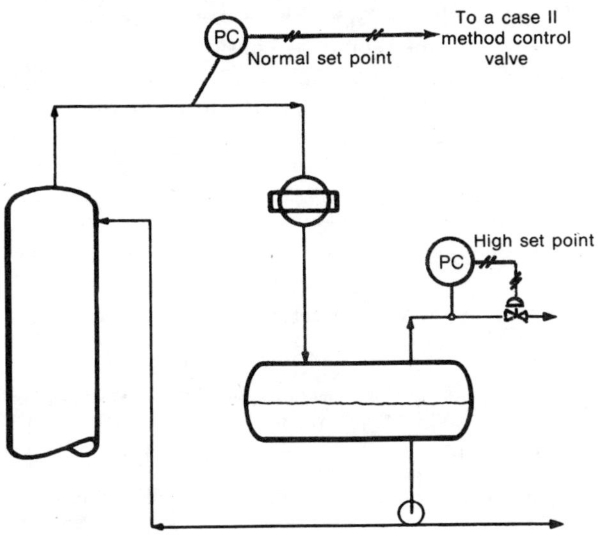

FIG. 20

Method D-3a. See discussion above.

Method E-1 (Fig. 20)

Method E-1 is a combination of Method A: PC on vent, and Method B, C, or D: Any variable condenser duty method except B-3, D-3a and D-3b.

Operation: For the configuration shown, if column pressure rises to the vent line PC set point, pressure will return to normal after vent rate decreases to zero. But return will occur gradually as noncondensibles in the reflux drum dissolve in liquid distillate and leave the system.

Variations: If net vapor rate is rarely greater than zero, the vent line PC may be replaced by a manual vent. A single split-range controller may be substituted for the two separate controllers if a single combination of controller settings will suit both loops. This will avoid possible problems with reset-windup in separate controllers.

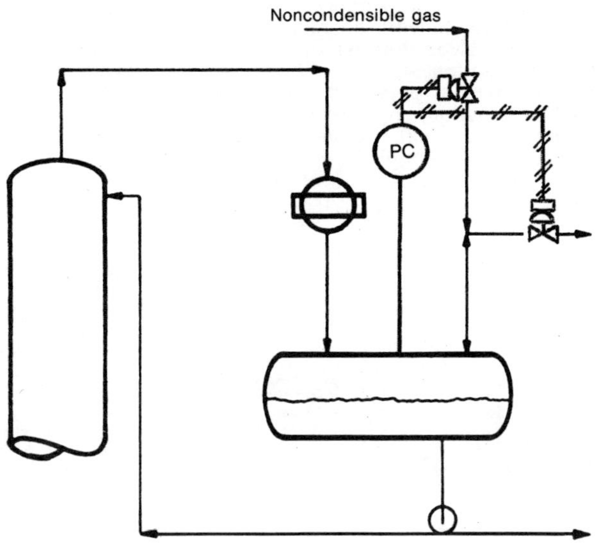

FIG. 21

Method E-2 (Fig. 21)

Operation: On falling pressure the make-up gas control valve opens to hold pressure; the bleedoff valve remains closed. On rising pressure the bleedoff control valve opens to hold pressure; the make-up gas control valve remains closed.

Advantages: Fast response, easy to understand.

Disadvantages: System increases the vent gas stream and requires a source of uncondensible gas compatible with the process and available at higher-than-column pressure. Although concentration of uncondensible gas dissolved in the distillate is less than equilibrium, it may still be objectionable. We have not found this a problem in petroleum refining. A two-valve system is harder to tune and requires careful calibration of the two positioners. May produce poorer control than a good single-valve system.

Alternative: This method may waste gas and in doing so increase the vent gas rate unnecessarily. This may result from improper controller tuning or from too much valve position overlap. A simpler system to operate is one where the make-up gas rate is fixed.

Comment: We have found that a continuous vapor path between the condenser outlet and the reflux drum is not required for this method.

LITERATURE CITED

[1] Shinsky, F. G., Advanced Control of Distillation Columns, *Chemical Engineering Progress,* Vol. 71, No. 6, June 1975, p. 53.

[2] Buckley, Page S., Techniques of Process Control, John Wiley and Sons, 1964, p. 265.

[3] Dodge, B. F., Chemical Engineering Thermodynamics, First Edition, McGraw Hill, 1944, p. 167.

[4] Hollander, Leonard, Pressure Control of Light Ends Fractionators, *ISA Journal,* May 1957.

[5] Christman, J. C., The Case for Air-cooled Exchangers, *Oil and Gas Journal,* November 13, 1961.

[6] Whistler, A. M., Locate Condensers at Ground Level, Petroleum Refiner, Vol. 33, No. 3, March 1954, pp. 173-174.

[7] Shinsky, F. G., Distillation Control for Productivity and Energy Conservation, McGraw Hill, 1977.

[8] Rademaker, O., Rjinsdorp, J. E., and Maarleveld, A., Dynamics and Control of Continuous Distillation Units, Elsevier Scientific Publishing Co., 1975.

[9] Boyd, D. M., Jr., Fractionation Instrumentation and Control, *Petroleum Refiner,* Vol. 27, No. 10, October 1948, pp. 115-118.

Multiplexing: Justifying, Specifying and Proposal Evaluation

Multiplexing systems are gaining in popularity due to increased reliability, capability and cost savings over direct wiring. Presented are considerations for justifying, specifying and evaluating proposals for multiplexing systems

L. T. Garrett, Fluor Engineers and Constructors Inc., Houston

SUCCESSFUL MULTIPLEXING application is highly dependent on proper design, specification and vendor selection. This article examines the technical justifications for multiplexing and discusses items to consider when developing specifications and selecting vendors.

Multiplexing means many things to many people, but in HPI plant areas, it is replacement of direct run wiring with devices that accomplish the same function. Depending on application and multiplexer manufacturer, devices will vary in design and operation. However, all multiplexer systems contain three basic elements: a remote multiplexer unit, a master controller and a man-machine interface (Fig. 1).

The remote multiplexer unit is normally located close to the field devices whose signals are being multiplexed. Signal conditioning and conversion may occur at this point but in some applications the signal is just switched into a master controller.

Functions of the master controller (normally located in the main control center) include scan control, signal conditioning, data conversion, data correction, man-machine interface control, auxiliary interface control and system communication control. The function of the man-machine interface is to display the scanned data in human consumable form and enable an operator to enter requests for data or control.

It should be pointed out that while these basic elements are present in all multiplexer systems, functions of each element may vary as well as the manner in which they are interconnected. There are many manufacturers of multiplexer equipment, and while they perform basically the same function, the difference between systems will range from subtle (yet important) to dramatic.

Justifying multiplexers. There are many reasons why multiplexer systems should be considered in lieu of direct wiring of field devices to main control centers (Table 1). However, in the final analysis, the question will be asked, "Is there any economic justification for multiplexing?" The answer can only be yes if all costs are realistically considered and if there are more than a few points to be multiplexed.

No doubt there is some magic number of points at which multiplexing is more economical than direct wiring, but it probably doesn't matter. Getting approval for multi-

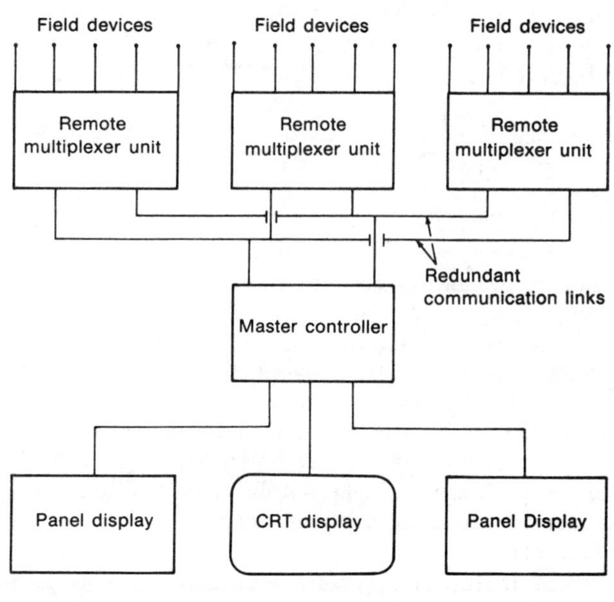

Fig. 1—Typical multiplexer system block diagram.

plexing without going through an economic evaluation is slim. To aid in this exercise (should it be necessary) there are several published articles on the subject of developing cost comparisions[1,2] so cost justification will not be covered here.

TABLE 1—Multiplexing benefits

- Reduced wiring errors
- Reduced wiring checkout
- Increased flexibility for changes
- Redundancy easily implemented
- Reduced noise and signal loss
- Increased flexibility for displaying data
- Easily interfaced to computers
- Capability exists for preprocessing data going to a computer
- Reduced control panel space

Once economics have been evaluated, there are many good technical reasons that justify multiplexer systems. For example, wiring errors should be reduced since there are fewer wires to handle, connect, document and check out. This is especially true when intermediate terminations are required.

Another plus is inherent flexibility in multiplexer systems for making additions or changes in the field. To add points to a direct wired plant, a cable with spare conductors must be located or in some instances an additional cable must be run. Multiplexer systems will usually have existing spare inputs or spare slots for adding new points to the system.

Redundancy, when required or justified, can be easily incorporated into multiplexer systems. For example, redundant data paths can be implemented between remote units and the master controller. These data paths can be run over separate routes so that if one cable becomes damaged (severance, fire, etc.) the system will continue to operate. With direct wired systems this redundancy is economically impractical. Furthermore, if a bundle of wires gets damaged in a direct wired plant, there is a considerable amount of time required to splice in or run and checkout a new set of cables. Since multiplexer systems use only one or two cables with a few twisted pairs in each cable, the problem is far less dramatic.

Data transmitted through multiplexer systems is normally less subject to signal degradation noise interference than direct wired inputs. This is especially true on systems where all data is transmitted in digital form. Special encoding of digital data can be utilized to virtually eliminate noise.

Multiplexer systems also offer considerably more flexibility in presenting data to the operator. For example, on a microprocessor or computer based system, a CRT could be used to display data in bar graph form and in groups that are functionally related. On hardwired systems, select displays can be custom built to enable an operator to display some values continuously while others could be selected and displayed by keying in a tag number or pushing a single button. Systems with logging capability can also printout periodic reports or alarm conditions.

Since multiplexer systems eventually convert data to digital form, transmitting data to a computer is fairly straightforward. In fact, most manufacturers have designed their system to connect to a computer system through industry standard connectors. Generally this throws a modest software development burden on the computer system, but again there are fairly standard industry protocols developed which most manufacturers follow.

While multiplexers are capable of transmitting data to computers, some multiplexer systems are capable of preprocessing data which enables the main plant process computer to shed some of its computing load. This preprocessing can include data limit checking; reporting data that exceed limits or a predetermined rate of change. Some systems can also use the data for making calculations with only the calculated results reported to the main plant process computer.

In some applications, multiplexing is self-justifying since distance from the field device to the control center precludes direct wiring. Distances up to 20,000 feet are usually standard with most multiplexers. If distances exceed the manufacturer's maximum, this maximum can be extended by adding modems and connecting into a communications circuit.

Another big advantage of multiplexing is conservation of control room panel space. If all points brought into a control room were to be continuously displayed by panel mounted instruments, meaningful data would often be masked by an abundance of other data and control rooms would be much larger.

In summary, there are applications where multiplexing is not economically justified. Additionally, there are many good technical reasons for multiplexing that should sway any marginally economical evaluation over to multiplexing.

Selection. In selecting a multiplexer system there are three main decisions that must be made. These are: (1) the system logic base, (2) how signals are to be handled and (3) operation of the man-machine interface.

System logic bases range from unintelligent hardwired systems to fairly complex computer based systems. There are many types available between the two extremes. These include "smart" hardwired systems, relay based systems, solid state systems, microprocessor based systems and microcomputer based systems.

The type of system logic used depends on application requirements. If, for example, the system is to be used for strictly wire replacement then a hardwired relay or solid state system will yield very satisfactory results.

However, if the application calls for preprocessing of data, simple logic decisions or distributed data scanning due to a large volume of points, then the microprocessor or microcomputer based systems would be more desirable.

Should the application call for checking interlocks, closed loop control, alarming logging, calculations or considerable flexibility, then a computer based system is warranted. In all cases, the application should be examined in terms of future requirements and the system logic base selected should reflect the ultimate, realistic requirements of the application. Other important considerations include the overall system size and scan rates required.

The second decision to be made, signal handling, is somewhat more involved and frequently is based on personal preference rather than actual requirements. Using thermocouples as an example, systems are available which transmit unamplified low level signals from the remote unit to the master controller through mercury-wetted or reed relays. At the master controller, the signal is amplified and digitized. This approach has the advantage that no power connections are required at the remote making them safe for a class D application. Also, since most of the system logic is in the master controller, requirement for maintenance at the remote sites is reduced.

The main drawbacks to this approach is that it requires special installation considerations and may be more subject to signal degradation and noise interference. Since the link from the remote units to the master controller is thermocouple wire, it must be handled accordingly. That is, intermediate terminals will require special terminals and shielding to insure signal integrity. Also, care must be taken when running the wire near other control and power cables. Further, when the application calls for long wire runs, there are more opportunities for signal degradation and noise interference.

When the signal is amplified at the remote unit, these problems tend to dissipate. However, the remote units must perform signal conversion so they become moderately complex. Also, there might be some tendency toward increased costs since the remote units contain more hard-wired logic and perform a few more functions.

Several manufacturers offer systems that amplify and digitize the low level signal at the remote unit. Once the signal is digitized, security information can be transmitted with the field data to ensure the data is free of noise.

In addition to security checking, scanning and transmission rates can be significantly increased. Also, by incorporating microprocessors or other logic systems into remote units, some preprocessing of data can take place with data transmitted by exception or in bursts.

Man-machine interface. The ultimate objective of any data gathering system is to present data in a human consumable form. While all manufacturers offer a standard man-machine interface, the ultimate decision of how the operator will interact with the multiplexer system will be dictated by the application. Frequently, the standard manufacturer man-machine interface will not satisfy all operating requirements and either functional, dimensional or hardware modifications will be required. Obviously there is considerably less flexibility in this area with "hardwired" systems than with microprocessor or computer based systems. At any rate operator interface with the system must be defined early in the systems design since it will impact costs, design and schedules.

Specifying multiplexers. Before development of a detail specification some decisions must be made on how multiplexers are to be used in the application under consideration. The application may range from bringing in a few points to transmitting a major portion of data and control points in a plant.

How multiplexers are applied will depend on previous experience and confidence level of engineers and managers who are considering multiplexers. If there is no previous experience with multiplexing, a "walk before you run" approach will usually be taken, resulting in a few "non-critical" points being transmitted. In fact, even though multiplexers have been available for more than 10 years, the predominant approach is to multiplex only those data points not critical to continuous plant operation.

In other operations, multiplexers are extended to include digital on-off control such as the start/stop control of pumps, auxiliary motors, opening/closing valves and others.

The next step in the hierarchy of multiplexers is some level of closed loop control. To make this step, the primary logic unit needs to be a micro or mini-computer. The controls tend to be somewhat specialized and frequently the user is sufficiently sophisticated in this area that the bulk of the required software is developed in-house.

One problem encountered in using multiplexers in this manner is that there is some overlap in the functions of the multiplexers and plant process computer. In some applications this overlap is desirable because the resulting disturbed data acquisition and control system provides for a graceful degradation when failures occur. At any rate, the decision on how multiplexers are to be applied should be considered in light of integration into and relationship with other computer and near-computer systems in the plant.

If there is a plant process computer, then some thought should be given to what data (if any) should be made available to the process computer. In making this decision, the communication link must be defined along with scan frequency, protocol, and a definition of how the computer will use the data or initiate controls (if provided).

Writing the specification. Before specification can be started, one early decision must be made. Should the specification be specific and contain considerable detail pertaining to the unique requirements of the application or should the specification outline general functional requirements of the system.

If the application is for a new plant, often a lot of information required for a detailed specification is not available. The tendency here is to write a functional specification outlining basic requirements such as number of points, scan rates, man-machine interfaces, logic base and others. Response from vendors solicited will most likely be for their standard systems which, from a maintenance standpoint, is preferable.

However, there is a major pitfall in using a functional specification. When the system becomes better defined, there is a high likelihood that the system quoted by a manufacturer will not provide capability to handle special application requirements. This means money budgeted for the multiplexer system will be overrun since items you need most are not included.

Another problem arising from a general functional specification is that interface with the vendor during manufacture is increased since the system design is continuously undergoing change to reflect and incorporate information and requirements not previously known. This increases the possibility of overlooking something and also increases paperwork.

On the positive side, a general functional specification allows vendors to quote standard equipment, which results in lower costs and easier maintenance.

Writing a detailed specification when possible will always yield more satisfactory results. All possible information should be included to clearly define all aspects of the multiplexer system. The end result will be a document that leaves little room for interpretation and increases the probability the system will be completed on schedule, within budget and perform as specified.

The negative aspect of a detailed specification is that frequently the responding vendors will not be able to quote as much standard equipment and in some applications experienced, reliable suppliers may not be able to respond at all. Also multiplexer suppliers approach multiplexing in many different ways and detailing a specification will read (whether true or not) as though the vendor was preselected. Another problem with detailed specifications is that special requirements may cause new component development costs to be included in the vendor quotation. This new development may not be required at all if the manufacturer offers standard equipment that will functionally satisfy the specification requirements.

The next major decision relates to the system logic base. That is, what type of electronic logic will be utilized to control scanning, conversions, switching, displays, logging, external interfaces, training, security checking and others. As previously discussed, types of electronic logic available include hardwired, solid state, microprocessor, microcomputer, minicomputer and process computer.

After the level of detail and system logic has been decided, there are several items that should be addressed in the system specification. These include factory acceptance testing and evaluation, documentation, training, field support, and wiring practices and standards.

System acceptance testing and evaluation at the factory is perhaps the single-most important function to be performed on the system. Testing should be as comprehensive and thorough as time and economics allow. A thorough test at the factory will not guarantee no problems will occur during field startup, but it will reduce startup problems, provide confidence in the system operation and shorten startup time.

The ideal factory test will include 100 percent testing of all inputs and outputs simultaneously and exercising all hardware logic and/or software. On very large applications, testing at this level may be impractical or very expensive. In those situations where testing cannot be 100 percent, the test should enable exercising all hardware and software in some random fashion since field inputs and outputs will operate or change randomly.

Signal simulation hardware should be provided to emulate signal levels that will be encountered in the field. That is, for low level signals millivolt sources should be applied to the input terminals. Oscilliscopes, digital voltmeters and manufacturer supplied system test equipment should be available to use during factory testing and a procedure defined for use of the equipment.

Defining documentation in the system specification in any meaningful detail is difficult. Generally, enough documentation should be supplied with the system to virtually enable the user to build his own system. This includes dimensional drawings, terminal wire lists, PCB logic diagrams along with detail descriptions of the circuit functions and timing, a listing of all software or microcode and logic flow diagrams, operator manuals, trouble shooting procedures and maintenance guides.

Directly associated with documentation is system training. Most vendors offer well defined training courses that enable the user to maintain, operate, expand and modify (if necessary) any part of the system in the field. The amount of training that should be specified depends primarily on the degree of self-sufficiency desired or required. For example, if there is a shortage of manpower for maintenance of this equipment perhaps vendor support of the equipment in the form of a maintenance agreement might be elected. This would minimize need for training. On the other hand, if the application is in a foreign plant, support by the manufacturer would normally take too long to be practical. This of course means a fairly extensive training requirement should be specified. Criticality of the application is another big factor determining training requirements.

In the area of wiring practices and standards, the specification should define as much detail about the wiring as possible. This includes defining signal sources, wire sizes, distances, numbers of wires, types of termination, power sources, grounding practices, commoning, wire labeling and tagging, shielding, requirements and any special wiring conventions. This is an important area to make known to potential vendors since this information will enable the vendor to price his system as low as his profit requirements will allow. By removing doubt about custom wiring requirements the vendor may elect to lower the contingency allowance in his quotation. This also reduces cost adders after the contract is awarded.

Vendor proposal evaluation. Vendor proposal solicitations should only be sent to those with the experience

TABLE 2—Itemized price list

One Time Charges
- Project Engineering
- System Engineering
- License Fees
- Software
- Specification Preparation
- Other (Please Specify)

Development Charges
- Software for Computer Interface
- Special Panel Insert Design
- Other (Please Specify)

Major Component Charges
- Remote Multiplexer #1
- Remote Multiplexer #2
- Remote Multiplexer #3
- Master Controller
- Display Unit #1
- Display Unit #2
- Cables (RMU to Master) Cost/Foot
- Diagnostic Test Unit
- Incremental Cost for Additional Points on RMU #1
- Incremental Cost for Additional Points on RMU #2
- Incremental Cost for Additional Points on RMU #3
- Cost for an Additional RMU with 8 Analogs Inputs, 8 Digital Inputs, 8 Digital Outputs
- Other (Please Specify)

Factory Testing
- Test Equipment
- Acceptance Testing

Field Support
- Startup Assistance
- Contract Maintenance
- Spare Parts

and/or expertise to offer systems that fully satisfy the application requirements. By limiting the vendor list in this manner, the job of quotation evaluation will be somewhat simplified.

Selecting the system that will offer the best cost/performance is not always easy. The single most difficult item to evaluate is how the vendor interpreted the specification. No matter how much detail is included in the specification, it will always be subject to interpretation.

Several other items that will aid in vendor evaluations should be included in the specification sent to vendors. These include an itemized price list of system hardware, software, engineering and project costs (Table 2), a statement that all exceptions, interpretations and clarifications be separately stated and that the vendor will respond to the specification paragraph by paragraph in the same sequence.

The itemized price list should be made out by the specification writer and should include a list of major system hardware components, services, one-time charges, system and project engineering, test equipment, factory testing, training and support.

An itemized price list is an invaluable aid in quotation comparisons. For example, if a vendor is very low or very high on any given item it is very likely a misinterpretation has occurred. A high price also sometimes indicates the vendor is incurring development costs to build that particular line item: information that might not otherwise be volunteered.

Having the vendor separately state any exceptions, clarifications or exceptions reduces the likelihood they will be overlooked. This can easily happen if they are noted only in the text. Also, a paragraph by paragraph response that follows the specification sequence will enable easier understanding of the system being offered.

Future multiplexers. It seems a new manufacturer enters the multiplexer market every week. Most new systems are offered by manufacturers who are established in the electronics field with products that are peripheral to the multiplexer field. Growth into multiplexing in those companies is natural and a reasonably well designed product can be expected. Between established manufacturers and those entering the field, an excellent range of very competitively priced products is available.

In the future, things are likely to change drastically. With availability of a microprocessor dedicated to an individual loop already a reality, change is inevitable. The trend toward distributed scan/control will continue with more and more capability placed in close proximity to the scanned/controlled devices. Contributing to this trend will be new memories that are cheap and physically small along with circuitry shrinking from large scale integration (LSI) to very large scale integration (VLSI) and beyond. While it would be naive to predict that every point in future plants will be multiplexed, it is clear that some form of multiplexing will provide scanning and control for most instrumentation in future and existing plants.

LITERATURE CITED

[1] Weideman, E. A., "Slash costs with multiplex systems," *Power Magazine*, Dec. 1975.
[2] Simon, Herbert, "Multiplex systems save multibucks in refinery and chemical plants," *ISA Proceedings*, 1970.

Change Controls to Save Energy

Simple control modifications to existing process units can reduce the energy used by the control systems. Eight examples show the savings that are possible

Norman P. Lieberman, Amoco Oil Co., Chicago

PROCESS CONTROLLERS don't normally save energy; they waste energy. They convert potential energy to low level heat; they restrict the flow of heat to a heat sink; they maintain high pressures which suppress vaporization. The engineer designing a process control scheme must seek to minimize this waste. The operator running a process unit needs to search for this type of energy inefficiency.

Efficient energy utilization is not always compatible with the smooth unit operation. Computers can help. Smart plant operators are even more important. Sometimes we must tolerate a few bumps in daily operations in order to achieve efficient energy use.

As energy costs continue to escalate, our rewards for saving energy will escalate, too. Plenty of energy can still be saved by low cost, unsophisticated changes to process instruments and controls. The prerequisite for success is a willingness to experiment and a tolerance toward an occasional failure.

INSTRUMENTATION FOR ENERGY SAVINGS

The following examples review several common forms of energy waste caused by instrumentation. In each instance, spending a little money resulted in a large energy savings. The examples cited are based on this writer's firsthand experiences. In some cases, simplifications are made as an aid to reader comprehension. In other cases, the specific services are changed to protect the identity of some operations. However, the essentials, as far as energy is concerned, are correct. Instrumentation symbols used herein are identified at the end of the article.

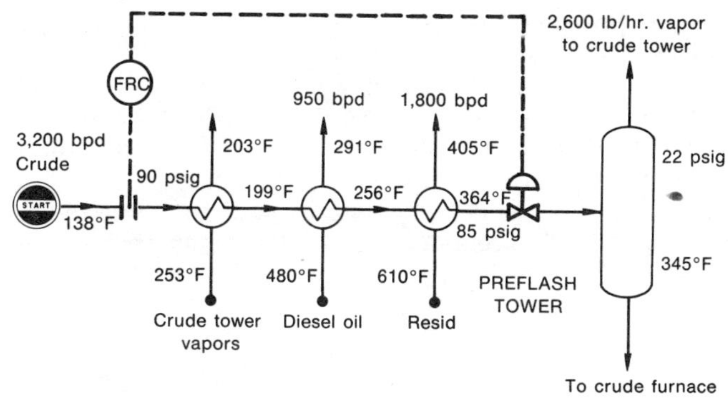

Fig. 1—Preheat exchanger train for crude topping.

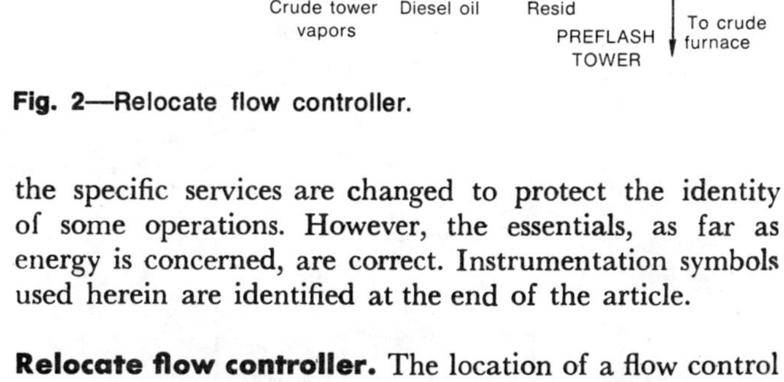

Fig. 2—Relocate flow controller.

Relocate flow controller. The location of a flow control valve in the preheat exchanger train of a crude topping plant was as shown in Fig. 1. Most of the pressure drop upstream of the preflash tower was taken through the flow control valve. This control valve was located immediately upstream of the preflash tower. The operating pressure on the crude side of the exchanger train was maintained at about 90 psig. This high pressure suppressed most of the vaporization in the crude preheat exchangers.

By relocating the control valve, as shown in Fig. 2, upstream of the exchanger train, vaporization in the exchangers was increased. Increased vaporization on the crude side of the exchanger train increased the amount of heat recovered to crude from products because of two factors:

- As crude vaporized in the exchangers, linear velocities increased, causing exchanger heat transfer coefficients to increase.
- Vaporization of crude in the exchangers occurred at lower temperatures when the pressure was reduced, thus increasing the temperature driving force between hot products and cooler crude.

Improved crude preheat resulted in a decrease of 4 percent in the furnace fuel required.

Use lower pressure steam. A naphtha debutanizer was supplied with reboiler heat from 18,000 lb./hr. of 100 psig

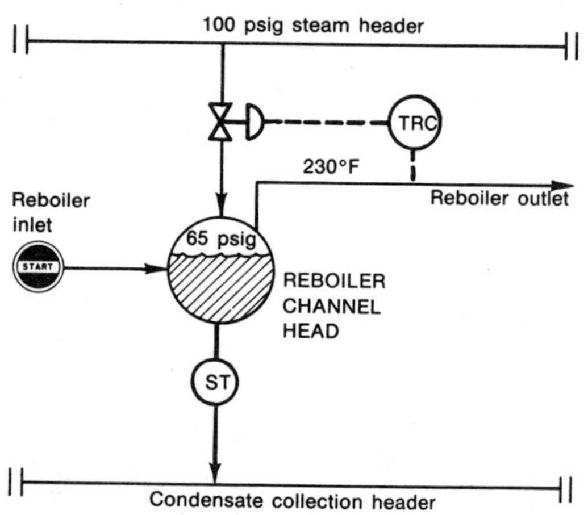

Fig. 3—Steam heated debutanizer reboiler.

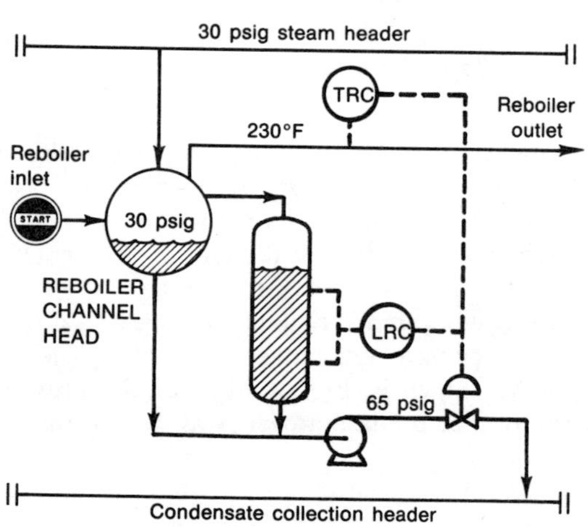

Fig. 4—Use lower pressure steam.

steam (condensing at 340° F) as shown in Fig. 3. The reboiler outlet temperature was running at 230° F. The 100 psig steam was valued at $3 per thousand pounds. Nearby, a large quantity of excess 30 psig steam was being vented to the atmosphere. Energy conservation suggested substituting the waste 30 psig steam, for the valuable 100 psig steam. Such a change could be made within the unit, without any complex study of the refinery steam system.

Process calculations of the required duty and the available heat transfer surface area indicated that it should be possible to reboil the tower with free 30 psig, instead of 100 psig steam. The calculations implied that the temperature control valve on the reboiler steam inlet was reducing the steam pressure in the reboiler channel head to 30 psig. To check these calculations, a pressure gage was put on the channel head of the reboiler. The pressure read 65 psig, instead of the expected 30 psig.

Something was wrong—either with the calculations or with the reboiler. With a wax stick having a melting point of 275° F, a vertical line was drawn on the channel head. The line remained unmelted for half the height of the channel head. This indicated that steam condensate was backing up into the channel head and water-logging the bottom half of the reboiler tube bundle.

To check this theory, the reboiler steam condensate drain line was temporarily disconnected from the steam condensate collection header. The steam condensate was manually drained directly on the ground. The rate of drainage was sufficient to prevent a water level from building up in the reboiler channel head. The control valve on the steam inlet line began to close. As expected, the reboiler channel head pressure dropped to 30 psig. This indicated that 30 psig steam could regularly be used in this reboiler, if the condensate drainage from the channel head could be improved.

The modifications that were made are shown in Fig. 4. A new steam line from the 30 psig steam header, was tied into the reboiler. There was no control valve in this line. The steam trap on the reboiler channel head drain line was replaced with a condensate drum. The level in the condensate drum was varied by a new control valve. This valve caused the reboiler tubes to be covered and uncovered by steam condensate, depending upon the required reboiler outlet temperature. At a low condensate drum liquid level, the reboiler outlet temperature controller (TRC) would be overridden by the condensate drum liquid level controller (LRC).

A small pump was also added to take suction on the condensate drum. This insured that the steam condensate would have plenty of head to reach the condensate collection header. The net result of these modifications were to save 18,000 lb./hr. of 100 psig steam with a value of $470,000/yr.

Control pump driver speed. A turbine driven pump was used to charge a crude distillation unit. The turbine was being driven by 110 psig steam exhausting to the atmosphere. As shown in Fig. 5, the turbine was run at constant speed. Excess pump discharge pressure was dissipated by a flow control valve (FRC) on the crude charge line. Since the crude charge rate was varied quite frequently, a great amount of the pumping energy supplied by the 110 psig steam was lost in the throttling action of the crude flow controller. It is difficult to justify the operating convenience of running turbine driven pumps in this manner when energy costs are high.

The modifications made to eliminate the energy wasted by the flow control valve are shown in Fig. 6. The crude flow transmitter is used to provide the signal that adjusts the turbine governor position. This forces the turbine and pump to speed up or slow down as the required charge rate changes. The turbine speed continues to be displayed for operator evaluation of pump performance. An average of 3,500 lb./hr. of 110 psig steam, valued at $120,000/yr., is saved.

This energy savings is achieved at the cost of some operator inconvenience. For one thing, the turbine could now be run-up over the design speed by increasing the setting on the crude flow controller. This could trip off the charge pump and cause a unit upset. Naturally, this problem should be prevented by operators checking the turbine speed indicator before increasing the crude rate. Also, there is an increased importance in keeping the turbine governor in good shape.

In general, energy conservation projects often make the operating man's job just a little bit harder. Some might even conclude than an alarm might be required. With the kind of savings we are talking about, the addition of an alarm would be a trivial matter. Incidentally, the above description clearly indicates that additional energy savings are possible. Can the reader spot this potential saving?

Change refrigerant fluid. Centrifugal compressors that are driven by constant speed motors are a likely source of great energy waste. This is because centrifugal compressors enter a regime of unstable operations called "surge." Surging happens when gas feed suction volume falls below a certain level. A compressor in surge sounds like a giant with terminal asthma. Surging is extremely hard on the rotor bearings.

To prevent the compressor from surging due to low suction volume flow, gas is bypassed from the discharge to the suction of the compressor. This wastes energy. The problem is most frequently encountered on motor driven centrifugal refrigeration compressors during the winter.

Fig. 7 details the problem of a refinery refrigeration compressor. It was observed that the minimum flow bypass was half open. The operators stated correctly that this was necessary to prevent the centrifugal compressor from surging due to low suction volume. Calculations showed that 33% of the compressed gas was being recircalculated through the minimum flow bypass. A check of the compressor manufacturer's operating curve indicated the machine really was operating close to its surge point. The calculated compressor horsepower lost through the minimum flow bypass flow control valve (FRC) was 284 hp.

To correct this problem, butane was added to the refrigerant fluid, which had been 100% propane. The result was the suction pressure of the compressor was reduced, without changing the refrigerant's vaporization temperature. The lower suction pressure naturally increased the suction volume. Fig. 8 illustrates the extreme case. Here

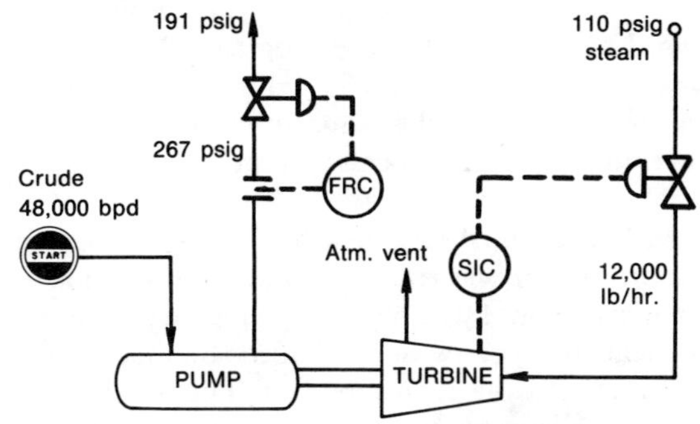

Fig. 5—Crude distillation unit charge pump.

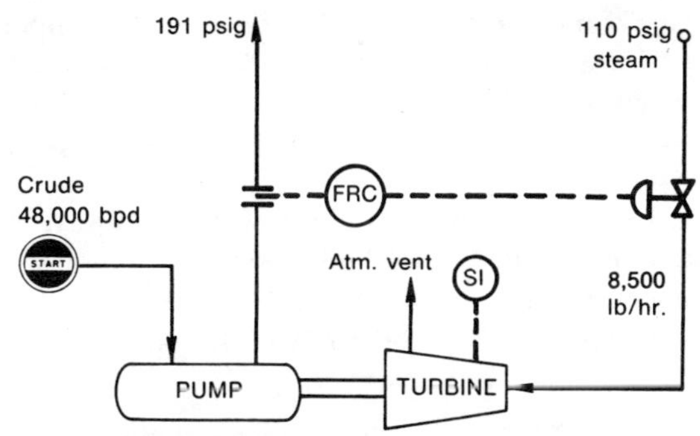

Fig. 6—Control pump driver speed.

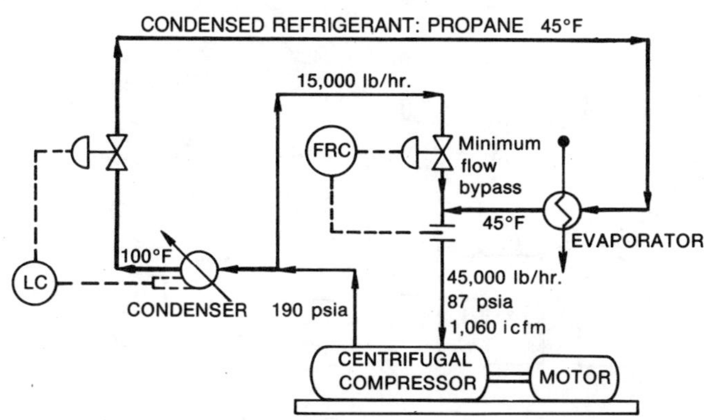

Fig. 7—Refrigeration compressor.

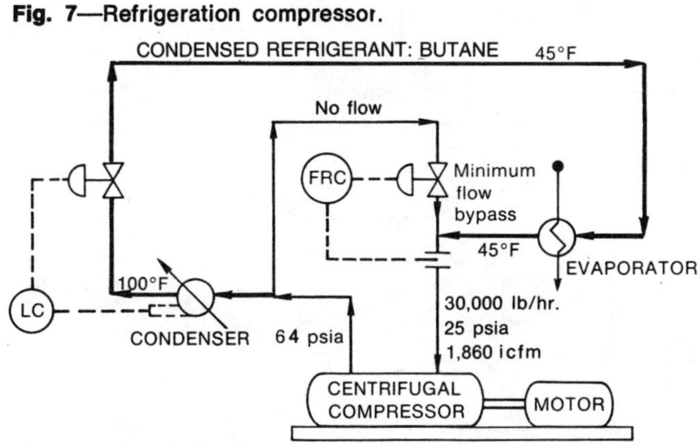

Fig. 8—Change refrigerant fluid.

the refrigerant has been switched to 100% butane.

The increase in the compressor's suction volume allowed the minimum flow bypass control valve to close. This reduced the number of pounds of refrigerant to be compressed. In practice, butane was added to the propane refrigerant until the compressor's suction volume increased enough to automatically close the bypass flow control valve. This change in refrigerant composition resulted in a wintertime energy savings of 226 kw worth $130 per day.

A word of caution: If you decide to spike your propane refrigerant with butane during the winter, don't forget to use dry butane. Otherwise, solid hydrates can form which could plug up process equipment.

Add product analyzer control. An alkylation unit debutanizer tower was required to make a bottoms product of not more than 1% isobutane. As shown in Fig. 9, the tower heat balance was controlled by flow controlling (FRC) steam to the twin reboilers. The varying composition of the alkylate (bottoms product), meant that the bottoms temperature could not be correlated with the isobutane concentration of alkylate. Since the operators did not have any quick way of checking the alkylate isobutane concentration, they always tried to minimize this concentration, so as not to get into trouble by going off-specification.

This practice meant that the tower was always run at its maximum reflux rate. This maximized reboiler steam requirements. Once a day a sample of alkylate was sent to the laboratory for an analysis by gas chromatography.

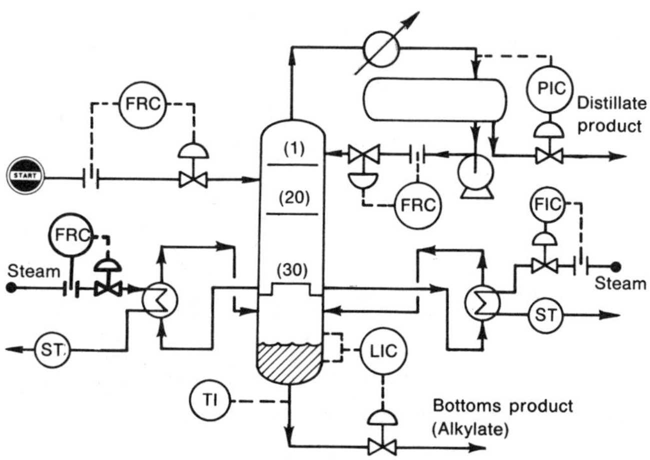

Fig. 9—Alkylation unit debutanizer.

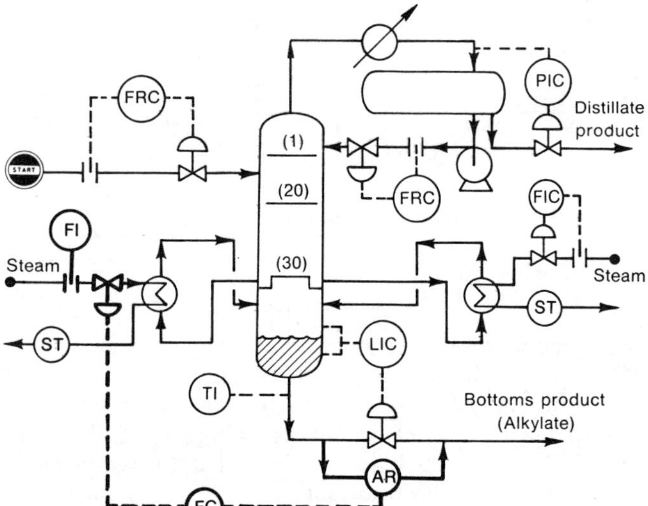

Fig. 10—Add product analyzer control.

If the result came back less than 1% isobutane, it was ignored. Furthermore, a 0.1% isobutane in alkylate analysis did not normally result in the operators cutting reflux to save steam.

To encourage operators to save reboiler steam, an on-stream gas chromatograph was purchased and installed to monitor continuously the isobutane content of alkylate. The cost was about $25,000. In general, most of the operators continued to run the tower at maximum reflux, regardless of the isobutane concentration of alkylate being shown by this expensive new instrument. Once again, human inertia took precedence to energy conservation.

To reduce the human factor, it was decided to purchase a small, feed forward, analog computer to control the reboiler steam flow. This computer would hold a constant 1% isobutane in the alkylate. Until the computer could be purchased, the gas chromatograph output was hooked to control directly the steam flow to one of the twin reboilers. By de-tuning the steam flow controller, a reasonably stable tower operation was obtained. The variation in the isobutane content of alkylate (0.8% to 1.2%) was evened out in the product tank. This control scheme is shown in Fig. 10. The results were so satisfactory that the analog computer was never purchased.

The point of this example is that energy savings were effected not by overly sophisticated instrumentation, but by reducing the human factor. This was accomplished by providing the minimum instrumentation required to evaluate and optimize performance. The reader should judge whether the operators used good judgment in ignoring lab analysis available on a once per day basis. Also, why would they first ignore the on-line gas chromatograph, and yet later depend on it to control steam flow directly?

Avoid overhead condenser flooding. The usual way to control the pressure in a distillation tower, where the total overhead is condensed, is the "flooded condenser" method. Fig. 11 shows how a depropanizer was being controlled with a flooded condenser. As the pressure in the depropanizer drops, the flow of distillate product is reduced. This forces liquid to back up into the condenser. As the liquid backs up, some of the tubes in the condenser are submerged. These covered tubes are then unavailable to condense any of the overhead vapors. As the rate of condensation falls, the depropanizer pressure will rise. Note that the reflux drum is always full of liquid. The flow of steam to the reboiler is used to control the depropanizer temperature (TRC).

In general, flooded condenser control is satisfactory. However, for this depropanizer, the condenser was much larger than it needed to be for the required duty. Consequently, much of the condenser's surface area was wasted, since most of the tubes were normally submerged. To use this excess condenser surface area, the condenser cooling water inlet valve was partially closed. This saved cooling water. However, since all distillation towers are subject to occasional upsets, it was necessary to increase the cooling water flow occasionally. After a few such incidents, the men running the unit decided to leave the cooling water valve open. Once again, the human factor defeated energy conservation.

The changes made to successfully save cooling water are shown in Fig. 12. The reboiler steam rate is controlling the depropanizer pressure. The distillate product rate is controlling the tower temperature (i.e., the split between the key components). The liquid level in the reflux drum is held by a new control valve (LRC) on the condenser cooling water supply. As the reflux drum liquid level drops, the flow of water to the condenser increases. This increases the rate of condensation. If the liquid level were to continue to fall, after the cooling water control valve is 100% open, the flow of reflux would be reduced automatically. Decreasing the reflux cuts the condenser duty.

This method of control insures that the minimum amount of cooling water is used, consistent with just condensing the overhead product to its bubble point. This method has the disadvantage that the rate of fouling on the water side of all coolers is increased at lower velocities. However, the cooling water saved permitted the shutdown of a 100 hp cooling water pump, and a 25 hp cooling water fan. Resulting power savings were valued at $26,000 per year.

Reduce cooling water bypass. Warm water from a box cooler was pumped back to a cooling tower by a motor driven centrifugal pump. The pump was instrumentated to maintain the water level in the box. This was done by allowing water from the pump discharge to flow back into the box through a bypass line. As shown in Fig. 13, a control valve (LRC) in the bypass line held the level in the box. This meant that the water return pump always operated far out on its pumping curve (i.e., it always was pumping lots of gallons). When a centrifugal pump operates at the end of its pumping curve, it requires the maximum motor horsepower characteristic of the particular pump. This wastes power.

To reduce waste of electricity, the pump controls were modified as shown in Fig. 14. The level controller was changed to a split range controller. Also, a new control valve was installed on the warm water return line downstream of the bypass line take-off.

During normal operations, the old control valve on the bypass line remained closed. The new control valve directly controlled the level in the box by throttling the pump discharge. When the new control valve was 90% closed, the split-range level control transmitter would signal the old control valve, in the bypass line, to start opening. This was necessary to prevent the water return pump from overheating due to insufficient flow.

The split-range level control normally forced the pump to operate well back on its pumping curve. As centrifugal pump horsepower requirements are reduced when less fluid is being pumped, the electric motor power requirements are reduced. This reduced the average motor power required to 70% of the maximum pump horsepower. Savings of about 50 kilowatts resulted.

One could argue that the instrumentation was all wrong to begin with. However, it is the job of operators and engineers to discover such energy hogs, which may have crept in by accident over the years.

Reduce compressed vapor bypass. This last example of energy saving through process control modifications

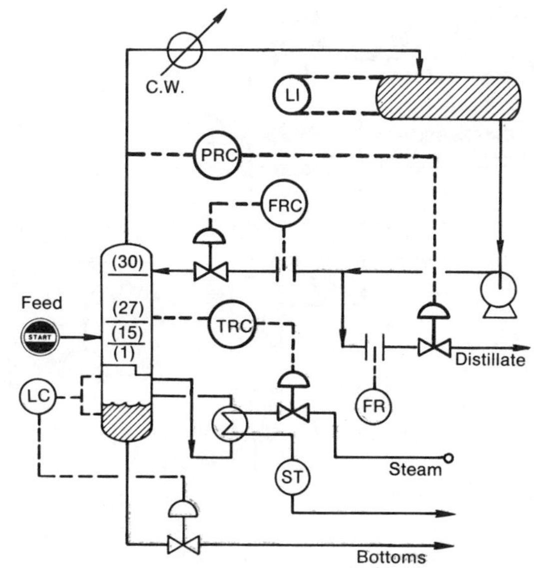

Fig. 11—Distillation using flooded condenser.

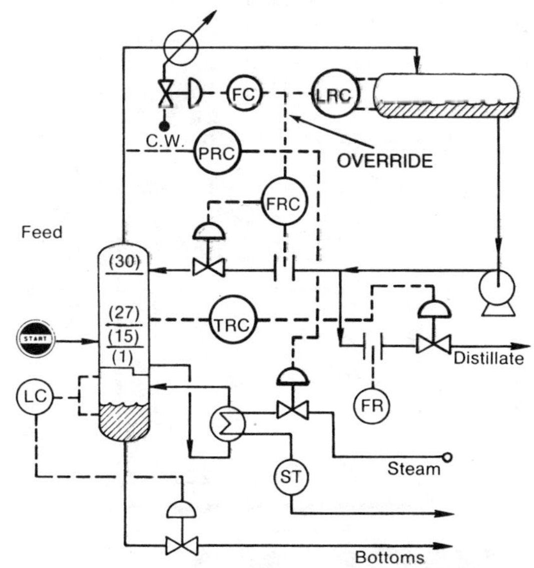

Fig. 12—Avoid overhead condenser flooding.

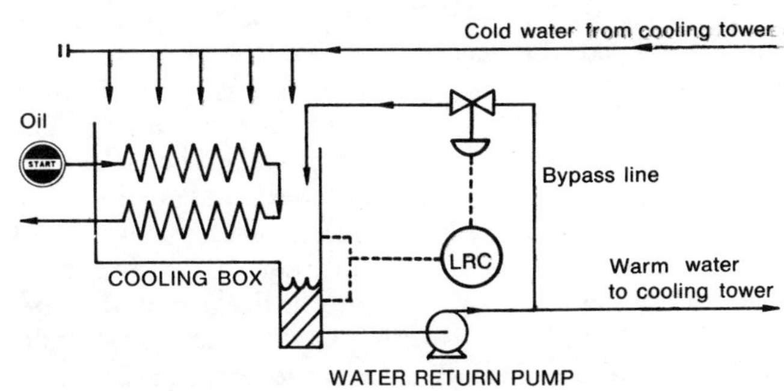

Fig. 13—Pumping water to a box cooler.

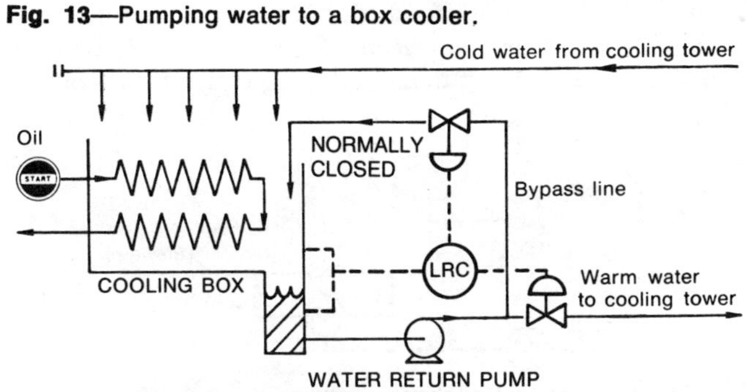

Fig. 14—Reduce cooling water bypass.

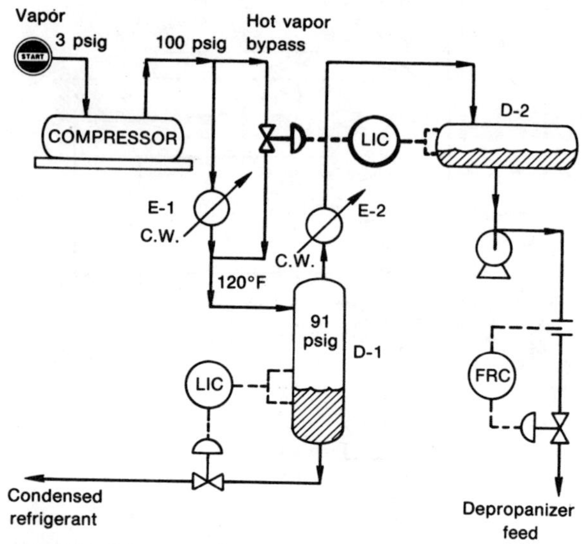

Fig. 15—Cooling vapor from a compressor.

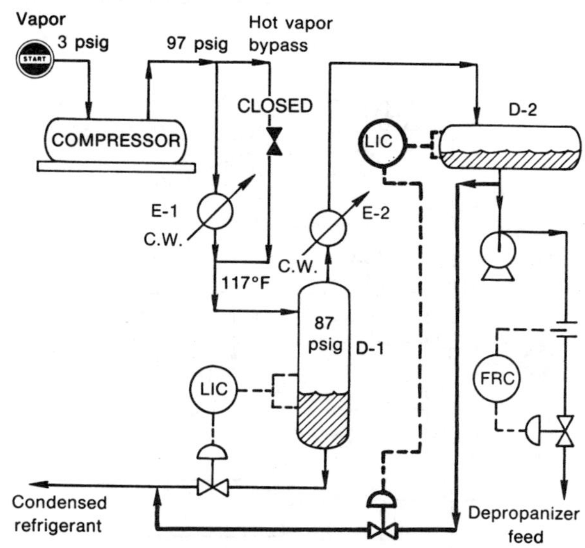

Fig. 16—Reduce compressed vapor bypass.

resulted in a savings of $180,000/yr. of 400 psig steam. A large (9,000 hp) centrifugal compressor was driven by 400 psig steam exhausting to a 28 psig steam header. This low pressure steam was excess in the refinery. The compressed fluid was a mixture of propane and isobutane.

The compressor discharge fluid was condensed in two stages as shown in Fig. 15. About half the refrigerant was condensed in the first stage condenser (E-1) and separated into vapor and liquid phases in the first stage separator drum (D-1). The resulting vapor stream was totally condensed in the second stage condenser (E-2). This lighter half of the compressor discharge fluid accumulated in the depropanizer feed drum (D-2). The depropanizer feed from this accumulator drum was on flow control (FRC).

The key point of the control scheme shown in Fig. 15 is how the liquid level is held in D-2. When the liquid level falls in D-2, its level control transmitter (LIC) signals the control valve in the hot vapor bypass line to open. When this control valve opens, hot vapors from the compressor discharge bypasses the first stage condenser (E-1). This bypassing reduces the total pounds of vapor condensed by E-1. Therefore, the flow of vapor from the first stage separator drum (D-1) to the second stage condenser (E-2) is increased. In turn, this increases the flow of liquid into the accumulator drum (D-2) and corrects for the falling level.

Any time the efficiency of a heat sink is impaired, energy is being wasted. Bypassing vapors around a compressor discharge condenser, for control purposes, is a typical example of energy waste. From the operating man's point of view, it's hard to understand why the designer would choose this route.

The modifications made to eliminate the horsepower wasted by the hot vapor bypass controller are shown in Fig. 16. A new line from the bottom of the accumulator drum (D-2) was installed. A control valve in this line maintained the liquid level in D-2. The effluent from this line joined with the liquid from D-1. The hot vapor bypass line was blocked in.

On occasion, the new control valve (LIC) would close all the way, and the liquid level in D-2 would continue to fall. When this happened, usually during cold weather, the depropanizer feed rate was reduced, until the LIC could again hold the level in D-2.

What happened when the depropanizer feed rate dropped too low? The only answer is that this never did happen.

In practice, closing the hot vapor bypass reduced the temperature in the first stage separator drum (D-1) by 3° F. This reduced the compressor discharge pressure from 100 psig to 97 psig. Compressor driver steam savings were 5,000 lb./hr. of 400 psig steam. (2.6% of 190,000 lb./hr. of total driver steam.) This modest change was worth $180,000/yr. to the refinery fuel budget.

INSTRUMENTATION ABBREVIATIONS USED

AR	Analyzer Recorder
C.W.	Cooling Water
FC	Flow Controller
FI	Flow Indicator
FIC	Flow Indicator Controller
FR	Flow Recorder
FRC	Flow Recorder Controller
LC	Level Controller
LI	Level Indicator
LIC	Level Indicator Controller
LRC	Level Recorder Controller
PIC	Pressure Indicator Controller
PRC	Pressure Recorder Controller
SI	Speed Indicator
SIC	Speed Indicator Controller
ST	Steam Trap
TI	Temperature Indicator
TRC	Temperature Recorder Control

Multi-temps Give Better Control

Product quality is more consistent when a distillation column is controlled by multiple temperature measurements

B. Joseph and **C. B. Brosilow,** Case Western Reserve University, Cleveland, Ohio, and **J. C. Howell** and **W. R. D. Kerr,** Marathon Oil Co., Findley, Ohio

DISTILLATION CONTROL is improved by using multiple temperature measurements to estimate product composition and to project changes for the manipulated variable. As a result, changes in product quality are reduced substantially compared to a constant temperature feedback control of a single tray.

Plant data collected over slightly more than a year for a debutanizer column operated by the Marathon Oil Co. are used to derive a computer model. Then an industry-university research project by Case Western Reserve University and Marathon Oil Co. compared estimation techniques and their anticipated control of product quality.

CONTROL THEORY

Product quality measurement is one of the major difficulties associated with the control of distillation columns. Gas chromatographs require large capital investment and are expensive to maintain. Also, chromatographs often introduce time delays in the process that are detrimental.

One of the common alternatives to chromatographic control is temperature control of a single distillation tray. This policy is good only if the temperature selected responds to various process disturbances in the same way as does the output composition. This is often not the case. As a result, there can be substantial variations in the product quality even though one is able to maintain the tray temperature constant.

Substantial improvement in control can be obtained by using multiple temperature measurements to predict output composition as shown by Weber and Brosilow.[1] Their approach was somewhat restrictive, however, since it required as many measurements as there were process disturbances. Easterday[2] extended Weber's approach to include cases where the number of measurements was less than the number of disturbances. The estimate of output quality is given by a linear combination of the measurements. Another approach to estimation using temperature measurements, proposed by Marathon Oil, is to relate the unmeasurable output to the measured variables by least squares regression.

This article examines both approaches of output estimation and compares their performance when applied to the control of a debutanizer tower. Attention will be focused on useful results rather than rigorous details. The interested reader can find a rigorous discussion of estimation in the theses of Easterday[2] and Joseph.[3]

Two types of estimators are discussed. The first, called the *projection estimator,* is based on a linear input-output model of the process. The second, called the *regression estimator,* is constructed by curvefitting the quantity to be estimated to the measurements, over predetermined ranges of process variables.

Projection estimator. Consider a process operating about a steadystate and subject to a vector of unknown disturbances u. Let θ denote the changes in the measured variables caused by the disturbances and y the change in the product composition. For small input disturbances, the variables θ and y can be linearly related to the disturbances u as follows:

$$\theta = A^T u \tag{1}$$
$$y = b^T u \tag{2}$$

where A and b are matrices of suitable size.

The case when there are as many measurements as input disturbances is discussed by Weber and Brosilow.[1] However, the number of measurements, generally, is less than the number of input disturbances and we will restrict our attention to this case. In this case, Easterday[2] has shown that an estimate which minimizes the expected value of the square of the error in estimation is given by

$$y = \alpha^T \theta \tag{3}$$

where $\alpha = [A^T A]^{-1} A^T b$

The vector α will be referred to as the estimator. Note that α projects the vector b onto the space spanned by the columns of A and hence the name projection estimator. Associated with each α are two numbers defined as projection error and condition number:

$$\text{Projection error} = \frac{||A\alpha\text{-}b||}{||b||}(100) \tag{4}$$

$$\text{Condition number} = \sqrt{\frac{\lambda_{\max}[AA^T]}{\lambda_{min}[AA^T]}} \tag{5}$$

where $||x||$ represents the length of the vector x, and $\lambda[M]$ represents an eigenvalue of the matrix M.

The projection error is an indication of how accurate the estimate will be. A projection error of zero indicates that perfect estimation is possible. As the number of measurements is increased, the projection error will generally decrease (it cannot increase).

The condition number is an indication of the sensitivity of the estimate to modelling errors. A large condition number indicates a high sensitivity. One can reduce the condition number by dropping some measurements or picking other measurements from the available set. The problem of measurement selection is discussed in detail by Joseph.[3]

The above development neglects the effect of dynamics

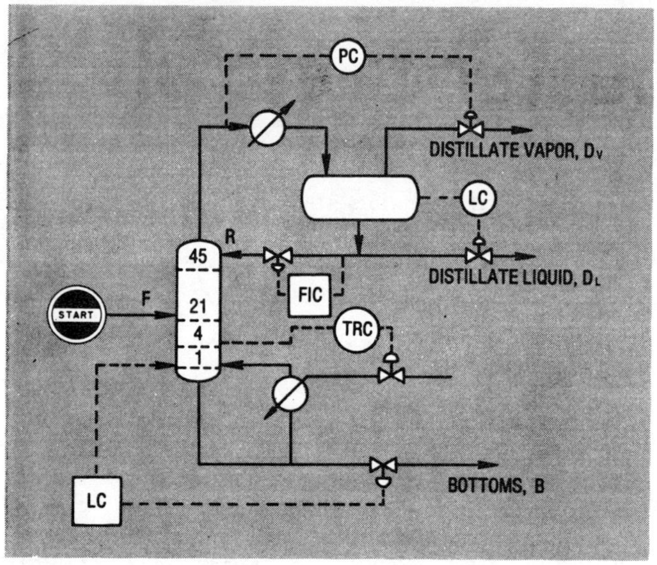

Fig. 1—How debutanizer was controlled.

introduced by the process. However, Tong and Brosilow[4] show how to add dynamics to the above estimator so as to minimize the transient response of the column to disturbances.

The parameters A and b in the linear model of Equations 1 and 2 can be determined in a number of ways. If a fast simulation model is available, the perturbation method can be used. Known disturbances are introduced into the process and the changes in θ and y are noted. From a sufficient number of such runs, the matrices A and b can be calculated. The perturbation technique can also be applied to the actual process itself, rather than the simulation model, provided all disturbances to the process can be measured. With an accurate mathematical model, A and b can also be calculated as the partial derivatives of θ and y with respect to u evaluated at the operating point. For a distillation column, this latter procedure is rather cumbersome and requires considerable computational effort.[2]

TABLE 1—Basis for Operating Point I.

	Concentration, mole fraction		
Component	Feed	Distillate liquid	Bottoms product
Hydrogen	0.0046	0.0006	......
Methane	0.0428	0.0429	......
Ethane	0.0405	0.0660	......
Propane	0.0900	0.2650	......
Isobutane	0.1033	0.4059	0.0007
N-Butane	0.0541	0.2256	0.0018
Isopentane	0.1078	0.0115	0.1588
N-Pentane	0.0130	0.0001	0.0196
9	0.0582	......	0.0877
10	0.0393	......	0.0591
11	0.0873	......	0.1314
12	0.1302	......	0.1962
13	0.0663	......	0.0999
14	0.0221	......	0.0332
15	0.0166	......	0.0250
16	0.0207	......	0.0311
17	0.0310	......	0.0467
18	0.0354	......	0.0533
19	0.0244	......	0.0367
20	0.0121	......	0.0182

Flow rates, lbs.-moles/day

Feed, F 90,000
Reflux, R 78,860
Distillate liquid, D_L 15,390
Bottoms, B 59,754

Conditions for Operating Point I

Feed temp, T_F = 340°F (approx. 50% vapor) on tray 21.
Column press. = 235 psia
Isobutane in bottoms = 0.0007 mole frac. (target)
N-Butane in bottoms = 0.0020 mole frac.

Temperature measurements are also affected by changes in measured disturbances or manipulated variables. The effect of such measured variables m, can be taken into account as follows.

The linear model which includes the effect of measured variables is written as:

$$\theta = A^T u + Pm \tag{6}$$

$$y = b^T u + Cm \tag{7}$$

The same estimation technique as before can be applied to the difference in the measured variables $(\theta - Pm)$. The resulting estimator is of the form

$$y = \alpha^T [\theta - Pm] + Cm \tag{8}$$

with α the same as before.

Regression estimator. This estimator is obtained by curve fitting the quantity to be estimated to the measurements. Let Θ denote a sample set of the measurement Θ, and Y the corresponding values of y. Then the linear regression estimator is constructed by solving the set of equations

$$\theta\alpha = Y \tag{9}$$

by the method of least squares. The solution is

$$\alpha = [\theta^T \theta]^{-1} \theta Y \tag{10}$$

The regression estimator can be designed to include nonlinear terms also. Easterday[2] gives a more detailed discussion of regression estimators.

THE DEBUTANIZER COLUMN

A schematic of a debutanizer column operated by the Marathon Oil Co. at their Robinson refinery is shown in Fig. 1. The tower has 47 stages and the feed enters (approximately 50 percent vapor) on tray 21. The feed has eight pure components and 12 pseudocomponents representing groups of heavier hydrocarbons. The overhead condensor has two products: a liquid distillate output and a smaller vapor withdraw. The column has a temperature feedback control with the temperature on tray 4 maintained constant by adjusting the fuel input to the reboiler which essentially adjusts the bottoms flow rate.

The tower instrumentation consists of flow measurements of the feed stream F, the reflux flow R, the bottoms flow B, the liquid distillate D_L and the vapor distillate D_V, the feed temperature T_F and the pressure of the column. The temperature of the feed is controlled by a feed preheater. Table 1 summarizes typical operating conditions of the column. The table also lists the mole fraction of isobutane and n-butane in the bottoms.

The simulation model of this tower was constructed by varying the number of stages in the model to achieve the best match of the calculated product compositions with measured data. The results showed that the 47-stage tower can be modeled by 35 theoretical stages.

As a first step in the construction of an estimator, a linear input-output model was generated by the perturbation technique around Operating Point I defined by Table 1. The input disturbances were taken to be the changes in flow rates of each of the 20 components. The measured variables were selected as the ratios of the flow rates of the reflux, liquid distillate and bottoms to the feed flow. These ratios take into account variations in the feed flow rates. They also reduce the nonlinearities in the process introduced by the interaction of input disturbances.

To obtain the linear model, the input flow rates of each component was perturbed by 10 percent. From the resulting steadystate changes in temperatures and composition the numbers in A and b of Equations 1 and 2 were calculated. It took approximately three minutes on a UNIVAC 1108 computer to generate the linear model of the process.

Preliminary studies using the linear model indicated the bottoms isobutane composition (the output composition we want to predict) is a highly nonlinear function of both the input disturbances as well as the manipulated variables. This is demonstrated in Fig. 2, which shows the behavior of both x_{ic4} and the logarithms of x_{ic4} as functions of the B/F ratio. Since the logarithm curve is more nearly linear, it was decided to construct an estimator to predict the changes in log x_{ic4} rather than x_{ic4} itself. The linear model now took the form

$$z = \Delta\ (\log x_{ic4}) = b^T u + Pm \qquad (11)$$

TABLE 2—Estimators compared

1. **Estimator using all six measurements:**

$$\begin{aligned}\Delta \log x_{ic4} = &- 0.0102\ \theta_1 + 0.00525\ \theta_4 - 0.0854\ \theta_{16} \\ &+ 0.00136\ \theta_{26} - 0.01092\ \theta_{34} - 0.000034\ \theta_{35} \\ &- 5.209\ \Delta\ (R/F) - 1.444\ \Delta\ (B/F) \\ &+ 0.065\ \Delta(D_V/F) + 0.06876\ \Delta(T_F)\end{aligned}$$

2. **Estimator using best five measurements:**

$$\begin{aligned}\Delta \log x_{ic4} = &- 0.00999\ \theta_1 - 0.07925\ \theta_{16} + 0.00111\ \theta_{26} \\ &- 0.01983\ \theta_{34} + 0.01\ \theta_{35} - 5.318\ \Delta(R/F) \\ &-0.9857\ \Delta(B/F) + 0.4069\ \Delta(D_V/F) \\ &+ 0.06778\ \Delta(T_F)\end{aligned}$$

3. **Estimator using best four measurements:**

$$\begin{aligned}\Delta \log x_{ic4} = &- 0.01172\ \theta_1 - 0.07088\ \theta_{16} - 0.02141\ \theta_{34} \\ &+ 0.01264\ \theta_{35} - 5.391\ \Delta(R/F) \\ &- 0.134\ \Delta(B/F) + 0.074\ \Delta(D_V/F) \\ &- 0.06522\ \Delta(T_F)\end{aligned}$$

4. **Estimator for n-butane using best five measurements, but effect of R/F, D_V/F, T_F not included:**

$$\begin{aligned}\Delta \log x_{nc4} = &\ 0.011\ \theta_1 - 0.0638\ \theta_{16} - 0.01017\ \theta_{26} \\ &-0.019\ \theta_{34} + 0.014\ \theta_{35} + 1.861\ \Delta(B/F)\end{aligned}$$

5. **Regression estimator for isobutane:**

$$\begin{aligned}\Delta x_{ic4} = &- 0.0000117\ \theta_1 + 0.0000211\ \theta_4 \\ &-0.00007746\ \theta_{16} - 0.0000053\ \theta_{35} \\ &+ 0.00325\ \Delta(B/F)\end{aligned}$$

6. **Nonlinear regression estimator for n-butane:**

$$\begin{aligned}\Delta x_{nc4} = &\ 0.00053\ \theta_1 - 0.00223\ \theta_4 + 0.00114\ \theta_{16} \\ &+ 0.0002\ T_F - 0.00000595\ \theta_1^{\,2} \\ &+ 0.0000733\ \theta_4^{\,2} + 0.000070\ \theta_{16}^{\,2} \\ &- 0.330\ \Delta(B/F)^2 - 0.0000157\ \theta_{34}\ \theta_1 \\ &+ 0.005\ \theta_{26}\ \Delta(B/F) - 0.051\ \theta_{11}\ \Delta(B/F)^2 \\ &+ 0.0000323\ \theta_{16}\ \theta_{26} - 0.080\ \theta_{16}\ \Delta(B/F)^2 \\ &+ 0.00001\ \theta_{34}\ \theta_{35} - 0.000128\ \theta_4\ \theta_{16} \\ &+ 0.03167\ \theta_4\ \Delta(B/F)^2\end{aligned}$$

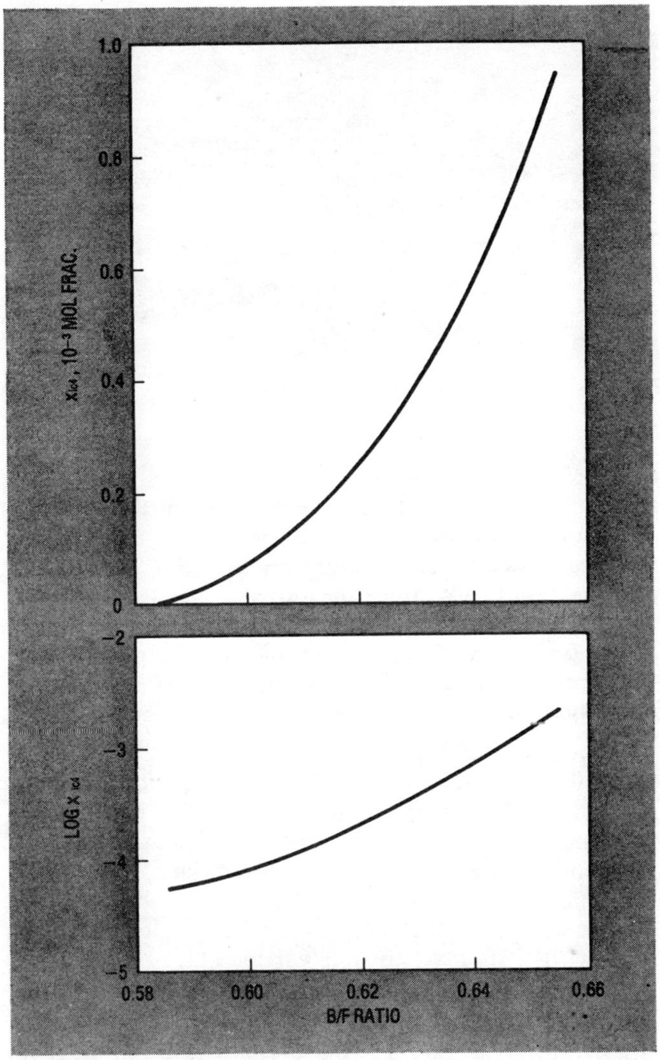

Fig. 2—Logarithm plot is linear.

Column estimators. The estimation principles presented in the Theory section can be applied to the new variable without any modification. Table 2 lists the various estimators constructed to predict the mole fraction of iso and normal butanes in the bottoms.

Projection estimators are the first four in Table 2 and, for reasons already discussed, predict changes in the logarithm of the product mole fraction. All were constructed based on a linear model about Operating Point I (Table 1), and can be expected to perform best when operating near these conditions.

The temperature measurements used in the estimators

TABLE 3—Factorial design to show effect of feed

Item	Base cond.	Range of independent variables: −2 (Minimum)	−1	0 (Average)	+1	+2 (Maximum)
B/F	0.6445	− 0.05	−0.025	0	0.025	0.05
R/F	0.87	− 0.0444	−0.0222	0	0.0222	0.0444
F_F, °F	340	−10	−5	0	5	10
D_L/F	0.18	− 0.0555	−0.0277	0	0.0222	0.0444
L/F	0.6393	− 0.1278	*	0	**	0.1882
		Sample No. 22		Sample No. 1		Sample No. 2

Joseph[3] lists the feed samples.
* Feed sample numbers: 23, 25, 26, 18, 16, 14, 17, 8, 15, 21, 27, 19, 13, 20.
**Feed sample numbers: 7, 9, 10, 11, 12, 28, 5, 6, 3, 4, 29, 30.

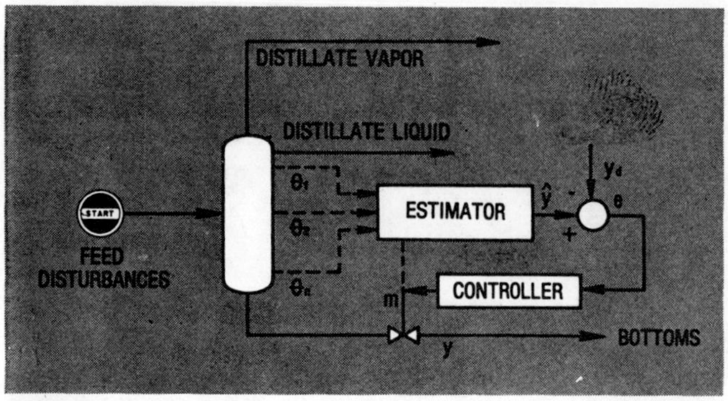

Fig. 3—An estimate based control system.

are those that were already available on the column. Estimator 1 uses all six of these measurements to predict the mole fraction of isobutane in bottoms. Estimator 2 uses the best five of these measurements and Estimator 3 uses the best four. The best measurements are used here in the sense of minimizing the projection error. Estimator 4 predicts the change in the mole fraction of *n*-butane in the bottoms.

Regression estimators are 5 and 6 in Table 2. These regression estimators were constructed by the operations research staff of the Marathon Oil Co. The data for the regression program was generated by simulating the column under a variety of operating conditions. A fractorial design procedure described by Cochran and Cox[5] was used to determine the values of the independent variables at which the simulations were carried out.

For the debutanizer column there are 23 independent variables, namely, the feed compositions [19 variables], B/F, R/F, D_L/F and T_F. Since it would be impractical to simulate every possible combination of variables at maximum, minimum and average values, the number of independent variables were reduced to five by treating only the liquid fraction L/F rather than the feed composition as an independent variable. This liquid fraction was calculated at an average temperature and pressure ($T_F = 340°F$, Press. = 235 psia). Feed samples taken on the column were used to determine the maximum, average and minimum ranges of the L/F ratio.

In the factorial design, the feed composition corresponding to the intermediate positions +1 and −1 were chosen randomly from the feed samples. This way the effect of all feed samples were included in the regression. Table 3 gives the ranges of variables used in the fractorial design procedure.

CONTROL STUDY SCHEME

Fig. 3 shows how an estimator can be used to control the output quality of the column. Consider the case when the process is subject to a set of new feed conditions. Recall that the estimator was constructed around a nominal operating point, all changes in variables are measured as the change from this operating point. The estimator measures the changes in temperatures from their nominal operating values to give an estimate $\bar{y}$ of the composition y. This value is compared against the desired value of y (also called the target value) denoted by y_d and the bottoms flow rate is adjusted till the error $\overline{y\text{-}y_d}$ becomes zero.

TABLE 4—Control Study I of isobutane around Operating Point I.

Feed sample no.	Steady state deviation of isobutane, % of target (0.0007 mole frac.)				
	Constant T_4 control	Estimator 1 (6 temp.)	Estimator 2 (Best 5 temp.)	Estimator 3 (Best 4 temp.)	Estimator 5 (Regression)
1	−244.81	−54.61	−50.06	−40.02	−77.59
2	−72.34	−46.85	−42.86	−41.83	−8.50
3	−11.49	−5.14	−1.50	0.08	30.50
4	−133.59	−16.49	−25.62	−33.90	−3.40
5	−144.60	1.59	−10.80	−47.93	−62.90
6	1.64	35.25	34.82	33.36	−12.90
7	−21.43	44.08	43.80	43.51	−59.00
8	−37.42	−72.48	−42.71	−2.37	8.58
9	−228.86	20.53	12.71	−0.73	−76.80
10	−288.49	−32.90	−56.82	−101.82	−104.10
11	−141.38	7.71	6.98	8.41	3.22
12	−26.95	−20.00	13.31	1.14	−10.79
13	28.49	15.99	16.62	27.99	−5.21
14	47.11	61.66	−30.04	−33.17	−92.00
15	11.23	40.62	36.42	24.27	−10.16
16	28.67	42.21	41.15	36.23	−19.08
17	33.06	38.59	36.05	23.83	−5.50
18	48.39	18.00	10.07	−9.63	−5.80
19	−425.15	6.03	−12.01	−37.37	−146.58
20	60.00	−39.78	−53.38	−77.00	−11.00
21	88.00	19.96	9.05	−16.89	−18.20
22	23.74	22.98	16.62	1.61	−17.40
23	−290.30	−20.07	15.00	−25.00	−82.88
24	71.00	21.31	14.76	0.44	−20.80
25	−93.17	29.62	29.83	29.48	0.97
26	84.00	−16.09	−9.23	22.18	91.59
27	57.69	12.83	14.29	13.27	6.31
28	−138.68	30.85	24.10	9.61	9.31
29	82.43	−50.00	66.00	−66.00	39.30
30	−53.94	3.63	−5.77	−24.92	20.19
Mean avg.	−55.56	3.30	3.36	−9.44	−21.29
Std. dev.	142.44	33.91	31.75	36.93	52.73
Cond. no.	......	275.40	46.80	7.67	......
Proj. error	......	5.10	5.11	5.57	......

TABLE 5—Control Study II of *n*-butane around Operating Point II

Feed sample no.	Deviation of n-butane, % of target		
	Constant T_4 control	Estimator 4 (Best 5 temp.)	Estimator 6 (Regression)
6	−2.7	−6.0	30.4
7	12.5	17.1	44.5
9	−28.7	−43.0	−3.7
11	−9.2	−26.5	6.4
13	13.1	−31.4	−215.4
14	24.3	2.6	26.6
19	−23.6	−42.6	−10.1
20	28.6	−29.5	66.5
22	38.6	3.1	−532.1
23	72.7	4.6	−276.8
24	15.8	−54.2	−123.7
25	19.4	−23.2	−2.5
26	13.9	−12.8	−31.0
Mean avg.	13.43	−18.6	−76.8
Mean, absolute	23.3	22.8	105.2
Std. dev.	28.8	28.0	181.8

Flow rates, lbs.-moles/day

Feed, F 90,000
Reflux, R 78,300
Distillate liquid, D_L 15,000
Bottoms, B 62,000

Conditions for Operating Point II

Feed temp, T_F = 340°F
Isobutane in bottoms = 0.01195 mole frac.
N-Butane in bottoms = 0.0232 mole frac. (target)

Due to estimation and measurement errors, the actual composition y will not be equal to the target value and one can define a percent error of control as $100\ (y\text{-}y_d)\ /y_d$. In order to provide a comparison, the test was repeated with the column under normal temperature feedback control and the steady state error in y was again calculated.

The study was done on the simulation model of the column using 34 different feed compositions, collected over a period of about one year from plant operating data.

Results of the control study are summarized in Tables 4, 5 and 6. As mentioned earlier, the projection estimators were constructed based on a linear model about Operating Point I (Table 1) and can be expected to per-

TABLE 6—Control Study III of isobutane around Operating Point III

Feed sample no.	T_F, °F	F, mole/day	R, mole/day	D_L, mole/day	Dev., % of target: Constant T_4	Dev., % of target: Estimator 4
31.......	−343.6	98,658	79,500	18,400	−49.94	25.87
32.......	343.6	93,096	86,000	16,300	5.96	−5.46
33.......	341.0	90,300	86,000	14,900	3.45	−3.05
Op. pt. III	340.7	93,367	78,800	15,900	(IC_4 target = 0.0059)	

Note: Bottoms rate = 58,000 lbs.-moles/day at Operating Point III.

form best when control is maintained around the target values of Operating Point I. For the studies summarized in Tables 5 and 6, the target values of y_d for isobutane were chosen at a different operating point. This will form a test of the robustness of the estimators.

The results of the Control Study I (Table 4) clearly indicate the superiority of estimator based control. The standard deviation of the percent error of control listed at the bottom is an indication of the ability of the control to keep the isobutane in the bottoms near its target (desired) value. Control using constant temperature feedback has a standard deviation of 142.4 compared to 31.75 for Estimator 2 (the best of all four estimators). Also, the estimator based control has much less bias as seen from the mean of the percent error of control over the 30 feeds. The regression estimator was slightly worse than all the projection estimators.

Some properties of the projection estimators are revealed by this study. Estimator 1, using all six measurements, gives a large condition number (275) indicating there is a strong coupling between the measurements. Hence, it can be expected to be very sensitive to modelling errors. This is indeed the case, since Estimator 2 using only five measurements is better. Note that Estimator 2 has a smaller condition number (46.8) but about the same projection error. Hence, adding a measurement to Estimator 2 deteriorated its performance. Removing a measurement from Estimator 2 gives Estimator 3 much less sensitivity (condition number of only 7.6) but causes a 10 percent larger projection error. The increase in projection error wiped out the benefits offered by the decreased sensitivity of the estimator. Thus, the condition number and the projection error can be used to choose the number of measurements and their locations. Joseph[3] presents a method for selecting a set of measurements with a low condition number and small projection errors.

Control Study II is summarized in Table 5 where the target value of n-butane was changed to Operating Point II as defined in Table 5. The major difference between Operating I and II is in the B/F ratio. This causes the new target value for n-butane to be 10 times the old value. As seen from the table, even under widely different operating conditions, the projection estimator is better than the constant temperature control. The regression estimator, in spite of the nonlinearities built into it, is worse than the constant temperature control. Since the regression estimator is constructed to fit over a wide range of input variables, it might not be expected to do well locally. The projection estimator takes advantage of the fact that the column is under control and will not deviate very far from the desired operating conditions. Some improvement in the performance of the projection estimator can be expected if a new linear model is constructed around the new operating point.

Control Study III is summarized in Table 6. Control Studies I and II concentrated on the effect of unknown feed disturbances only, and the measured variables R/F, D_v/F, T_F and F were kept constant. Control Study III tested the effect of variations in the measured variables in the control of isobutane. The operating point was again changed and the new target value of isobutane was 0.0059 compared to 0.0007 for operating Point I. In two out of three cases studied, both the estimator and temperature control do a good job of maintaining the isobutane in the bottoms constant. In the third case, the estimator is twice as good. Although this study is not conclusive, it brings out the point that the estimator can handle variations in measured variables and operating conditions well.

In Control Studies II and III, the set point of Temperature 4 was changed so that, with no input disturbances, the desired mole fraction of isobutane or n-butane is obtained in the bottoms. In actual operation, the new temperature set point is not known *a priori* and must be obtained by trial and error, based on product analyses obtained from the laboratory. This highlights another important advantage of the estimator based control scheme: In order to change the mole fraction of isobutane or n-butane in the bottoms when the column is under estimator based control mode, one simply adjusts the setpoint. On the other hand, adjusting the set point of the temperature to get a desired mole fraction of isobutane or n-butane in the bottoms requires considerably more effort.

NOMENCLATURE

A An (n) (k) matrix relating θ to u
A^T Transpose of A
B Bottoms flow rate, moles/day
b A vector of size k relating y to u
C Matrix relating θ to m
D_L Liquid distillate flow rate, moles/day
D_V Vapor distillate flow rate, moles/day
e Error in estimation $= y - \bar{y}$
F Feed flow rate, moles/day
k Number of input disturbances, u
L Liquid flow rate in the feed
m Vector of manipulated variables $(B/F, R/F, D_v/F, T_F)$
n Number of temperature measurements
P Vector relating y to m
R Reflux flow rate, moles/day
T_i Temperature on stage i, °F
T_F Feed temperature, °F
u Vector of input disturbances of size K
x_{ic4} Mole fraction of isobutane in bottoms
x_{nc4} Mole fraction of n-butane in bottoms
y Change in mole fraction of the product composition
$\bar{y}$ Estimated change in the mole fraction of product quality
y_d Desired value of the change in the product composition

GREEK LETTERS

α A vector of size n when multiplied by θ gives the estimate of y
θ Change in temperature of stage i
Δ Denotes a change

LITERATURE CITED

1 Weber, R. and Brosilow, C. B., "The Use of Secondary Measurements to Improve Control," *AIChE Journal*, Vol. 18, No. 3, May 1972, pp. 614-623.
2 Easterday, D. E., "The Use of Secondary Measurements in Estimating Unmeasurable Process Output," Ph.D. Thesis, Case Western Reserve University, 1973.
3 Joseph, B., "Estimation of Unmeasurable Outputs from Secondary Measurements with application to Control of Multicomponent Distillation Processes," Ph.D. Thesis, Case Western Reserve University, 1975.
4 Tong, M. and Brosilow, C. B., "Inferential (Feedforward) Control Applied to a Multicomponent Distillation Column," presented at a meeting of AIChE, Houston, March 17-20, 1975.
5 Cochran, C. G. and Cox, G. M., Experimental Designs, John Wiley and Sons, New York, 1957.

Better Control for C_3/C_4 Split

Better control assures on-specification product without purity giveaway and without energy waste

Jorge A. Bonilla,
Foster Wheeler Energy Corp., Livingston, N.J.

SPLITTERS for C_3/C_4 mixtures are frequently operated so that their products are substantially purer than required to minimize the possibility of production losses caused by the products being off-specification. This situation occurs because of the difficulties involved in holding the product purity constant during periods of disturbances.[1,2] The usual control scheme of Fig. 1 is revised as shown in Fig. 2. The revised control scheme uses the temperature difference between two trays, one in the rectifying section and one in the stripping section, to achieve control of produce purity that is independent of fluctuations in both system pressure and in feed composition.

Better control needed. A typical control scheme (see Fig. 1) which one might consider for a C_3/C_4 splitter lo-

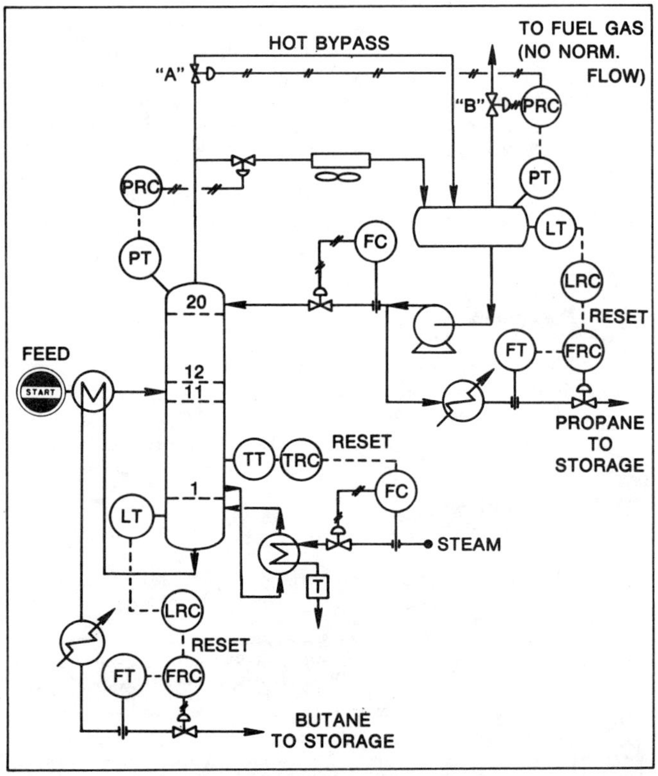

Fig. 1—A typical control scheme.

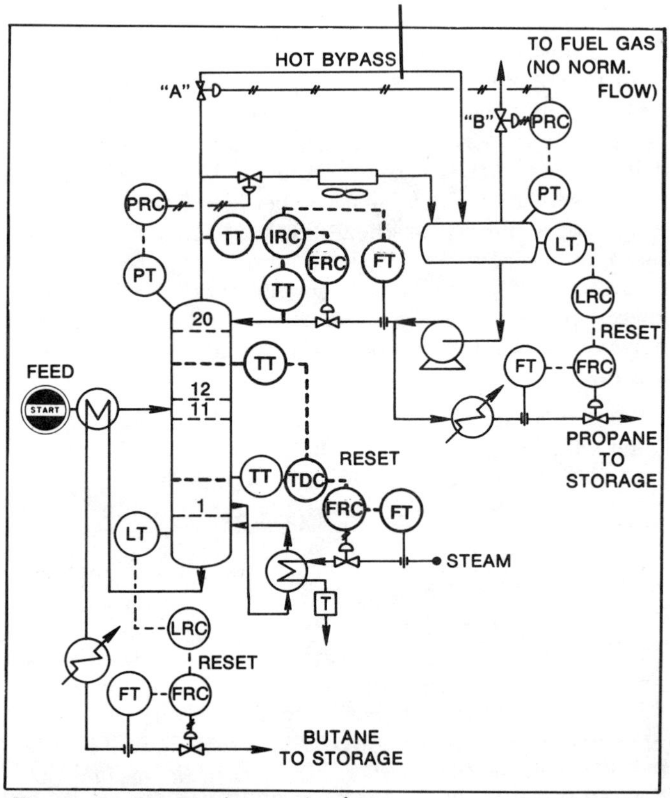

Fig. 2—An improved control scheme.

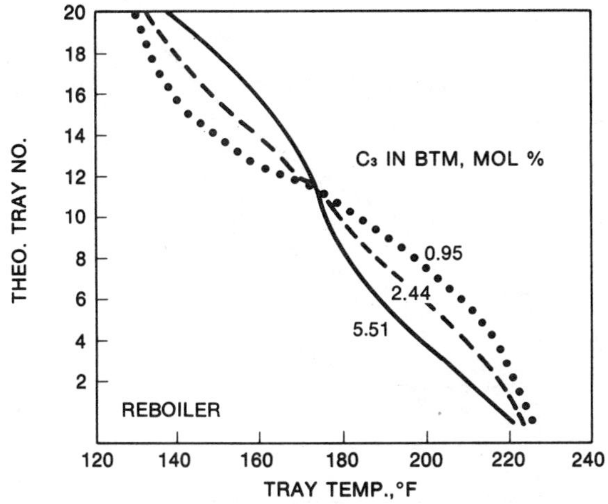

Fig. 3—Temperature profile of theoretical trays for an overhead accumulator pressure of 260 psia.

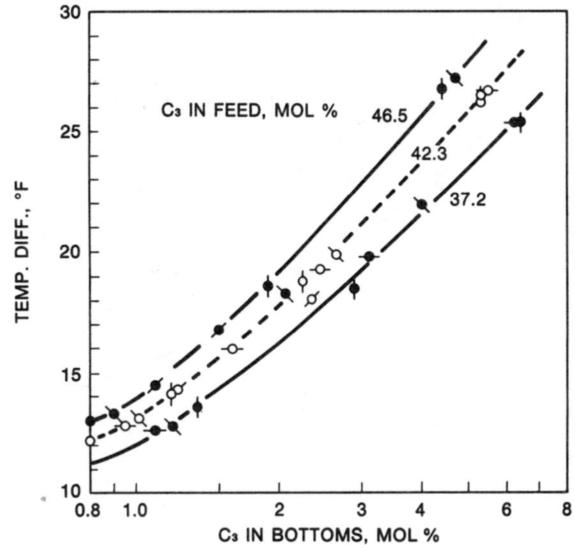

Fig. 4—Temperature difference (bottoms minus tray 5) for various bottom product purity. Parameters: C_3 in feed.

cates the temperature controller bulb several actual trays above the reboiler return or at the reboiler outlet. This temperature controller resets the flow controller on the steam to the reboiler.

The problems associated with this type of control were studied to explore different control schemes. For example, consider a tower having 9 theoretical rectifying trays and 11 theoretical stripping trays. The overhead condenser is a total condenser and the overhead accumulator pressure is 260 psia. The overhead system pressure drop is 7 psi and tower pressure drop 5 psi. Tower feed is assumed to have the composition shown in Table 1.

The propane overhead product must contain less than 2 mol % isobutane and heavier and the butane bottoms product a maximum of 2 mol % propane and lighter.

In order to study the influence of the different parameters, this problem was solved on a digital computer for several closely related conditions. Vapor-liquid equilibrium data were generated using Chao Seader K values[3] and enthalpies from The API data book.[4]

For 260 psia overhead accumulator pressure the reboiler outlet temperatures shown in Table 2 change only from 224.0 to 225.2° F when the C_3 content in the tower bottom product is varied from 1 to 2 mol %. Thus, if the normal bottoms purity is 1.5 mol % C_3, and if this purity is allowed to fluctuate between 1 and 2%, a temperature transmitter located at the reboiler outlet must respond to a change of 1.2° F for the whole proportional band. Technically this is not a problem since it is possible to keep a temperature within ± 0.02° F using a resistance thermometer and an electronic controlled.[5] Table 2 also shows that a fluctuation of ± 2% in the overhead accumulator pressure, due to atmospheric pressure changes and to the pressure controller itself, gives a bottom temperature change of 3.1° F at a constant bottom purity of 1.5 mol % C_3.

This fact makes it difficult to use the reboiler outlet temperature to control the product purity because the changes in temperature are more sensitive to the pressure fluctuations than to the purity of the product.

The temperature profile in the tower is shown in Fig. 3 for three cases, 0.95, 2.44 and 5.51 mol % C_3 in the bottoms product, at a 260 psia overhead accumulator pressure. This figure shows that the temperature changes that occur when the bottoms product purity varies are much larger at an intermediate tray between the reboiler return and the feed tray than at the reboiler return itself. Thus, by placing the temperature controller at theoretical tray No. 7, the temperature changes are substantially larger. Table 3 shows that at 260 psia a bottoms product purity change from 1 to 2 mol % corresponds to a change in tray No. 7 temperature of 6.1° F compared with 1.2° F at the reboiler return. Temperature changes due to pressure fluctuations are also increased, but to a lesser extent. For 1.5 mol % C_3 purity (see Table 3), a ± 2% pressure fluctuations gives a 3.4° F temperature change. Thus, a temperature controller at this location allows for better control than one at the reboiler outlet would. When product purity is not critical, and when steady process conditions are expected, this arrangement would be acceptable.

When steady process conditions are not expected, compensation for pressure fluctuations can be achieved by controlling the temperature difference between two points in the tower instead of a single tray temperature. The temperature difference between the bottoms product and theoretical tray No. 5, for example, is independent of fluctuations in tower pressure, as is shown in Fig. 4. However, this established temperature difference changes for each feed composition. Fig. 4 shows that a fluctuation of

TABLE 1—Splitter feed composition

Component	Mol %
C_2	0.8
C_3	42.3
iC_4	17.2
nC_4	38.5
$C_5{+}$	1.2
Total	100.0

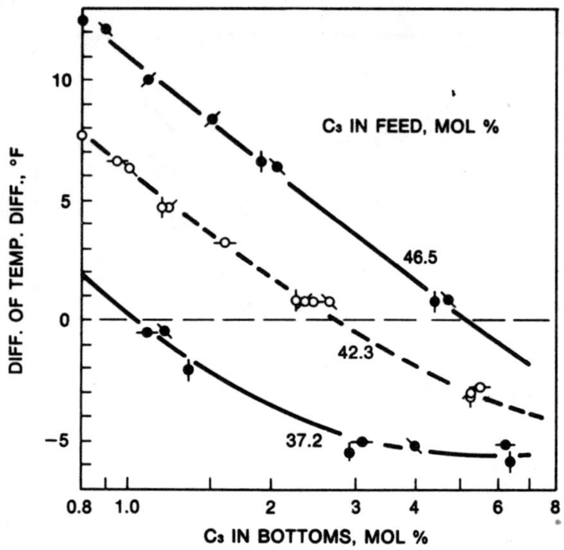

Fig. 5—Temperature difference (tray 5 minus tray 9) minus (tray 14 minus tray 18).

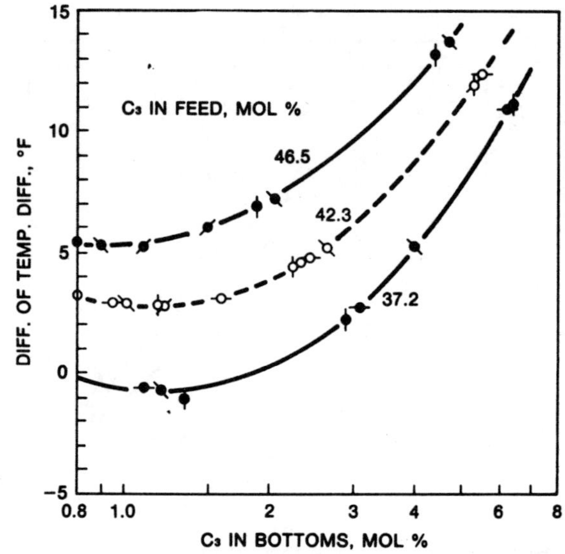

Fig. 6—Temperature difference (bottoms minus tray 5) minus (tray 15 minus tray 18).

C_3 content in the feed of about ± 5% gives a temperature difference change of 2.4° F for a constant bottoms product purity of 1.5%. This change corresponds to a purity fluctuation of 1.2 to 1.8%. In many cases, this fluctuation can be considered acceptable. By taking temperature differences between two intermediate trays in the stripping section, plots similar to Fig. 4 are obtained. All are dependent upon feed composition.

Another method tested. D. M. Boyd[6] reported a control scheme using four temperature points, two to give a temperature difference between two trays in the rectifying section and two more to give a temperature difference between two trays in the stripping section. By subtracting these differences it was possible to get control of a split between two almost pure components.

We studied this type of control for a C_3/C_4 splitter. It was found that for this particular system, the temperature differential obtained was independent of pressure fluctuations, but not of feed composition (see Fig. 5). It was also found that the function may pass through a minimum (see Fig. 6), thus indicating that for two different product purities, the same differential temperature could be found which would not give practical control. It appears then that while this method of control is very useful for high purity products, the control may not be suitable in the case of a C_3/C_4 splitter where product purity is not so high—the impurities usually being from 1.5 to 2 mol %. Thus a careful analysis would be required if this type of control were applied to a C_3/C_4 splitter.

A better method. Our studies show that it is possible to achieve a control independent of both pressure and feed composition fluctuations for a C_3/C_4 splitter by controlling the temperature difference between only two trays, one above and one below the feed point. By choosing these trays, it is possible to find an almost logarithmic function between the temperature difference and the

CODE FOR DATA POINTS IN FIGURES:

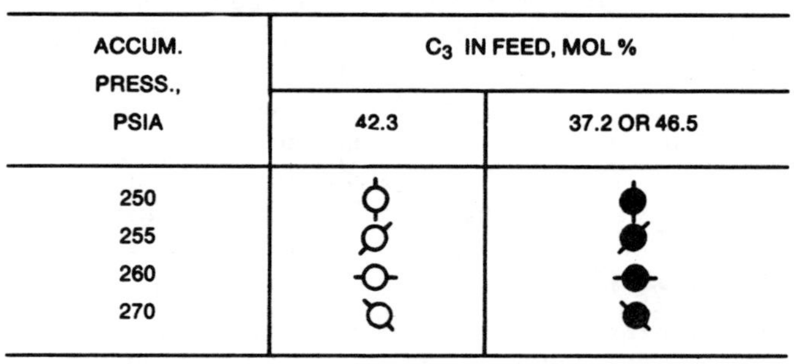

product purity, independent of both pressure and feed composition, as shown on Fig. 7. This phenomena is due firstly to the similarity in the vapor pressure curve slope of the overhead and bottoms products that compensate for the pressure fluctuations. For example, a pressure increment gives similar bubble point temperature increments of the two products or a mixture of them at any tray in the tower that cancels out when a differ-

TABLE 2—Reboiler outlet temperature

C_3 in btm. prod., mol %	Reboiler outlet temp., °F. for overhead accumulator press. shown		
	255 psia	260 psia	265 psia
1.0	222.3	224.0	225.4
1.5	222.8	224.5	225.9
2.0	223.6	225.2	226.6

TABLE 3—Theoretical tray No. 7 temperature

C_3 in btm. prod., mol %	Theo. tray No. 7 temp., °F, for overhead accumulator press. shown		
	255 psia	260 psia	265 psia
1.0	200.1	201.7	203.7
1.5	196.7	198.7	200.1
2.0	193.8	195.6	197.2

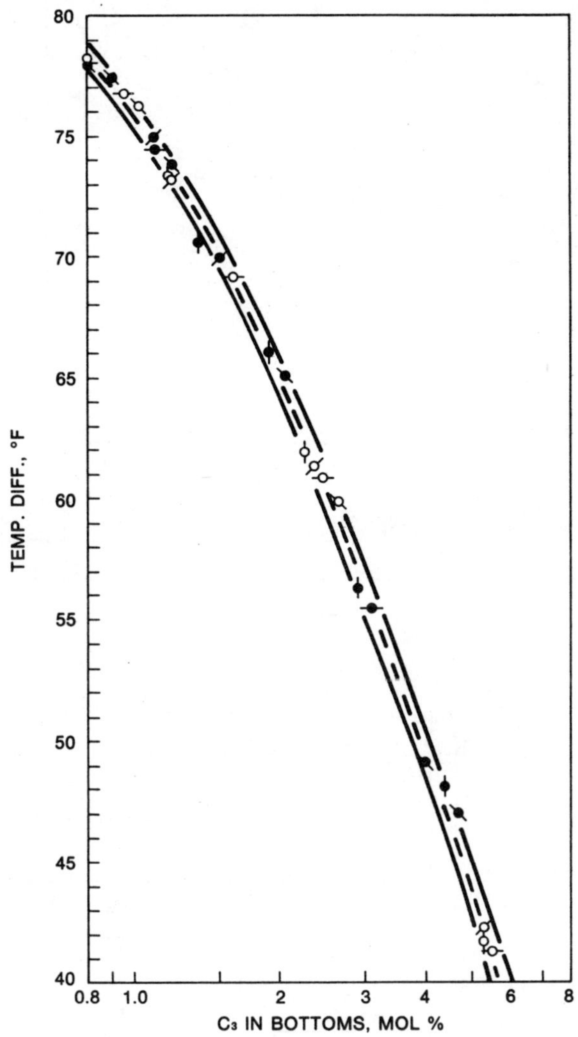

Fig. 7—Temperature difference (tray 5 minus tray 17) for improved control.

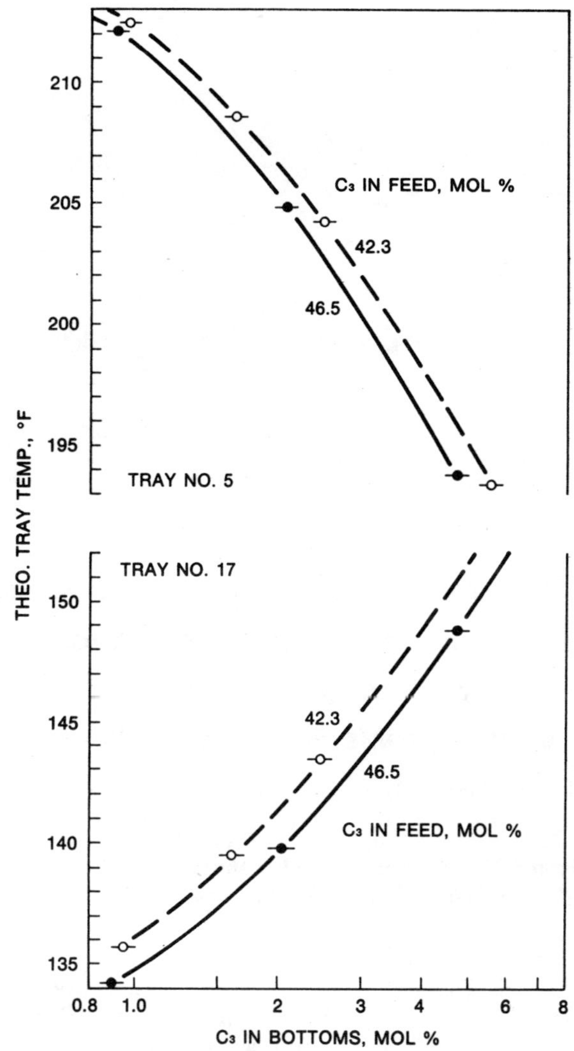

Fig. 8—Change in feed composition gives approximately same temperature change for both trays 5 and 17.

ential is considered. Secondly, because the rectifying and stripping tray temperature change in the same direction when the feed composition is changed (see Fig. 8). The proposed control was shown in Fig. 2. An internal reflux control is preferred to the constant reflux flow to compensate for the reboiler charge changes.

It is important to observe that the proposed temperature difference changes from 76° F to 65° F (a 11° F change) when the bottoms product purity varies from 1 to 2 mol % C_3, compared with a change of 1.2° F at the reboiler outlet or 6.1° F at the theoretical tray No. 7.

Contrary to the method of control proposed by Boyd,[6] this method is not independent of the pressure drop in the column that depends on the unit load; but this fluctuation, for a C_3/C_4 splitter, is negligible compared with the magnitude of the controlled difference. The fluctuation in the bottoms product purity resulting from a change in unit capacity of 60 to 100% is expected to be lower than 0.05% mol C_3.

The simple control scheme of using a temperature difference between the stripping and the rectifying sections allows for very steady and reliable control of C_3/C_4 splitters. This method could be extended to other systems in which a mixture of two similar components are split in a column into relatively pure products.

LITERATURE CITED

1 Fauth, G. F. and Shinskey, F. G., "Advanced Control of Distillation Columns," *Chemical Engineering Progress*, Vol. 71, No. 6, June 1975, pp. 49-54.
2 Rademaker, O. and Rynsdays, J. E., "Dynamics and Control of Continuous Distillation Columns," Fifth World Petroleum Congress, June 1959.
3 Chao, K. C. and Seader, J. D., "A General Correlation of Vapor-Liquid Equilibria in Hydrocarbon Mixtures," *AIChE Journal*, Vol. 7, No. 4, December 1961, pp. 598-605.
4 American Petroleum Institute, "Technical Data Book—Petroleum Refining."
5 Harriott, Peter, *Process Control*, McGraw-Hill Book Co., 1964.
6 Boyd, D. M., "Fractionation Column Control," *Chemical Engineering Progress*, Vol. 71, No. 6, June 1975, pp. 55-60.

Consider Reactor Control Lags

Dynamic lags caused by temperature measurement and heat transfer give control problems with open-loop unstable reactors

William L. Luyben and **Marko Melcic***,
Lehigh University, Bethlehem, Pa.

IN AN industrial reactor, the control lags caused by temperature measurement and heat transfer are often significant. The effects of these lags are identified in this study of open-loop unstable reactors. The results suggest a way to tune controllers for optimum gain and an acceptable control stability.

These types of reactors have been the subject of many studies. Ellingsten and Cealgske[1] used root locus methods to examine the effects of various types of controllers on closed-loop stability. Koppel[2] studied sampled-data control of a first-order open-loop unstable process. Luyben[3] extended this to include second-order open-loop unstable processes. Later, Luyben[4] explored the effect of reaction rate on open-loop stability.

These authors have considered the case where temperature measurement lags and heat-removal lags are negligible. In a real industrial-scale reactor, the dynamics of these parts of the control loop are often quite significant. The purpose of this communication is to show quantitatively the effect of these lags on the stability of an open-loop unstable reactor.

System. The process is assumed to have a simple first-order open-loop transfer function, relating heat removal rate Q to reactor temperature T with a positive pole at $+1/\tau$.

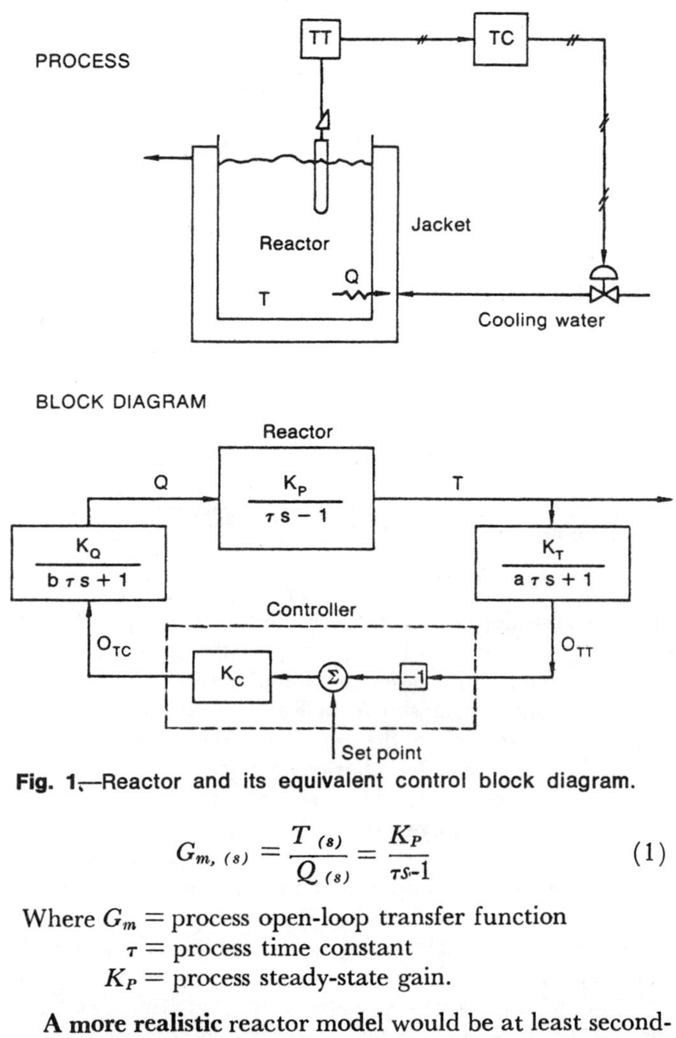

Fig. 1—Reactor and its equivalent control block diagram.

$$G_{m,(s)} = \frac{T_{(s)}}{Q_{(s)}} = \frac{K_P}{\tau s - 1} \qquad (1)$$

Where G_m = process open-loop transfer function
τ = process time constant
K_P = process steady-state gain.

A more realistic reactor model would be at least second-order if changes in reactor composition as well as temperature were considered. This more complex model was not used in this study so that the effects of the additional lags could be illustrated more easily using the simplest possible reactor model.

* Present address: Pliva (Pharmaceutical & Chemical Works Research Institute) Zagreb, Yugoslavia.

The temperature measurement and heat-removal lags are assumed to be first order.

$$\frac{O_{TT,(s)}}{T_{(s)}} = \frac{K_T}{a\tau s + 1} \quad (2)$$

$$\frac{Q_{(s)}}{O_{TC,(s)}} = \frac{K_Q}{b\tau s + 1} \quad (3)$$

where O_{TT} = Temperature transmitter output signal, % of control signal range

K_T = Temperature transmitter gain, 100% of control signal range/transmitter span

O_{TC} = Temperature controller output signal, % of control signal range

K_Q = Heat-removal system gain, maximum heat-removal rate/100% of control signal range

a = Ratio of the temperature-measurement lag time constant to the process time constant

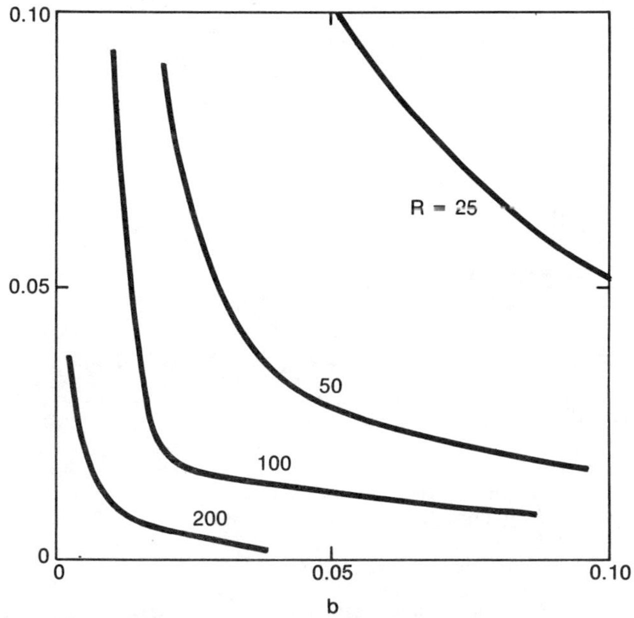

Fig. 2—Stability ratio, *R*, decreases for increase in *a* or *b*.

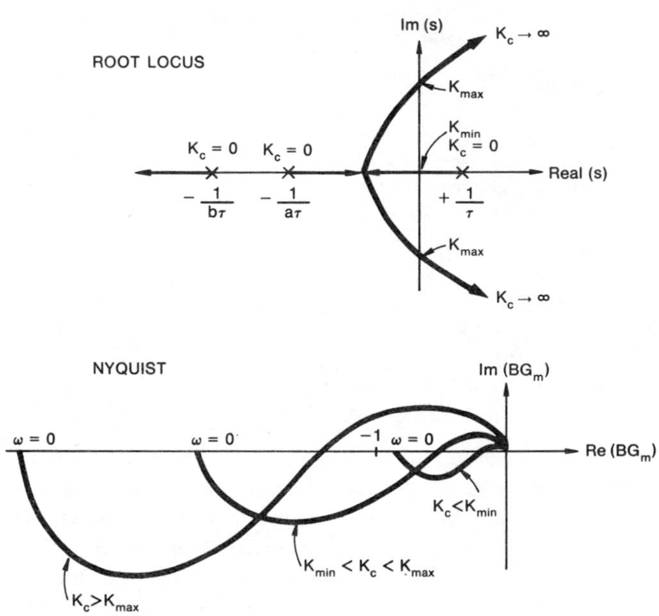

Fig. 3—Two ways to depict controller gain.

b = Ratio of the heat-removal lag time constant to the process time constant.

The closed-loop characteristic equation of the system with a proportional-only feedback controller ($B_{(s)} = K_c$) is:

$$1 + B_{(s)}\,G_{m,(s)} = 0 \quad (4)$$

$$1 + K_C\left(\frac{K_P}{\tau s - 1}\right)\left(\frac{K_T}{a\tau s + 1}\right)\left(\frac{K_Q}{b\tau s + 1}\right) = 0$$

$$(ab\tau^3)s^3 + [(a+b)\tau^2 - (ab)\tau^2]s^2 + [\tau - \tau(a+b)]s + (K_P K_C K_T K_Q - 1) = 0 \quad (5)$$

Maximum and minimum values of controller gain K_c can be found by substituting $i\omega$ for s in Equation 5 and solving for the resulting two values of gain.

$$K_{\min} = \frac{1}{K_P K_T K_Q} \quad (6)$$

$$K_{\max} = [1 + (1 - a - b)\,(\frac{a+b}{ab} - 1)]/K_P K_T K_Q \quad (7)$$

The ratio of the maximum to minimum gain is defined as the "stability ratio," R:

$$R = 1 + (1 - a - b)\,(\frac{a+b}{ab} - 1) \quad (8)$$

From Equation 5 it can be seen that a proportional-only feedback controller cannot stabilize this open-loop unstable system if:

$$a + b \geq 1 \quad (9)$$

The value of a and b must be a fairly small fraction for the system time constant to be stable. This qualitatively shows the importance of minimizing heat removal lags and temperature-measurement lags in reactor control systems.

The stability ratio R varies with a and b as shown in Fig. 2, where R decreases as either a or b increases.

Root locus and Nyquist plots of the system are shown in Fig. 3 identifying the occurrence of maximum and minimum gains. Note the Nyquist plot must encircle the point $(-1, 0)$, once in a negative direction—counterclockwise—since there is one pole in the right-half of the s-plane. The general Nyquist stability criterion is

$$N = Z - P \quad (10)$$

where N = Number of clockwise encirclements of point $(-1, 0)$

Z = Number of roots (zeros) of the closed-loop characteristic equation that lie in the right-half of the s-plane

P = Number of poles of the closed-loop characteristic equation that lie in the right-half of the s-plane.

Then Z must be zero for closed-loop stability. Since P is one (there is a pole at $s = +1/\tau$), N must be negative one.

In theory any system can be stabilized by a proportional controller if the stability ratio R is greater than one. In practice stability ratios of approximately 15 to 20 are

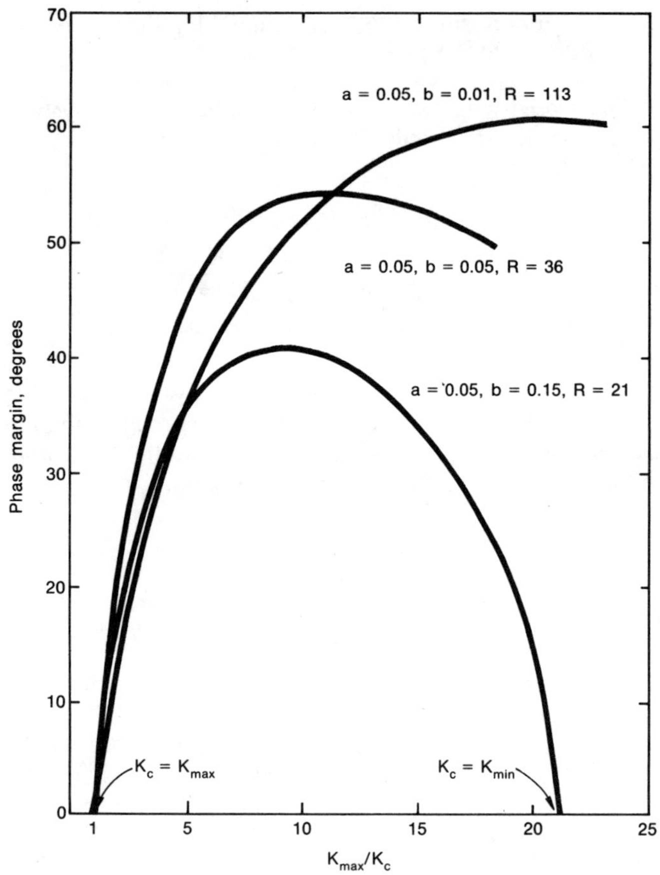

Fig. 4—Phase margins for the stable control region.

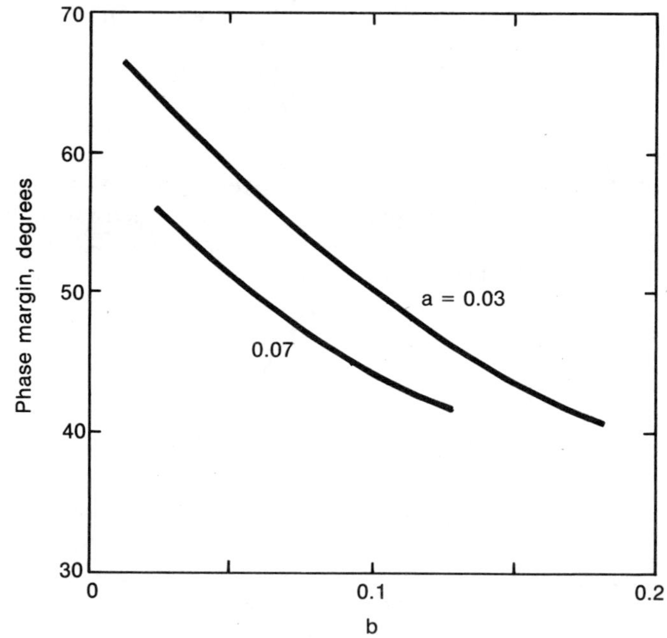

Fig. 5—Maximum phase margins for optimum gains.

probably about as low as one can go. Fig. 2 shows that these values of R occur when the sum of the two lags is about 20% of the negative process time constant.

The tuning of the controller, even when R is in the stable region, is not a trivial problem. Fig. 4 shows the phase margins, for various values of a and b, as controller gain is varied over the stable region. At both ends (where $K_c = K_{min}$ and where $K_c = K_{max}$) the phase margin is zero. Between these values, there is an optimum gain setting which maximizes phase margin for the system.

Notice that this optimum gain is 1/10th to 1/20th of the maximum or ultimate gain for the a and b values shown. The normal Ziegler-Nichols setting of 1/2 the maximum gain gives very low phase margins.

Finally, Fig. 5 shows the maximum phase margins obtainable, using the optimum gains, for various values of a and b. Stability decreases as a and b increase.

NOMENCLATURE

- a Ratio of temperature measurement lag to process time constant
- b Ratio of heat-removal lag to process time constant
- B Feedback controller transfer function
- G_m Process open-loop transfer function
- Im Imaginary part of a complex parameter
- K_c Feedback controller gain
- K_{max} Maximum feedback controller gain
- K_{min} Minimum feedback controller gain
- K_P Process steady-state gain
- K_Q Heat-removal system gain
- K_T Temperature transmitter gain
- N Number of clockwise encirclements of point $(-1, 0)$ made by $B_{(i\omega)} G_{m,(i\omega)}$
- O_{TC} Temperature controller output signal
- O_{TT} Temperature transmitter output signal
- P Number of poles
- Q Heat-removal rate
- R Ratio of maximum to minimum stable feedback controller gains
- Re Real part of a complex parameter
- s Laplace transform variable
- T Reactor temperature
- Z Number of zeros
- τ Process time constant
- ω Frequency

LITERATURE CITED

[1] Ellingsen, W. R. and Ceaglske, N. H., *AIChE J.*, Vol. 5, 1959, p. 30.
[2] Koppel, L. B., *Ind. Eng. Chem. Fund.*, Vol. 5, 1966, p. 396.
[3] Luyben, W. L., *Ind. Eng. Chem. Fund.*, Vol. 11, 1972, p. 557.
[4] Luyben, W. L., *AIChE J.*, Vol. 20, 1974, p. 175.

P D Control Improves Reactor Stability

Controllability of exothermic reactors decreases as reactor size increases. Presented is an analysis of the controllability of several reactors indicating proportional plus derivative control is the best solution to this problem

T. F. Cheung and **W. L. Luyben,** Department of Chemical Engineering, Lehigh University, Bethlehem, Pa.

Devia and Luyben (1977) have quantitatively shown that controllability of an exothermic, perfectly mixed, chemical reactor (Fig. 1) decreases when reactor size increases.[1] This presents a serious problem in scaling up chemical reactors. One way to avoid the problem is by increasing heat removal capacity of the reactor. This may be done by increasing heat transfer area or the overall heat transfer coefficient. However, these process modifications are usually expensive and sometimes impractical. An alternative is to improve the control system.

Ellingsen and Ceaglske (1959), in a pioneering paper,[2] explored the effects of changing controller modes on reactor stability.

In this article, a proportional-derivative (PD) controller is studied for several reactor sizes. Improvements on the control characteristics are shown, and optimum tuning of the PD controller is examined.

Proportional-only control. The reactor system studied by Devia and Luyben had an open-loop transfer function of

$$G_{M(s)} = \frac{T_{(s)}}{F_{J(s)}} = \frac{a_{26}\, a_{37}\, (s - a_{11})}{s^3 + B_2 s^2 + B_1 s + B_0}$$

Table 1 gives the definitions of the coefficients and Table 2 shows the assumptions and base case conditions.

This system is open-loop unstable. After adding a temperature measurement lag and a proportional-only (P-only) controller ($B_{(s)} = K_c$), the total open-loop transfer function becomes

$$B_{(s)}\, G_{M(s)} = K_c \left(\frac{1}{\tau s + 1}\right) \left(\frac{a_{26}\, a_{37}\, (s - a_{11})}{s^3 + B_2 s^2 + B_1 s + B_0}\right)$$

This system will be stable within a range of controller gains. Maximum and minimum gains (K_{max} and K_{min}) of the stable range can be determined by applying the stability criteria in the root locus plot or the Nyquist plot. Luyben and Melcic (1977) have shown that systems with stability ratios (K_{max}/K_{min}) less than about 15 are probably not controllable in plant type conditions.[3] They proposed that this ratio (R) could be used as a measure of controllability.

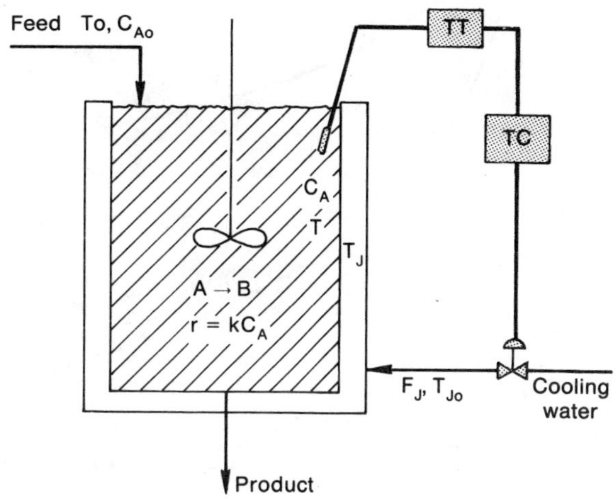

Fig. 1—Temperature controlled reactor.

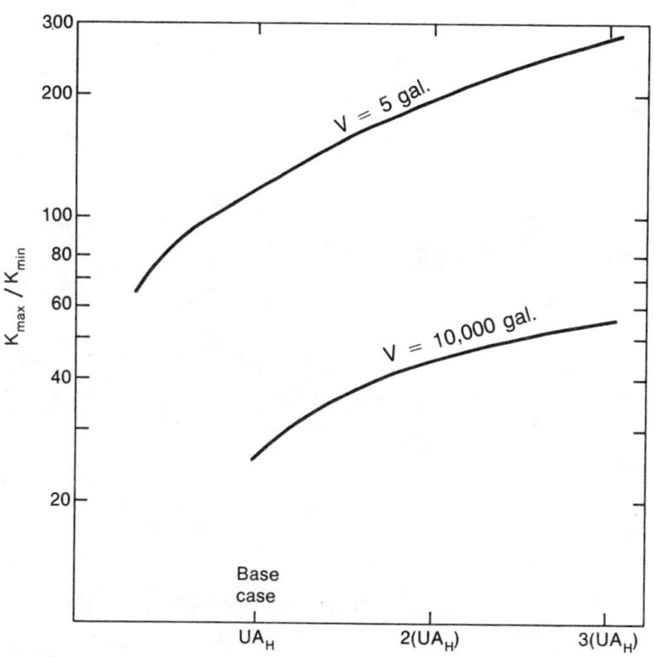

Fig. 2—Controllability improvement through increased heat transfer (P-only control).

Devia and Luyben have shown that with a P-only controlled reactor, R decreases as reactor size increases. This is due to the fact that heat removal capacity of a jacketed reactor does not increase proportionately with increasing

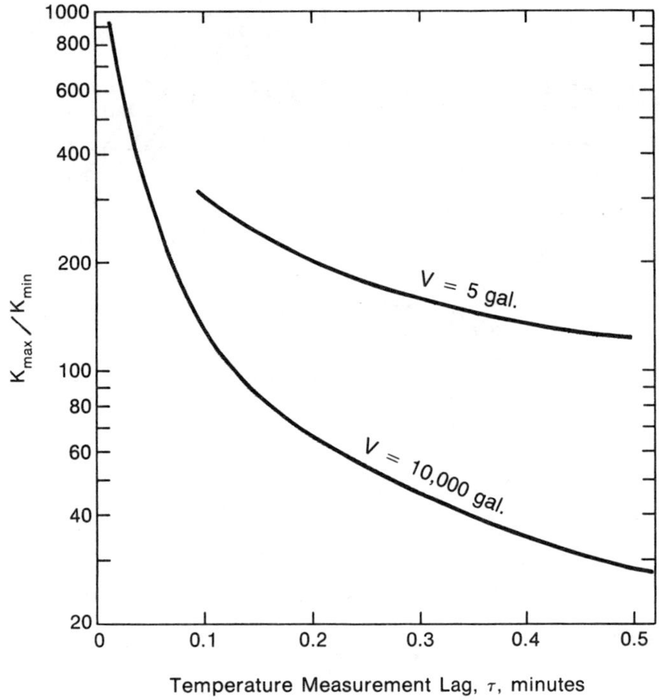

Fig. 3—Controllability improvement through reduced transmission lag (P-only control).

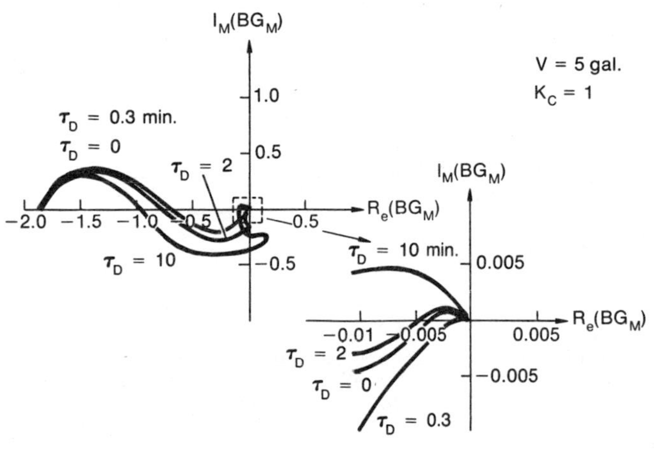

Fig. 4—Nyquist plot of PD controlled reactor.

TABLE 1—Coefficients in transfer function

$a_{11} = -F/V - \bar{k}$

$a_{12} = -(\bar{C}_A\, E\, \bar{k})/(R\, \bar{T}^2)$

$a_{13} = F/V$

$a_{15} = (\bar{C}_{A_o} - \bar{C}_A)/V$

$a_{21} = -(\lambda\, \bar{k})/(\rho\, C_p)$

$a_{22} = -(\lambda\, \bar{k}\, E\, C_A)/(\rho\, C_p\, R\, \bar{T}^2) - F/V - (U\, A_H)/(\rho\, V\, C_p)$

$a_{24} = F/V$

$a_{25} = (\bar{T}_o - \bar{T})/V$

$a_{26} = (U\, A_H)/(\rho\, V\, C_p)$

$a_{32} = (U\, A_H)/(\rho_J\, V_J\, C_J)$

$a_{36} = -(\bar{F}_J/V_J) - (U\, A_H)\, (\rho_J\, V_J\, C_J)$

$a_{37} = (T_{J_o} - \bar{T}_J)/V_J$

$B_2 = -a_{11} - a_{22} - a_{36}$

$B_1 = a_{11}\, a_{22} + a_{11}\, a_{36} + a_{22}\, a_{36} - a_{12}\, a_{21} - a_{26}\, a_{32}$

$B_0 = a_{12}\, a_{21}\, a_{36} - a_{11}\, a_{22}\, a_{36} + a_{11}\, a_{26}\, a_{32}$

TABLE 2

Assumptions

1) Reaction is first order.
2) Both the reactor liquid and the jacket cooling water are perfectly mixed.
3) The reactor is a vertical cylinder with height to diameter ratio of two.
4) All physical properties, the overall heat transfer coefficient, and the reactor and jacket hold-up times are held constant.
5) The temperature transmitter span is 100°F. The control valve has linear trim and constant pressure drop and a span of twice the steady-state flow.

Reactor parameters	
Reactor hold-up time, V/F	1.2 hr.
Jacket hold-up time, V_J/F_J	0.077 hr.
Overall heat transfer coefficient, U	150 Btu/hr.-°F-ft².
Heat capacity of feed and product, C_p	0.75 Btu/lb.-°F
Heat capacity of cooling water, C_J	1.0 Btu/lb.-°F
Density of feed and product, ρ	50 lb./ft.³
Density of cooling water, ρ_J	62.3 lb./ft.³
Inlet cooling water temperature, T_{J_o}	70°F
Temperature measurement lag, τ	30 sec.
Feed concentration, C_{A_o}	0.5 moles A/ft.³
Feed temperature, T_o	70°F
Base case conditions	
Steady-state reaction temperature, T	140°F
Preexponential factor, α	7.08×10^{10} hr.$^{-1}$
Activation energy, E	30,000 Btu/mole
Heat of reaction, λ	−30,000 Btu/mole
Steady-state reaction concentration, C_A	0.245 moles A/ft.³
Specific reaction rate, k	0.8672 hr.$^{-1}$

reactor volume, if the height to diameter ratio is kept constant. The problem may be rectified by improving heat transfer. For example, the reactor may be redesigned and cooling coils or external heat exchanger may be installed to provide more heat transfer surface. Also, agitation may be increased to raise the overall heat transfer coefficient. Fig. 2 shows how much controllability can be improved through these means. However, besides being expensive, these process modifications are sometimes impractical due to physical limitations and process restrictions.

Another potential route to improvement lies in reducing the temperature measurement lag (τ). Fig. 3 shows the effect this has on R. Controllability shows marked improvement when τ is reduced, especially for the 10,000-gallon reactor. However, temperature measurement lags often cannot be reduced because of severer reactor environments and fouling problems.

Proportional-derivative control. An alternative to changing the process or the sensor is to change the controller. Control can be improved by adding a derivative term.

The PD controller has a transfer function:

$$B_{(s)} = K_c \left(\frac{\tau_D\, s + 1}{(\tau_D/\alpha_D)\, s + 1} \right)$$

The normal rule of thumb for tuning derivative action in a temperature controller is to set τ_D equal to the temperature measurement lag. This heuristic will be shown not to apply to open-loop unstable reactors.

In this study, $\alpha_D = 20$ is used unless otherwise stated. With the PD controller, the total open-loop transfer function is

$$B_{(s)}\, G_{M(s)} = K_c \left(\frac{\tau_D\, s + 1}{(\tau_D/\alpha_D\, s + 1)} \right) \left(\frac{1}{\tau\, s + 1} \right) \left(\frac{a_{26}\, a_{37}\, (s - a_{11})}{s^3 + B_2\, s^2 + B_1\, s + B_0} \right)$$

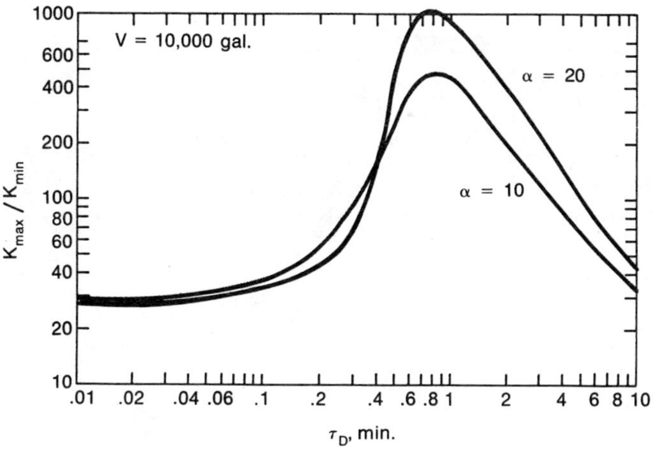

Fig. 5—K_{max}/K_{min} of PD controlled reactor.

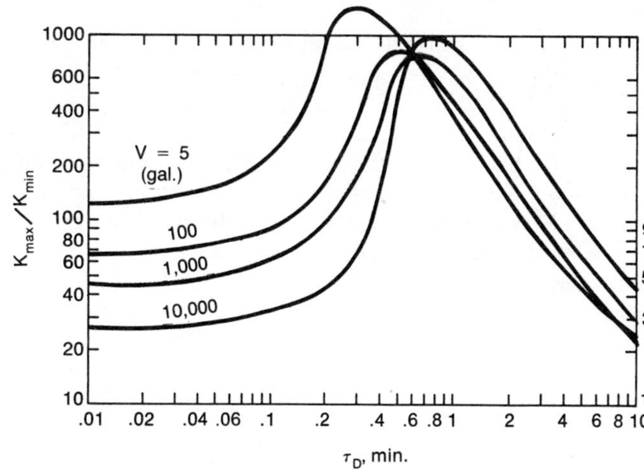

Fig. 6—K_{max}/K_{min} of PD controlled reactor.

For the open-loop unstable system to be closed-loop stable, the Nyquist plot must have one negative (counterclockwise) encirclement of the $(-1, 0)$ point in the complex $B\,G_M$ plane. Therefore, using this stability criterion, K_{max} and K_{min} can be calculated.

Fig. 4 shows the Nyquist plot of a five-gallon reactor with $K_c = 1$ and different values of derivative time constant (τ_D). K_{max} and K_{min} can be calculated from the intersections of the $B\,G_M$ curve and the negative real axis.

As τ_D is increased from 0, K_{min} is not affected much but K_{max} is increased. However, as τ_D is increased further, K_{min} begins to decrease and the $B\,G_M$ contour is moved away from the $(-1, 0)$ point in the low frequency part of the contour. However, the higher frequency part of the contour moves towards the $(-1, 0)$ point. Clearly there is an optimum τ_D value for maximum stability.

Fig. 5 shows the change in the K_{max}/K_{min} ratio for a 10,000-gallon reactor for various values of τ_D. It shows that R increases to a maximum before falling off. The optimum τ_D for this system, based on maximum K_{max}/K_{min} ratio, is about 0.75 minutes. Since a temperature measurement lag of 0.5 minutes was used, these results would tend to support the commonly employed tuning technique of setting τ_D equal to the temperature measurement lag. However, maximizing the K_{max}/K_{min} ratio does not necessarily give optimum controller tuning.

Fig. 5 shows that at high τ_D (>10 min.), the system is less stable than the P-only control case $(\tau_D = 0)$. This shows the importance of correctly specifying the amount of derivative action. Also shown in the same graph is the effect of changing α_D. Decreasing α_D seems to lower the R peak but has little influence on the optimum τ_D.

Fig. 6 gives the R vs. τ_D curves for reactors with sizes ranging from five to 10,000 gallons. For all size reactors, controllability, as measured by R, peaks at τ_D's of 0.3 to 0.75 minutes. All these optimum τ_D's are very close to the 0.5 minutes measurement lag. These results suggest that derivative action makes the process more stable by compensating for the measurement lag.

It should be noted that although R can be used as a measure of reactor stability, it cannot be used as a tool for tuning controllers. The "optimum" τ_D given by the R plot is not necessarily the optimum in the sense of system response.

The most general controller tuning criterion is to achieve a reasonable maximum closed-loop log modulus (LM) of the servo function $(B\,G_M/(1 + B\,G_M))$. Other

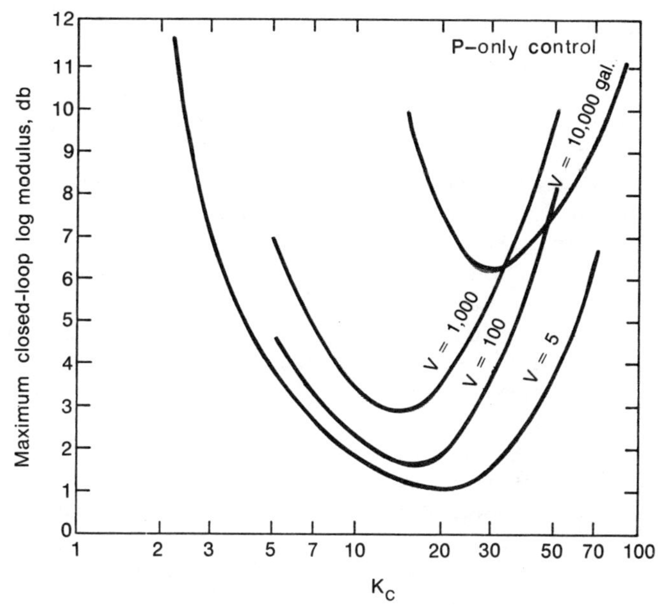

Fig. 7—Closed loop response of P-only controlled reactor.

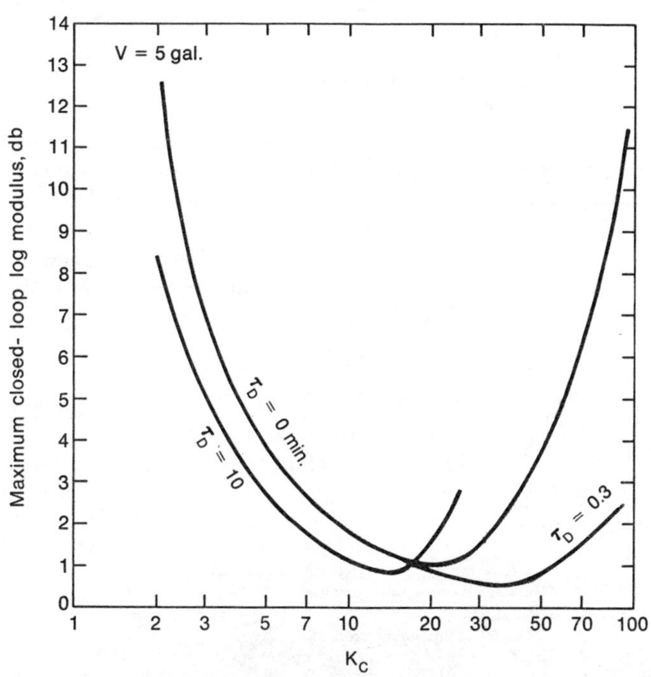

Fig. 8—Closed loop response of PD controlled reactor.

commonly used criteria, phase margin (PM) and gain margin (GM) of the open-loop function $B\,G_M$, are less reliable for non-standard systems.

Fig. 7 is an LM vs. controller gain plot for P-only controlled reactors. It shows that the minimum of the LM curve increases with increasing reactor size. An LM of +2 db is usually considered the limit of system response specification. Therefore, responses of the P-only controlled reactors larger than 100 gallons are undesirable, even when controller gain is set at the optimum. Note also that the curves of the larger reactors are narrower, due to the smaller range between K_{max} and K_{min}. As a result, if the controller setting drifts from optimum, system response deteriorates very rapidly.

Fig. 8 gives the LM vs. K_c plot of a five-gallon reactor with PD control. In this example, a small amount of derivative action ($\tau_D = 0.3$ min.) seems to broaden and lower the LM curve. Therefore, the system is more stable over a wider range of K_c. However, too much derivative action ($\tau_D = 10$ min.) has the opposite effect.

Benefits of derivative action can be seen more clearly in a larger reactor, e.g., the 10,000 gallons (Fig. 9). Increasing τ_D seems to lower and broaden the LM curve. At $\tau_D = 2$ minutes the curve near the minimum point becomes very flat. Further increase of τ_D to 5 minutes continues to lower the curve but now the curve becomes narrower. Increasing τ_D further would raise and narrow the LM curve. Therefore, with this plot, the PD controller can be tuned to a maximum closed-loop log modulus specification. For the 10,000-gallon reactor shown, a K_c of 130 and τ_D of 2 minutes would give an LM of +2.2 db. These are acceptable settings on the PD controller. At this value of τ_D, K_{max} is 6850.

If P-only control were used, the optimum K_e is 30 and K_{max} is 239. Gain calculated from the Ziegler-Nichols method is considerably higher at 119.5. PD control not only raises K_{max} but also allows a higher optimum gain. This helps reduce steady-state offset and gives faster speed of response.

With τ_D of 2 minutes, the range of K_e that gives LM less than 3 db is from 30 to 600. For τ_D of 25 minutes, the K_c range giving the same LM is 18 to 95. Therefore, too much derivative action reduces the range of K_c that gives desirable system response.

It is interesting to note that the optimum PD controller settings suggested for the 10,000-gallon reactor are very close to the settings of $K_c = 141$ and $\tau_D = 1.15$ minutes calculated from the Ziegler-Nichols method for PID controllers. However, the Ziegler-Nichols method cannot be used with confidence to tune PD controllers, as shown by the five-gallon reactor example. Calculated Ziegler-Nichols settings for the five-gallon reactor are $K_c = 104$ and $\tau_D =$ 0.3 minutes. These settings are not close to the optimum given by LM tuning.

Fig. 10 shows the effects of K_c and τ_D on phase margin. Optimum gains given by PM tuning are lower than those by LM. Since LM is more reliable, PM tuning should be used with caution.

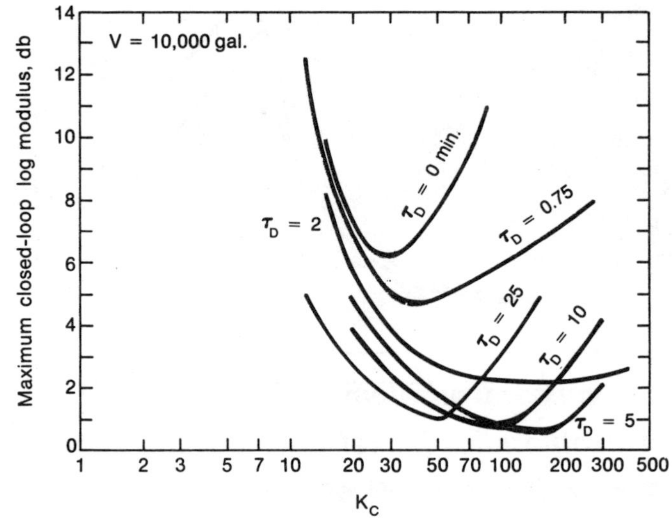

Fig. 9—Closed loop response of PD controlled reactor.

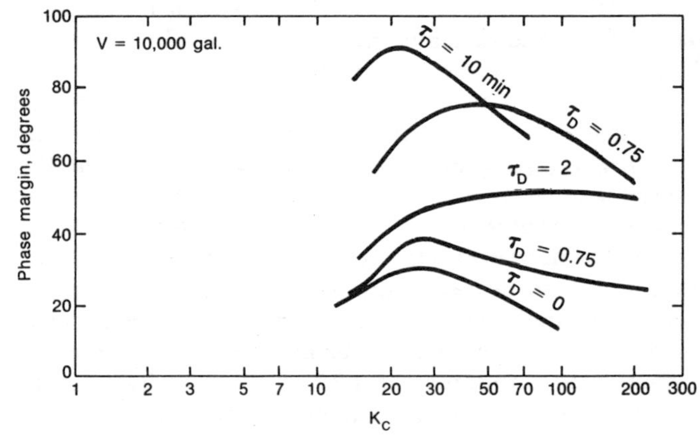

Fig. 10—Closed loop response of PD controlled reactor.

NOMENCLATURE

- a_{ij} Constant coefficients in linearized equations
- A Chemical component, reactant
- A_H Heat transfer area, ft.^{2}
- B Chemical component, product
- B_i Coefficient in denominator of open-loop transfer function
- $B_{(s)}$ Feedback controller transfer function
- C_A Reactant concentration in reactor and product, moles/ft.^{3}
- C_{A_0} Initial concentration of A in feed, moles/ft.^{3}
- C_J Heat capacity of cooling medium, Btu/lb.-°F
- C_p Heat capacity of reactants, Btu/lb.-°F
- E Activation energy, Btu/mole
- F Feed rate, ft.^{3}/hr.
- F_J Cooling water rate, ft.^{3}/hr.
- $G_{M(s)}$ Process open-loop transfer function
- k Specific reaction rate, hr.$^{-1}$
- Q Heat transfer rate, Btu/hr.
- R Gas constant, 1.987 Btu/mole-°F
- s Laplace transform variable
- t Time
- T Temperature of reactor, °R
- T_J Jacket cooling water temperature, °R
- T_{J_0} Inlet cooling water temperature, °R
- T_0 Feed temperature, °R
- U Overall heat transfer coefficient, Btu/hr-°F-ft.^{2}
- V Reactor volume, ft.^{3}
- V_J Jacket volume, ft.^{3}
- α Pre-exponential factor, hr.$^{-1}$
- α_D Measure of derivative action
- ρ Density of reactor liquid, lb./ft.^{3}
- ρ_J Density of cooling water, lb./ft.^{3}
- τ Temperature transmission time constant, min.
- τ_D Derivative time constant, min.
- ω Frequency, radians/min.

ACKNOWLEDGMENT

Based on material originally presented at the *JACC Meeting*, Philadelphia, October 1978.

LITERATURE CITED

[1] Devia, N. and Luyben, W. L., "Effect of Scale-Up on Reactor Stability," *Hydrocarbon Processing*, Vol. 57, No. 6, 1978, p. 119.
[2] Ellingsen, W. and Ceaglske, N., "Control System Design for a Chemical Process by the Root Locus Method," AIChE J., 5, p. 30, 1959.
[3] Luyben, W. L. and Melcic, M., "Effect of Temperature Measurement and Heat Removal Lags on the Stability of Open-Loop Unstable Reactors," *Hydrocarbon Processing*, Vol. 57, No. 3, p. 115.

Reactors: Size Versus Stability

A significant problem of scale-up for an exothermic, jacket-cooled, chemical reactor is the effect of reduced heat removal on dynamic stability

N. Devia and **William L. Luyben,**
Lehigh University, Bethlehem, Pa.

FOR EXOTHERMIC REACTIONS in jacket-cooled reactors, the region of conditional closed-loop stability is decreased as reactor size increases. This report explores quantitative changes in dynamic stability and controllability of exothermic perfectly mixed chemical reactors. Controllability is measured as a ratio of the maximum to minimum gain for the feedback controller. The reactors have volumes in the range of 5 to 10,000 gallons, while reactor hold-up time is constant at 1.2 hours.

Scale-up of reactor data from one size to a larger size must consider these narrower ranges of control stability. The ranges of stability are related to reactor dimensions as follows:

Reactor surface area is related directly to reactor diameter, while reactor volume is related to the square of the diameter. Increasing a reactor size causes a reduction in the ratio of area to volume. Since heat transfer for a jacket-cooled reactor is related to reactor surface area, scaling-up a reactor size gives a relative reduction in heat-transfer capacity per unit volume.

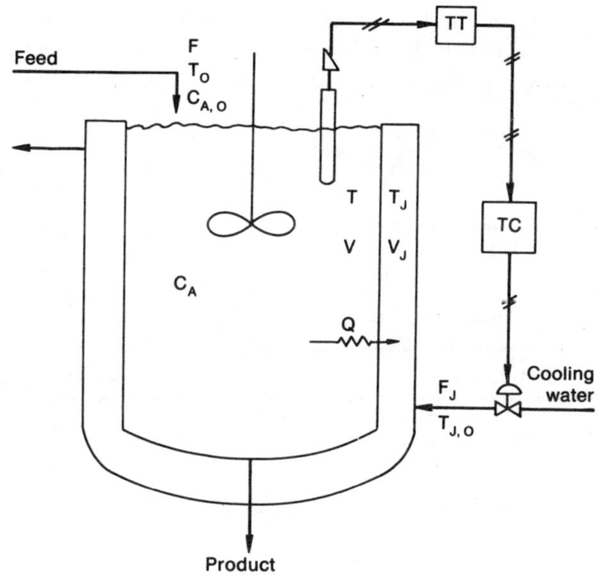

Fig. 1—The liquid mixture is assumed perfectly mixed.

TABLE 1—Reactor parameter values

Reactor hold-up time, V/F	1.2 hr.
Jacket hold-up time, V_J/F_J	0.077 hr.
Overall heat transfer coefficient, U	150 Btu/(hr.) (sq. ft.) (° F)
Heat capacity of feed and products, C_P	0.75 Btu/(lb.) (° F)
Heat capacity of jacket water, C_J	1.0 Btu/(lb.) (° F)
Density of feed and products, ρ	50 lb/cu. ft.
Density of jacket water, ρ_J	62.3 lb./cu. ft.
Make up cooling water temperature, $T_{J,O}$	70° F
Temperature measurement lag	30 sec.
Feed concentration, $C_{A,O}$	0.50 moles A/cu. ft.
Feed temperature, T_O	70° F
Base case conditions	
Steady state reaction temperature, T	140° F
Pre-exponential factor, α	7.08 x 10^{10} (1/hr.)
Activation energy, E	30,000 Btu/mole
Heat of reaction, λ	-30,000 Btu/mole
Steady state reactant concentration, C_A	0.245 moles A/cu. ft.
Specific reaction rate	0.8672 (1/hr.)

An example. A single, continuous, perfectly-mixed reactor is considered as depicted in Fig. 1 in which an exothermic, irreversible reaction takes place in the liquid phase as symbolized by

$$A \rightarrow B$$

The reaction is assumed to be first-order in the concentration of reactant, C_A.

Feed and product streams are at a constant rate. Reactor volume is assumed constant, as are all physical properties. The metal walls of the reactor and jacket and the agitator are neglected.

Heat of reaction is removed by heat transfer to cooling water in a jacket surrounding the reactor. Perfect mixing is assumed in the reactor and in the jacket. The reactor is a vertical cylindrical vessel with a height to diameter ratio of two. Reactor hold-up time (V/F) and jacket hold-up time (V_J/F_J) are held constant as reactor size is increased.

The overall heat transfer coefficient between the reactor liquid at temperature T and the jacket at temperature T_J is assumed constant at 150 Btu/(hr.) (sq. ft.) (°F).

A temperature measurement lag of 30 seconds is assumed for all cases. Table 1 gives parameter values and base case conditions. The temperature of the make-up cooling water to the jacket is 70° F.

A reactor temperature transmitter span of 100° F is

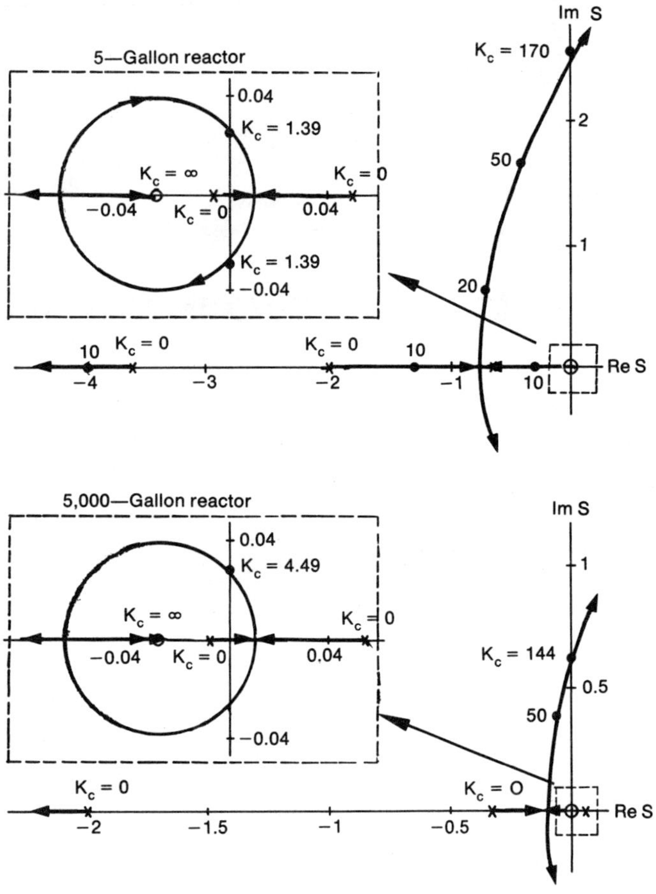

Fig. 2—Root locus plots for two reactor sizes.

used. Jacket water make-up control valve is sized to give twice the normal flow rate when wide open. Linear trim and constant control valve pressure drop are assumed. Feedback controller gains are reported as dimensionless (psi/psi in pneumatics).

Steady-state conditions and controller gains for reactors from 5 to 10,000 gallons are shown in Table 2 for the same reaction.

Control stability is explored using the root locus and Nyquist plots of the 5 and 5,000-gallon reactors as shown in Figs. 2 and 3. The systems are both open-loop unstable (a positive root when $K_C = 0$). Therefore, a minimum value of feedback controller gain ($K_C = K_{min}$) is required to stabilize the system.

There is also a maximum value of controller gain ($K_C = K_{max}$) since the system is net third order. These maximum and minimum gains are clearly seen on the root locus plots. Associated equations are shown in a later section.

There must be one negative (i.e., counterclockwise) encirclement of the $(-1,0)$ point since there is one pole in the right-half of the s-plane.

The general Nyquist stability criterion is

$$N = Z - P$$

where

N = number of clockwise encirclements of $(-1,0)$ point

Z = number of roots (zeros) of the closed-loop characteristic equation that lie in the right-half of the s-plane

P = number of poles of the closed-loop characteristic equation that lie in the right-half of the s-plane.

For closed-loop stability, Z must be zero. Since P is one for an open-loop unstable process (there is a pole in the right-half of the s-plane), N must be negative one.

Thus, the conditional stablility of the system can be seen in both the frequency and the Laplace domains.

Maximum and minimum gains for various sized reactors are also shown in Table 2. Maximum gain decreases from 5 to 500 gallons, and then increases for larger reactors. Minimum gain increases slowly with reactor

TABLE 2—Design conditions for various sized reactors

Volume, gallons	5	50	500	1,000	2,000	5,000	10,000
Feed rate, lb/hr	27.8	278	2,780	5,560	11,120	27,800	55,600
Heat transfer rate, 10^6 Btu/hr	0.0028	0.028	0.280	0.560	1.120	2.800	5.600
Reactor height, ft	1.504	3.24	6.98	8.80	11.08	15.04	18.95
Reactor diameter, ft	0.752	1.62	3.49	4.40	5.54	7.52	9.476
Heat transfer area, sq. ft	3.99	18.56	86.15	136.8	217.1	400.0	634.8
Cooling water flow, gpm	0.086	0.934	11.58	26.22	62.9	240.0	998.4
Jacket temperature, °F	135.3	129.9	118.3	112.7	105.6	93.3	81.2
Controller gains							
Max	169.8	116.5	100.3	102.5	110.6	144.2	259.3
Min	1.37	1.52	1.95	2.25	2.76	4.41	9.65
Ratio	124	76.6	51.3	45.6	40.1	32.7	26.9

TABLE 3—Effect of kinetic parameters

Case	Base					Higher T					Higher λ					Higher E				
Reactor temp., T, °F	140					200					140					140				
Specific reaction rate, k, 1/min	0.0144					0.0144					0.0144					0.0144				
Heat of reaction, λ, Btu/mole	−30,000					−30,000					−40.000					−30,000				
Activation Energy, E, Btu/mole	30,000					30,000					30,000					40,000				
Reactor concentration, C_A, moles A/cu. ft	0.245					0.245					0.245					0.245				
	T_J, °F	F_J, gpm	K_{max}	K_{min}	Ratio	T_J, °F	F_J, gpm	K_{max}	K_{min}	Ratio	T_J, °F	F_J, gpm	K_{max}	K_{min}	Ratio	T_J, °F	F_J, gpm	K_{max}	K_{min}	Ratio
Reactor volume, gal.																				
5	135.3	0.086	170	1.34	123.8	197.4	0.024	654	2.37	276	133.0	0.13	95.1	1.45	65.6	135.3	0.086	165	2.54	64.9
50	129.9	0.93	116	1.52	76.7	194.4	0.25	361	2.43	148	124.8	1.54	70.6	1.70	41.4	129.9	0.93	111	2.80	39.7
500	118.3	11.6	100	1.95	51.3	188.0	2.6	228	2.58	88.3	107.4	22.6	69.3	2.66	26.1	118.3	11.6	92.5	3.59	25.8
1,000	112.7	26.2	102	2.25	45.6	184.9	5.4	206	2.67	77.3	98.9	58.5	77.5	3.55	21.8	112.7	26.2	92.6	4.13	22.4
2,000	105.6	62.9	111	2.76	40.0	181.0	11.1	189	2.77	68.3	88.2	186	103	5.92	17.4	105.6	62.9	97.1	5.07	19.1
5,000	93.3	240	144	4.41	32.7	174.2	29.7	175	2.98	58.7	.....	......	*	....	....	93.3	240	119	8.11	14.7
10,000	81.2	998	259	9.65	26.9	167.5	63.4	169	3.21	52.7	.....	......	*	....	....	81.2	998	198	17.9	11.0

* Unattainable

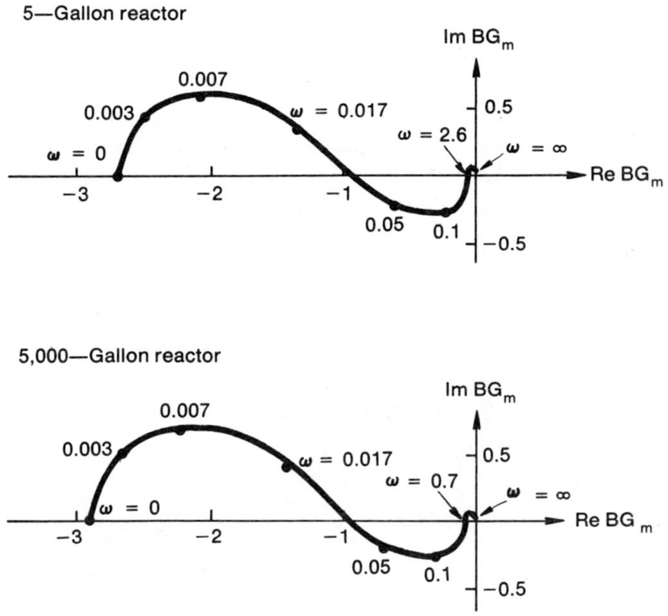

Fig. 3—Nyquist plots for the two reactor sizes.

size until about 2,000 gallons and then begins to increase sharply. The important parameter from the standpoint of controllability is the ratio of K_{max} to K_{min}. This ratio decreases as reactor size increases. The relative rate of this decrease is compared with other reactor parameters in Fig. 4 in which the 5-gallon reactor is the basis for comparison.

A small value of the ratio K_{max}/K_{min} signifies a narrow range of conditional stability. Such a system will be more difficult to control and more sensitive to changes in process parameters and conditions. This is shown in the root locus plots of Fig. 2. The path of the 5,000-gallon reactor is much closer to the unstable right-half of the *s*-plane than is the path of the 5-gallon reactor.

Luyben and Melcic have shown[1] that systems with stability ratios (K_{max}/K_{min}) lower than about 15 to 20 have phase margins less than 30 to 40 degrees at optimum controller settings. Therefore, reactors with ratios less than about 15 are probably not controllable in plant-type conditions.

Changing kinetic parameters. The effects of various parameters are shown in Table 3. The first group of data are for the base case. Subsequent groupings are for the following:

- Raising reactor operating temperature assuming reaction rate does not change. This is equivalent to assuming a lower temperature coolant is used. Stability is increased.
- Higher heat of reaction gives decreased stability. It is impossible to remove enough heat with 70° F cooling water for the 5,000-gallon and 10,000-gallon reactors.
- Increasing activation energy decreases stability. The 10,000-gallon reactor is probably not controllable because of its low stability ratio.

The effect of raising reactor operating temperature with reaction rate also increasing is shown in Table 4. The stability ratio decreases as temperature is raised from 140° F to 160° F. But then it begins to increase for higher temperatures. This occurs because the minimum gain decreases. In fact, the reactor becomes open-loop stable at 200° F, meaning there is no minimum gain. The increasing temperature raises the reaction rate and reduces reactant concentration which stabilizes the system as shown[2] by Luyben.

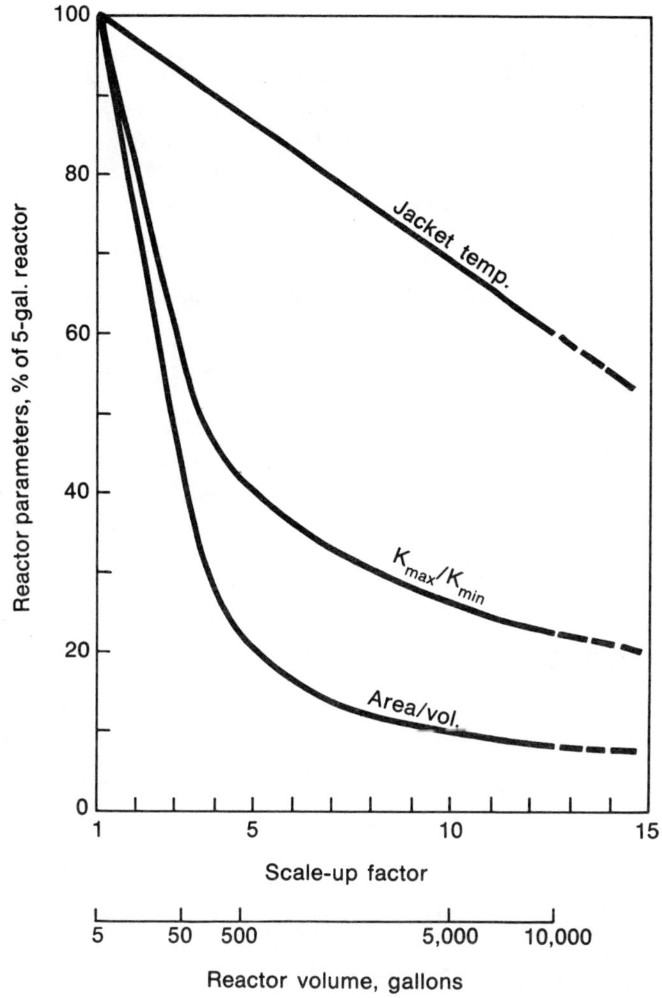

Fig. 4—Controllability decreases for larger reactors.

Equations. The nonlinear equations describing the open-loop system are:

$$\frac{dC_A}{dt} = \left(\frac{F}{V}\right) C_{A,O} - \left(\frac{F}{V}\right) C_A - C_A \, \alpha \, e^{-E/RT} \tag{1}$$

TABLE 4—Temperature effect

Reactor temp., °F	Parameters	Reactor volume, gallons 5	500	5,000
140	C_A, moles/cu. ft.	0.245	0.245	0.245
	k, 1/min.	0.0144	0.0144	0.0144
	T_J, °F	135.3	118.3	93.4
	F_J, gpm	0.086	11.6	240
	K_{max}	170	100	144
	K_{min}	1.34	1.95	4.41
	Ratio	123.8	51.3	32.7
160	C_A, moles/cu. ft.	0.150	0.150	0.150
	k, 1/min.	0.0325	0.0325	0.0325
	T_J, °F	153.4	129.2	93.8
	F_J, gpm	0.095	13.4	335
	K_{max}	112	66.7	102
	K_{min}	1.21	1.88	5.68
	Ratio	92.1	35.5	17.9
180	C_A, moles/cu. ft.	0.083	0.083	0.083
	k, 1/min.	0.0695	0.0695	0.0695
	T_J, °F	172.2	143.9	102.2
	F_J, gpm	0.091	12.6	290
	K_{max}	95.3	53.9	70.7
	K_{min}	0.65	1.09	3.65
	Ratio	147.3	49.6	19.4

$$\frac{dT}{dt} = \left(\frac{F}{V}\right) T_o - \left(\frac{F}{V}\right) T - \frac{\lambda C_A \alpha}{\rho C_p} e^{-E/RT} - \frac{U A_H}{\rho V C_p} (T - T_J) \quad (2)$$

$$\frac{dT_J}{dt} = \left(\frac{F_J}{V_J}\right) (T_{J,o} - T_J) + \frac{U A_H}{\rho_J C_J V_J} (T - T_J) \quad (3)$$

Linearization and conversion to perturbation variables gives:

$$\frac{dC_A}{dt} = a_{11} C_A + a_{12} T + a_{13} C_{A,O} + a_{15} F \quad (4)$$

$$\frac{dT}{dt} = a_{21} C_A + a_{22} T + a_{24} T_o + a_{25} F + a_{26} T_J \quad (5)$$

$$\frac{dT_J}{dt} = a_{32} T + a_{36} T_J + a_{37} F_J \quad (6)$$

where

$$a_{11} = -\frac{F}{V} - \bar{k} \quad (7)$$

$$a_{12} = -\frac{\bar{C}_A E \bar{k}}{R \bar{T}^2} \quad (8)$$

$$a_{13} = \frac{F}{V} \quad (9)$$

$$a_{15} = \frac{\bar{C}_{A,O} - \bar{C}_A}{V} \quad (10)$$

$$a_{21} = \frac{-\lambda \bar{k}}{\rho C_p} \quad (11)$$

$$a_{22} = \frac{-\lambda \bar{k} E \bar{C}_A}{\rho C_p R \bar{T}^2} - \frac{F}{V} - \frac{U A_H}{\rho V C_p} \quad (12)$$

$$a_{24} = \frac{F}{V} \quad (13)$$

$$a_{25} = \frac{\bar{T}_o - \bar{T}}{V} \quad (14)$$

$$a_{26} = \frac{U A_H}{V \rho C_p} \quad (15)$$

$$a_{32} = + \frac{U A_H}{\rho_J V_J C_J} \quad (16)$$

$$a_{36} = -\frac{\bar{F}_J}{V_J} - \frac{U A_H}{\rho_J V_J C_J} \quad (17)$$

$$a_{37} = \frac{T_J - \bar{T}_J}{V_J} \quad (18)$$

Equations 4, 5 and 6 can be combined to yield the transfer function between reactor temperature and cooling water flow rate.

$$\frac{T_{(s)}}{F_{J,(s)}} = \frac{a_{26} a_{37} (S - a_{11})}{S^3 + B_2 S^2 + B_1 S + B_0} \quad (19)$$

where

$$B_2 = -a_{11} - a_{22} - a_{36} \quad (20)$$

$$B_1 = a_{11} a_{22} + a_{11} a_{36} + a_{22} a_{36} - a_{12} a_{21} - a_{26} a_{32} \quad (21)$$

$$B_0 = a_{12} a_{21} a_{36} - a_{11} a_{22} a_{36} + a_{11} a_{26} a_{32} \quad (22)$$

A temperature measurement lag, τ, of 30 seconds is assumed. If a proportional-only feedback controller is used, the closed-loop characteristic equation of the system is: $1 + B_{(s)} G_{m,(s)} = 0$

$$1 + (K_c) \left(\frac{1}{\tau s + 1}\right) \left(\frac{a_{26} a_{37} (S - a_{11})}{S^3 + B_2 S^2 + B_1 S + B_0}\right) = 0 \quad (23)$$

$$(\tau) S^4 + (\tau B_2 + 1) S^3 + (\tau B_1 + B_2) S^2 + (\tau B_0 + B_1 + K_c a_{26} a_{37}) S + (B_o - K_c a_{26} a_{37} a_{11}) = 0 \quad (24)$$

The Nyquist plot of Fig. 3 gives the total open-loop system transfer function $B_{(s)} G_{m,(s)}$:

$$B_{(s)} G_{m,(s)} = K_c \left(\frac{1}{\tau s + 1}\right) \left(\frac{a_{26} a_{37} (S - a_{11})}{S^3 + B_2 S^2 + B_1 S + B_0}\right) \quad (25)$$

Summary. The dynamic stability and controllability of exothermic perfectly mixed chemical reactors are functions of the physical size of the reactor. The ratio of the maximum feedback controller gain to the minimum can be used as a measure of the controllability and is found to decrease as reactors are scaled up to larger sizes for constant volume to feed rate ratios.

NOMENCLATURE

- a_{ij} Constant coefficients in linearized equations.
- A Chemical component, reactant.
- A_H Heat transfer area, sq. ft.
- B Chemical component, product.
- B_i Coefficients in denominator of open-loop transfer function.
- $B_{(s)}$ Feedback controller transfer function.
- C_A Reactant concentration in reactor liquid, moles A/cu. ft.
- $C_{A,o}$ Concentration of reactant in feed, moles A/cu. ft.
- C_J Heat capacity of jacket water, Btu/(lb.)(°F).
- C_p Heat capacity of reactor liquid, Btu/(lb.)(°F).
- E Activation energy, Btu/mole.
- F Feed rate, cu. ft./hr.
- F_J Jacket cooling water rate, cu. ft./hr.
- $G_{m,(s)}$ Process open-loop transfer function.
- Im Imaginary part of a complex number.
- k Specific reaction rate, 1/hours.
- N Number of encirclements.
- P Number of poles.
- Q Heat transfer rate, Btu/hr.
- R Perfect gas constant, Btu/(mole)(°F).
- Re Real part of a complex number.
- S Laplace transform variable, 1/minutes.
- t Time, hours.
- T Temperature of reactor, °R.
- T_J Jacket cooling water temperature, °R.
- $T_{J,o}$ Inlet cooling water temperature, °R.
- T_o Feed temperature, °R.
- U Overall heat transfer coefficient between reactor liquid and jacket water, Btu/(hr.)(sq. ft.)(°F).
- V Reactor volume, cu. ft.
- V_J Jacket volume, cu. ft.
- Z Number of zeros.
- α Pre-exponential factor, 1/hours.
- ρ Density of reactor liquid, lb./cu. ft.
- ρ_J Density of jacket cooling water, lb./cu. ft.
- ω Frequency, radians/minutes.

ACKNOWLEDGMENT

Original presentation was before the American Institute of Chemical Engineers meeting, New York, Nov. 15, 1977.

LITERATURE CITED

[1] Luyben, William L. and Melcic, Marko, "Consider reactor control lags," HYDROCARBON PROCESSING, Vol. 57, No. 3, March 1978, pp. 115-117.

[2] Luyben, William L., "Effect of reaction rate on the open loop stability of of chemical reactors," *AIChE Journal*, Vol. 20, No. 1, January 1974, pp. 175-177.

Maximize Orifice Meter Performance

Presented is a computer program for more accurate orifice sizing, checking flowrate and differential for existing orifices and determining power losses for operating cost comparisons with other type meters

Hung Xuan Nguyen, Phillip Morris, Inc Richmond, Va.

ORIFICE METERS are still the most prevalent method of measuring flow in spite of significant developments in other types of flow meters. Successful application of an orifice meter requires accurate sizing of the orifice for a specific fluid, under desired operating conditions. Presented is a computer program, using time sharing, to calculate not only the required orifice diameter but also the flow rate for a known orifice or the pressure differential for checking the d/p cell setting for gases, steam and liquid. In addition, the permanent loss and power consumption of the meter are also computed. Furthermore, the recommended drain hole sizes and orifice plate thickness are also printed out.

Generally, the engineer notes the operating temperature, inside pipe diameter, rate of flow, desired pressure differential and properties of the fluid under flowing conditions. He then sizes the required orifice diameter by a

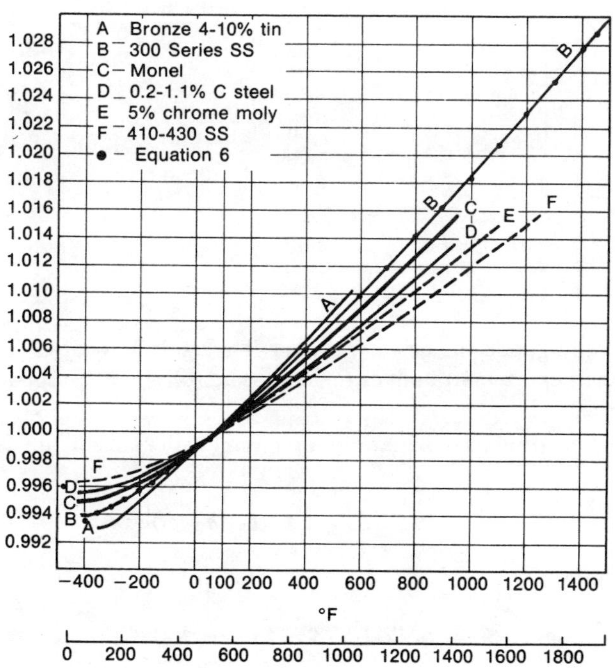

Fig. 1—Area factor, F_a, for the thermal expansion of primary elements.

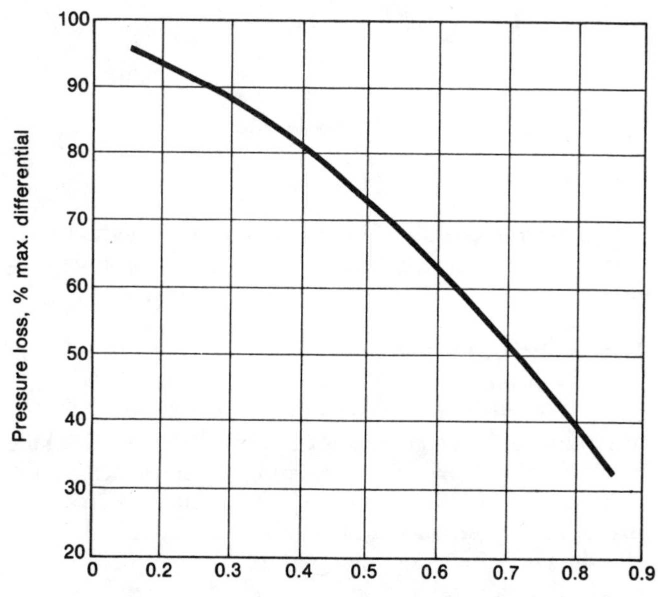

Fig. 2—Over-all pressure loss in orifice meter.

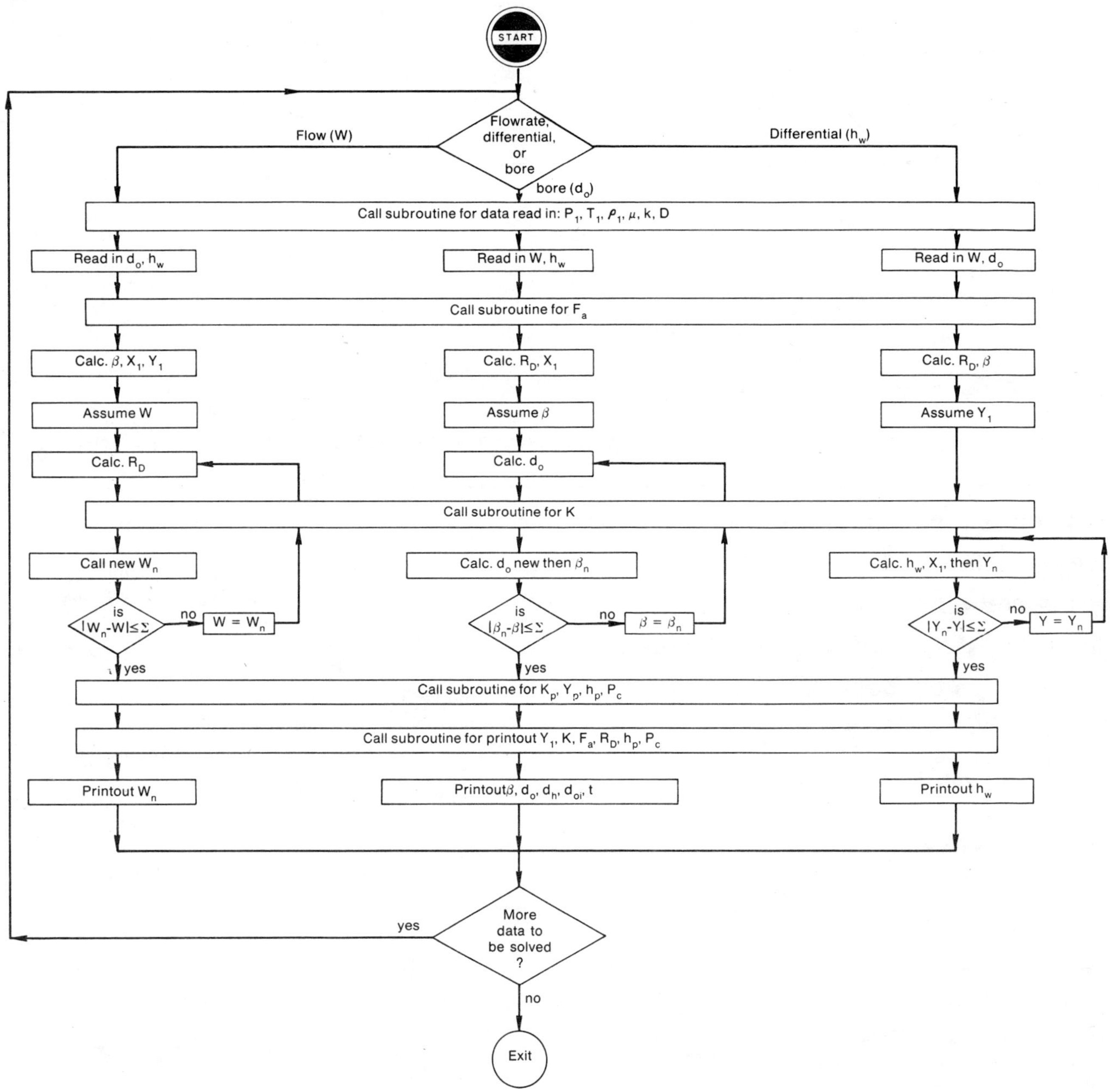

Fig. 3—Computer program flow diagram.

trial-and-error technique or by a short-cut method with charts. This makes orifice design tedious and time-consuming, and the final result is still questionable.

Orifice calculations for gases, vapors or steam. Most engineers have had experiences in calculating orifice size for a particular flowrate and orifice differential. However, another problem he often encounters is checking the d/p cell setting for an existing orifice meter in the plant, or in determining the power consumed by an orifice meter. The latter calculation enables an engineer to evaluate the cost of average pumping per year comparing the orifice plate against the lo-loss tube, the venturi tube or the magnetic meter (low power loss results) to determine whether an extra investment is paid off in a reasonable time.

The universal equation for gases, vapors or steam is:[1]

$$W = 359\ K F_a\ F_m\ d_o^2\ Y_1\ (h_W\ \rho_1)^{1/2} \tag{1}$$

where:

$$\rho_1 = M P_1 / 10.73\ Z\ (T_1 + 460) \tag{2}$$

$$K_o = K_e\ [(10^6\ d_o + 15\ E_f)/10^6\ d_o]^{-1} \tag{3}$$

(a)

```
SUBROUTINE CALCFA(FA,T1)
IF(T1.LE.(-200.)) GO TO 51
IF(T1.LE.70.) GO TO 52
IF(T1.LE.500.) GO TO 53
IF(T1.LE.1000.) GO TO 54
FA=.00002275*T1+.996012
GO TO 55
51 FA=.00001133*T1+.9981667
GO TO 55
52 FA=.000014067*T1+.998713
GO TO 55
53 FA=.00001919054*T1+.9984648
GO TO 55
54 FA=.000021*T1+.9975
55 RETURN
END
```

(b)

```
SUBROUTINE CALCK(REN,DP,XK,D,BETA)
XK1=.5993+.007/DP+(.364+(.076/SQRT(DP)))*(BETA**4)
ONE=.07+.5/DP-BETA
IF(ONE) 6,6,5
5 XK1=XK1+.4*((1.6-(1./DP))**5)*(ONE**2.5)
6 TWO=.5-BETA
IF(TWO) 8,8,7
7 XK1=XK1-(.009+.034/DP)*(TWO**1.5)
8 THREE=BETA-.7
IF (THREE) 10,10,9
9 XK1=XK1+((65./(DP*DP))+3.)*(THREE*2.5)
10 E=(830.-5000.*BETA+9000.*(BETA**2)-4200.*BETA**3+530./&
SQRT(DP))*D
XK2=XK1*((1000000.*D)/(1000000.*D+15.*E))
XK=XK2*(1.+(BETA*E/REN))
RETURN
END
```

(c)

```
SUBROUTINE PLOSS(REN,DP,XP,D,BETA,YP,YX,XL&
,PMLOSS,FA,RHO1,W,RATIO,PSIA1,HP)
XKOP=.5925+.0182/DP+(.44-(.06/DP))*(BETA**2)
XKOP=XKOP+(.935+(.225/DP))*(BETA**5)+1.35*(BETA**14)
BB=.25-BETA
IF(BB)70,70,71
71 XKOP=XKOP+(1.43/SQRT(DP))*(BB*2.5)
70 EE=(905.-5000.*BETA+9000.*BETA**2-4200.*BETA**3+
  875./DP)*D
XKOP=XKOP*((1000000.*D)/(1000000.*D+15.*EE))
XP=XKOP*(1.+(BETA*EE/REN))
YX=1.
5 PMLOSS=(W/(359.*XP*D*D*FA*YX*SQRT(RHO1)))**2
XL=.0361*PMLOSS/PSIA1
YP=1.-(.333+1.145*(BETA*BETA+.7*(BETA**5)+12.*(BETA
  **13&
)))*(XL/RATIO)
IF(ABS(YP-YX)-1.D-9)300,300,301
301 YX=YP
GOTO 5
300 HP=4.20875D-8*W*PMLOSS
RETURN
END
```

Fig. 4—Subroutines for F_a, (a), K (b), and K_p, h_p, P_c (c).

$$Y_1 = 1 - (0.41 + 0.35\,\beta^4)\,(X_1/k) \quad (4)$$

$$X_1 = 0.0361\; h_W/P_1 \quad (5)$$

$$F_m = (1 - 0.0737\;G)^{1/2}$$

$$K = K_o\,(1 + (\beta\,E_F/R_D))$$

For flange taps:

$$K_e = \Bigg[\left(0.5993 + \frac{0.007}{D}\right) + \left(0.364 + \frac{0.076}{\sqrt{D}}\right)\beta^4$$

$$+\,0.4\left(1.6 - \frac{1}{D}\right)^5\left(0.07 + \frac{0.5}{D} - \beta\right)^{5/2}$$

$$-\left(0.009 + \frac{0.034}{D}\right)(0.5 - \beta)^{3/2} + \left(\frac{65}{D^2} + 3\right)$$

$$(\beta - 0.7)^{5/2}\Bigg]$$

$$E_f = d_o\,(830 - 5000\,\beta + 9000\,\beta^2 - 4200\,\beta^3 + (530/\sqrt{D}))$$

F_a is the correction factor for thermal expansion of the primary device and can be obtained from Fig. 1.[2] However, for computer application, equations for F_a as a function of operating temperature were developed, using the curve fitting method, for orifice-type material 300 Series SS, the type usually used for orifice plate material. They are as follows:

$T_1 \leq -\,200°$ F then $F_a = 0.00001133\;T_1 + 0.9981667$

$-200 < T_1 \leq 70°$ F then $F_a = 0.000014067\;T_1$

$+\;0.998713$

$70°$ F $< T_1 \leq 500°$ F then $F_a = 0.00001919054\;T_1$

$+\;0.9984648$

$500°$ F $< T_1 \leq 1000°$ F then $F_a = 0.000021\;T_1 + 0.9975$

$1000°$ F $< T_1$ then $F_a = 0.00002275\;T_1 + 0.996012$ (6)

For any gas, Equation 2 can be used to find its specific weight at flowing condition. However, when the flowing fluid is wet steam, the specific weight of the mixture under the operating conditions must be corrected by dividing ρ_1 for dry steam by the quality (percent dry steam/100), or:

$$\rho_{fc} = \rho_1/(\text{percent dry steam}/100)$$

TABLE 1—Recommended orifice plate thicknesses*

Pipe size, inches	Thickness (in.) below 600° F	Thickness (in.) above 600° F
Up to 3	3/32 ± 1/32	1/4
4 to 6	5/32 ± 1/32	3/8
7 to 8	1/4 ± 1/32	1/2
10 and over	3/8 ± 1/8	1/2 to 3/4

*From the ASME Flowmeter Computation Handbook, 1961.

TABLE 2—Recommended drain hole sizes for various orifice diameters*

Orifice dia. d_o, inches	Drain hole dia., inches
1—3½	3/32
3½—4¼	1/8
4¼—5	5/32
5—6	3/16
6—6¾	7/32
6¾—7½	1/4
7½—8⅜	9/32
8⅜—9¼	5/16
9¼—10	11/32
10—10⅞	3/8
10⅞—11⅝	13/32
11⅝—12½	7/16
12½—13¼	15/32
13¼ and up	1/2

*From Fischer and Porter Handbook, Flowmeter Orifice Size.

and ρ_{fc} must be substituted for ρ_1 in Equation 1.

Design procedure. Equation 1, Equation 6 and the empirical equations for correction factors are used in the computer program to solve for either the rate of flow, the orifice diameter or the pressure differential.

For all dry or diaphragm, bellows or spring-balanced meters, F_m is equal to unity.

If flowrate and operating pressure are known for an existing orifice and the differential is to be found for checking the d/p cell setting, the computer program starts with an assumed value for Y_1, then calculates h_w via Equation 1. After h_w is found, X_1 is determined from Equation 5, thus new Y_1 is computed from Equation 4 and compared with the assumed Y_1. Successive trials of Y_1 are applied until the error is less than 10^{-9}.

The same procedure is employed to find flowrate or orifice size, except successive trials of W are applied for solving the rate of flow and that of β are employed for computing the orifice diameter.

When the required unknown is obtained, the permanent loss and the power consumed by the orifice meter are also determined. The power consumption of the orifice meter is determined from

$$P_c = 4.2087 \times 10^{-8}\,(W)\,(h_p)$$

where h_p is the permanent loss in inches of H_2O.

The permanent loss, rather than the orifice differential h_w, is measured by pipe taps where the downstream tap is eight pipe diameters below the orifice plate. The empirical equations used for discharge coefficients, for pipe taps are:[1]

$$K_o = K_e\,[(10^6\,d_o + 15\,E_p)/10^6\,d_o]^{-1}$$

$$K_p = K_o\,(1 + \beta\,E_p/R_D)$$

$$K_e = \left[0.5925 + \frac{0.0182}{D} + \left(0.440 - \frac{0.06}{D}\right)\beta^2 + \left(0.935 + \frac{0.225}{D}\right)\beta^5 + 1.35\,\beta^{14} + \frac{1.43}{\sqrt{D}}\,(0.25 - \beta)^{5/2}\right]$$

$$E_p = d_o\,(905 - 5000\,\beta + 9000\,\beta^2 - 4200\,\beta^3 + (875/D))$$

$$Y_{p1} = 1 - (0.333 + 1.145\,(\beta^2 + 0.7\,\beta^5 + 12\,\beta^{13}))\,(X_{p1}/k) \tag{7}$$

$$X_{p1} = 0.0361\,h_p/P_1 \tag{8}$$

$$R_D = 6.32\,W/\mu D$$

and h_p is determined as:

$$h_p = (W/359\,F_a\,K_p\,Y_{p1}\,d_o^2\,\sqrt{\rho_1})^2$$

The percent differential that is permanently lost depends on the value of β, and the relationship between the percent differential and β is shown in Fig. 2.

For solving h_p, the successive trials of Y_{p1} are also employed until the error is less than a given tolerance.

Three separate computer programs may be written for determining orifice size, rate of flow and differential pressure. However, one main program with a number of subprograms for correction factors can be developed to save storage space and time. Each subprogram can be "called" wherever necessary.

Such a main program was developed and its flow diagram is shown in Fig. 3.

Three subprograms for the orifice calculation were subprogram F_a, K and K_p, and they are shown in Fig. 4. In addition, a subprogram for data read-in and a subprogram for results printout may be written since input values other than W, h_w and d_o are always required, and the resulting values of K, R_D, F_a, and Y_1 are always printed out in calculation of orifice size, flowrate or differential.

Orifice calculation for liquid. The equations used for liquid flow are:[1]

$$W = 2835\,KF_a\,F_m\,d_o^2\,\sqrt{G_f\,h_w}$$

$$h_p = (W/2835\,K_p\,F_a\,F_m\,d_o^2\,\sqrt{G_f})^2$$

```
THIS COMPUTER PROGRAM CAN BE USED TO MAKE THREE TYPES OF
CALCULATIONS:

        1) RATE OF FLOW (LB/HR)
        2) BORE DIAMETER (INC.)
        3) PRESSURE DIFFERENTIAL (INC. H2O)

SELECT YOUR TYPE OF FLUID:
LIQUID OR GAS..........?GAS

      ♦♦♦♦♦ FLANGED ORIFICE CALCULATIONS ♦♦♦♦♦
          ♦♦♦♦ FOR GASES OR STEAM ♦♦♦♦

SELECT YOUR TYPE OF GASES:
GASES OR STEAM..........?STEAM

SELECT YOUR TYPE OF CALCULATION:
FLOW , BORE , OR PRES..........?BORE

      ♦♦♦♦♦ YOUR TYPE OF CALCULATION IS ♦♦♦♦♦
            ♦♦♦♦♦ BORE DIAMETER ♦♦♦♦♦
               ♦♦♦ FOR STEAM ♦♦♦

♦♦♦♦♦ INPUT THE FOLLOWING DATA:

CP/CV RATIO.......................................?1.3

FLUID TEMPERATURE................F.................?298

FLUID VISCOSITY...................CP................?.014

UPSTREAM PRESSURE...............PSIA.............?64.7

SPECIFI WT. OF DRY STEAM.........LB/FT3...........?.14928

QUALITY OF STEAM.................%................?98

INSIDE PIPE DIAMETER.............INC..............?8.125

FLOW RATE........................LB/HR............?20000

PRESS. DIFFERENTIAL..............INC.H2O..........?120

          ♦♦♦♦♦ THE RESULTS ARE AS FOLLOWS ♦♦♦♦♦

     TYPE OF METER                              DRY
     PRESSURE TAP LOCATION                 FLANGED
     TYPE OF ORIFICE                    CONCENTRIC
     ORIFICE PLATE MATERIAL          300 SERIES SS

REYNOLDS NUMBER                    =   1111208.50000
FA                                 =         1.00418
K                                  =          .63671
Y                                  =          .97708
BETA                               =          .56210

BORE DIA WITHOUT DRAIN HOLE (INC.) =         4.56704
DRAIN HOLE DIA.             (INC.) =          .15625
BORE DIA. WITH DRAIN HOLE   (INC.) =         4.56436

     1/4 < ORIFICE PLATE THICKNESS (INC.) < 1/2

PERMANENT LOSS                (INC.H2O)  =  79.80962
POWER CONSUMPTION OF THE METER  (HP)     =    .06718
```

Fig. 5—Example 1 printout.

where G_f is specific gravity of liquid at flowing conditions, and K, K_p, F_a and F_m are the same as above.

Note that, for liquid flow, Y_1 or Y_{p1} is equal to unity; thus, the calculation of h_w or h_p is straightforward without trial-and-error iterations. However, successive trials of W or β are still applied, and the design procedures are similar to those employed for gases, steam or vapor.

In calculating for orifice diameter the computer program also determines the orifice plate thickness based on Table 1 and the size of the drain hole based on Table 2, then corrects the proper size for the orifice to be cut in the center of the plate, using equation:

$$d_{oi} = (d_o^2 - d_h^2)^{1/2}$$

The drain or weep hole is placed at the bottom of the horizontal pipe ahead of the orifice plate when wet gases or saturated steam are measured, and the vent hole is placed at the top when a liquid carrying gas bubbles is metered.

In practice, it is difficult to keep drain holes from plugging with dirt or scale. If this is a problem, the orifice meter should be installed in a vertical line with flow vertically upwards to carry gas bubbles through the orifice or downwards to purge condensate.

Illustrations. The following examples illustrate the use of the computer program:

Example 1. Size an orifice meter to measure the steam flowing through a pipe line of 8.125 inches ID. Other operating conditions are:

$h_w = 120$ in. of water

$T_1 = 298°F$

$P_1 = 64.7$ psia

$\rho_1 = 0.14928$ lb./ft.^{3}

$\mu = 0.014$

$k = 1.3$

$W = 20{,}000$ lb./hr.

```
♦♦♦♦♦ FLANGED ORIFICE CALCULATION ♦♦♦♦♦
♦♦♦♦♦ FOR LIQUID ♦♦♦♦♦

SELECT YOUR TYPE OF CALCULATION:
FLOW , BORE , OR PRES............?PRES

♦♦♦♦♦ YOUR TYPE OF CALCULATION IS ♦♦♦♦♦
♦♦♦♦♦ PRESSURE DIFFERENTIAL ♦♦♦♦♦

♦♦♦♦ INPUT THE FOLLOWING DATA:

SP.GR. OF LIQUID AT FLOWING TEMP................?.954

FLUID TEMPERATURE..................F.............?225

FLUID VISCOSITY....................CP............?.25

DESIGNED FLOWRATE..................LB/HR........?62500

INSIDE PIPE DIAMETER...............INC...........?3.068

BORE DIAMETER......................INC...........?1.8573

♦♦♦♦♦ THE RESULTS ARE AS FOLLOWS ♦♦♦♦♦

      TYPE OF METER                        DRY
      PRESSURE TAP LOCATION            FLANGED
      TYPE OF ORIFICE               CONCENTRIC
      ORIFICE PLATE MATERIAL     300 SERIES SS

REYNOLDS NUMBER                     =  514993.50000
FA                                  =       1.00278
K                                   =        .65235
PRESS. DIFFERENTIAL      (INC.H20)  =     100.04544

PERMANENT LOSS           (INC.H20)  =      61.79204
POWER CONSUMPTION OF THE METER (HP) =        .16254
```

Fig. 6—Example 2 printout.

The orifice plate is 304SS and the manometer type is dry. Steam quantity is 98 percent.

Example 2. Determine the d/p cell setting for a given orifice meter if the flowrate of water at 225°F is 62,500 lb./hr. Water line is 3.068 inches ID, the orifice diameter is 1.8573 inches. Water specific gravity and viscosity are 0.954 and 0.25 cps, respectively.

Also, determine the permanent loss and power required to operate the meters in both examples at full load.

The results of the computer run for the examples are shown in Figs. 5 and 6, respectively.

Note that when calculations for gases, vapors or steam are made based on the downstream operating pressure (P_2), the gas expansion factor should be corrected as follows:

For flange taps

$$Y_2 = Y_1\ (1/(1 - X_1))^{1/2} \tag{9}$$

For pipe taps

$$Y_{p2} = Y_{p1}\ (1/(1 - X_{p1})^{1/2} \tag{10}$$

where

$$X_1 = (h_w/P_2)/(27.7 + (h_w/P_2))$$

and

$$X_{p1} = (h_p/P_2)/(27.7 + (h_p/P_2))$$

X_1 and X_{p1} must be used in calculations of Y_1 and Y_{p1} by Equations 4 and 7, and the conversion to Y_2 and Y_{p2} by Equations 9 and 10, respectively.

NOMENCLATURE

- D Inside diameter of pipe, inches
- d_h Drain hole diameter, inches
- d_o Orifice diameter without drain hole, inches
- d_{oi} Orifice diameter corrected for drain hole, inches
- F_a Area ratio, for thermal expansion of primary device
- F_m Manometer correction factor
- G Specific gravity of the fluid on the mercury surface of the mercury type meter
- G_f Specific gravity of liquid at flowing conditions
- h_p Permanent differential pressure, inches of water
- h_w Operating differential pressure, inches of water
- k Specific heat ratio, c_p/c_v
- K Coefficient of discharge including velocity approach for flange taps
- K_p Coefficient discharge including velocity approach for pipe taps
- M Molecular weight, lb./lb. mole
- P_c Power consumption of the meter, hp
- P_1 Upstream operating pressure, psia
- R_D Reynolds number based on pipe diameter, 6.32 $W/D\mu$
- t Orifice plate thickness, inches
- T_1 Upstream operating temperature, °F
- W Rate of flow, lb./hr.
- X_1, X_{p1} Differential divided by upstream pressure in same unit
- Y_1, Y_{p1} Gas expansion factor based on upstream operating pressure, flange taps and pipe taps, respectively
- Y_2, Y_{p2} Gas expansion factor based on downstream operating pressure, flange taps and pipe taps, respectively
- Z Compressibility ratio at flowing conditions
- β Diameter ratio, d_o/D
- ρ_1 Specific weight of flowing fluid (or dry steam) at upstream conditions, lb./ft.^{3}
- ρ_{fc} Specific weight of wet stream, lb./ft.^{3}
- μ Fluid viscosity, cps

LITERATURE CITED

[1] Spink, L. K., Principles and Practice of Flow Meter Engineering, 9th Edit., The Foxboro Co., Foxboro, Mass. (1967).
[2] Bean, H. S., Edit., Fluid Meter, Their Theory and Application, 6th Edit., ASME, New York City (1971).

Use Mass Measurement for NGL

Hard to measure mixtures such as those containing ethanes need special attention

W. J. Templeton, Cities Service Oil Co., Tulsa, Okla.

Mass measurement of liquefied petroleum gases provides an accurate measuring system which does not depend upon unknown physical properties for correcting the measurement to standard conditions. Mass measurement is applicable to mixtures containing ethane or ethylene and to streams being measured under extreme temperature and pressure conditions where nonideal mixing occurs.

In the measurement of liquefied petroleum gases, it is necessary to consider the physical properties of the components and the mixture composition before selecting a measurement system utilizing either volumetric or mass determination methods.

Volumetric methods of measurement are utilized where physical properties are known and correction factors can be applied in correcting the measurement of the component to standard conditions. Volumetric determinations are applicable to pure components or commercial grade products containing less than 10 percent ethane.

Mass measurement methods are utilized where properties in addition to temperature and pressure will affect the measurement such as intermolecular adhesion or volumetric changes caused by solution mixing. Accurate physical correction factors cannot be determined to correct for solution mixing which occurs in mixed streams containing ethane or ethylene. The higher the percentage of ethane or ethylene in the mixture, the greater the effect of solution mixing. An example of solution mixing is when one gallon of ethane is added to one gallon of propane and the volume of the combined mixture is less than two gallons. Mass measurement thus provides an accurate means of determining flow because one pound of one component added to one pound of another component always gives two pounds of mixture.

MASS MEASUREMENT

Mass measurement of natural gas liquids is accomplished by either static or dynamic methods and the resulting mass is converted to volume by individual components at standard conditions by the use of a component weight analysis of the mixed stream. Mass measurement eliminates the need to compensate for the effects of compressibility and nonideal mixing because physical properties and correction factors are not required to convert the measurement to standard conditions. Improvement in measurement accuracy by 4 to 9 percent is thus accomplished.

Mass measurement in a dynamic state utilizes

- A volumetric determination at flowing conditions
- A density or gravity determination at flowing conditions
- A composite sample of the fluid flowing through the measuring device.

Mass measurement is obtained by multiplying the measured volume at flowing condition times flowing density using consistent units. The equivalent volume at standard conditions of each component in the mixture is obtained by a component analysis and the density of the pure component at standard conditions.

Density of the flowing fluid in mass per unit volume at operating conditions is equal to the specific gravity of the fluid in the line times the density of water in vacuum at standard conditions. Densitometers can be used which provide the density of the fluid directly in mass per unit volume.

Conventional volumetric measuring methods are applicable to mass measurement when the volumetric measurement is accomplished at flowing temperature and pressure conditions and no correction factors are applied.

The dynamic volumetric measurement is accomplished by using one of the industry standard methods such as

- Orifice meter
- Positive displacement meter
- Turbine meter.

These devices are installed, maintained and calibrated in accordance with industry standards such as the AGA Report No. 3 and API 1101, 2534 and 2531. Temperature compensated metering devices are not to be used in mass measurement.

Special precautions are required to ensure that the volumetric measurement and density determination are accomplished at the same flowing conditions. When positive displacement meters or turbine meters are proved, the meter proving factor obtained must provide a factor for flowing conditions. Corrections are not to be applied to the liquid in the prover; corrections for temperature and pressure are to be applied to the steel in the prover only. This is important.

An analysis of a sample which is representative of the fluid being measured is required in order to convert the total mass of the mixture into individual component masses. Sample increments proportional to the flowing rate must be taken to provide a composite sample representative of the fluid being measured. A good sampling procedure recommended is the procedure outlined in the Gas Processors Association Procedure GPA 2174-74. Pure component physical properties are used to convert the individual component masses to volumetric units.

METHOD OF CALCULATION

Positive displacement or turbine meters. The calculation of total mass flowing must be performed continuously on-line by a suitable device or by off-line integration of charts on which measured volume and density are continuously recorded so that at all times the density corresponds to the volume measured. Multiplying the total metered volume by the density averaged over a period of time is permissible when either density or volumetric flow rate remains constant.

$$\text{Total mass} = \text{Meter factor} \times \begin{matrix}\text{Metered}\\ \text{vol. at}\\ \text{oper.}\\ \text{cond.}\end{matrix} \times \begin{matrix}\text{Specific}\\ \text{gravity at}\\ \text{operating}\\ \text{conditions}\end{matrix} \times \begin{matrix}\text{Density of}\\ \text{water in}\\ \text{vacuum at}\\ \text{standard}\\ \text{conditions}\end{matrix}$$

The analysis of the composite sample is obtained in weight percent and this analysis is multiplied by the total mass of the fluid to determine the mass of each component.

The calculation of the volume of each component at equilibrium vapor pressure at standard temperature condition is performed as follows:

$$Q = \text{Mass} \div \text{Density}$$

where:

Q = Volume measured at equilibrium pressure at standard temperature conditions for each component
Mass = Weight of fluid per unit of time for each component
Density = Weight per unit of volume at standard condition for the pure component
Total volume of all components = Sum of individual component volumes.

EXAMPLE CALCULATION

Mass measurement using orifice meters. The orifice meter used in conjunction with a densitometer for obtaining a density at flowing conditions provides a means of measuring mass without using the factors for temperature and compressibility. The basic orifice equation for mass flow is given in the American Gas Association Gas Measurement Committee Report No. 3 when the specific weight of the flowing fluid is obtained. The equation is as follows:

$$W = 1.0618\, F_b\, F_r\, F_m\, F_a\, (h_w r)^{1/2}$$

where:

W = Mass, lb./hr.
F_b = Basic orifice factor
F_r = Reynolds number factor
F_m = Manometer factor
F_a = Orifice thermal expansion factor
h_w = Differential pressure, inches of water
r = Actual specific weight of fluid at flowing condition, lb./cu. ft.

This equation and factors are explained more fully in AGA Report No. 3.

When the orifice differential and gravity of the fluid are recorded on the same chart, both can be integrated with a conventional gas integrator for L-10 charts and zero setback. The integrator will provide the $\Sigma(h_w r)^{1/2}$ for the period integrated.

The mass units obtained would be applied to the sample analysis weight percent for determining the component gallons at standard conditions.

Other volumetric measuring devices. Other types of flow meters can be used to obtain the mass of the fluid directly or in conjunction with a densitometer as outlined above. The preferential use of the metering equipment used should be by mutual agreement of the contractual parties involved.

Static mass measurement. Static mass measurement is determined by weighing the container or vessel before and after product has been added to or removed from the vessel. The difference in weight provides the total mass of the product contained in the vessel. Weight determination in this way accounts for the portion of the product which vaporizes and remains in the vapor state in the vessel.

Calculation of volume by using the difference in weight is determined by dividing the density of the liquid product at standard conditions into the mass.

SPECIAL PRECAUTIONS

Special precautions which pertain to dynamic mass measurement must be observed and adhered to as follows:

- Volume measurement must be obtained at flowing conditions.
- Proving of the measuring device is required at flowing conditions. Special precautions to prevent deviation of operating temperatures and pressures from the proving temperatures and pressures are required.
- When proving the meter, corrections are not to be applied to the liquid in the prover; corrections for temperature and pressure are to be applied to the steel in the prover only.
- Density measurement must be obtained at flowing conditions and must be made at the same pressure and temperature conditions as the volume measurement is made.
- Calibration and proving of the densitometer is required

EXAMPLE CALCULATION

1. Calculation of Total Mass Flowing

Total Mass = Meter Factor x Meter Volume at Operating Conditions U.S. gallons x Specific Gravity at Operating Conditions Vacuum x Density Water at 60°F lb/U.S. gal. = Total lb./Mass

= .9995 x 204,702 x .450 x 8.337 = 767,586

2. Convert Analysis to Weight Fraction

Composite Sample Comp	Mol%	GPA 2145-75 Constant Lb Mass/Lb Mole	Lb Mass x 100 Mol% x Lb/Lb Mole	Weight Fraction of Component
C_1	1.335	16.043	21.42	.00584
C_2	50.991	30.070	1533.30	.41788
C_3	46.805	44.097	2063.96	.56251
iC_4	.869	58.124	50.51	.01377
Total	100.000		3669.19	1.00000

3. Calculate the Mass of Each Component As Follows:

Wt fraction comp x total lb mass = lb mass each comp

Comp	Weight Fraction of Component	Total Lb Mass	Lb Mass of Component
C_1	0.00584	767,586	4,483
C_2	0.41788	767,586	320,759
C_3	0.56251	767,586	431,775
iC_4	0.01377	767,586	10,569
Total	1.00000		767,586

4. Calculate the Volume of Each Component at Equilibrium Vapor Pressure and 60°F As Follows:

$$\text{U.S. Gal at } 60^\circ\text{F and equilibrium pressure} = \frac{\text{lb of each component}}{\text{density, lb mass/U.S. gal at } 60^\circ\text{F and equilibrium pressure}}$$

Comp	Lb Mass of Component	GPA-2145-75 Density (Vacuum) Lb Mass/U.S. Gal	U. S. Gal of Component
C_1	4,483	2.5	1,793
C_2	320,759	2.971	107,963
C_3	431,775	4.223	102,002
iC_4	10,569	4.695	2,251
Total	767,586		214,009

The GPA Publication 8173-75 is the standard for converting natural gas liquids and vapors to equivalent liquid volumes and provides additional explanation of converting mass to equivalent liquid volume.

at flowing conditions. Proving the densitometer in addition to the calibration is an additional feature which increases measurement accuracy.

► A composite sample must be obtained which is representative of the fluid measured and must provide a weighted average sample of the fluid measured.

► Temperature compensated metering devices shall not be used.

► General requirements to ensure good measurement such as single phase conditions, prevention of turbulence through the measuring device, elimination of pulsation or vibration and elimination of electrical noise voltages in the flow meter pickup coil must be maintained.

Presented to the Gas Processors Association meeting at San Antonio, Texas, March 22-24, 1976; API Pipeline Measurement Seminar, Houston, April 5, 1976; 1976 International School of Hydrocarbon Measurement, Norman, Okla., April 15, 1976.

LITERATURE CITED

[1] Gas Processors Association, GPA Publication 8173-75; 2145-75; 2174-74.
[2] American Petroleum Institute, API Standards 1101; 2534; 2531.
[3] American Gas Association, Gas Measurement Committee Report No. 3.
[4] "Principles & Practices of Flow Meter Engineering," by L. K. Spink.

LNG Density Determination

Test program is determining the best of the available approaches

Dwain E. Diller, Cryogenics Division, Institute for Basic Standards, National Bureau of Standards, Boulder, Colo.

LNG DENSITY determination must be very accurate to avoid large mischarges in normal transactions. Incorrect charges of as much as $40,000 per 125,000 m^3 tanker cargo can easily result from systematic errors of 1% in density determination. Consequently, new data and improved measurement techniques are needed that can provide acceptable accuracies.

Liquid density determination is fundamental to the calculation of heating value which is used for buying and selling LNG. The basis for determining density is usually agreed to in custody transfer contracts which may involve several companies and countries. LNG density can be either calculated from a mathematical model as a function of composition and temperature (or pressure) or measured directly in a tank or pipe line. Both approaches are being systematically examined and revised to try to meet the goal of determining a density which is within 0.1% of the true value.

Accurate orthobaric liquid density data have been obtained for six LNG components and their mixtures. These data are being used to optimize and evaluate four promising calculation methods. Within various composition limitations these models have correlated most data within estimated inaccuracies. Several significant differences between measured and calculated densities still remain. Efforts are continuing to resolve these differences.

A density reference system is being used to examine the performance of commercially available densimeters for use on LNG type mixtures. Several types look promising for direct density measurements with an imprecision less than three tenths of 1%.

CALCULATION METHODS

Use of mathematical models has advantages of providing coverage of a large range of mixture compositions, temperatures and pressures while utilizing the most accurate laboratory determined densities for the components of LNG and their mixtures. On the other hand, some models are fairly difficult to use and depend on knowing the mixture composition, which may be difficult or inconvenient to obtain with sufficient accuracy.

The excess volume approach provides sensitive and relatively simple means of correlating the densities of liquefied gas mixtures.

The molar volume V_{mix} of a fluid mixture can be written

$$V_{mix} = \Sigma\, x_i\, V_i + V^E \qquad (1)$$

where x_i, V_i and V^E are conventionally the mole fractions, molar volumes of the constituents and the excess volume, respectively, all at the temperature and pressure of the mixture. For LNG type mixtures V^E is typically negative and 1-2% of V_{mix}. Therefore an accurate mathematical model for V_{mix} (or V^E) is required to achieve an inaccuracy of less than 0.1% in the density. The mathematical model must be based on and tested against accurate density measurements for LNG components and their mixtures. For contractual purposes it is desirable that measurements and evaluation be carried out by an independent third party.

Until recently there have been few accurate ($\pm$0.1%) density measurements available for liquefied LNG components and their mixtures in the normal boiling temperature range (110-115 K) of LNG type mixtures. Additional data are being determined using a magnetic suspension densimeter constructed and tested for measurements in the temperature range 90-300 K at pressures to 5 MPa (725 psi)[5] in a comprehensive program of orthobaric liquid density measurements. Determinations have been made on six LNG components (methane, ethane, propane, isobutane, normal butane and nitrogen),[5,6] 29 compositions of 13 binary mixtures[7] and 13 multicomponent mixture compositions[8] including several LNG type mixtures. Measurements on the pure components mainly at temperatures between 95 and 160 K have an estimated imprecision less than 0.02% and an estimated inaccuracy of approximately 0.1%. The estimated imprecision is primarily based on the standard deviation obtained for 123

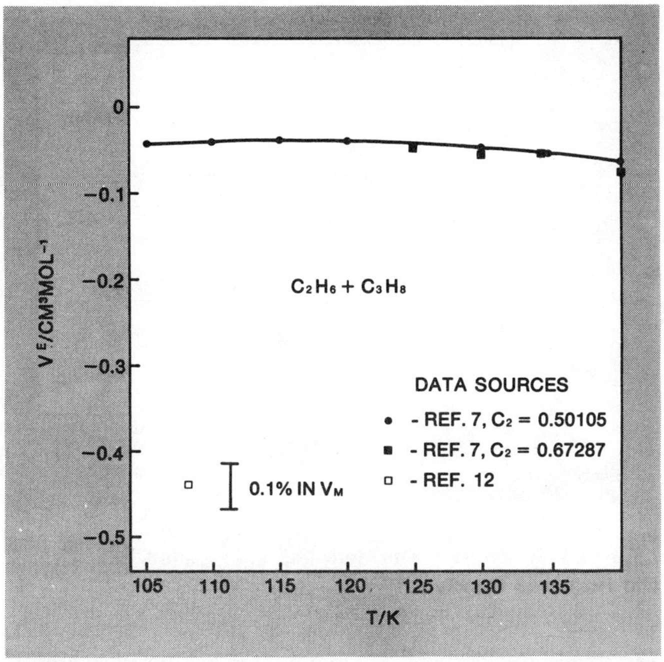

Fig. 1—Dependence of excess volumes of ethane-propane mixtures on temperature at fixed composition (nearly equimolar).

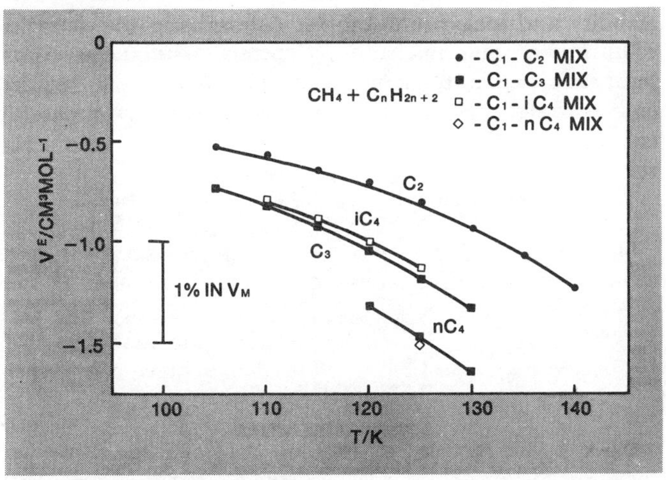

Fig. 2—Dependence of excess volumes of nearly equimolar alkane mixtures containing methane on temperature at fixed composition. 0, methane-ethane mixture; □, methane-propane; △, methane-isobutane; ◊, methane-normal butane.

measurements on liquid methane. The estimated inaccuracy of a single density measurement on liquid methane was taken as the square root of the sum of the squares of the systematic error plus an allowance for random error of three times the standard deviation. Measurements on binary and multicomponent mixtures mainly at temperatures between 105 and 140 K have a maximum estimated inaccuracy of 0.16%. The derived excess volumes V^E of the binary mixtures in the temperature range of these measurements can be classified into several quantitatively distinguishable groups.

- Alkane mixtures without methane have excess volumes which are less than 0.1% of V_{mix} (Fig. 1).

- Alkane mixtures containing methane have excess volumes up to 3% of V_{mix} (Fig. 2).

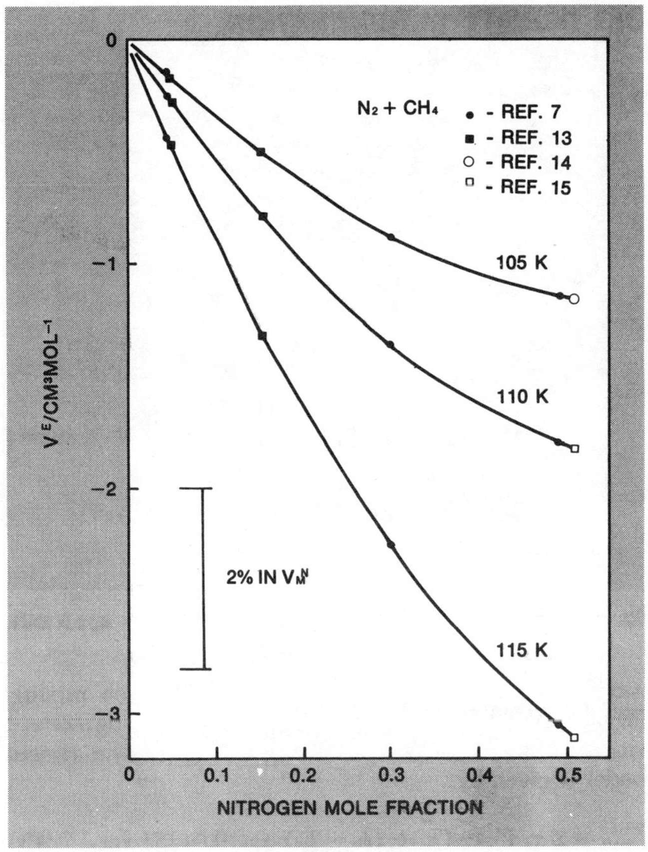

Fig. 3—Dependence of excess volumes of nitrogen-methane mixtures on composition at fixed temperatures (105, 110 and 115 K). 0, reference [7]; □ Rodosevich and Miller [13]; △ Liu and Miller [14]; ▽ Massengill and Miller [15].

- Nitrogen-alkane mixtures have excess volumes up to 10% of V_{mix} (Fig. 3).

Several promising models for calculating accurate densities of LNG type mixtures have been published previously. They include the Klosek and McKinley excess volume correlation,[1] the modified cell model,[2] the modified corresponding states model,[3] and the modified hard sphere equation of state model.[4]

The modified Klosek and McKinley correlation,[1] the modified corresponding states model[3] and the modified hard sphere equation of state model[4] are being optimized and evaluated at the National Bureau of Standards and the modified cell model[2] is being optimized and evaluated at the Phillips Petroleum Co.

Within limited composition, temperature and pressure ranges, all of the models examined look promising for calculating the densities of commercial LNG mixtures with an inaccuracy less than several tenths of 1%. The empirical excess volume approach, of which the Klosek and McKinley correlation is typical, provides one of the simplest methods for calculating the densities of commercial LNG mixtures. Satisfactory calculations can be obtained without the use of a computer program. The Klosek and McKinley correlation, as originally proposed,[1] is given by

$$V_{mix} = \Sigma\, x_i\, V_i - k\, x_{CH_4}, \tag{2}$$

where x_{CH_4} is the mole fraction of methane in the mixture and k is the coefficient which corrects the mole fraction

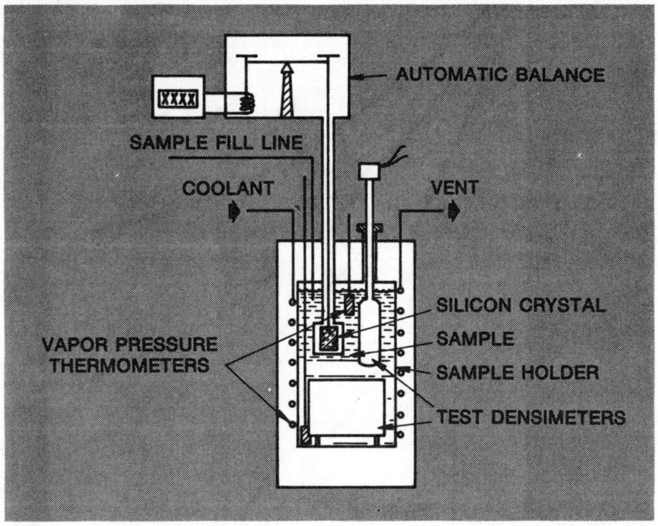

Fig. 4—Cryogenic fluids density reference system apparatus.

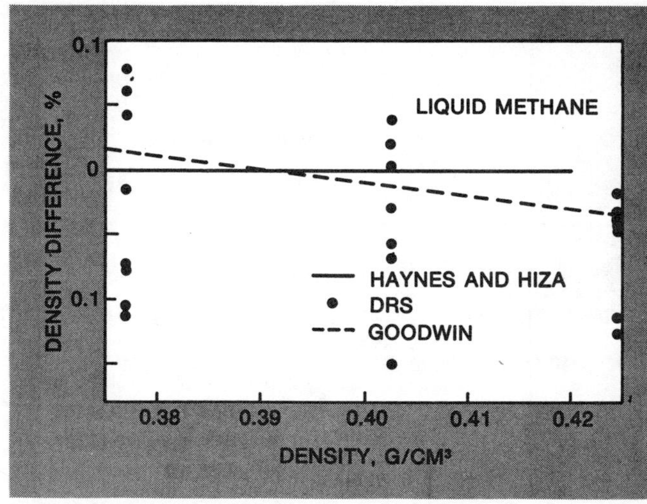

Fig. 5—Comparison of densities, 0, measured with the DRS reference densimeter with densities interpolated from Haynes and Hiza[5] and Goodwin.[16]

average, $\Sigma x_i V_i$, for the volume contraction on mixing. The model has been revised[9] to improve the agreement with data for mixtures containing nitrogen. The revised model is given by

$$V_{mix} = \Sigma x_i V_i - [k_1 + (k_1 - k_2) x_{N_2}/0.0425] x_{CH_4}, \quad (3)$$

where x_{N_2} is the mole fraction of nitrogen in the mixture and k_1 and k_2 are coefficients which correct for the volume contraction on mixing. This model is limited to the calculation of orthobaric liquid densities of methane rich mixtures in their normal boiling temperature range. Its predictive capability is also limited by the empirical basis of the method.

DIRECT READING DENSIMETERS

Direct reading densimeters are relatively simple to use (knowledge of the mixture composition is not required) and may be conveniently used for continuous density monitoring. They have the disadvantages of needing to be calibrated and of requiring assurance that the calibration and performance of a particular densimeter is stable. For contractual purposes it is desirable that the performance of types of available densimeters be examined by an independent third party. Accordingly, an apparatus[10] for examining the performance of direct reading densimeters on LNG components and their mixtures has been constructed and tested at the National Bureau of Standards. The densimeters are being examined in a density reference system (DRS) (Fig. 4) containing carefully characterized liquefied gases and an accurate absolute densimeter based on Archimedes' principle. The accessible temperature range of the DRS is 80 to 300 K and the pressure limit is 7 bars (~ 100 psi). The estimated inaccuracy of the reference densimeter is less than 0.21%, based on the sum of the known sources of systematic error, ±0.026%, plus three times the standard deviation. ±0.062%, based on 71 independent measurements using eight different samples of liquid methane (99.97% pure) (Fig. 5). The commercial direct reading densimeters are also being compared with densities calculated from a mathematical density model. Tests are being performed on commercially available densimeters based on several different principles of operation.[11] Densimeter types being examined include an Archimedes' principle or displacement type, capacitance or dielectric type, vibrating cylinder type and a vibrating plate type. Preliminary tests indicate that several direct reading densimeters have good stability and look promising for determining the densities of commercial LNG type mixtures with an imprecision less than three tenths of 1%. As all of them require calibration, their inaccuracy depends mainly on careful calibration against accurate reference data or an accurate reference densimeter.

ACKNOWLEDGMENT

This work was carried out at the National Bureau of Standards under the sponsorship of British Gas Corp., Chicago Bridge and Iron Co., Columbia Gas Service Corp., Distrigas Corp., Easco Gas LNG, Inc., El Paso Natural Gas, Gaz de France, Marathon Oil Co., Mobil R&D Corp., Natural Gas Pipeline Co., Phillips Petroleum Co., Shell International Gas, Ltd., Sonatrach, Southern California Gas Co., Tennessee Gas Pipeline, Texas Eastern Transmission Co., Tokyo Gas Co., Ltd., and Transcontinental Gas Pipe Line Corp., through grants administered by the American Gas Association, Inc.

LITERATURE CITED

1 Klosek, J., and McKinley, C., "Densities of Liquefied Natural Gas and of Low Molecular Weight Hydrocarbons," Proc. First Int. Conf. on LNG, IGT, Chicago, 1968, Session 5, Paper 22.
2 Albright, M. A., "A Model for the Precise Calculation of Liquefied Natural Gas Densities," Technical Publication TP-3, 1973, NGPA, Tulsa, Okla.
3 Mollerup, J., and Rowlinson, J. S., "The Prediction of Densities of Liquefied Natural Gas and of Lower Molecular Weight Hydrocarbons," *Chem. Eng. Sci.*, Vol. 29, 1974, p. 1373.
4 Rodosevich, J. B., and Miller, R. C., "Calculation of LNG Excess Volumes by a Modified Hard-Sphere Model," *Adv. Cryog. Eng.*, Vol. 19, 1974, p. 339.
5 Haynes, W. M., Hiza, M. J., and Frederick, N. V., "A Magnetic Suspension Densimeter for Measurements on Fluids of Cryogenic Interest," *Rev. Sci. Instr.*, Vol. 47, 1976, p. 1237.
6 Haynes, W. M., and Hiza, M. J., "Measurements of the Orthobaric Liquid Densities of Methane, Ethane, Propane, Isobutane and Normal Butane," *J. Chem. Thermo.* (in press), 1977.
7 Hiza, M J., Haynes, W. M., and Parrish, W. R., "Orthobaric Liquid Densities and Excess Volumes for Binary Mixtures of Low Molecular Weight Alkanes and Nitrogen Between 105 and 140 K," submitted to *J. Chem. Thermo.*, 1977.
8 Hiza, M. J., and Haynes, W. M., unpublished data.
9 McCarty, R. D., "A Comparison of Mathematical Models for the Prediction of LNG Densities," *Nat. Bur. Stand. (U.S.), Internal Report* (in press), 1977.
10 Younglove, B. A., and Siegwarth, J. D., "Cryogenic Fluids Density Reference System Provisional Accuracy Statement," *Nat. Bur. Stand. (U.S.), Internal Report 77-852*, January 1977.
11 Younglove, B. A., and Siegwarth, J. D., unpublished data.
12 Shana'a, M. Y., and Canfield, F. B., *Trans. Faraday Soc.*, Vol. 64, 1968, p. 2281.
13 Rodosevich, J. B., and Miller, R. C., *A.I.Ch.E. J.*, Vol. 19, 1973, p. 729.
14 Liu, Y. P., and Miller, R. C., *J. Chem. Thermodynamics*, Vol. 4, 1972, p. 85.
15 Massengill, D. R., and Miller, R. C., *J. Chem. Thermodynamics*, Vol. 5, 1973, p. 207.
16 Goodwin, R. D., "Thermophysical Properties of Methane, From 90 to 500 K at Pressures to 700 Bar," *Nat. Bur. Stand. (U.S.), Tech. Note 653*, April 1974.

Crude Unit Computer Control . . . How Good Is It?

Here is what can realistically be expected based on data from several installations. Suggestions are also given on how to make it better.

L. D. Van Horn, Setpoint, Inc., Houston

CRUDE UNIT computer control is providing large cost savings for many refiners. As with most refining projects, these attractive economic benefits don't come automatically. Computer control functions that generate benefits in one place, may not do so well elsewhere. Operating environments vary from plant to plant and practical problems stand in the way. Some evaluation and selection is necessary, just as it is with processes.

This discussion will draw from our experience to relate to the question, "How good is it?," and to offer some suggestions on how to make it good.

Crude unit computer control functions. Many of the functions in a crude unit computer project have been presented before.[4,5,6,7,8,9] They are summarized below:

►Engineering calculations

Part of a control computer's job is to read process data—temperatures, flows, and pressures (Fig. 1). Once these variables are read, the computer can calculate useful operating parameters, like an engineer would from test run data. These variables include heat duties, efficiencies, volume balances, yields, liquid and vapor flows inside the column, approach to flood. These calculations are made from measured process data on a frequent basis—say every five minutes. For example, an internal reflux flow is calculated often enough that it appears to be measured. Normally the liquid traffic in the column is calculated only for test runs. These engineering variables are used as operating guides, or as a basis for control.

►Operator assistance controls

Some of the routine and bothersome operator control functions have been automated; e.g., balancing of furnace passes, control of crude charge rate and ramping of crude rate changes, and control of the crude column bottom level.

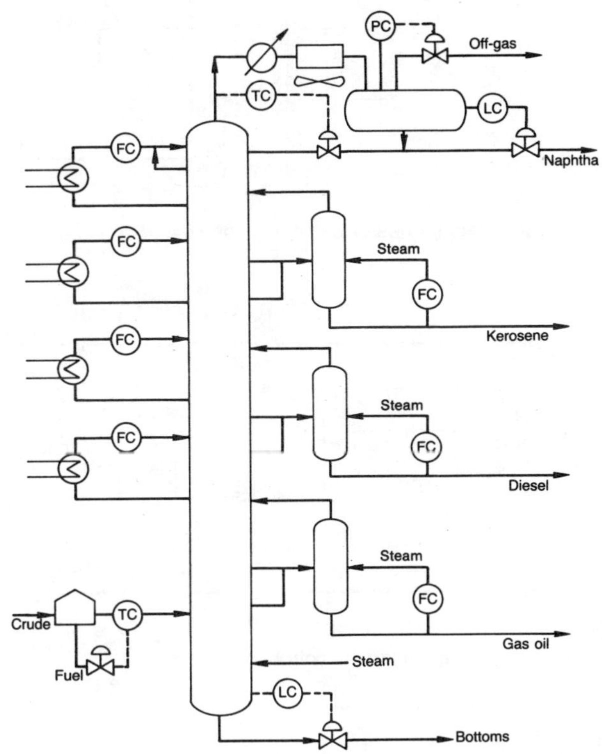

Fig. 1—Crude distillation column.

►Quality controls

The computer adjusts the crude column top temperature and side draw flow rates to control distillation properties, or properties related to distillation. Examples are endpoints, ASTM 10 percent or 90 percent points, for naphtha, light distillate, and heavy distillate cuts.

►Heat balance controls

The purpose of these controls is to save heat, generally by maximizing heat recovery to process streams, and minimizing heat loss to air-cooled or water-cooled exchangers. Examples are:

- Computer adjustments to side pumparound heats to maximize heat recovery to crude
- Adjustment of flows through parallel heat exchanger networks to maximize heat recovery
- Adjustment of atmospheric gas oil draw to minimize excess reflux or overflash.

The adjustment of pumparound heats also influences reflux or separation in the column, and heat

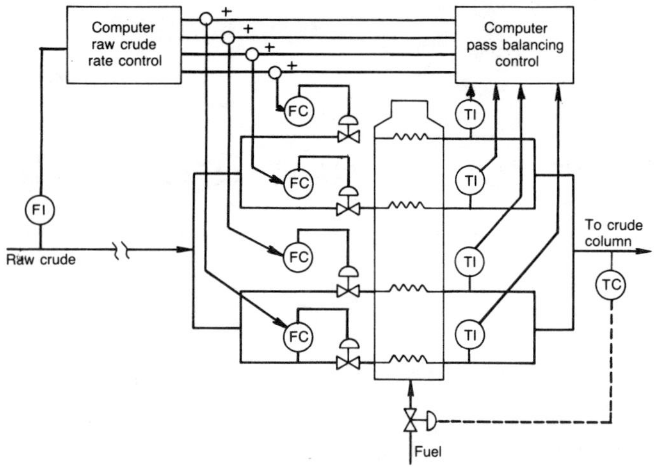

Fig. 2—Computerized heater balancing and charge rate.

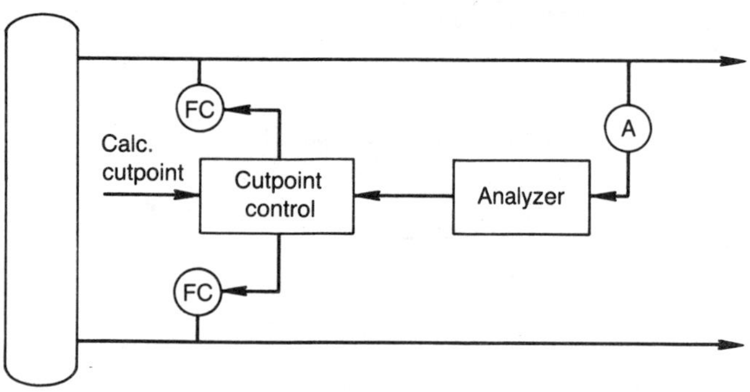

Fig. 3—Analyzer and cutpoint control.

balance controls have been used to control reflux ratios and separation when that is desired.[8]

Engineering calculations. Many of the desired engineering variables are calculated from heat and material balances around envelopes of pertinent tower sections.[8,9] Some of the key variables computed are

- Heat duties (MMBtuh)
- Internal liquid flows (Mpph, bpd)
- Internal vapor flows (Mpph, bpd)
- Tray loadings (percent of flood).

These calculations have proven to be very helpful as operator guides and as input for closed loop control. They are

- Aids in understanding traffic in the column, especially effects of changes in heat balance.
- Basis for heat balance adjustments when crude changes or operating conditions change.
- Guides for monitoring approach to flooding.
- Indicators of low reflux conditions and approach to pulling trays dry.
- Basis for optimizing heat recovery.

Calculations are based on steady state concepts. Therefore, it is necessary to average the input signals for the calculations, and that slows response. Typical time constants for response of the calculations to process changes are about 20 minutes. Normally, that is fast enough, but short term transients are not handled well.

Accumulation of errors affects the accuracy of internal reflux estimates in the lower sections of the column. Since overflash is normally the lowest calculated internal reflux in the column, it has the highest accuracy problem. It will reflect errors in flows, temperatures, API gravities, etc.

Calculated overflash by itself is not suitable for control. However, it has been used in engineering studies where the change was important, not the absolute value.[8]

Engineering calculations are combined with measured process variables in CRT operating displays, operating logs, and other reports. Together, these variables make a powerful operating information system. One refiner has said that the information system in his crude unit computer control project "probably paid for the project many times over."[2]

Operator assistance controls. Some of the easier control functions have proven to be the most popular with the operators.

► Balancing of heater pass outlet temperatures on multipass heaters (Fig. 2)

► Crude charge rate control by adjustment of pass flows (Fig. 2)

► Feedforward adjustment of draws and pumparounds from crude charge rate changes

► Control of crude column bottom level by adjustments to vacuum heater pass flows

► Control of overhead receiver level.

In most cases, control requires adjustment of several flow controllers at once. This was a bothersome task for the operators. The computer helps the operators a great deal by performing these controls automatically.

Control performance is excellent. Heater pass temperatures control within ±2°F, even through transients. All of these controls operate through most upsets. On stream time for closed loop heater balancing controls of about 95 percent has been achieved.[11] These controls help win operator confidence for the entire computer control system.

Quality controls. For purposes of this discussion, quality means boiling range control of the ASTM 90 percent point or endpoint on the naphtha and side draws.

Quality controls are accomplished by[8]:

- TBP cutpoint controls
- Analyzer controls.

TBP cutpoints are calculated by the computer and analyzer controls are based on a boiling point analyzer.

TBP cutpoints. The calculated cutpoint includes effects of tray temperature, local column pressure, local reflux ratio and steam partial pressure.[8,10] Temperature, pressure and steam flows are all measured variables. The local reflux ratio comes from the internal liquid and vapor flow calculations mentioned earlier.

Calculated TBP cutpoints are well tested now. They are numerically close to ASTM 95 percent points, and some operators understand them as such. They reflect dynamic changes well, and respond quickly to process changes with time constants of about 10 minutes.

Analyzers. Analyzer performance is improved over earlier experience, but maintenance problems still remain. We don't find too many boiling point analyzers on closed loop control. There are some that work well. However, noise and drift seem to keep their closed loop control time low.

There is a considerable lag between process changes and analyzer response. It comes from sampling system lags and from mixing volumes in the overhead condensers and accumulators for a naphtha analyzer, and from strippers for side draws. Lag times range from 45 minutes to two or three hours, and it may vary when accumulator or stripper levels vary.

Control. The calculated TBP cutpoints have been used as the primary control variables. They are used by themselves if there is no analyzer—or as a fast inner loop with slower analyzer feedback trim (Fig. 3). When cutpoints are on by themselves, the laboratory ASTM distillation data are used to trim the targets for the controllers. Computer control of quality is especially worthwhile where crude quality changes often. Typical variations for ASTM 90 percent points without the computer are ± 10 to 20°F. Computer control will hold these variations within 3 to 5°F. This allows operating management to set operating targets closer to real limits and improve yields of desired products. Examples are maximizing naphtha to an endpoint specification at the expense of kerosene, or maximizing distillate fuel oil at the expense of FCC conversion feed.

A sample of cutpoint response compared to analyzer response is shown in Fig. 4. This is not precise data, but represents observations. In this example, a specialty run required adjustment of the naphtha/kerosene cutpoint by 30°F. The change was made directly to the target of the naphtha/kerosene TBP cutpoint controller in the computer. The controller adjusted top temperature and kerosene draw to reach the new target. It was lined out at the new target in about 90 minutes. The other cutpoints in the column remained on target. Meanwhile, the naphtha 90 percent point analyzer moved slowly to its new steady state value, reaching it roughly two hours later. The transition was made quickly and accurately by cutpoint control. It would have been very slow if the analyzer had been used.

Yield improvements are the most obvious benefits from good quality control.[3] Assume computer control reduced naphtha 90 percent variation from ± 10°F to ± 3°F, a reasonable case. During gasoline season when naphtha is maximized, the average naphtha ASTM 90 percent point could be increased 7°F without violating the naphtha endpoint specification. This would produce an increase in naphtha yield of approximately 0.7 percent basis crude. On a 50,000 bpd unit that is

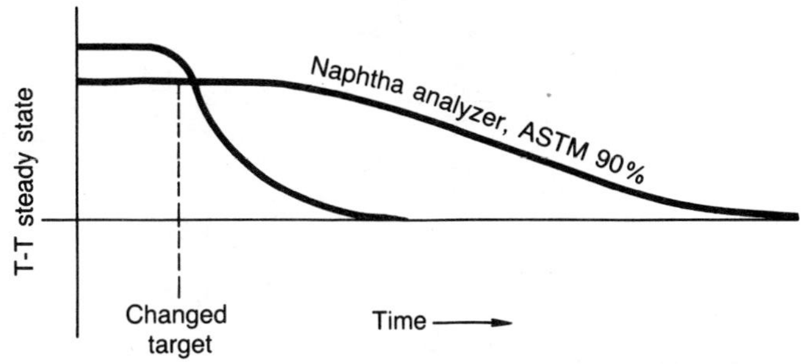

Fig. 4—Cutpoint vs. analyzer response.

350 bpd more naphtha and less kerosene. These types of benefits have been realized in practice.

Onstream time for the cutpoint controls can be expected to average between 80 and 90 percent. Most of the downtime results from process operating problems and computer tests. One refiner experienced closed loop performance of 87 percent over a three month test period.[11] This came after about two years of crude unit computer control experience in the refinery.

Heat balance controls. The purpose of these controls is to maximize heat recovery to process streams, subject to limits.[5] There are a number of possible heat balance controls, but this discussion will concentrate on the adjustment of pumparounds (Fig. 5).

The variable heat exchangers in pumparounds are usually to raw or desalted crude. The goal is to maximize heat exchange to crude, subject to limits. This can and has been handled rigorously with a small linear program. It requires many temperature inputs from exchangers, in this case.

The problem often simplifies to recovering the heat at the highest possible temperature.[1] That means recovering it as low as possible in the column, subject to the operating limits. This strategy is easier to implement than a full optimization and requires fewer input signals.

The physical limits often govern what can be done. Limits are:

- Maximum heat transfer on pumparounds, as indicated by a wide open flow control valve, or by insensitivity of heat removal to added pumparound flow
- Minimum liquid reflux in the column—must watch stripper levels so strippers don't pull dry during transients
- Heat exchange to pre-desalted crude is sometimes limited by maximum desalter temperature.

The physical limits can and often do override everything the computer wants to do to maximize heat recovery. When the pumparound heat exchangers are dirty, the pumparounds will run wide open most of the time.

Heat balance controls will save heat. When the computer has room to work, an increase in crude preheat temperature of about 10°F is achievable. For a 50,000 bpd unit, that is 3 to 4 MMBtuh.

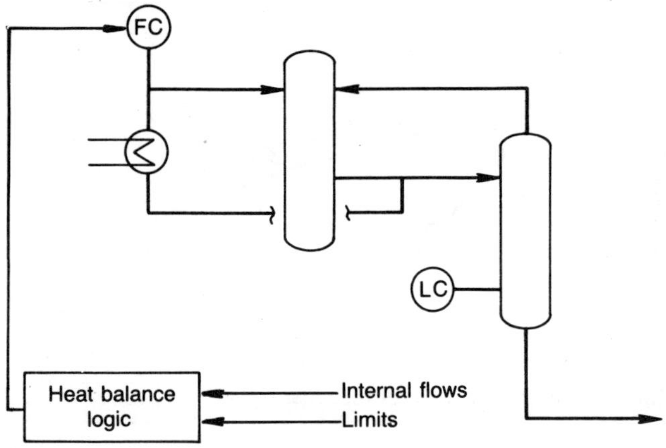

Fig. 5—Pumparound adjustments.

Overrides. A big problem with computer control a few years ago was operator acceptance. At least part of that problem was failure to include enough limit checks and overrides to keep computer control within safe, practical operating limits. As a result, operators had to watch carefully and take the computer off control for most upsets. We now include several practical limit checks and overrides in all control logic. These checks and overrides enforce operating limits regardless of what else happens. So, if a process upset occurs, the computer will help the situation instead of aggravating it. The result is improved operator confidence, and more closed loop control time. This gives the computer controls a chance to achieve the control benefits they were designed for.

Project organization. The way a project is planned and executed has a lot to do with "how good it is."[7] Crude unit computer control projects are usually complex. Loosely planned or poorly managed projects are likely to grow in cost and take more time than desired.

Start a project with a good definition phase.[7] Tasks in this work include:

- ►Evaluate and select cost effective computer functions
- ►Set the project scope
- ►Estimate benefits for the functions
- ►Define instrument work to be done
- ►Define computer system requirements
- ►Outline training requirements
- ►Estimate project costs
- ►Plan the project. List the tasks to be performed and determine how they will be performed
- ►Prepare the project schedule.

The definition work is the basis for project profitability analysis and for a decision to proceed or not. If the project is approved, the definition work is the blueprint for project execution.

Given an approved project, then implement the project within the established scope with sound project management. This is just as important for a computer project as it is for refinery construction projects.

Involve operating personnel and refinery engineers in the project to help assure success. They are the ones who will use the computer control system. They have to be comfortable with it, and if they are not, the chances for success are slim. They also contribute greatly to the project.

- Needed practical and operating information for the control system design
- Advice for operator display and console design
- Operator training functions
- Coordination of work within the plant
- Assistance in commissioning

The large companies may form a task force of operating and technical personnel to work with the consultants on these projects.[4,6] Smaller independents can't do that, but usually participate to a significant extent. A refiner who did not have operations participation encouraged others to "promote maximum plant personnel and operator input right from the start."[2]

Crude unit computer control has reached a degree of maturity. There are proven functions available, and economic benefits have been demonstrated. Overall benefits of $1000 to $2000/day are reasonable for a 50 Mbpd crude unit.

A successful installation can be achieved, with appropriate planning and execution. It requires a careful analysis of the feasibility of computer control functions for the particular case, and selection of a project scope with good payout potential. It requires adequate planning and effective project management.

Originally presented at the NPRA Computer Conference, St. Louis, Mo., Oct. 28-31, 1979.

LITERATURE CITED

[1] Bannon, R., et al, "Heat Recovery in Hydrocarbon Distillation," *CEP*, July 1978, pp. 41-45.
[2] Carbonneau, A., et al., "Golden Eagle's Experience with Process Computer Control," presented to E.I.C. (Petroleum Section), Toronto Branch, Toronto, April 12, 1978.
[3] Latour, P.R., "The Hidden Benefits from Better Process Control," presented to ISA-76, Houston, Oct. 11, 1976.
[4] McClean, G., "Gulf's Edmonton Refinery Computer System," presented to Canadian Chemical Engineer's Conference, Calgary, Oct. 23, 1977.
[5] Ryskamp, C.J., et al., "Improve Crude Unit Operation," *Hydrocarbon Processing*, May 1976, pp.81-86.
[6] Terris, J.S., et al., "Supervisory Process Computer Control of the Crude and Vacuum Units," Digital Computer Applications to Process Control, IFAC and North-Holland Publishing Co., 1977.
[7] Van Horn, L.D., "Computer Control—How to get started," *Hydrocarbon Processing*, Sept. 1979.
[8] Van Horn, L.D., and Latour, P.R., "Computer Control of a Crude Still," *Instrumentation Technology*, Nov. 1976.
[9] Wade, H.L., et al., "Tray Flooding Sets Crude Throughput," *Hydrocarbon Processing*, Nov. 1977, pp. 281-285.
[10] Watkins, R.N., Petroleum Refinery Distillation, Gulf Publishing Co., Houston (1973).
[11] Van Horn, L.D., and Latour, P.R., "Crude Distillation Computer Control Experience," presented at ISA-76, Houston, Oct. 11-14, 1976.

Understand Minicomputer Control Systems

Minicomputer systems, the most common form of digital control, are presented in simple terms to provide a working knowledge of these systems for process and plant engineers.

G.C. Shah, American Hoechst Corp., Bayport, Texas

WITH STEEPLY RISING raw material and energy costs, computer control has now become almost a necessity for many processes. Usually, medium or large computer projects for data acquisition and process control are managed jointly by groups such as process engineering, plant engineering, data management and instrument systems. Thus, a process or a plant engineer does not usually get deeply involved in the intricate details of a computer system and consequently needs to know little about high level electronics or systems engineering. Conversely, working knowledge of these systems is becoming increasingly important to intelligently evaluate and use them. The purpose of this article is to discuss key features that are crucial to understanding minicomputers.

SYSTEMS OVERVIEW

Various definitions of the term "minicomputer" are found in literature. They are based on parameters such as processing power price or size. From a user's point of view, however, precise definitions are irrelevant. Computer vendors name and classify their products. It is sufficient to recognize that minicomputers are a subclass between large computers and microcomputers. All the computers are functionally the same, i.e., to perform and store calculations. These computations are done digitally. The logic to perform the digital computations is binary logic which is discussed later.

Before discussing major functional blocks of a minicomputer system, let us review different architectures or arrangements in which computer systems are available (Fig. 1).

Glossary of Selected Terms*

Access time: Time required to retrieve information from memory.

Acoustic coupler: A device to convert digital signals for transmission on telephone lines.

Address: Location of information in memory.

Baud rate: Number of bits transmitted per second.

Bootstrap program: An initializing or a loader program.

Buffer storage: Temporary data storage.

Host: A central supervisory computer in a distributed system.

Interpreter: A program resident in some computers to translate high level programs into machine instructions.

Interrupt: A computer based command to start work on a higher priority item while stopping work on the current task.

Modem: A device to transmit digital data to/from telephone lines.

Protocol: Several integrated circuits that define data formats, their meanings, feed-back etc. necessary for communication between two devices. e.g. ASCII, BASIC, CAMAC.

Register: A circuit to temporarily store a "digital word."

Serial transmission: The term refers to sending one bit at a time as opposed to parallel transmission which is multiple bits at a time.

Volatile memory: Memory which loses its contents upon loss of power. This term refers to the main memory in contrast to disk memory which is non-volatile.

(*Terms discussed in the article are not repeated in the table.)

Supervisory control: This arrangement is analogous to cascade control in purely analog systems. A computer is used to change the set point of analog controllers. The changed set point then serves as the basis for control. Thus, the control is done by analog controllers (slaves) with the computer in a supervisor's position. Since process signals are simultaneously fed to the computer and analog controllers, failure of the computer does not cause immediate process upset. In highly critical services, backup computers may be considered.

Direct digital control (DDC): As the name implies, process control is done directly by a digital computer in DDC. Transmitter signals go to a computer. Since a computer operates digitally, process signals are condi-

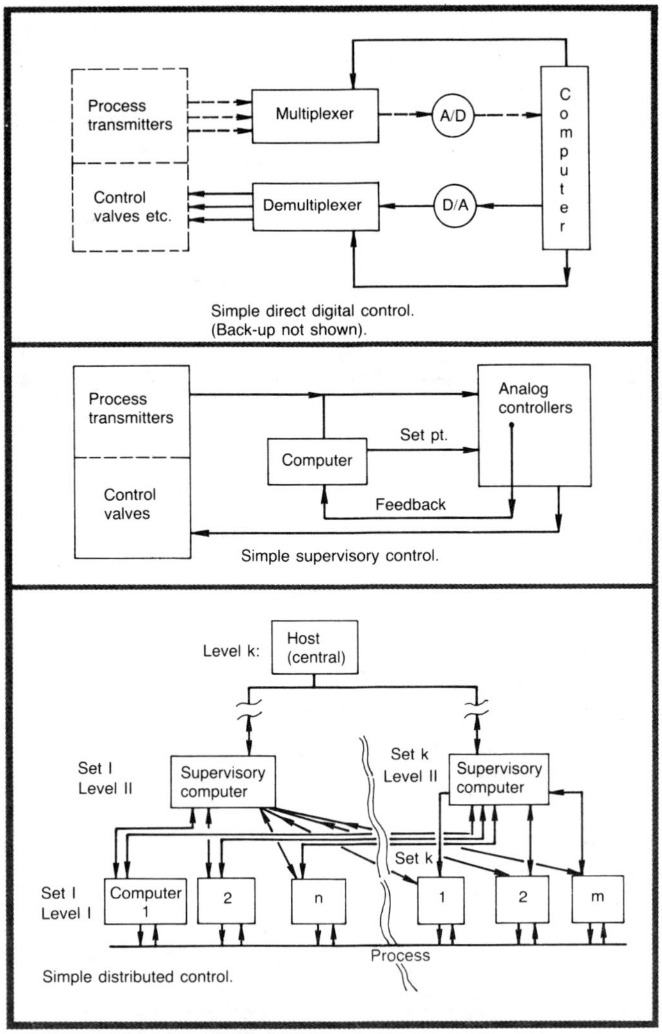

Fig. 1—Different computer system arrangements.

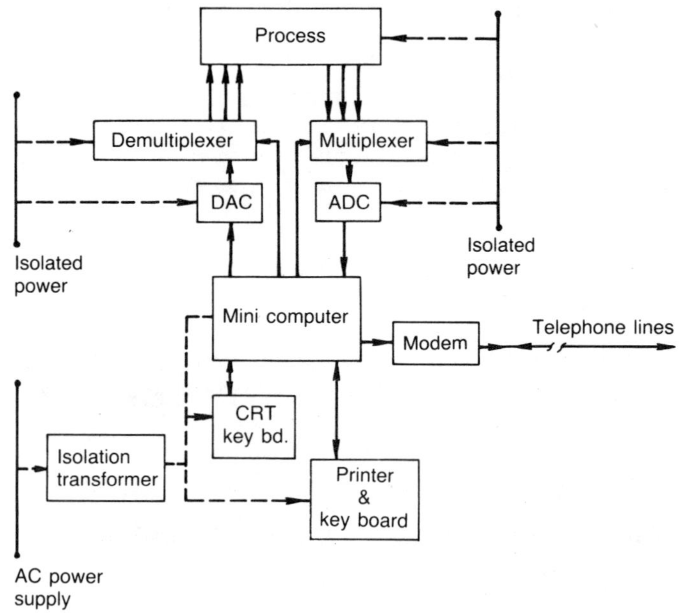

Fig. 2—Major functional blocks of a minicomputer system.

tioned and converted into digital form before entering the computer. Conversely, the output signals from the computer are converted to analog form suitable for control valves. Control valves usually accept pneumatic signals. These systems, in addition to control, often involve a variety of display and record functions. Although analog controllers are absent from DDC, they may be considered for backup. Sometimes backup computers are provided. In case of a computer failure, bumpless transfers to backup controllers are relatively complex in DDC systems.

Distributed controls: It is normally understood that distributed control is all digital. When plant control is divided or distributed among several computers, or data processing devices such as microprocessors, the architecture is known as distributed control. The computers can communicate to each other by means of a central computer. Microprocessors generally handle several control loops. The number of loops that can be handled by a microprocessor will depend on its capacity and complexity of the control loops. Microprocessors are often provided with backup units. These units are either analog controllers or identical microprocessors. Since plant control is shared by several computing devices, failure of one of them does not affect the whole process. The impact of a failure of a computer on the plant will depend on the number of loops handled by that computer.

Distributed systems are available in a variety of arrangements or architectures. Important ones include hierarchal arrangement, resource sharing systems and process defined systems.[1] Computer systems offered by most vendors are hybrid and do not fit neatly into these classifications.

In a major computer network, one computer system will communicate with another. Different computers in a network are often from different vendors. "Emulators" are used to establish communications between computers of different makes. Thus they are computer interfaces.

BASIC FUNCTIONAL BLOCKS

A minicomputer system can be represented by its basic functional blocks shown in Fig. 2.

In the discussion that follows, we are looking from outside in, i.e., we begin with process signals then go to the computer and finally to the associated components. In a typical process plant, a variety of process signals are available. The most commonly encountered process signals are listed in Table I. Most of the signals are analog. Since a computer works digitally, these signals must be first conditioned (linearized, magnified, filtered, etc.) and converted to digital form before entering the computer.

Multiplexers: Analog to digital converters (A/D) can be used to accomplish the A/D conversion described previously. Simply stated, A/D converters consist essentially of amplifiers (transistors), resistors and capacitors. One A/D converter converts one analog signal into its digital equivalent. A typical plant, however, contains a large number of signals. Converting all the signals by individual A/D converters thus involves extensive wiring and is expensive. Using multiplexers considerably reduces the wiring required. A multiplexer handles several analog signals in a serial manner. It is like a selector switch which, with one A/D converter, performs A/D conversions in a specific sampling order. To convert pneumatic signals into corresponding binary coded signals, pneumatic A/D converters and multiplexers are also available. Several vendors offer modular designs.

For the computers to communicate to the outside world, their digital signals must be converted to compatible analog outputs. This is done by digital to analog (D/A) converters. Again, multiplexers are used to simplify wiring from the computer back to the process.

The computer: The digital outputs from the A/D converter enter the input section of a computer. Complex calculations of industrial control equations are done swiftly in computers by binary logic. Stated in simple terms, binary logic involves two states: presence (state 1) or absence (state 2) of voltage or current levels.

The numbers 1 or 0 are individually known as bits. A digital word is a collection of bits. For example, 00100101 is a word containing eight bits. Often "word" is interchangeably used with "byte" in specifying sizes or processing capacity of computers. (Byte = 8 bits) Word size depends on the internal construction of a computer. As shown in Fig. 4, a computer can be subdivided into three major functional sections: Input/Output devices, main memory and central processing unit (CPU).

Input/Output devices (I/O) are used to establish communications between the outside (or peripheral devices) and the computer. These devices are connected to I/O buses. I/O buses are composed of sets of wires or conductors that carry current. In the context of I/O bus operation, it is important to note two basic factors:

►Computers operate at astronomically faster speeds than the peripheral devices. One consequence of this difference in speeds is that there is a time lag in transmitting the results to the peripherals. The computer system must synchronize its output with the peripherals and accomplishes this function by the use of I/O devices.

►A computer has to communicate to several peripheral devices and these devices are often from different vendors and therefore differ structurally. The I/O bus must be able to identify (address) the peripheral devices and adapt its output signals to these differing peripherals. The I/O bus accomplishes these primary functions and consists of three groups of wires: address bus, control bus and data bus.[2]

Main memory or core memory is a computer's storehouse for data and instruction sets. Common forms of memory for computers are ferrite cores and integrated circuits (IC). Although, over the years, substantial gains in access speeds have been made by reducing the size of the individual cores, compact and faster IC's are replacing ferrite memories. IC's however lose their storage instantly once power is lost. Battery backups are available for memory as well as other critical components of most computer systems and should be considered. The time required to retrieve data from storage is known as access time. Data and instructions (executive programs) are transferred between memory and the CPU by buses or conductors similar to I/O described earlier, i.e., the communica-

TABLE 1—Typical process inputs.

Variable	Electronic or pneumatic signal	Comments
Pressure; dp	4-20 mA; 10-50 mA; 3-15 psig	Signals have to be converted to digital by A/D converters or multiplexers & A/D
Temperature; t/c, RTD	Volt or millivolt 4-20 mA; 10-50 mA 3-15 psig	Same as above.
Level	4-20 mA; 10-50 mA; 3-15 psig	" "
Flow; dp cell; pitot,	4-20 mA; 10-50 mA; 3-15 psig	" "
Target meter,	" " "	" "
Mag meter,	" " "	" "
Gear & turbine meters	pulse (If pulse/current converter absent.)	A/D conversion not required
Alarms	On or off	A/D conversion not required.
Operator inputs	Commands, real or integer numbers (lab-analyses, level etc.)	A/D conversion not required.
Inter-computer-Communications	All digital through protocols.	A/D conversion not required.
Continuous-process-analyzers	Volts or amperes	A/D conversion necessary by A/D converters or multiplexers & A/D
Viscosity	" "	" " " " "

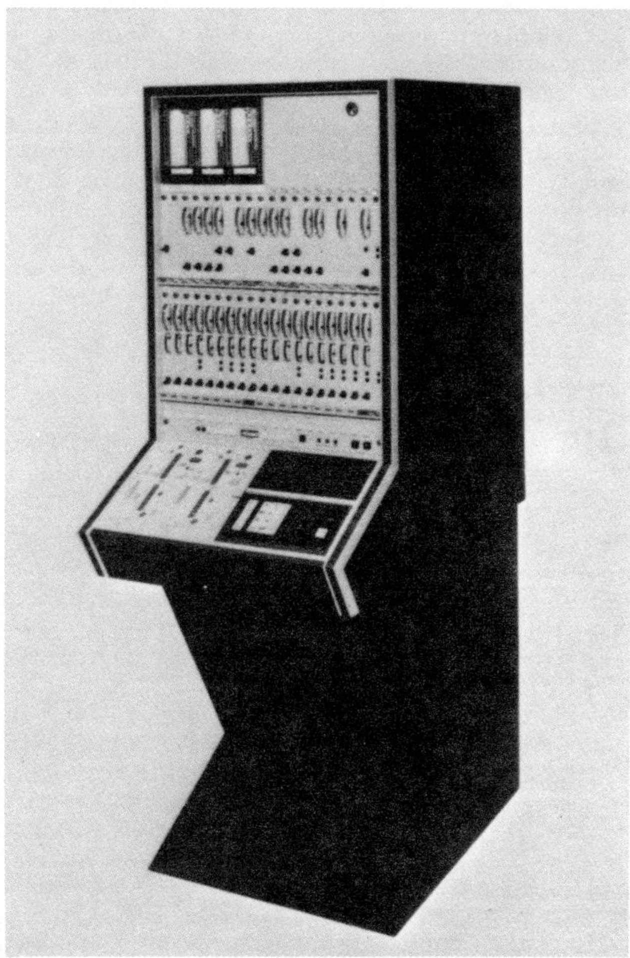

Fig. 3—A minicomputer for process control. (Courtesy CATCO, Inc., Houston)

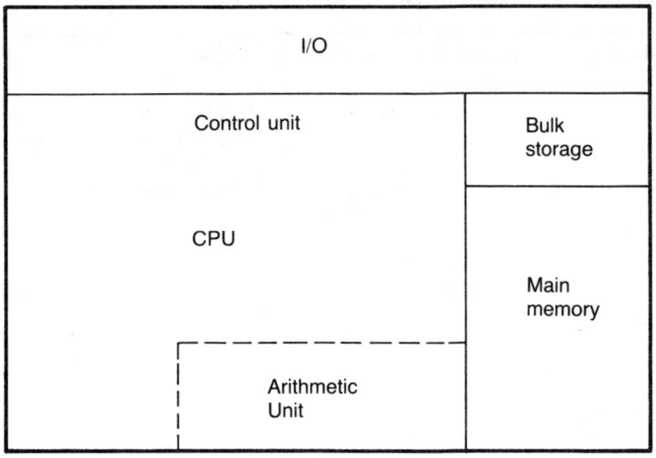

Fig. 4—Major functional blocks of a minicomputer.

tion bus, which consists of the data bus, control bus and address bus. Data can be stored and retrieved at any memory location in the main memory (i.e. randomly). The memory is therefore called random-access memory (RAM). This is direct data acquisition and does not require address search through the file.

Quite often, several segments of data or instructions are used more frequently than others. To speed up retrieval, some computers place these data at special locations called "cache" (cache memory). Taking this a step further, several data functions have definite relationships between input and output variables and never require changing. To ensure the information is not altered or lost during power failure, it is placed into a special type of memory called Read Only Memory (ROM). Function access in ROM is provided by either hard wiring or special "microprograms." Functional relations like logarithms, trigonometric functions, etc. are conveniently handled by ROM. It should be noted that memory systems offered by most vendors are modular type, i.e., memory can be changed by adding or removing memory boards. Memories, whether main or bulk as discussed later, are specified by total word storage capacity, e.g., storage capacity of 64k bytes (words).

A computer's memory can be significantly increased by using external (auxiliary) storage or bulk storage. This is accomplished by disks, drums or tapes. Disks are more common in the minicomputer systems. Disks look just like ordinary soundtract records and are coated with magnetic material. Information transfer to/from a disk takes place by write/read heads. Disks rotate at a fixed speed and transfer of information does not involve physical contact between heads and a disk. Thus, the transfer works by electromagnetism. A disk has several tracks and the proper track must be found by the write/read head before information transfer takes place. These disks are called Moving Head Disks (MHD). Fixed Head Disks (FHD) are also available where multiple heads on a disk do not require their physical movement (to respective tracks) for data transfer. Disks can thus be considered close to random access devices for their data acquisition. FHD's have obviously shorter access times than MHD's since the former does not involve read/write head movement. Smaller versions of MHD are called diskettes or floppy-disks. Here the read/write head touches the diskette for information transfer. Because of this contact wear, the diskette loses the magnetic coating and has to be replaced periodically.

The central processing unit (CPU) is the heart of a computer system. It consists of two major functional blocks: control unit and arithmatic unit. The control unit, as the name implies, controls various operations such as data transfer, computations, timing etc. It is therefore the CPU "supervisor." The arithmatic unit is comparable to an accountant. Arithmatic operations such as addition, subtraction, multiplication and division are performed by this unit. Temporary storage of data during computations is accomplished by an accumulator which resides in the arithmatic unit. Like memory and I/O devices, the CPU usually consists of several integrated circuits. Some computers incorporate a single Large Scale Integrated Circuit (LSIC) CPU. This is the familiar microprocessor. Combinations of the IC's or LSIC's depend on the computer make.

Computers, from a user's viewpoint, are therefore combinations of IC's (or LSIC's) to perform arithmatic operations. Four basic elements can be identified in the countless array of IC's in different computers:[3] gates, inverters, delays and flip-flops (Fig. 5).

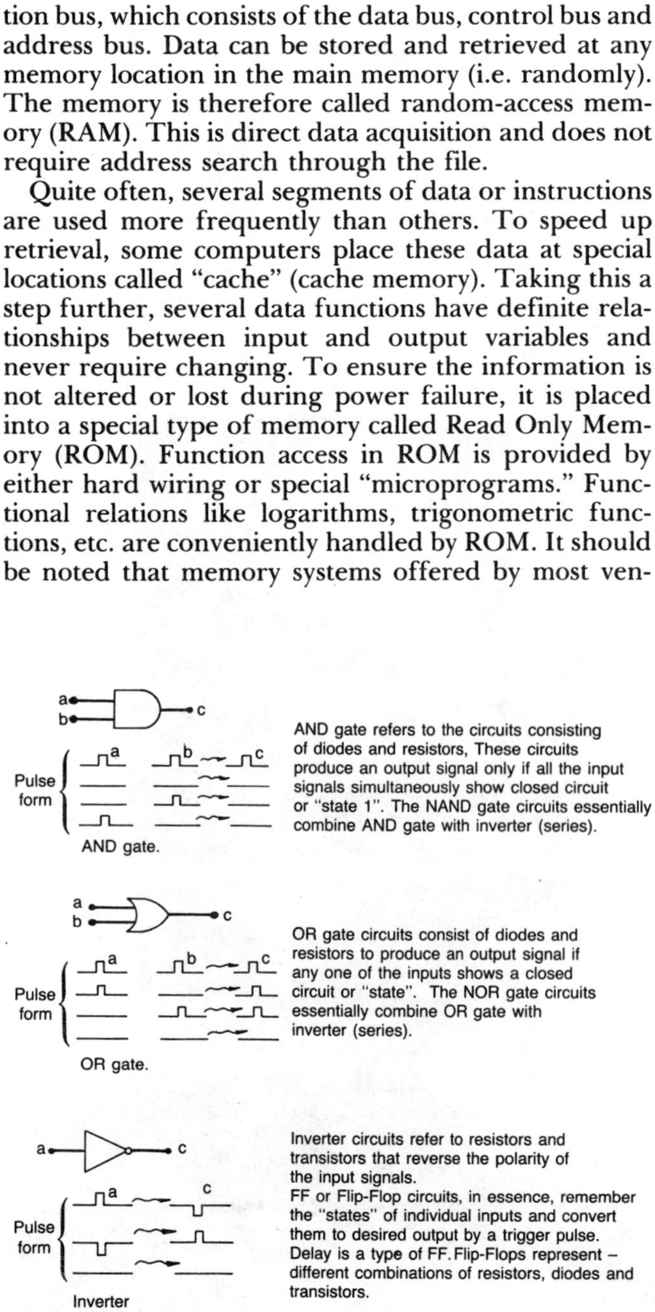

Fig. 5—Symbols for computer logic circuits.

Peripherals: This term collectively describes devices used with the computer unit to form a "computer system." These devices include printers, CRT (cathode ray tube) displays and card or tape readers. Sometimes disks, tapes and drum storage units are also grouped as peripherals. Disks have been discussed earlier. For minicomputers in industrial control, printers and CRT's are the more commonly used operator interface items.

Printers are used to communicate with the computer by means of a printed paper called hard copy.

Printers look just like ordinary typewriters, i.e., typing head, roll of paper and usually a keyboard. When the signals come from the computer, the printer receives the translated outputs from the I/O interface discussed earlier. Printers are electromechanical devices and they operate much slower than the computer. The I/O devices, besides decoding the computer signals for the printer, synchronize the computer output with the printer's speed. There are two types of printers available: character printers and line printers. As the names suggest, a character printer prints one character at a time and is obviously much slower than a line printer which prints one line (row) at a time.

Typing on paper can be done either by conventional means (metal impact) or electrostatically.[4] Electrostatic printers require special kinds of paper that will hold electric charge. Since they do not involve mechanical movement of an impact head, the electrostatic printers are extremely fast.

CRT displays are television-like display screens commonly used as operator interfaces in minicomputer systems. The CRTs may be monochromatic (green or white on black background) or color. CRTs used for minicomputers are electromagnetic type and operate by electron beam impingement on a phosphor-coated screen that emits light. Although the persistance by these screens is relatively short, most vendors provide special registers to sustain display as long as desired. Most CRT's have keyboard terminals for operator inputs.

By incorporating a microprocessor in a CRT, a computer's overall processing capacity can be enhanced. CRT's without intelligence (dumb CRT's) use the computer's CPU for on-line editing. Software necessary for displays, such as trend representation, equipment display (column temperature profile, reactor temperature profile, etc.) is supplied by the computer vendor.

Computer software forms two main groups: user software and supervisory programs. The user is mainly concerned with the "user software" which is available in high level languages like BASIC, FORTRAN or COBOL. Supervisory programs are supplied by the computer vendor and perform such tasks as editing, scheduling, assigning priorities, communication control, etc. Compiler programs that translate high level languages into machine instructions are obviously an example of a supervisory program. The proportion of processing capacity necessary for supervisory programs depends on the number and complexity of the tasks. This is an important factor to consider during the early sizing phase of a computer system.

WIRING AND INSTALLATION

Wiring between the components of a computer system is as important as the components themselves. While loose connections and moisture are to be avoided, immunity from stray electromagnetic fields is crucial to the accuracy of the system and personnel safety. Techniques to reduce noise from electrical sources include twisting the pair of wires, providing shields (steel conduits), insulation and proper grounds. Ground potentials must be minimized. Fluctuations in voltage or frequency are, of course, harmful to computer operation. Vendors do specify allowable surge levels in power. Isolation transformers (installed between the power supply and the system components) are probably the most common means of protection against power transients.

Computers should be installed in clean, dust-free, air conditioned and humidity controlled rooms. The components to be installed in the plant area must be intrinsically safe and should be installed at clean, vibration-free locations.

PROJECT MANAGEMENT

Planning is the most important phase of a computer project. Economic justification is based on the savings that computer control will generate by virtue of its "finer control" and improved data gathering and manipulation capabilities. Improved data gathering enables a better understanding of the process and hence, better control strategies. Sizing the system (word capacity) starts out with counting the number and type of inputs, number and nature of control loops, amount of storage required, peripherals and desired spare capacity. With the modular designs available, it is good engineering practice to provide for spare capacity rather than purchase extra capacity at the outset.

Economic justification is followed by preparation of specifications and bids evaluation. Criteria for selection include total price, system configuration and its versatility, vendor's image in terms of stability, reliability, competence and his past performance with his customers. Consideration, at this stage, is also given to system maintenance. Most vendors offer comprehensive maintenance programs of their systems and the decision to select such programs is governed by several factors, such as the company's maintenance policy, type of contract and price and complexity of maintenance.

Once a vendor is selected, the customers and the vendor jointly work on the tasks of data base preparation, system details, site preparation, installation and start-up. To help improve communications, a glossary of some key computer-related terms is presented.

Understanding minicomputer systems is the key to their effective application. The bottom line is, of course, profitability.

ACKNOWLEDGMENT

I would like to thank Mr. John Engel of CATCO, Inc., Houston for providing Fig. 3.

LITERATURE CITED

[1] Merrit, R., "Large Systems are in Control," Instruments and Control Systems, Nov. 1978.
[2] Leibson, S. "What is I/O?" Instruments and Control Systems, Oct. 1979.
[3] Lenk, J.D. and Marcus, A., Computers for Technicians, Prentice Hall, Englewood Cliffs (1973).
[4] Davis, S., Computer Data Displays, Prentice Hall, Engelwood Cliffs (1969).